Vibrational Spectroscopy of Molecular Liquids and Solids

NATO ADVANCED STUDY INSTITUTES SERIES

A series of edited volumes comprising multifaceted studies of contemporary scientific issues by some of the best scientific minds in the world, assembled in cooperation with NATO Scientific Affairs Division.

Series B: Physics

RECENT VOLUMES IN THIS SERIES

Volume 46 – Nondestructive Evaluation of Semiconductor Materials and Devices
edited by Jay N. Zemel

Volume 47 – Site Characterization and Aggregation of Implanted Atoms in Materials
edited by A. Perez and R. Coussement

Volume 48 – Electron and Magnetization Densities in Molecules and Crystals
edited by P. Becker

Volume 49 – New Phenomena in Lepton-Hadron Physics
edited by Dietrich E. C. Fries and Julius Wess

Volume 50 – Ordering in Strongly Fluctuating Condensed Matter Systems
edited by Tormod Riste

Volume 51 – Phase Transitions in Surface Films
edited by J. G. Dash and J. Ruvalds

Volume 52 – Physics of Nonlinear Transport in Semiconductors
edited by David K. Ferry, J. R. Barker, and C. Jacoboni

Volume 53 – Atomic and Molecular Processes in Controlled Thermonuclear Fusion
edited by M. R. C. McDowell and A. M. Ferendeci

Volume 54 – Quantum Flavordynamics, Quantum Chromodynamics, and Unified Theories
edited by K. T. Mahanthappa and James Randa

Volume 55 – Field Theoretical Methods in Particle Physics
edited by Werner Rühl

Volume 56 – Vibrational Spectroscopy of Molecular Liquids and Solids
edited by S. Bratos and R. M. Pick

This series is published by an international board of publishers in conjunction with NATO Scientific Affairs Division

A Life Sciences B Physics	Plenum Publishing Corporation London and New York
C Mathematical and Physical Sciences	D. Reidel Publishing Company Dordrecht, Boston and London
D Behavioral and Social Sciences E Applied Sciences	Sijthoff & Noordhoff International Publishers Alphen aan den Rijn, The Netherlands, and Germantown, U.S.A.

Vibrational Spectroscopy of Molecular Liquids and Solids

Edited by
S. Bratos
and
R. M. Pick
Université Pierre et Marie Curie
Paris, France

PLENUM PRESS • NEW YORK AND LONDON
Published in cooperation with NATO Scientific Affairs Division

Library of Congress Cataloging in Publication Data

NATO Advanced Study Institute on Vibrational Spectroscopy of Molecular Liquids and Solids, Menton, 1979.
Vibrational spectroscopy of molecular liquids and solids.

(NATO advanced study institutes series: Series B, Physics; v. 56)
"Lectures presented at the NATO Advanced Study Institute on Vibrational Spectroscopy of Molecular Liquids and Solids, held in Menton, France, June 25–July 7, 1979."
Includes indexes.
1. Vibrational spectra–Congresses. I. Bratos, S. II. Pick, R. M. III. Title. IV. Series.
QD96.V53N37 1979 539'.6 80-12174

ISBN-13: 978-1-4613-3113-1 e-ISBN-13: 978-1-4613-3111-7

DOI: 10.1007/978-1-4613-3111-7

Lectures presented at the NATO Advanced Study Institute on Vibrational Spectroscopy of Molecular Liquids and Solids, held in Menton, France, June 25–July 7, 1979.

Softcover reprint of the hardcover 1st edition 1980

A Division of Plenum Publishing Corporation
227 West 17th Street, New York, N.Y. 10011

PREFACE

This book has its origin in a NATO Summer School organized from June 25 to July 7 1979, in Menton, France. The purpose of this School was a comparative study of the various aspects of vibrational spectroscopy in molecular liquids and solids. This field has been rapidly expanding in the last decade; unfortunately, its development took place independently for liquids and for solids. In these circumstances, the comparison of the basic concepts and techniques used in these two branches of physics appeared as a necessity. The lectures given at the Menton Advanced Study Institute, as well as the exceptionally fruitful and lively discussions which followed them confirmed this point of view.

The need of putting together these lectures, in the form of a monograph, clearly appeared during the ASI and the lecturers accepted to write down the material they presented at the Institute, improved thanks to the remarks of the participants. It is the result of this collective work which appears in the familiar Plenum Series.

The Editors would like to thank the Scientific Affairs Division of NATO for the financial support it offered to the Institute, and Dr. M. di Lullo, its director, for its constant help before and during the School. They are also very happy to acknowledge the very useful support of the Direction des Relations Universitaires Internationales du Ministère des Universités, as well as the Conseil Municipal of Menton who kindly accepted to have our Summer School organized in his town. The Editors are grateful to all the authors for their cooperation, as well as to Mrs. M. Debeau, E. Dervil, J. Vincent and M. Yvinec for their scientific collaboration to this project. Finally they would like to thank their constant collaborators, N. Claquin and C. Lanceron, who took a very important part either in the preparation or in the every day life of the School and in the publication of this book.

S. Bratos and R.M. Pick

PREFACE

CONTENTS

PART I INTRODUCTION

Intermolecular Forces 1
A.D. Buckingham

PART II LIQUIDS

Infrared and Raman Study of Vibrational Relaxation in Liquids 43
S. Bratos

Rotational Spectral Band Shapes in Dense Fluids 61
W. A. Steele

Computer Calculation of Vibrational Band Profiles in Liquids 91
D. W. Oxtoby

Vibrational Relaxation of Hydrogen-bonded Species in Solution: Fine Structure of the ν_s(XH) Band 101
G. N. Robertson

Experimental Study of Rotational and Vibrational Relaxation in Liquids from Investigation of the IR and Raman Vibrational Profiles . . 117
J. Vincent-Geisse

Collision-Induced Vibrational Spectroscopy in Liquids 147
G. Birnbaum

Picosecond Vibrational Spectroscopy 167
A. Laubereau

Vibrational Bandshapes in Viscous Liquids and Glasses 187
C. A. Angell

The Study of Vibrational Relaxation by Ultrasonic and Light Scattering Techniques 203
D. Sette

PART III SOLIDS

The Dynamics of Molecular Crystals: I. General Theory 221
S. Califano

The Dynamics of Molecular Crystals: II. Intermolecular Potentials 231
S. Califano

The Dynamics of Molecular Crystals: III. Infrared and Raman Intensity of Lattice Bands 243
S. Califano

The Dynamics of Molecular Crystals: IV. Hamiltonian Perturbation Treatment 253
S. Califano

Large Amplitude Motions in Molecular Crystals 263
K. H. Michel

Raman and Infrared Lineshapes of Internal Modes in ODIC Phases 305
R. M. Pick

Vibrational Band Shapes of Amorphous Solids 341
M. F. Thorpe

Order-Disorder Phase Transitions in Solids 367
C. H. Perry

The Effect of Multiphonon Processes on Vibrational Band Shapes in Molecular Solids 405
V. Schettino

Neutron Spectroscopy Studies on Rotational States in Molecular Solids 415
H. Stiller

Rotation-translation Coupling in Collective Modes of Molecular Solids and Liquids 431
G. Jaccuci

Participants 443

Author Index 451

Subject Index 459

Substance Index 463

INTERMOLECULAR FORCES

A.D. Buckingham

University Chemical Laboratory
Lensfield Road
Cambridge, CB2 1EW, United Kingdom

LECTURE 1

GENERAL SURVEY

1.1 Introduction

Molecules attract one another when they are far apart, since liquids and solids exist. They repel one another when close, since the densities of liquids and solids have the values they do under normal conditions of temperature and pressure. Figure 1.1 illustrates this important truth and shows a typical interaction energy u(R) of two spherical molecules as a function of their separation R. For two argon atoms, the well-depth ε is 0.198×10^{-20} J (ε/k = 143 K) and the equilibrium separation R_e is 3.76×10^{-10} m (1,2).

Since we shall be concerned with intermolecular forces, we should first consider what we mean by a molecule and by a force (3). Two argon atoms form a bound diatomic Ar_2 but we do not normally consider the species Ar_2 as a molecule, since the binding energy is only about one-half kT at room temperature. Collisions may easily dissociate Ar_2, and there are many thermally populated vibration-rotation states, ψ_{vJ}, each with a different mean bond length $\bar{R} = \langle\psi_{vJ}|R|\psi_{vJ}\rangle/\langle\psi_{vJ}|\psi_{vJ}\rangle$ and a large uncertainty $[\langle\psi_{vJ}|(R-\bar{R})^2|\psi_{vJ}\rangle/\langle\psi_{vJ}|\psi_{vJ}\rangle]^{\frac{1}{2}}$ in R. We prefer to speak of Ar_2 as a dimer of argon atoms. Similarly H_4, formed on cooling gaseous hydrogen to about 20 K at 1 atmosphere, is an infrared-active species in which two bonds are very

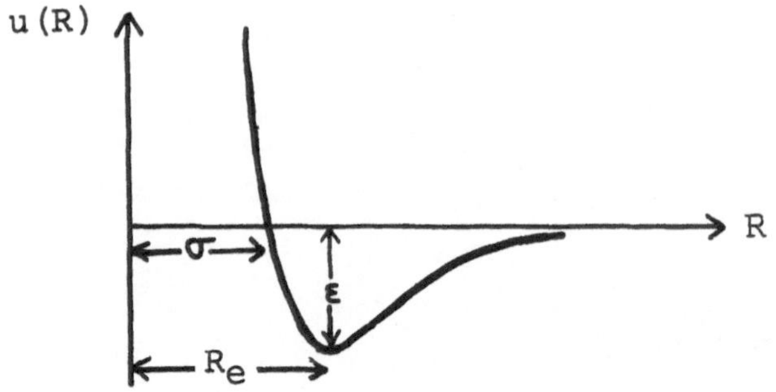

Figure 1.1. The interaction energy u(R) of two spherical molecules as a function of their separation R. The distance σ is known as the collision diameter.

similar to that in H_2 (4); we prefer to think of H_4 as $(H_2)_2$, i.e. as the hydrogen molecule dimer. So by a molecule we mean a group of atoms (or a single atom) whose binding energy is large compared to kT. A molecule therefore interacts with its environment without losing its identity. In some non-rigid molecules, such as NH_3, 1,2-dichloroethane ($ClCH_2$-CH_2Cl), or a polypeptide, there may be only a small change in energy with a large change in an internal coordinate; the influence of the environment in producing changes in the energy surface involving this coordinate may be of interest (5).

And what do we mean by a force? In Figure 1.1 the force is $-du/dR$ and there is no difficulty here. The concept may easily be extended to a many-dimensional surface, as in Figure 1.2 where the torque acting on a diatomic molecule in collision with an atom is $-\partial u/\partial\theta$. But what

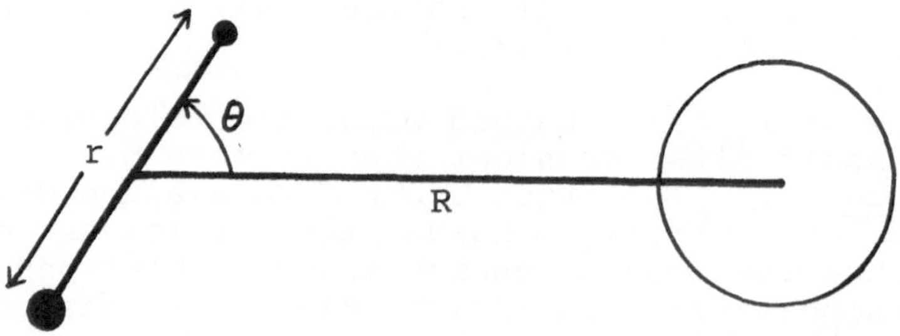

Figure 1.2. The three coordinates R, θ, r appropriate to the interaction of a diatomic molecule and a spherical atom.

is the effective force on two ions in aqueous solution? It is convenient to consider the potential of average force A(R) which is a Helmholtz free energy and is the mean interaction energy of the two ions at a fixed separation R, averaged over all configurations of all the other molecules and ions in the solution. A(R) is the sum of u(R) and -TS(R) where both u(R) and the entropy S(R) are functions of the temperature T. The entropic contribution may be supposed to arise from the change in the order in the environment resulting from the interaction of the pair. The attractive force in a stretched rubber band is attributable to a decrease in entropy on stretching (6); and the hydrophobic effect that appears to produce an attractive force between hydrocarbon chains in aqueous media (7) depends on S(R), for the decrease in entropy in forming a cage of water molecules (8) is presumably less in the case of a close pair of chains than when they are far apart.

1.2 The Born-Oppenheimer Approximation

The concept of a potential energy function u(R) is dependent upon the Born-Oppenheimer approximation (9,10). The potential energy u(R, θ, r) is the interaction energy for fixed positions of all the nuclei, i.e. it is the difference between the energy of the system in that particular configuration and its value when the intermolecular separation $R \to \infty$. There are interesting effects resulting from the breakdown of the Born-Oppenheimer approximation (particularly when there are electronic degeneracies (11)), but for the purposes of studying liquids and solids we may safely employ it. The accuracy of the approximation may be gauged from the following:

(i) The Rydberg constant for the H atom is reduced by 0.054 per cent on changing the nuclear mass from infinity to that of the proton. This energy change of 59.8 cm^{-1} is equal to the mean kinetic energy of the nucleus.

(ii) The clamped-nuclei non-relativistic dissociation energy D_e for H_2 is 38292.83 cm^{-1} (12). The relativistic correction takes D_e to 38292.30 cm^{-1} (13) and the experimental value is 38295.6 $cm^{-1} < D_e < 38297.6$ cm^{-1} (14).

(iii) The dipole moment of HD, as determined from pure-rotational absorption intensities in the far infrared (15), is 5.85×10^{-4} D = 19.5×10^{-34} C m, and it arises entirely from the breakdown of the approximation, since HD is electrically centrosymmetric in the clamped-nuclei approximation. The dipole may have the sense H^+D^- (16); it is an order of magnitude smaller than the dipole of CH_3D which is 5.64×10^{-3} D (17) and is attributable to the different mean internuclear distances in CH_3D and CH_4 (18).

1.3 The Number of Variables Determining u(R)

The number of independent variables upon which the intermolecular energy depends increases as the molecular size increases. For two atoms there is only one variable R (Figure 1.1), and for an atom interacting with a diatomic there are the three variables R, θ, r (Figure 1.2). For two diatomics there are six (R, θ_1, θ_2, ϕ, r_1, r_2), where ϕ is the angle between the planes containing the line of centres and the internuclear axis of each molecule. In the general case, for molecules containing N_1 and N_2 nuclei, there are $3(N_1+N_2)-6$ independent variables of which $3N_1-6$ and $3N_2-6$ are vibrational coordinates in each molecule and the remaining six (R, θ_1, χ_1, θ_2, χ_2, ϕ) determine the relative positions and orientations of the molecules; χ_1 and χ_2 determine the orientation of molecules 1 and 2 about their axes at angles θ_1 and θ_2 to the line of centres. Thus the intermolecular potential surface of the water dimer $(H_2O)_2$ has twelve variables, six of which are related to the vibrational coordinates of the two H_2O molecules.

The six translational and orientational degrees of freedom of an interacting pair of non-linear polyatomic molecules generally fluctuate slowly compared to the intramolecular vibrations. For some purposes, such as rotational relaxation, it may be sufficient to average u over the vibrational motion, thereby reducing the number of variables upon which u depends to just six. And for vibrational relaxation of a particular mode, it may sometimes be reasonable to average over the other vibrational modes, thus reducing the effective dimensionality of the problem.

Intramolecular vibrational motion is too rapid to permit adjustment of the relative positions and orientations of neighbouring molecules, so their contribution to the entropic force T $\partial S(R)/\partial r_i$ may normally be neglected.

1.4 Classification of Intermolecular Forces

The significant forces between molecules have an electric origin. It is true that there are also magnetic and gravitational interactions, but these can normally be neglected. In considering the nature of intermolecular potentials, it can be helpful to separate various contributions.

The primary separation is to divide the interactions into two classes, long-range and short-range. The former decrease as R^{-m} at large R where m is a positive integer. Thus the interaction energy of two ions varies as R^{-1} and that of two dipoles as R^{-3} at large R; the corresponding forces vary as R^{-2} and R^{-4}. Short-range interactions decrease approximately as e^{-aR} times a polynomial in R and result from overlap of the electronic wavefunctions describing the isolated molecules. At large separations, this overlap is negligible and it is possible to consider the electrons as belonging to one molecule or another, and the n-electron wavefunction, where $n = n_1+n_2$, need not be antisymmetrized with respect to exchange of electrons between molecules 1 and 2; such antisymmetrization leads to short-range forces. Long-range forces can be related to properties of the free molecules such as charge densities and polarizabilities.

Short-range forces may be attractive or repulsive, although for small R they are always repulsive. They arise (see Lecture 4, section 4.2) from the Coulomb and exchange energies (19). Long-range forces can also be attractive or repulsive but for pairs of inert-gas atoms in their ground states, the long-range force is attractive.

The Hellmann-Feynman theorem (20) requires that the forces on the nuclei may be evaluated by classical electrostatics from the charge distribution. The attractive force between two inert-gas atoms at long-range is associated with a slight build-up of electron charge in the region between the nuclei. Each atom acquires a dipole moment proportional to R^{-7} at large R but the dipoles cancel in a homonuclear pair such as Ar_2. The attractive force varying as R^{-7} results from the force exerted on each nucleus by the distorted electron cloud of its own atom, but evaluation of the interaction energy does not require such a detailed knowledge of the charge distribution (21). The interaction at long-range results from intermolecular electron correlation. In the short-range overlap region, it is not necessary to invoke a redistribution of charge to explain the force, although such a redistribution does occur and tends to reduce the strength of the repulsion.

A secondary classification of long-range interactions into several distinct types can be helpful. Table 1 shows these interactions and whether they are additive in the sense that $u_{123} = u_{12} + u_{23} + u_{31}$; it also shows whether the forces are attractive or repulsive.

Table 1. Classification of molecular interaction energies.

Range	Type	Attractive (-) or Repulsive (+)	Additive or non-additive
short	overlap (Coulomb & exchange)	$\mp$	non-additive
long	electrostatic	$\mp$	additive
	induction	-	non-additive
	dispersion	-	nearly additive
	resonance	$\mp$	non-additive
	magnetic	$\mp$	(weak)

1.5 Electrostatic Energy

The simplest, and for systems such as polar gases or electrolyte solutions, the most important, long-range interaction is the electrostatic energy. It is the interaction energy of the unperturbed charge distribution of the molecules, and may be evaluated by performing an integration over the space of each molecule. If the separation between the molecules is large compared to their dimensions, the multipole expansion may conveniently be employed (see Lecture 2).

The electrostatic energy has a major role in hydrogen bonding (see Lecture 4).

1.6 Induction Energy

The induction energy is the energy resulting from the distortion of one molecule by the mean electric field due to the other molecules. Like the electrostatic energy, it is absent in the case of a pair of inert-gas atoms. The main contribution to the induction energy is due to the electric dipole induced in the ith molecule by the field $\underset{\sim}{F}^{(i)}$ resulting from the charge distribution of the other molecules:

$$u_{\text{induction}} = -\tfrac{1}{2}\sum_i \underset{\sim}{\alpha}^{(i)} : \underset{\sim}{F}^{(i)2} - \ldots\ldots \qquad (1.1)$$

where $\underset{\sim}{\alpha}^{(i)}$ is the static polarizability tensor of molecule i. Thus, in the interaction of an ion of charge q with a spherical atom, the induction energy is $-\tfrac{1}{2}\alpha q^2 R^{-4}(4\pi\varepsilon_0)^{-2}$ where $q\, R^{-2}(4\pi\varepsilon_0)^{-1}$ is the field strength at the atom distant R from the ion. If $\underset{\sim}{\alpha}^{(i)}$ is isotropic, $\alpha_{\alpha\beta}^{(i)} = \alpha^{(i)}\delta_{\alpha\beta}$ and

$$u_{\text{induction}} = -\tfrac{1}{2}\sum_i \alpha^{(i)}(F^{(i)})^2 - \ldots\ldots \qquad (1.2)$$

$\alpha^{(i)}$ is positive for a molecule in its ground electronic state, so $u_{\text{induction}} < 0$.

The induction energy is not additive since

$$\underset{\sim}{F}^{(i)} = \sum_{j \neq i} \underset{\sim}{F}^{(ij)}$$

$$(F^{(i)})^2 = \sum_{j \neq i} (F^{(ij)})^2 + \sum_{j \neq i}\sum_{k \neq i,j} \underset{\sim}{F}^{(ij)} \cdot \underset{\sim}{F}^{(ik)} \qquad (1.3)$$

the second term in (1.3) being responsible for the non-addivity. Thus the dipole induction energy of an atom midway between two ions of charge q is zero (since the field vanishes at that point), although the induction energy of the atom with each of the ions separately is $-\tfrac{1}{2}\alpha q^2 R^{-4}(4\pi\varepsilon_0)^{-2}$, where α is the polarizability of the atom and 2R the separation of the ions.

1.7 Dispersion Energy

Dispersion forces act between all molecules, although they are absent in the interaction of a proton and an atom. They result from intermolecular correlations in the fluctuations of the electronic coordinates of the molecules, and are a consequence of the quantum-mechanical nature of the electron (see Lecture 2).

If the electron were a classical particle, its position could be specified and there would be an electrostatic energy for each electronic configuration. For two spherical atoms, this classical electrostatic energy would average to zero in first order but would lead to a temperature-dependent attractive energy in second order because of the Boltzmann favouring of the configurations of lower energy. Temperature-dependent forces of this nature were discussed by Keesom in 1921 (22). The origin of the binding energy of the liquid and solid inert gases remained a mystery until Wang (23) and London (24) showed that there is an attraction due to an energy varying as R^{-6} between two spherical atoms. London (24) pointed out a link between his second-order perturbation theory for this energy and optical dispersion and hence introduced the name dispersion energy.

1.8 Resonance Energy

The resonance energy is the additional energy that results from the lifting of degeneracy by the interaction. The degeneracy may arise because one of the molecules is in a degenerate state, as in the interaction of an H atom with principal quantum number 2 with an ion or polar molecule. The degeneracy might also result from the exchange of excitation between identical molecules, as in the case of a vibrating molecule having one quantum of excitation in its ith mode ($v_i = 1$) near an identical molecule with $v_i = 0$. The lifting of the degeneracy by the interaction produces two or more potential surfaces which lie above and below zero; a sum over all the surfaces produces zero in the long-range limit, although in any particular collision the resonance energy produces either an attractive or repulsive interaction, according to the quantum numbers describing the state of the pair.

Resonance energy may be a significant contributor to vibrational relaxation. Thus the lifetime of a vibrationally excited C_6H_6 molecule was found to be significantly shorter in liquid C_6H_6 than in C_6D_6 (25,26).

1.9 Magnetic Interactions

Since magnetic dipoles are of the order of 1 Bohr magneton = 0.9274 x 10^{-20} e.m.u. = 0.9274 x 10^{-23} A m^2, while electric dipoles are ≈ 1D = 10^{-18} e.s.u. = 3.336 x 10^{-30} C m, magnetostatic energies are typically 10^{-4} of electrostatic energies. If the magnetic moments are transitory, as in a diamagnetic molecule, then the magnetic energies are smaller still and can normally be neglected. In optically active species, where the molecules exist in right- and left-handed forms, there is a coupling of the fluctuating electric and magnetic moments giving rise to a weak dispersion energy that is dependent on the handedness of the molecules. This weak dispersion force varies as R^{-7} and is attractive between similar species (i.e. left with left, and right with right) but is repulsive between dissimilar species (left with right) (27). However, it is probable that this difference is negligible and that the important discriminatory forces are of short range.

1.10 Short-range Interactions

When the overlap of the electron clouds is significant it is essential that the total wavefunction be antisymmetric with respect to exchange of all pairs of electrons, in accord with the Pauli principle (28).

One important route to short-range interaction energies is through applications of self-consistent-field theory to the interacting system at fixed nuclear positions and to the free molecules. The interaction energy is then the difference between the calculated energies, but unlike the total energy it is not in general bound. This approach can give useful results for short-range energies but if only a single configuration is employed there can be no electron correlation and hence no dispersion energy at long-range. Since the dispersion energy is the sole source of attraction between inert-gas atoms, it is to be expected that the Hartree-Fock potential curve for these systems should have no minimum. Minima have sometimes been obtained but these result from a basis-set extension effect and can be eliminated by the addition of "ghost orbitals" in the calculations on the separate molecules to compensate for the extension of the basis set of the pair (29,30).

In the region of electron overlap, the identity of the interacting molecules is lost and they are merged into a "supermolecule". It is therefore unlikely to be helpful to seek a theory of short- and intermediate-range forces which relates the interaction to the properties of the free molecules. However, in the long-range region such a general theory can be formulated and will be considered in Lecture 2.

LECTURE 2

LONG-RANGE FORCES

2.1 Introduction

In Lecture 1 we divided molecular interactions into short-range and long-range effects and we subdivided the long-range interactions into several distinct categories. In this Lecture we use quantum-mechanical perturbation theory to formulate a general theory of long-range interaction energies. The objective is to relate the interactions to properties of the free molecules.

But first we consider a simple coupled-oscillator model which can be solved exactly. It illuminates some general features of long-range interactions (3).

2.2 A Pair of Coupled Oscillators

Consider two identical one-dimensional harmonic oscillators of mass m and force constant K (Figure 2.1).

Figure 2.1. A pair of identical linear oscillators. The interaction potential energy is proportional to z_1z_2 and the Hamiltonian is given by equation (2.1).

The Hamiltonian for each free oscillator is $(1/2m)\ p_z^2 + \frac{1}{2} Kz^2$ where p_z is the momentum operator and z the displacement from equilibrium. The Hamiltonian for the coupled pair is

$$\mathcal{H} = \frac{1}{2m}\left[p_{1z}^2 + p_{2z}^2\right] + \frac{1}{2}K\left[z_1^2 + z_2^2 + 2Cz_1z_2\right] \qquad (2.1)$$

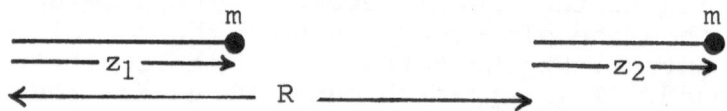

If the oscillators are considered to be electric dipoles of moments qz_1 and qz_2, the coupling constant C in (2.1) is

$$C = -2\,(q^2/KR^3)\,(4\pi\varepsilon_0)^{-1} \tag{2.2}$$

The Hamiltonian $\mathcal{H}$ can be rearranged to the separated form

$$\mathcal{H} = \frac{1}{4m}\left[(p_{1z}+p_{2z})^2+(p_{1z}-p_{2z})^2\right] + \frac{1}{4}K\left[(1+C)(z_1+z_2)^2+(1-C)(z_1-z_2)^2\right] \tag{2.3}$$

so that the oscillations split into two normal vibrations of angular frequency ω_+ and ω_- where

$$\omega_{\pm} = \omega_0\,(1 \pm C)^{\frac{1}{2}} \tag{2.4}$$

and $\omega_0 = (K/m)^{\frac{1}{2}}$ is the angular frequency of an isolated oscillator.

The equilibrium thermodynamic properties of an ensemble of harmonic oscillators of angular frequency ω are obtained through the partition function Q.

$$\begin{aligned} Q &= \sum_{v=0}^{\infty} \exp\left[-(v+\tfrac{1}{2})\hbar\omega/kT\right] \\ &= \exp(-\hbar\omega/kT)\left[1-\exp(-\hbar\omega/kT)\right]^{-1} \end{aligned} \tag{2.5}$$

In the classical limit $\hbar\omega/kT \to 0$ and Q in (2.5) becomes $kT/\hbar\omega$. Hence for the coupled oscillators

$$Q_{\text{classical}} = \frac{kT}{\hbar\omega_-}\,\frac{kT}{\hbar\omega_+} = \left(\frac{kT}{\hbar\omega_0}\right)^2 (1-C^2)^{-\frac{1}{2}} \tag{2.6}$$

The corresponding Helmholtz free energy is

$$A = -kT \ln Q_{\text{classical}} = A_0 + \tfrac{1}{2}\,kT \ln\,(1-C^2) \tag{2.7}$$

The change $\Delta A = A - A_0$ on coupling the oscillators is negative, giving an attractive force $-dA/dR$. This force is purely entropic, for the entropy change on coupling is

$$\Delta S = -\frac{\partial \Delta A}{\partial T} = -\tfrac{1}{2}\,k \ln(1-C^2) = -\Delta A/T \tag{2.8}$$

The change in internal energy is therefore zero in this classical limit.

In reality, the oscillators possess zero-point energy which changes on coupling the pair. For the ground state

$$u-u_o = \frac{1}{2}\hbar\left(\omega_- + \omega_+ - 2\omega_o\right) = \frac{1}{2}\hbar\omega_o\left[(1-C)^{\frac{1}{2}}+(1+C)^{\frac{1}{2}}-2\right] \quad (2.9)$$

This is negative for either positive or negative C and vanishes when C is zero. For small C

$$u-u_o = -\frac{1}{8}\hbar\omega_o\left[C^2 + \frac{5}{16}C^4 + \ldots\right] \quad (2.10)$$

and the interaction energy $u-u_o$ varies quadratically with the strength C of the coupling. This is the dispersion energy and it is due to correlation in the motion of the quantum oscillators. If the oscillators are dissimilar, $u-u_o$ is again negative and proportional to C^2.

If there is a vibrational quantum in one of the identical oscillators, the coupling lifts the degeneracy and gives a resonance energy linear in C. The energy levels are

$$u'_- = \frac{3}{2}\hbar\omega_- + \frac{1}{2}\hbar\omega_+ - 2\hbar\omega_o = \hbar\omega_o\left[\frac{3}{2}(1-C)^{\frac{1}{2}}+\frac{1}{2}(1+C)^{\frac{1}{2}}-2\right] \quad (2.11)$$

$$u'_+ = \frac{1}{2}\hbar\omega_- + \frac{3}{2}\hbar\omega_+ - 2\hbar\omega_o = \hbar\omega_o\left[\frac{1}{2}(1-C)^{\frac{1}{2}}+\frac{3}{2}(1+C)^{\frac{1}{2}}-2\right] \quad (2.12)$$

The mean interaction energy $\frac{1}{2}(u'_- + u'_+) = 2(u-u_o)$ and is twice that in the ground state. There is a splitting

$$u'_+ - u'_- = \hbar\omega_o\left[(1+C)^{\frac{1}{2}} - (1-C)^{\frac{1}{2}}\right] \quad (2.13)$$

For small C this reduces to

$$u'_+ - u'_- = \hbar\omega_o\left[C + \frac{1}{8}C^3 + \ldots\right] \quad (2.14)$$

The resonance energy is $\pm\frac{1}{2}(u'_+ - u'_-)$ and results from the lifting of the degeneracy of the excited states by the coupling.

2.3 General Theory of Long-range Two-body Forces (31-33)

Consider the interaction of molecule a in the electronic state m_a with molecule b in state m_b. We suppose that the effects of electron exchange are negligible, so that this treatment is limited to the long-range regime. The electrons may therefore be associated with one molecule or the other. The effects of electron exchange lead to additional interactions which decrease exponentially with R at large R.

The Hamiltonian of the pair is

$$\mathcal{H} = \mathcal{H}^{(a)} + \mathcal{H}^{(b)} + \mathcal{H}' \tag{2.15}$$

where $\mathcal{H}^{(a)}$ and $\mathcal{H}^{(b)}$ are the clamped-nuclei Hamiltonians of the free molecules a and b and $\mathcal{H}'$ is the interaction.

$$\mathcal{H}' = (4\pi\varepsilon_o)^{-1} \sum_{i,j} e_i^{(a)} e_j^{(b)} R_{ji}^{-1} \tag{2.16}$$

where R_{ji} is the distance from the ith charge $e_i^{(a)}$ (either an electron or a nucleus) in molecule a to the jth charge $e_j^{(b)}$ in b. A general theory of long-range intermolecular forces is obtained by treating $\mathcal{H}'$ as a perturbation to $\mathcal{H}^{(a)} + \mathcal{H}^{(b)}$. The unperturbed wavefunctions are the eigenfunctions of $\mathcal{H}^{(a)} + \mathcal{H}^{(b)}$ and are simple products of the wavefunctions of the free molecules a and b. The perturbed wavefunction may be expressed in terms of these unperturbed eigenfunctions by standard quantum-mechanical theory and is

$$|\psi_{m_a m_b}\rangle = |m_a m_b\rangle - \sum_{p_a,p_b}{}' \frac{\langle p_a p_b|\mathcal{H}'|m_a m_b\rangle}{W_{p_a} - W_{m_a} + W_{p_b} - W_{m_b}} |p_a p_b\rangle + \ldots \tag{2.17}$$

where the symbol $\sum'_{p_a,p_b}$ denotes a sum over all the unperturbed states with the exception of $m_a m_b$, and where $W_{p_a} - W_{m_a}$ is the energy difference between the unperturbed states of a. The perturbed and unperturbed states are taken to be orthonormal. The energy of the pair of molecules in the perturbed state is

$$W_{m_a m_b} = \langle \psi_{m_a m_b} | \mathcal{H} | \psi_{m_a m_b} \rangle$$

$$= W_{m_a} + W_{m_b} + \langle m_a m_b|\mathcal{H}'|m_a m_b\rangle - \sum_{p_a,p_b}{}' \frac{|\langle p_a p_b|\mathcal{H}'|m_a m_b\rangle|^2}{W_{p_a} - W_{m_a} + W_{p_b} - W_{m_b}} + \ldots \tag{2.18}$$

If $m_a m_b$ is degenerate and if this degeneracy is lifted by $\mathcal{H}'$, then the zero-order wavefunction $m_a m_b$ is chosen so that $\mathcal{H}$ is diagonal. The first-order energy $\langle m_a m_b|\mathcal{H}'|m_a m_b\rangle$ is the expectation value of $\mathcal{H}'$ for the unperturbed system and is the sum of the electrostatic and resonance energies:

$$u_{\text{electrostatic}} + u_{\text{resonance}} = \langle m_a m_b | \mathcal{H}' | m_a m_b \rangle \qquad (2.19)$$

If $m_a m_b$ is non-degenerate, or if $\mathcal{H}'$ does not lift the degeneracy to first-order (and this is the normal situation for molecules in liquids and solids), then the electrostatic energy is $\langle m_a m_b | \mathcal{H}' | m_a m_b \rangle$. It can be interpreted as the coulombic interaction of the unperturbed charge distributions of the two molecules.

The second-order energy in equation (2.18) may be separated into two types. The first is the induction energy and is comprised of all those terms in which either $p_a = m_a$ (with $p_b \neq m_b$) or $p_b = m_b$ (with $p_a \neq m_a$):

$$u_{\text{induction}} = u^{(a)}_{\text{induction}} + u^{(b)}_{\text{induction}} \qquad (2.20)$$

where

$$u^{(a)}_{\text{induction}} = -\sum_{p_a \neq m_a} \frac{|\langle p_a m_b | \mathcal{H}' | m_a m_b \rangle|^2}{W_{p_a} - W_{m_a}} \qquad (2.21)$$

$$u^{(b)}_{\text{induction}} = -\sum_{p_b \neq m_b} \frac{|\langle m_a p_b | \mathcal{H}' | m_a m_b \rangle|^2}{W_{p_b} - W_{m_b}} \qquad (2.22)$$

$u_{\text{induction}}$ is due to distortion of the electronic charge distribution of each molecule by the electric field resulting from the static charge distribution of its neighbours. If m_a is the ground state of molecule a, $W_{p_a} - W_{m_a} \geqslant 0$ and $u^{(a)}_{\text{induction}} \leqslant 0$. This is in accord with the variation principle which ensures that if a system in its ground state is given freedom to distort, it may utilize that freedom to lower its energy. However, if m_a is an excited electronic state of molecule a, it is possible that $u^{(a)}_{\text{induction}} > 0$.

The other type of second-order interaction energy is $u_{\text{dispersion}}$.

$$u_{\text{dispersion}} = -\sum_{\substack{p_a \neq m_a \\ p_b \neq m_b}} \frac{|\langle p_a p_b | \mathcal{H}' | m_a m_b \rangle|^2}{W_{p_a} - W_{m_a} + W_{p_b} - W_{m_b}} \qquad (2.23)$$

The dispersion energy results from matrix elements of $\mathcal{H}'$ which are off-diagonal in the electronic eigenfunctions of both molecules and is attributable to correlations in the fluctuations in the charge distributions of different molecules. It also is negative when both molecules are in their ground states.

We need to relate $u_{electrostatic}$, $u_{induction}$ and $u_{dispersion}$ to properties of the free molecules a and b. We could follow Longuet-Higgins (32) and use the diagonal and off-diagonal matrix elements of the charge-density operator $\rho(\underset{\sim}{r})$.

$$\rho(\underset{\sim}{r}) = e\left[\sum_N Z_N \delta(\underset{\sim}{r}-\underset{\sim}{r}_N) - \sum_n \delta(\underset{\sim}{r}-\underset{\sim}{r}_n)\right] \tag{2.24}$$

where the summation $\sum_N$ is over all the nuclei, whose charges are $Z_N e$ and positions $\underset{\sim}{r}_N$, and $\sum_n$ is a sum over all electrons of the system. $\delta(\underset{\sim}{r}-\underset{\sim}{r}')$ is the Dirac delta function, which is zero unless $\underset{\sim}{r}=\underset{\sim}{r}'$ and is such that

$$\int f(r')\, \delta(\underset{\sim}{r}-\underset{\sim}{r}')\, d\underset{\sim}{r}' = f(\underset{\sim}{r}) \tag{2.25}$$

The diagonal matrix elements of $\rho_a(\underset{\sim}{r})$, $\langle p_a|\rho_a(\underset{\sim}{r})|p_a\rangle$, are the total charge density of molecule a in the state p_a. The off-diagonal elements of $\rho_a(\underset{\sim}{r})$, $\langle p_a|\rho(\underset{\sim}{r})|m_a\rangle$, are called <u>transition densities</u> (32) and relate to the change in charge density when the molecule undergoes the transition from state m_a to p_a.

The interaction Hamiltonian of equation (2.16) may be written

$$\mathcal{H}' = (4\pi\epsilon_0)^{-1} \iint \frac{\rho_a(\underset{\sim}{r}_a)\,\rho_b(\underset{\sim}{r}_b)}{|\underset{\sim}{r}_b - \underset{\sim}{r}_a|}\, d\underset{\sim}{r}_a\, d\underset{\sim}{r}_b \tag{2.26}$$

The electrostatic energy for the non-degenerate state $m_a m_b$ may be written

$$u_{electrostatic} = (4\pi\epsilon_0)^{-1} \iint \frac{\langle m_a|\rho_a(\underset{\sim}{r}_a)|m_a\rangle\langle m_b|\rho_b(\underset{\sim}{r}_b)|m_b\rangle}{|\underset{\sim}{r}_b - \underset{\sim}{r}_a|}\, d\underset{\sim}{r}_a\, d\underset{\sim}{r}_b \tag{2.27}$$

and is therefore determined by the charge densities of the molecules in their unperturbed states. The second-order energies involve transition densities as well as charge-density response functions (32); since little is known of these properties, we shall not pursue this approach.

2.4 The Multipole Expansion

The standard means of expressing the various long-range interaction energies in terms of the properties of the free molecules is through the multipole expansion of $\mathcal{H}'$. Now

$$\mathcal{H}' = \sum_i e_i^{(a)} \phi_i^{(a)} = \sum_j e_j^{(b)} \phi_j^{(b)} \tag{2.28}$$

where $\phi_j^{(b)} = (4\pi\varepsilon_0)^{-1}\Sigma e_i^{(a)} R_{ji}^{-1}$ is the potential at the jth charge of molecule b due to all the charges in molecule a. This potential may be expanded in a Taylor series about an origin O_b fixed in molecule b (see Figure 2.2). This origin may be the centre-of-mass of molecule b but it could be any other point, such as the electrical centre in the molecule HD, which is midway between the two nuclei. $\mathcal{H}'$ is independent of the choice of origin.

$$\mathcal{H}' = \sum_j e_j^{(b)} \phi_j^{(b)} = \sum_j e_j^{(b)} \left[\phi_0^{(b)} + (\nabla_\alpha \phi^{(b)})_0 r_{j\alpha} + \tfrac{1}{2}(\nabla_\alpha \nabla_\beta \phi^{(b)})_0 r_{j\alpha} r_{j\beta} + \ldots\right] \tag{2.29}$$

We use the tensor summation convention in which a repeated tensor suffix implies a summation over all three components; thus

$$(\nabla_\alpha \phi)_0 r_{j\alpha} = (\nabla_x \phi)_0 r_{jx} + (\nabla_y \phi)_0 r_{jy} + (\nabla_z \phi)_0 r_{jz}.$$

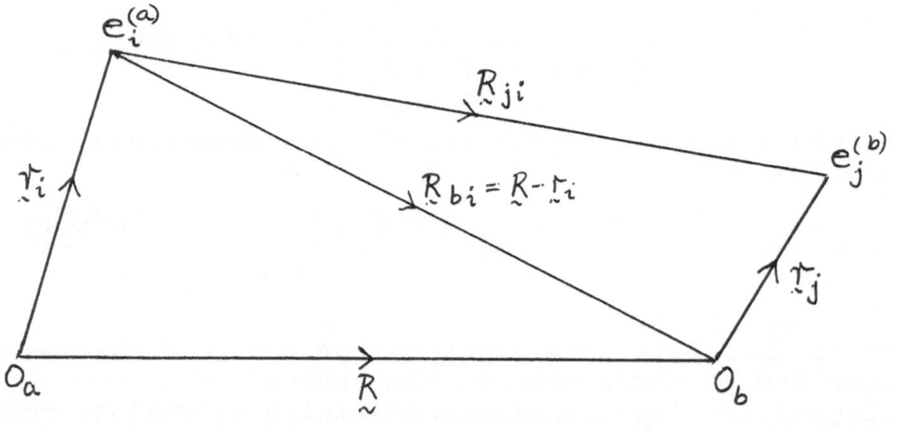

Figure 2.2. The relative positions of the point charges $e_i^{(a)}$ and $e_j^{(b)}$ in molecules a and b. O_a and O_b are origins fixed in a and b.

Equation (2.29) can be expressed in terms of the electric multipole moment operators of molecule b. These moments of the charge distribution are defined as follows:

$$\text{zeroth moment} = \sum_j e_j = q = \text{total charge} \tag{2.30}$$

$$\text{first moment} = \sum_j e_j r_{j\alpha} = \mu_\alpha = \text{dipole moment} \tag{3.21}$$

$$\text{second moment} = \sum_j e_j r_{j\alpha} r_{j\beta} = Q_{\alpha\beta} \tag{2.32}$$

The second moment is normally defined as a traceless quantity called the quadrupole moment,

$$\Theta_{\alpha\beta} = \sum_j e_j \left(\tfrac{3}{2} r_{j\alpha} r_{j\beta} - \tfrac{1}{2} r_j^2 \delta_{\alpha\beta}\right) = \tfrac{3}{2} Q_{\alpha\beta} - \tfrac{1}{2} Q_{\gamma\gamma}\delta_{\alpha\beta} \tag{2.33}$$

The nth-order multipole moment is

$$\xi^{(n)}_{\alpha\beta\ldots\nu} = (-1)^n (n!)^{-1} \sum_j e_j r_j^{(2n+1)} \frac{\partial}{\partial r_{j\alpha}} \frac{\partial}{\partial r_{j\beta}} \cdots \frac{\partial}{\partial r_{j\nu}} r_j^{-1} \tag{2.34}$$

$\xi^{(n)}$ is symmetric with respect to interchange of any pair of suffixes (thus $\Theta_{\alpha\beta} = \Theta_{\beta\alpha}$) and is reduced to zero on contraction since $\frac{\partial}{\partial r_\alpha}\frac{\partial}{\partial r_\alpha} r^{-1} = 0$ if $r \neq 0$ (thus $\Theta_{\alpha\alpha} = 0$, $\xi^{(n)}_{\alpha\alpha\gamma\cdots\nu} = 0$). Only the first non-vanishing moment is independent of the choice of origin. For example, the dipole moment of an ion varies linearly with the origin. On moving the origin through r' from O to O′, the dipole and quadrupole moments become

$$\mu'_\alpha = \sum_j e_j r'_{j\alpha} = \sum_j e_j (r_{j\alpha} - r'_\alpha) = \mu_\alpha - q r'_\alpha \tag{2.35}$$

$$\Theta'_{\alpha\beta} = \Theta_{\alpha\beta} - \tfrac{3}{2}\mu_\alpha r'_\beta - \tfrac{3}{2}\mu_\beta r'_\alpha + \mu_\gamma r'_\gamma \delta_{\alpha\beta} + q\left(\tfrac{3}{2} r'_\alpha r'_\beta - r'^2 \delta_{\alpha\beta}\right) \tag{2.36}$$

The permanent multipole moment of a molecule in state m is the expectation value $\langle m|\xi^{(n)}|m\rangle$, and in general there are 2n+1 independent components of the nth moment. However, symmetry may reduce this number; for example, there is just one independent moment of each order for a linear molecule ($C_{\infty v}$ symmetry) (34); all odd multipoles $\xi^{(2n+1)}$ about a centre of symmetry must vanish. All ions have a centre of charge at which the dipole moment $\langle m|\mu'|m\rangle$ vanishes. However, only linear molecules have a centre of dipole where the quadrupole moment is zero, for in general there is no way movement of the origin could simultaneously eliminate all five independent quadrupoles. Thus in an uncharged molecule of C_{2v} symmetry (such as H_2O), in which the two-fold rotation axis is the z-axis, the combination $\Theta_{xx}-\Theta_{yy}$ is independent of origin (see equation 2.36)).

Equation (2.29) may be written in terms of the multipole moments of molecule b,

$$\mathcal{H}' = q^{(b)}\phi_0^{(b)} - \mu_\alpha^{(b)}F_\alpha^{(b)} - \frac{1}{3}\Theta_{\alpha\beta}^{(b)}F'^{(b)}_{\alpha\beta} \ldots - \frac{1}{1.3.5\ldots(2n-1)}\xi^{(n)(b)}_{\alpha\beta\ldots\nu}F^{(b)}_{\alpha\beta\ldots\nu} \ldots \quad (2.37)$$

where $F_\alpha^{(b)} = -(\nabla_\alpha\phi^{(b)})_0$ is the field at the origin of molecule b due to molecule a, and $F'^{(b)}_{\alpha\beta} = -(\nabla_\alpha\nabla_\beta\phi^{(b)})_0$ is the field-gradient.

To complete the multipolar expression for $\mathcal{H}'$ it is necessary to relate $\phi_0^{(b)}$ to the electric moments of a. This may be done by expanding $R_{bi} = -r_i + R$ in Figure 2.2 as a Taylor series:

$$\phi_0^{(b)} = (4\pi\epsilon_0)^{-1}\sum_i e_i^{(a)} R_{bi}^{-1}$$

$$= (4\pi\epsilon_0)^{-1}\sum_i e_i^{(a)}\left[R^{-1} + \left(\frac{\partial}{\partial r_{i\alpha}}R_{bi}^{-1}\right)_0 r_{i\alpha} + \frac{1}{2}\left(\frac{\partial}{\partial r_{i\alpha}}\frac{\partial}{\partial r_{i\beta}}R_{bi}^{-1}\right)_0 r_{i\alpha}r_{i\beta} + \ldots\right]$$

$$= (4\pi\epsilon_0)^{-1}\sum_i e_i^{(a)}\left[R^{-1} - r_{i\alpha}\nabla_\alpha R^{-1} + \frac{1}{2}r_{i\alpha}r_{i\beta}\nabla_\alpha\nabla_\beta R^{-1} - \ldots\right]$$

$$= q^{(a)}T - \mu_\alpha^{(a)}T_\alpha + \frac{1}{3}\Theta_{\alpha\beta}^{(a)}T_{\alpha\beta} - \ldots + (-1)^n\frac{1}{1.3.5\ldots(2n-1)}\xi^{(n)}_{\alpha\beta\ldots\nu}T_{\alpha\beta\ldots\nu} + \ldots \quad (2.$$

where $T_{\alpha\beta\ldots\nu} = (4\pi\varepsilon_0)^{-1}\nabla_\alpha\nabla_\beta\ldots\nabla_\nu R^{-1}$ is an nth rank tensor proportional to $R^{-(n+1)}$ which is symmetric under interchange of any pair of suffixes and is reduced to zero on contraction ($T_{\alpha\alpha\gamma\ldots\nu}=0$). The first four $\underset{\sim}{T}$ tensors are:

$$T = (4\pi\varepsilon_0)^{-1} R^{-1} \tag{2.39}$$

$$T_\alpha = -(4\pi\varepsilon_0)^{-1} R_\alpha R^{-3} \tag{2.40}$$

$$T_{\alpha\beta} = (4\pi\varepsilon_0)^{-1} (3R_\alpha R_\beta - R^2\delta_{\alpha\beta}) R^{-5} \tag{2.41}$$

$$T_{\alpha\beta\gamma} = -3(4\pi\varepsilon_0)^{-1} \left[5R_\alpha R_\beta R_\gamma - R^2(R_\alpha\delta_{\beta\gamma} + R_\beta\delta_{\gamma\alpha} + R_\gamma\delta_{\alpha\beta})\right] R^{-7} \tag{2.42}$$

The electric field and field-gradients at b may easily be obtained from equation (2.38):

$$F_\alpha^{(b)} = -\nabla_\alpha \phi_0^{(b)} = -q^{(a)} T_\alpha + \mu_\beta^{(a)} T_{\alpha\beta} - \tfrac{1}{3}\Theta_{\alpha\beta}^{(a)} T_{\alpha\beta\gamma} + \ldots \tag{2.43}$$

$$F'^{(b)}_{\alpha\beta} = -\nabla_\alpha \nabla_\beta \phi_0^{(b)} = -q^{(a)} T_{\alpha\beta} + \mu_\gamma^{(a)} T_{\alpha\beta\gamma} - \tfrac{1}{3}\Theta_{\gamma\delta}^{(a)} T_{\alpha\beta\gamma\delta} + \ldots \tag{2.44}$$

Combining equations (2.37)-(2.44) gives the multipole expansion for $\mathcal{H}'$,

$$\mathcal{H}' = Tq^{(a)}q^{(b)} + T_\alpha\left(q^{(a)}\mu_\alpha^{(b)} - q^{(b)}\mu_\alpha^{(a)}\right) + T_{\alpha\beta}\left(\tfrac{1}{3}q^{(a)}\Theta_{\alpha\beta}^{(b)} + \tfrac{1}{3}q^{(b)}\Theta_{\alpha\beta}^{(a)} - \mu_\alpha^{(a)}\mu_\beta^{(b)}\right) + \ldots$$

$$= \sum_{n=0}^{\infty}\sum_{n'=0}^{\infty} (-1)^{n'} \frac{1}{1.3.5\ldots(2n-1)} \frac{1}{1.3.5\ldots(2n'-1)} T_{\alpha\beta\ldots\nu\alpha'\beta'\ldots\nu'} \xi_{\alpha\beta\ldots\nu}^{(n)(b)} \xi_{\alpha'\beta'\ldots\nu'}^{(n')(a)} \tag{2.45}$$

Movement of either origin O_a or O_b changes each term in $\mathcal{H}'$ but leaves $\mathcal{H}'$ unchanged. The multipole expansion is useful provided the intermolecular distance R is large compared to the dimensions of each molecule. In many applications only the first non-vanishing contribution is evaluated and one exploits the asymptotic behaviour of the series. Kreek et al (35) discussed the validity of the expansion in R^{-2} of the dispersion energy of atoms and showed that outside the region of validity of the expansion, inclusion of more terms leads to a poorer approximation.

LECTURE 3

EVALUATION OF INTERMOLECULAR FORCES

3.1 Introduction

In the previous lecture a general perturbation theory of long-range intermolecular forces was presented. We now use the multipole expansion for $\mathcal{H}'$ to express the long-range interaction energy in terms of the properties of the free molecules.

3.2 The Electrostatic Energy

If molecules a and b are in non-degenerate states m_a and m_b, the electrostatic energy is just the expectation value of $\mathcal{H}'$ and, from equation (2.45), can be written in terms of the permanent electric moments, $\underset{\sim}{\xi}^{(n)(m_b)}$ and $\underset{\sim}{\xi}^{(n')(m_a)}$:

$$u_{\text{electrostatic}} = \sum_{n=0}^{\infty}\sum_{n'=0}^{\infty}(-1)^{n'}\frac{2^{n+n'}n!\,n'!}{(2n)!\,(2n')!}T_{\alpha\beta\ldots\nu\alpha'\beta'\ldots\nu'}\xi^{(n)(m_b)}_{\alpha\beta\ldots\nu}\xi^{(n')(m_a)}_{\alpha'\beta'\ldots\nu'} \quad (3.1)$$

For n=0, 1 and 2 there are sound experimental techniques for determining these moments for molecules in their ground electronic states (36). The charge q is independent of state. The dipole moment of a molecule in an excited state may be measured through the optical Stark effect (37).

If the two molecules are axially symmetric, and in the relative orientations in Figure 3.1, the first few terms in the electrostatic energy are

$$\begin{aligned}
u_{\text{electrostatic}} = {} & (4\pi\epsilon_0)^{-1}q^{(a)}\left[q^{(b)}R^{-1} + \mu^{(m_b)}\cos\theta_b R^{-2} + \Theta^{(m_b)}\left(\tfrac{3}{2}\cos^2\theta_b - \tfrac{1}{2}\right)R^{-3} + \ldots\right] \\
& + (4\pi\epsilon_0)^{-1}\mu^{(m_a)}\Big[q^{(b)}\cos\theta_a R^{-2} + \mu^{(m_b)}(2\cos\theta_a\cos\theta_b + \sin\theta_a\sin\theta_b\cos\phi)R^{-3} \\
& \qquad + 3\,\Theta^{(m_b)}\left\{\cos\theta_a\left(\tfrac{3}{2}\cos^2\theta_b - \tfrac{1}{2}\right) + \sin\theta_a\sin\theta_b\cos\theta_b\cos\phi\right\}R^{-4} + \ldots\Big] \\
& + (4\pi\epsilon_0)^{-1}\Theta^{(m_a)}\Big[q^{(b)}\left(\tfrac{3}{2}\cos^2\theta_a - \tfrac{1}{2}\right) \\
& \qquad + 3\mu^{(m_b)}\left\{\cos\theta_b\left(\tfrac{3}{2}\cos^2\theta_a - \tfrac{1}{2}\right) + \sin\theta_a\cos\theta_a\sin\theta_b\cos\phi\right\}R^{-4} \\
& \qquad + \tfrac{3}{4}\Theta^{(m_b)}\{1 - 5\cos^2\theta_a - 5\cos^2\theta_b + 17\cos^2\theta_a\cos^2\theta_b + \\
& \qquad\qquad 2\sin^2\theta_a\sin^2\theta_b\cos^2\phi + \\
& \qquad\qquad 16\sin\theta_a\cos\theta_a\sin\theta_b\cos\theta_b\cos\phi\}R^{-5} + \ldots\Big] \\
& + \ldots
\end{aligned} \quad (3.2)$$

Some of the configurations favoured by electrostatic forces are shown in Figure 3.2.

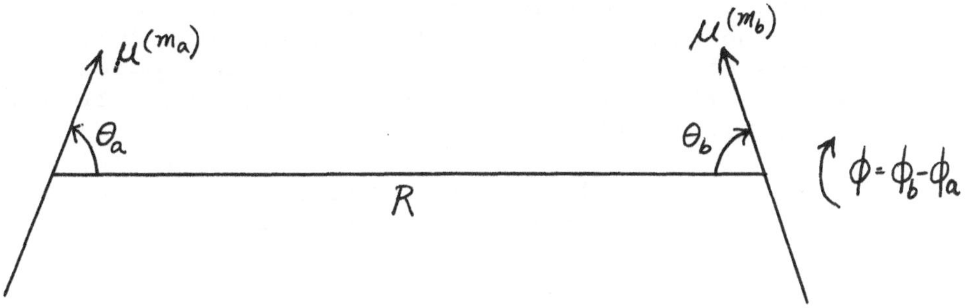

Figure 3.1. The relative position and orientation of a pair of axially symmetric molecules a and b. ϕ is the angle between the planes formed by the molecular axes with the line of centres.

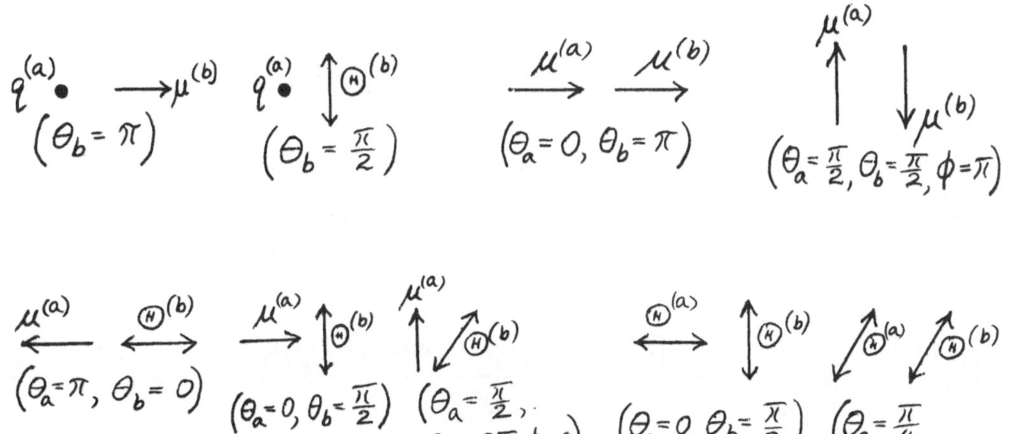

Figure 3.2. Favoured orientations for a pair of linear molecules with charge-dipole and charge-quadrupole, dipole-dipole, dipole-quadrupole, and quadrupole-quadrupole, interactions. The moments q, μ, Θ are taken to be positive.

3.3 Static Polarizabilities

The induction and dispersion energies result from the second-order perturbed energy and therefore depend upon off-diagonal matrix elements of $\mathcal{H}'$ (equations (2.21)-(2.23)). When the multipole expansion of $\mathcal{H}'$ is used, the longest-range contributions involve the square of the transition dipole moment (since there are no off-diagonal elements of the charge q). Transition dipoles also determine the optical properties of matter, including the polarizability α.

For a molecule fixed in orientation in a uniform electric field $\underset{\sim}{F}$, the Hamiltonian is

$$\mathcal{H}(\underset{\sim}{F}) = \mathcal{H} - \mu_\alpha F_\alpha \tag{3.3}$$

and the energy can be written as a power series in $\underset{\sim}{F}$ (38)

$$W_m(\underset{\sim}{F}) = W_m - \mu_\alpha^{(m)} F_\alpha - \tfrac{1}{2}\alpha_{\alpha\beta}^{(m)} F_\alpha F_\beta - \frac{1}{6}\beta_{\alpha\beta\gamma}^{(m)} F_\alpha F_\beta F_\gamma - \frac{1}{24}\gamma_{\alpha\beta\gamma\delta}^{(m)} F_\alpha F_\beta F_\gamma F_\delta - \ldots \tag{3.4}$$

where $\mu_\alpha^{(m)} = \langle m|\mu_\alpha|m\rangle$ is the permanent dipole moment, $\alpha_{\alpha\beta}^{(m)}$ is the polarizability, and $\beta_{\alpha\beta\gamma}^{(m)}$ and $\gamma_{\alpha\beta\gamma\delta}^{(m)}$ the first and second hyperpolarizabilities of the molecule in state m. Perturbation theory leads to the expressions

$$\alpha_{\alpha\beta}^{(m)} = \sum_{p\neq m} \frac{\langle m|\mu_\alpha|p\rangle\langle p|\mu_\beta|m\rangle + \langle m|\mu_\beta|p\rangle\langle p|\mu_\alpha|m\rangle}{W_p - W_m} = \alpha_{\beta\alpha}^{(m)} \tag{3.5}$$

$$\beta_{\alpha\beta\gamma}^{(m)} = S(\alpha\beta\gamma)\sum_{p\neq m}\left[\sum_{q\neq m}\frac{\langle m|\mu_\alpha|p\rangle\langle p|\mu_\beta|q\rangle\langle q|\mu_\gamma|m\rangle}{(W_p-W_m)(W_q-W_m)} - \frac{\langle m|\mu_\alpha|m\rangle\langle m|\mu_\beta|p\rangle\langle p|\mu_\gamma|m\rangle}{(W_p-W_m)^2}\right]$$

$$= \beta_{\beta\gamma\alpha}^{(m)} = \beta_{\gamma\alpha\beta}^{(m)} = \beta_{\alpha\gamma\beta}^{(m)} = \beta_{\beta\alpha\gamma}^{(m)} = \beta_{\gamma\beta\alpha}^{(m)} \tag{3.6}$$

$S(\alpha\beta\gamma)$ indicates a sum of all permutations of the tensor components - thus $S(\alpha\beta\gamma)X_{\alpha\beta\gamma}=X_{\alpha\beta\gamma}+X_{\beta\gamma\alpha}+X_{\gamma\alpha\beta}+X_{\alpha\gamma\beta}+X_{\beta\alpha\gamma}+X_{\gamma\beta\alpha}$. The static polarizability and hyperpolarizabilities are symmetric with respect to interchange of any pair of suffixes. The dipole moment in the field is obtained by differentiating the energy $W_m(\underset{\sim}{F})$ with respect to $\underset{\sim}{F}$,

$$\mu_\alpha^{(m)}(\underset{\sim}{F}) = -\frac{\partial W_m(\underset{\sim}{F})}{\partial F_\alpha} = \mu_\alpha^{(m)} + \alpha_{\alpha\beta}^{(m)} F_\beta + \frac{1}{2}\beta_{\alpha\beta\gamma}^{(m)} F_\beta F_\gamma + \frac{1}{6}\gamma_{\alpha\beta\gamma\delta}^{(m)} F_\beta F_\gamma F_\delta + \ldots \quad (3.7)$$

Thus $\underset{\sim}{\alpha}^{(m)}$ determines the linear polarization of a molecule by an external field; the hyperpolarizabilities determine the departure from a linear polarization law. If the molecule is centrosymmetric, $\mu_\alpha^{(m)}$, $\beta_{\alpha\beta\gamma}^{(m)}$, and all coefficients of odd powers of the field in equation (3.4), must vanish.

Typical values for these properties of a small polar molecule are as follows:

$$\mu \approx 1\text{D} = 10^{-18}\ \text{e.s.u.} = 3.3356\times10^{-30}\ \text{C m}$$

$$(4\pi\varepsilon_0)^{-1}\alpha \approx 1\text{Å}^3 = 10^{-24}\ \text{e.s.u.} = 10^{-30}\ \text{m}^3$$

$$(4\pi\varepsilon_0)^{-2}\beta \approx \pm 10^{-30}\ \text{e.s.u.} = \pm0.29979\times10^{-30}\ \text{C}^{-1}\ \text{m}^5$$

$$(4\pi\varepsilon_0)^{-3}\gamma \approx 10^{-36}\ \text{e.s.u.} = 0.089876\times10^{-30}\text{C}^{-1}\ \text{m}^7$$

The quantity $(4\pi\varepsilon_0)^{-1}\alpha$ may be compared to the volume of the molecule. The various induced dipoles in equation (3.7) are of the same order of magnitude when the field strength $F \approx 1$ atomic unit $= ea_0^{-2}(4\pi\varepsilon_0)^{-1} = 1.7153\times10^7$ e.s.u.$=5.1423\times10^{11}$ V m^{-1}.

When the external field is non-uniform, as in the vicinity of a polar molecule, the Hamiltonian becomes

$$\mathcal{H}(\underset{\sim}{F},\underset{\sim}{F}') = \mathcal{H} - \mu_\alpha F_\alpha - \frac{1}{3}\Theta_{\alpha\beta} F'_{\alpha\beta} - \ldots\ldots \quad (3.8)$$

and the energy

$$W_m(\underset{\sim}{F},\underset{\sim}{F}') = W_m - \mu_\alpha^{(m)} F_\alpha - \frac{1}{3}\Theta_{\alpha\beta}^{(m)} F'_{\alpha\beta} - \ldots - \frac{1}{2}\alpha_{\alpha\beta}^{(m)} F_\alpha F_\beta - \frac{1}{3}A_{\alpha\beta\gamma}^{(m)} F_\alpha F_{\beta\gamma} - \frac{1}{6}C_{\alpha\beta\gamma\delta}^{(m)} F'_{\alpha\beta} F'_{\gamma\delta} - \ldots \quad (3.9)$$

The total dipole and quadrupole moments are, for linear distortion,

$$\mu_\alpha^{(m)}(\underset{\sim}{F},\underset{\sim}{F}') = \mu_\alpha^{(m)} + \alpha_{\alpha\beta}^{(m)} F_\beta + \frac{1}{3} A_{\alpha\beta\gamma}^{(m)} F'_{\beta\gamma} + \ldots \tag{3.10}$$

$$\Theta_{\alpha\beta}^{(m)}(\underset{\sim}{F},\underset{\sim}{F}') = -3\frac{\partial W_m(\underset{\sim}{F},\underset{\sim}{F}')}{\partial F'_{\alpha\beta}} = \Theta_{\alpha\beta}^{(m)} + A_{\gamma\alpha\beta}^{(m)} F_\gamma + C_{\alpha\beta\gamma\delta}^{(m)} F'_{\gamma\delta} + \ldots \tag{3.11}$$

$\underset{\sim}{C}^{(m)}$ is the quadrupole polarizability; atoms and molecules of all symmetries have a non-vanishing $\underset{\sim}{\alpha}^{(m)}$ and $\underset{\sim}{C}^{(m)}$. The dipole-quadrupole polarizability $\underset{\sim}{A}^{(m)}$ determines both the dipole induced by a field-gradient <u>and</u> the quadrupole induced by a field; it is origin-dependent and vanishes for centrosymmetric molecules if the origin is at the centre. From perturbation theory

$$A_{\alpha\beta\gamma}^{(m)} = \sum_{p\neq m} \frac{\langle m|\mu_\alpha|p\rangle\langle p|\Theta_{\beta\gamma}|m\rangle + \langle m|\Theta_{\beta\gamma}|p\rangle\langle p|\mu_\alpha|m\rangle}{W_p - W_m} = A_{\alpha\gamma\beta}^{(m)} \tag{3.12}$$

$$C_{\alpha\beta\gamma\delta}^{(m)} = \frac{1}{3}\sum_{p\neq m} \frac{\langle m|\Theta_{\alpha\beta}|p\rangle\langle p|\Theta_{\gamma\delta}|m\rangle + \langle m|\Theta_{\gamma\delta}|p\rangle\langle p|\Theta_{\alpha\beta}|m\rangle}{W_p - W_m} = C_{\gamma\delta\alpha\beta}^{(m)} \tag{3.13}$$

From equations (2.35), (2.36), (3.5), (3.6) and (3.12), it may be shown that on moving the origin through $\underset{\sim}{r}'$ from O to O',

$$\alpha'_{\alpha\beta} = \alpha_{\alpha\beta}, \quad \beta'_{\alpha\beta\gamma} = \beta_{\alpha\beta\gamma}, \quad \gamma'_{\alpha\beta\gamma\delta} = \gamma_{\alpha\beta\gamma\delta}, \ \ldots \tag{3.14}$$

$$A'_{\alpha\beta\gamma} = A_{\alpha\beta\gamma} - \frac{3}{2} r'_\beta \alpha_{\alpha\gamma} - \frac{3}{2} r'_\gamma \alpha_{\alpha\beta} + r'_\delta \alpha_{\alpha\delta} \delta_{\beta\gamma} \tag{3.15}$$

For a tetrahedron such as CH_4, $\underset{\sim}{\alpha}$ is isotropic and there is only one independent component of $\underset{\sim}{A}$, viz $A_{xyz} = A_{yzx} = A_{zxy} = A_{xzy} = A_{yxz} = A_{zyx}$, where x,y,z axes are perpendicular to the faces of a cube in which the corners of the tetrahedron are at the points (1,1,1), (1,-1,-1), (-1,1,-1), (-1,-1,1) (see Figure 3.3). $A_{xyz} = A$ is independent of the origin in a tetrahedron (see (3.15)). If the tetrahedral molecule can be represented by four anisotropically polarizable atoms, distant R_o from the centre, then (38)

$$A = A_{xyz} = \frac{4}{\sqrt{3}} (\alpha_{\parallel} - \alpha_{\perp}) R_o \qquad (3.16)$$

where $\alpha_{\parallel} - \alpha_{\perp}$ is the difference between the polarizabilities of one of the atoms in the directions parallel and perpendicular to its bond to the centre. Thus for a small tetrahedral molecule, $(4\pi\varepsilon_o)^{-1}A \sim 1\ \mathring{A}^4 = 10^{-32}\ cm^4$. For CH_4, an accurate ab initio computation (39) yielded $(4\pi\varepsilon_o)^{-1}\alpha = 2.45 \times 10^{-24}\ cm^3$ and $(4\pi\varepsilon_o)^{-1} A = 0.79 \times 10^{-32}\ cm^4$.

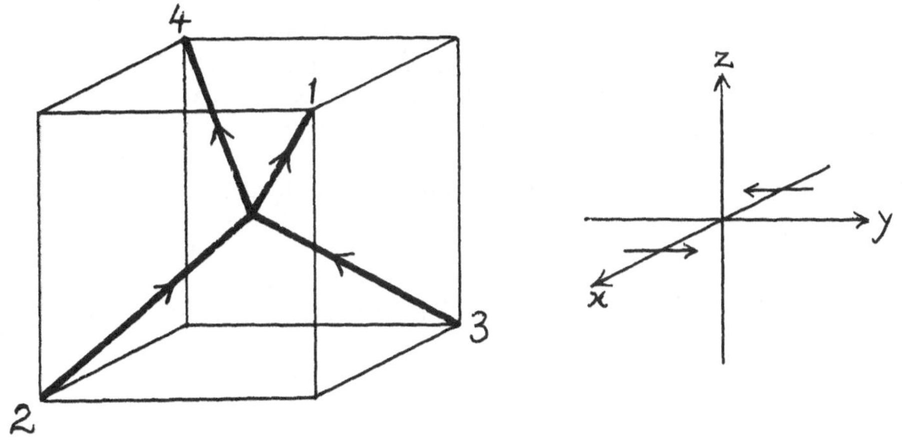

Figure 3.3. A tetrahedron enclosed in a cube. In the field-gradient F'_{xy} illustrated on the right, the anisotropically polarizable atoms at the four corners of the tetrahedron contribute in the same sense to the induced dipole μ_z. Atoms 1 and 2 experience positive fields F_y while 3 and 4 are in negative fields in the y direction.

3.4 Frequency-dependent Polarizabilities

If the external field is time-dependent, the molecule undergoes a time-dependent distortion which can be described by frequency-dependent polarizabilities. If the external field oscillates at an angular frequency ω

$$F_\beta = F_\beta^{(o)} \exp(-i\omega t) \qquad (3.17)$$

then the complex induced dipole proportional to F_β is

$$\mu_\alpha = \alpha_{\alpha\beta}^{(m)}(\omega) F_\beta \qquad (3.18)$$

From time-dependent perturbation theory (40, 33, 38)

$$\alpha_{\alpha\beta}^{(m)}(\omega) = \sum_p \frac{\omega_{pm}\left[\langle m|\mu_\alpha|p\rangle\langle p|\mu_\beta|m\rangle + \langle m|\mu_\beta|p\rangle\langle p|\mu_\alpha|m\rangle\right]}{\hbar\left(\omega_{pm}^2 - \omega^2 - i\omega\Gamma_{pm}\right)} = \alpha_{\beta\alpha}^{(m)}(\omega) \tag{3.19}$$

where $\hbar\omega_{pm}=W_p-W_m$ and Γ_{pm} is the width at half the maximum height of the absorption centred at ω_{pm}. The real part of $\alpha_{\alpha\beta}^{(m)}(\omega)$ determines the optical refractivity of the system, and the imaginary part determines the absorption coefficient. If the molecules have an electronic angular momentum (as in a sodium atom in its ^2S ground state), or if they are in a magnetostatic field, there is also an antisymmetric contribution to $\tilde{\alpha}_{\alpha\beta}^{(m)}(\omega)$ depending on $\langle m|\mu_\alpha|p\rangle\langle p|\mu_\beta|m\rangle - \langle m|\mu_\beta|p\rangle\langle p|\mu_\alpha|m\rangle$. This exists if the molecules lack symmetry under time reversal (33):

$$\tilde{\alpha}_{\alpha\beta}^{(m)}(\omega) = \alpha_{\alpha\beta}^{(m)}(\omega) - i\,\alpha'^{(m)}_{\alpha\beta}(\omega) = \alpha_{\beta\alpha}^{(m)}(\omega) + i\,\alpha'^{(m)}_{\beta\alpha} \tag{3.20}$$

3.5 The Induction Energy

The induction energy results from distortion of the electron clouds by the electric field and field-gradients due to the charge distribution of neighbouring molecules. It is therefore dependent upon the static polarizabilities. For molecule b in the state m_b

$$u_{\text{induction}}^{(b)} = -\tfrac{1}{2}\alpha_{\alpha\beta}^{(m_b)}F_\alpha^{(b)}F_\beta^{(b)} - \tfrac{1}{3}A_{\alpha\beta\gamma}^{(m)}F_\alpha^{(b)}F'^{(b)}_{\beta\gamma} - \tfrac{1}{6}C_{\alpha\beta\gamma\delta}^{(m)}F'^{(b)}_{\alpha\beta}F'^{(b)}_{\gamma\delta} - \ldots \tag{3.21}$$

where $F_\alpha^{(b)}$ and $F'^{(b)}_{\alpha\beta}$ are the field and field-gradient at the origin of molecule b due to the unperturbed charge distributions of neighbouring molecules. These fields may be expressed in terms of the permanent multipole moments of the neighbours by means of equations (2.43) and (2.44).

Induction energy is easy to evaluate, since static polarizabilities are readily accessible. However, $u_{induction}$ is rarely predominant and in many cases it can be neglected in comparison to the electrostatic and dispersion energies.

3.6 The Dispersion Energy

Unlike the induction energy, the dispersion energy does not separate easily into single-centre contributions like equation (3.21). For a precise formulation, it is necessary to consider the frequency-dependent polarizabilities at imaginary frequencies. Their relationship to dispersion energy depends upon the identities (for A > 0, B > 0)

$$\frac{1}{A+B} = \frac{2}{\pi}\int_0^\infty \frac{AB}{(A^2+u^2)(B^2+u^2)}\,du = \frac{2}{\pi}\int_0^\infty \frac{u^2}{(A^2+u^2)(B^2+u^2)}\,du \quad (3.22)$$

The polarizability at the imaginary frequency $\omega = iu$ (where u is real) is

$$\tilde{\alpha}^{(m)}_{\alpha\beta}(iu) = \sum_p \frac{2\omega_{pm}\,\mathrm{Re}\{\langle m|\mu_\alpha|p\rangle\langle p|\mu_\beta|m\rangle\} - 2u\,\mathrm{Im}\{\langle m|\mu_\alpha|p\rangle\langle p|\mu_\beta|m\rangle\}}{\hbar(\omega_{pm}^2 + u^2)}$$

$$= \alpha^{(m)}_{\alpha\beta}(iu) - i\alpha'^{(m)}_{\alpha\beta}(iu) \quad (3.23)$$

which is real ($\alpha'^{(m)}_{\alpha\beta}(iu)$ is pure imaginary). The dependence of $\tilde{\alpha}^{(m)}_{\alpha\beta}(iu)$ on the line width Γ_{pm} (see equation (3.19)) has been dropped since there are no poles in $\alpha^{(m)}_{\alpha\beta}(iu)$; actually $\alpha^{(m)}_{\alpha\beta}(iu)$ is a positive, monotonically decreasing, function of u - the singular behaviour of $\alpha^{(m)}_{\alpha\beta}(\omega)$ at $\omega = \omega_{pm}$ is removed on going to imaginary frequencies.

From equations (2.23) and (2.45), the dipole dispersion energy is

$$u_{dispersion}(R^{-6}) = -T_{\alpha\beta}T_{\gamma\delta}\sum_{\substack{p_a \neq m_a \\ p_b \neq m_b}} \frac{\langle m_a m_b|\mu^{(a)}_\alpha \mu^{(b)}_\beta|p_a p_b\rangle\langle p_a p_b|\mu^{(a)}_\gamma \mu^{(b)}_\delta|m_a m_b\rangle}{\hbar(\omega_{p_a m_a} + \omega_{p_b m_b})}$$

(3.24)

Applying the identities (3.22) to (3.24) gives

$$u_{dispersion}(R^{-6}) = -\frac{\hbar}{2\pi} T_{\alpha\beta} T_{\gamma\delta} \int_0^\infty \left\{ \alpha_{\alpha\gamma}^{(m_a)}(iu)\alpha_{\beta\delta}^{(m_b)}(iu) + \alpha'^{(m_a)}_{\alpha\gamma}(iu)\alpha'^{(m_b)}_{\beta\delta}(iu) \right\} du \quad (3.25)$$

which has achieved the representation of the dispersion energy in terms of properties of the separate molecules.

Equation (3.25) is precise and it reflects the fact that the dispersion energy results from a dynamic coupling of the electronic fluctuations in the interacting molecules. The contribution in $\int_0^\infty \alpha'^{(m_a)}_{\alpha\beta}(iu)\alpha'^{(m_b)}_{\beta\delta}(iu)du$ is small and negative for a pair of alkali-metal atoms having the same spin, and is positive when the spins are opposed. For isotropically polarizable molecules, (3.25) simplifies to

$$u_{dispersion}(R^{-6}) = -(4\pi\varepsilon_0)^{-2} \frac{3\hbar}{\pi} R^{-6} \int_0^\infty \alpha^{(m_a)}(iu)\alpha^{(m_b)}(iu)du$$
$$= C_6 R^{-6} \quad (3.26)$$

Very reliable values of C_6 are now available (41-44), and the higher coefficients C_8 and C_{10} are approximately known for some simple atoms (44).

In many cases, the dispersion of molecular polarizabilities is not available, and (3.26) is not appropriate for evaluating the dispersion energy. Then one may use approximate formulae involving the <u>static</u> polarizabilities and mean electronic excitation energies U (45, 38).

$$u_{dispersion} = -\frac{U_a U_b}{4(U_a + U_b)} \Big[T_{\alpha\beta} T_{\gamma\delta} \alpha_{\alpha\gamma}^{(m_a)} \alpha_{\beta\delta}^{(m_b)} + \frac{2}{3} T_{\alpha\beta} T_{\gamma\delta\varepsilon} \left(\alpha_{\alpha\gamma}^{(m_a)} A_{\beta\delta\varepsilon}^{(m_b)} - A_{\beta\delta\varepsilon}^{(m_a)} \alpha_{\alpha\gamma}^{(m_b)} \right)$$
$$+ \frac{1}{3} T_{\alpha\beta\gamma} T_{\delta\varepsilon\phi} \Big(\alpha_{\alpha\delta}^{(m_a)} C_{\beta\gamma\varepsilon\phi}^{(m_b)} + C_{\beta\gamma\varepsilon\phi}^{(m_a)} \alpha_{\alpha\delta}^{(m_b)} -$$
$$\frac{2}{3} A_{\alpha\varepsilon\phi}^{(m_a)} A_{\delta\beta\gamma}^{(m_b)} \Big) - \frac{2}{9} T_{\alpha\beta} T_{\gamma\delta\varepsilon\phi} A_{\alpha\gamma\delta}^{(m_a)} A_{\beta\varepsilon\phi}^{(m_b)}$$
$$+ \ldots \Big] \quad (3.27)$$

In practice U_a and U_b are normally set equal to the first ionization potentials of molecules a and b, but if the first dipole-allowed excitation energy is used for U, then the dispersion energy given by (3.27) is a lower bound. For spherical atoms, (3.27) reduces to

$$u_{dispersion} = -(4\pi\varepsilon_0)^{-2} \frac{3U\alpha^2}{4R^6} \left(1 + \frac{10C}{\alpha R^2} + \ldots\right) \qquad (3.28)$$

where $C = \frac{1}{5} C_{\alpha\beta\alpha\beta}$ is the static quadrupole polarizability. For the H atom, the first ionization energy is $\frac{1}{2}$, $\alpha=\frac{9}{2}$, $C=\frac{15}{2}$ (in atomic units), and the dispersion coefficients given by (3.28) are $C_6=-7.59$, $C_8=-123$, and the accurate values are $C_6=-6.499$, $C_8=-124.4$ (46).

LECTURE 4

SHORT-RANGE INTERACTIONS AND HYDROGEN BONDING

4.1 Introduction

In this final lecture we shall be concerned with short-range intermolecular forces, with non-additivity and the effects of a medium, with the hydrogen bond, and finally with vibrational contributions to interaction energies.

4.2 Short-range Intermolecular Forces

As noted in Lecture 1, short-range forces result from overlap of the electron clouds of the interacting molecules. In this region it is essential that the wave-function be antisymmetric with respect to exchange of all pairs of electrons. One way to ensure this antisymmetry is to choose a basis set for a variational calculation that is antisymmetric, and such a basis set for the interacting pair ab could be the antisymmetrized product functions $\mathcal{A}p_a p_b$, where $\mathcal{A}$ is the operator which antisymmetrizes with respect to intermolecular exchange of electrons. The set of functions $\mathcal{A}p_a p_b$ are not orthogonal at separations at which overlap is significant and cannot therefore be eigenfunctions of a Hamiltonian. Normal quantum-mechanical perturbation theory is therefore not applicable, and because the basis is overcomplete, there is no unique transformation to an orthogonal set. It is possible to perform a variational calculation with a trial function which is a sum of a finite number of terms of the set $\mathcal{A}p_a p_b$. A simpler technique (Lecture 1, § 1.10) is to use the usual self-consistent-field theory (47) to obtain the best one-electron wavefunctions for the interacting molecules and to evaluate the interaction energy by substracting the energy computed for the separate molecules.

Accurate variational calculations, involving extensive configuration interaction, have been performed on He_2 (48-50). The well-depth ε is 15×10^{-23} J (ε/k = 11 K) and the equilibrium separation $R_e = 3.0\times10^{-10}$ m. A variational calculation, including intermolecular but not intramolecular correlation, has been performed on Ne_2 (51), yielding $\varepsilon = 54\times10^{-23}$ J and $R_e = 3.08\times10^{-10}$ m, and long-range interaction closely following C_6R^{-6}. However, if there is no intra-atomic correlation, one might expect that the computed long-range behaviour should correspond to a Hartree-Fock C_6, which would differ from the true C_6 by a factor (see 3.26) of approximately $(\alpha_{\text{Hartree-Fock}}/\alpha_{\text{true}})^2 = 0.77$ (52).

In the region of small overlap, the interaction energy is small compared to the total energy and it is tempting to seek a perturbation scheme for calculating the interaction. Unfortunately, there are difficulties stemming from electron exchange. The problems have been considered by Hirschfelder (53), and the perturbation approach has been reviewed by Certain and Bruch (54). Gerratt has applied his spin-coupled formalism to the problem (55).

If the unperturbed wavefunction is taken to be the antisymmetrized product $\psi_o = \mathcal{A}m_a m_b$, where m_a and m_b are the unperturbed states of the isolated molecules a and b, the first-order energy $W^{(1)}_{m_a m_b} = \langle\psi_o|\mathcal{H}-W_{m_a}-W_{m_b}|\psi_o\rangle/\langle\psi_o|\psi_o\rangle$ separates as follows into a Coulomb and an exchange contribution. $\mathcal{A}$ may be written

$$\mathcal{A} = \left[\frac{(n_i)!\ \ (n_j)!}{(n_i + n_j)!}\right]^{\frac{1}{2}} (1 + P) \tag{4.1}$$

where

$$P = -\sum_{i,j} P_{ij} + \sum_{i,j} \sum_{\substack{i'>i \\ j'>j}} P_{ij} P_{i'j'} - \dots \tag{4.2}$$

n_i and n_j are the numbers of electrons in the separate molecules a and b, and P_{ij} exchanges the ith electron of a with the jth of b. With $\mathcal{H} = \mathcal{H}^{(a)} + \mathcal{H}^{(b)} + \mathcal{H}'$ as in (2.15), the first-order energy is therefore

$$W^{(1)}_{m_a m_b} = \frac{\langle m_a m_b|\mathcal{H}'|m_a m_b\rangle + \langle Pm_a m_b|\mathcal{H}'|m_a m_a\rangle + \Delta}{1 + \langle Pm_a m_b|m_a m_b\rangle} \tag{4.3}$$

where

$$\Delta = \frac{\langle Pm_a m_b|\mathcal{H}^{(a)} + \mathcal{H}^{(b)} - W_{m_a} - W_{m_b}|m_a m_b\rangle}{1 + \langle Pm_a m_b|m_a m_b\rangle} \tag{4.4}$$

is a correction which vanishes when m_a and m_b are eigenfunctions of $\mathcal{H}^{(a)}$ and $\mathcal{H}^{(b)}$. The first term in the numerator of (4.3) is the Coulomb or electrostatic energy of interaction of the unperturbed charge distributions of a and b. For spherical atoms, it vanishes in the absence of overlap and is negative when the overlap is small. The second term in the numerator of (4.3) is the exchange interaction, and in collisions of inert-gas atoms it is positive and several times larger in magnitude than the Coulomb interaction, giving short-range repulsion (56).

In going to the second order of perturbation, there appears to be no natural way to proceed in the short-range region. Murrell and Shaw (57) used the perturbed wavefunction

$$\psi = \mathcal{A} m_a m_b + \sum_{p_a, p_b}{}' c_{p_a p_b} p_a p_b \tag{4.5}$$

in which the perturbation to ψ_o is a simple product function. This approach introduces various second-order energies, such as those due to induction, dispersion, and exchange-polarization (58).

4.3 Non-additivity

In many theories of liquids and solids, the total interaction energy is assumed to be pairwise additive, that is, $u = \sum_{i<j} u_{ij}$. However, this is not precise (see Lecture 1) and many-body interactions may be important. For example, the transient electric dipole moment in liquid argon, giving a weak absorption in the far-infrared, must be associated with at least three atoms, since single atoms and pairs of atoms are centrosymmetric. One can

obtain an estimate of the importance of many-body forces in determining the binding energy of a liquid or solid by calculating the sublimation energy of crystalline argon at 0 K (1). The experimental value is 7.74±0.04 kJ mol^{-1} (59). Using the known pair potential for Ar_2, Barker calculated that the twelve nearest neighbours contribute 97% of this energy and that more distant neighbours contribute 10%; many-body effects must therefore account for -7% of the binding energy of solid argon (1).

The earliest calculation of three-body forces was that of Axilrod and Teller (60) who evaluated the long-range dipole-dipole-dipole dispersion energy of three atoms. It comes from the third-order perturbed energy and takes the form

$$\Delta u_{abc}=D_{abc}(1+3\cos\theta_a\cos\theta_b\cos\theta_c)R_a^{-3}R_b^{-3}R_c^{-3} \qquad (4.6)$$

where R_a, R_b, R_c are the sides and θ_a, θ_b, θ_c the interior angles of the triangle formed by the atoms. The constant D_{abc} can be expressed in terms of the polarizabilities of the free atoms a, b, c at imaginary frequencies (see Lecture 3) (41):

$$D_{abc}=(4\pi\varepsilon_o)^{-3}\,\frac{3\hbar}{\pi}\int_0^\infty \alpha_a(iu)\alpha_b(iu)\alpha_c(iu)du \qquad (4.7)$$

For atomic hydrogen, D_{HHH} = 21.64 a.u., while C_6 = -6.499 a.u. For three identical atoms at the vertices of an equilateral triangle, the ratio of the three-body to the two-body energy is

$$\frac{\Delta u_{abc}}{u_{ab}+u_{ac}+u_{bc}} = \frac{(11/8)\ D\ R^{-9}}{3\ C_6\ R^{-6}} = \frac{11\ D\ R^{-3}}{24\ C_6} \qquad (4.8)$$

For atomic hydrogen at R = 7 a.u. = 3.70×10^{-10} m, $\Delta u/u$ = -0.00445. For argon at R = 3.76×10^{-10} m, $\Delta u/u$ = -0.010. For crystalline argon, the dipole-dipole-dipole dispersion energy contributes -7.5% of the sublimation energy (1), which is approximately equal to the many-body contribution. From this one example, it should not be assumed that the dipole-dipole-dipole long-range interaction is the only important non-additive contribution. There are also non-additive short-range forces; the strong short-range repulsion between overlapping molecules must be affected by the distortion due to interaction with a third molecule, whether it be in long- or short-range interaction with one of the others. There are also quadrupole and other higher terms in the three-body long-range energy. These additional many-body interactions apparently approximately cancel in crystalline argon, but may not in other materials or in other arrangements of clusters of Ar atoms.

4.4 The Effect of a Medium

A medium of relative permittivity (or dielectric constant) ε_r reduces the electrostatic energy of interaction of two molecules immersed in it by a factor of ε_r. McLachlan (61) considered the influence of a medium on the interaction of two fixed molecules in a liquid and identified five effects relating to the presence of the medium: (i) the force is determined by the free energy $A(R) = U(R) - T\,S(R)$, as discussed in Lecture 1; (ii) the fixed molecules disturb the local order of the liquid; (iii) the long-range electromagnetic forces which would exist in free space are modified; (iv) Archimedes' Principle is involved because a molecule can move only by displacing the liquid from its path; (v) the electron clouds overlap those of the medium and change the properties of the molecules.

The problem of molecular conformation in solution has been discussed by Sinanoğlu (62).

The presence of polarizable matter between interacting molecules may <u>increase</u> their mutual potential energy. For example, if a spherical atom b of polarizability α_b is at the point midway between a pair of charges q and -q at a large separation R, the energy is

$$u_{abc} = -(4\pi\varepsilon_o)^{-1}\, q^2\, R^{-1}\left[1+32(4\pi\varepsilon_o)^{-1}\,\alpha_b\, R^{-3}\right] \quad (4.9)$$

If the charges were of the same sign, the dipole induced in b would vanish and the repulsive forces between the charges would not be affected by α_b (there is a field-gradient $F'_{zz} = -2F'_{xx} = -2F'_{yy} = -32(4\pi\varepsilon_o)^{-1}\, q\, R^{-3}$ and it induces a quadrupole moment $C_b\, F'_{zz}$, giving an induction energy $-\frac{1}{4}\, C_b (F'_{zz})^2 = -256(4\pi\varepsilon_o)^{-2}\, C_b\, q^2\, R^{-6}$).

There is much interest in solvent effects on the properties of molecules. Theoretical approaches vary from those employing the reaction field of Onsager (63-65) to the "supermolecule" approach (66) in which the interacting molecules are treated as a single large quantum-mechanical system.

4.5 The Hydrogen Bond

The intermolecular, or intramolecular, interaction between a hydrogen atom bonded to an electronegative atom X and another electronegative atom or group of atoms Y, known as the hydrogen bond, has long been of special interest. It has a dissociation energy of the order of 10 kJ mol^{-1} and is therefore intermediate in strength between that of a typical chemical bond (~500 kJ mol^{-1}) and a 'van der Waals' interaction (~1 kJ mol^{-1}).

The hydrogen bond X-H ··· Y is a localised interaction which may be detected in a number of ways (67):

(i) through molecular association in liquids and solutions, leading to abnormal molecular weights and transport properties;

(ii) through crystal-structure determinations, particularly by neutron diffraction, in which a proton is found near the electronegative atoms X and Y;

(iii) through abnormally high heats of vapourization of solids and liquids;

(iv) through a large red shift and broadening in the XH stretching vibrational frequency;

(v) through a blue shift in the electronic spectra associated with $\pi^* \leftarrow n$ transitions (68);

(vi) through a large downfield shift in the proton magnetic resonance spectrum.

In most cases the potential has two minima. Thus in the dimer $(HF)_2$, two isoenergetic conformations exist with either molecule acting as the proton donor (see Figure 4.1). The proton tunnelling through the hydrogen-bond barrier causes a splitting of the vibration-rotation energy levels, which may be interpreted as the rate of barrier penetration. In $(HF)_2$ this rate is 19.8 GHz, while in $(DF)_2$ it is 1.58 GHz; tunnelling is not seen in HFDF because H and D are distinguishable (69). Only the deuterium-bonded species HF...DF was observed, indicating that it is more stable than DF...HF.

The dipole moment of the HF dimer along the F-F axis is 3.0 D = 1.0×10^{-30} C m. This is substantially larger than the vector sum of the monomer moments (2.38 D), indicating that induction and short-range distortion effects are important in this case.

Ab initio computations have thrown much light on the hydrogen bond. The single linear hydrogen bond in the water dimer is substantially favoured over the bifurcated (i) and cyclic (ii) structures (72).

(i) 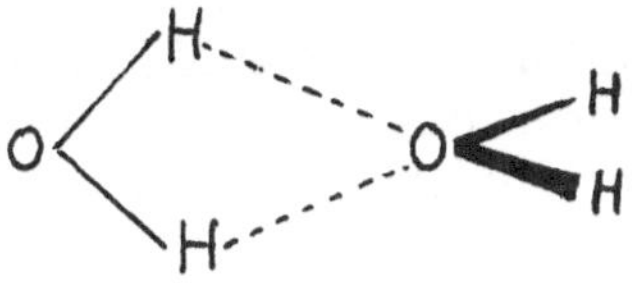

(ii) 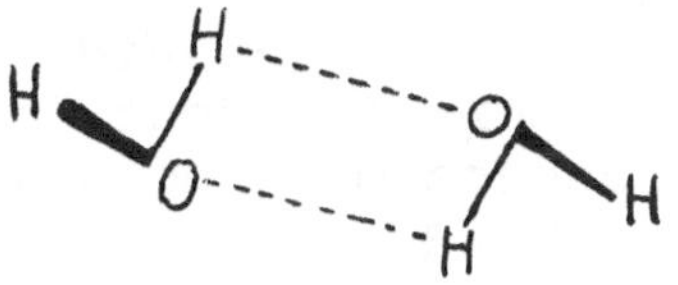

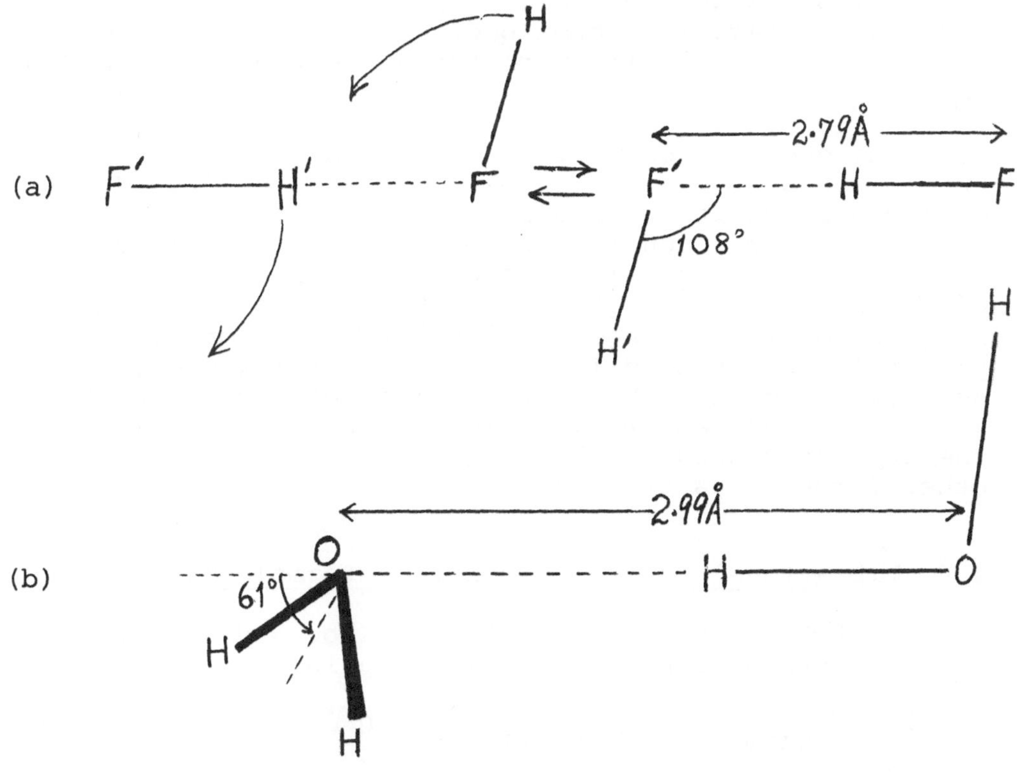

Figure 4.1. (a) Proton tunnelling in $(HF)_2$ has the effect of interchanging the proton donor and acceptor HF molecules. The structure is shown on the right (69,70). (b) The structure of the water dimer (71).

The hydrogen bond results from interatomic forces that probably should not be divided into components, although no doubt electrostatic and short-range interactions are the principal ingredients. It is not easy to study experimentally the energy and electronic properties of hydrogen-bonded species as a function of the important distances, namely X-H and H···Y. However, it is feasible to compute the effect of changes in these and other variables and thereby to throw much light on the hydrogen bond (73,74).

There is evidence for an important non-additive and co-operative contribution to hydrogen bonding. The presence of a negative charge near the proton in an H-X bond enchances the charge separation H^+-X^- in the bond and therefore makes the X atom more attractive to protons in other molecules. This co-operative effect was investigated by Del Bene and Pople (75) and Hankins et al (76) who found a negative three-body energy of -4.4 kJ mol^{-1} for a trimer $(H_3O)_3$ in which there are two linear hydrogen bonds.

The question of what it is, if anything, that is special about the hydrogen bond is of interest. Morokuma (74) considers it to be a special case of an intermediate to weak electron donor-acceptor complex, with linear bonding and appropriate directionality. Perhaps the main cause of the relative strength of the hydrogen bond is the weakness of the short-range overlap repulsive force; the proton in an X-H bond is in a region of low electron density (there are no inner-shell electrons on the H) so it can approach abnormally close to electronegative atoms, thereby increasing the binding due to electrostatic and other forces of attraction.

4.6 Vibrational Contributions to Intermolecular Forces

The vibrational Hamiltonian for a diatomic molecule in interaction with a fixed neighbouring atom is

$$\mathcal{H} = \mathcal{H}_{harmonic} + \mathcal{H}_{anharmonic} + U$$

$$= - \hbar c B_e \frac{d^2}{d\xi^2} + \frac{\hbar c \omega_e^2}{4B_e} (\xi^2 + a\xi^3 + \ldots) + U \qquad (4.10)$$

where B_e and ω_e are the rotational and vibrational constants in wavenumbers; the anharmonic constant a is negative for all diatomics. If U is expanded as a series in the relative displacement $\xi = (r-r_e)/r_e$ of the diatomic molecule from its equilibrium separation r_e,

$$U = U_e + U'_e\,\xi + \tfrac{1}{2}U''_e\,\xi^2 + \ldots \qquad (4.11)$$

then by treating $\mathcal{H}_{\text{anharmonic}} + U$ as a perturbation to $\mathcal{H}_{\text{harmonic}}$, one obtains the change in the vth vibrational energy level due to U (77):

$$\Delta W_v = U_e + (v+\tfrac{1}{2})\,\frac{B_e}{\omega_e}\,(U''_e - 3aU'_e) + O(B_e/\omega_e)^2 \qquad (4.12)$$

The shift in the fundamental vibrational energy due to the presence of the environment is

$$\Delta = \langle \Delta W_1 - \Delta W_0 \rangle = \frac{B_e}{\omega_e}\langle U''_e - 3aU'_e \rangle \qquad (4.13)$$

where the angular brackets denote a statistical average over all configurations of the neighbouring molecules. The equation may be generalised to polyatomics (77).

At very high densities ($R < R_{\text{equilibrium}}$), $U'_e > 0$ and $U''_e > 0$, so the shift Δ is blue. At normal densities ($R \approx R_{\text{equilibrium}}$), $U'_e < 0$ and $U''_e > 0$, the shift due to anharmonicity is red and that due to U''_e is blue; the former is usually the larger.

From (4.12) and (4.13), we can obtain a measure of the vibrational contribution to the interaction energy of the diatomic molecule with the atom. Figure 4.2 shows vibrational red shifts in HX and DX due to the environment. The vibrational contribution to the interaction energy of the ground state (v = 0) is $\tfrac{1}{2}\Delta_H$ or about -100 cm^{-1} = -1.2 kJ mol^{-1} for HX and $2^{-\frac{1}{2}}$ times this for DX. One might deduce from this that a deuterium bond is weaker than the corresponding hydrogen bond. However, while it is true that the molar volume of water at 20°C is smaller by 0.074 cm^3 than that of heavy water (implying a tighter binding, since anharmonicity makes the XD bond length shorter than the XH; the molar volume of CH_4 is greater than that of CD_4 by 0.46 cm^3, but in this case there is no hydrogen bonding), the heat of vapourization of D_2O is 1.032 times that of H_2O. And we saw that the HF···DF dimer is favoured

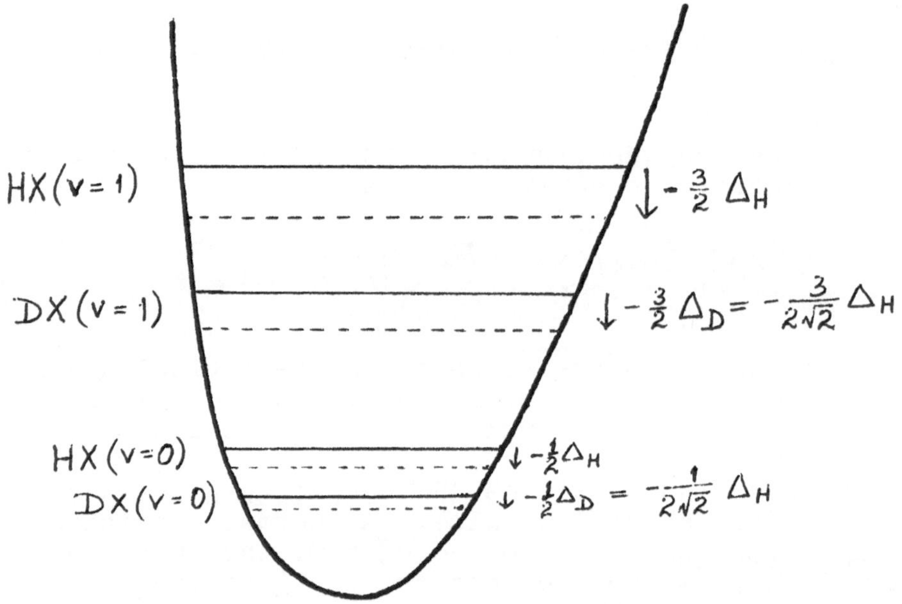

Figure 4.2. The vibrational energy changes of HX and DX due to a neighbouring atom for V=0 and 1.

over DF...HF. Apparently, it is necessary to consider the effects of vibrations other than that of the XH "diatomic" to gain a full understanding of the effect of vibration on the hydrogen bond.

REFERENCES

(1) J.A. Barker in Rare Gas Solids, M.L. Klein and J.A. Venables, eds., Academic Press, New York (1976), Chap. 4.
(2) R.A. Aziz and H.H. Chen, J. Chem. Phys. 67, 5719 (1977).
(3) H.C. Longuet-Higgins, Discussions Faraday Soc. 40, 7 (1965).
(4) A. Watanabe and H.L. Welsh, Canadian J. Phys. 43, 818 (1965); 45, 2859 (1967). A.R.W. McKeller and H.L. Welsh, Can. J. Phys. 52, 1082 (1974).
(5) S. Mizushima and H. Okazaki, J. Amer. Chem. Soc. 71, 3411 (1949).
(6) A Discussion on Rubber Elasticity, Proc. Roy. Soc. A, 351, 295-406 (1976).
(7) W. Kauzmann, Adv. Protein Chemistry 14, 1 (1959). G. Némethy and H.A. Scheraga, J. Chem. Phys. 36, 3401 (1962). C. Tanford, The Hydrophobic Effect, Wiley, New York (1973)
(8) J.A.V. Butler, Trans. Faraday Soc. 33, 229 (1937).
(9) M. Born and R. Oppenheimer, Ann. Phys. 84, 457 (1927).
(10) M. Born and K. Huang, Dynamical Theory of Crystal Lattices, Oxford University Press (1954) pp. 166 and 402.
(11) H.C. Longuet-Higgins, Adv. in Spectrosc. 2, 429 (1961).
(12) W. Kołos and L. Wolniewicz, J. Chem. Phys. 49, 404 (1968).
(13) W. Kołos and L. Wolniewicz, J. Chem. Phys. 41, 3663 (1964).
(14) G. Herzberg, J. Mol. Spectrosc. 33, 147 (1970)
(15) M. Trefler and H.P. Gush, Phys. Rev. Letters 20, 703 (1968).
(16) P.R. Bunker, J. Mol. Spectrosc. 46, 119 (1973).
(17) S.C. Wofsy, J.S. Muenter and W. Klemperer, J. Chem. Phys. 53, 4005 (1970).
(18) F.A. Gangemi, J. Chem. Phys. 39, 3490 (1963).
(19) H. Margenau and N.R. Kestner, Theory of Intermolecular Forces, 2nd edn., Pergamon Press, Oxford (1971).
(20) H. Hellmann, Einführung in die Quantenchemie, Deuticke, Leipzig (1937), p. 285. R.P. Feynman, Phys. Rev. 56, 340 (1939).
(21) J.O. Hirschfelder and M.A. Eliason, J. Chem. Phys. 47, 1164 (1967).
(22) W.H. Keesom, Physik. Z. 22, 129 (1921).
(23) S.C. Wang, Physik. Z. 28, 663 (1927).
(24) F. London, Z. Physik. 63, 245 (1930); Z. Physik. Chem. B 11, 222 (1930); J. Phys. Chem. 46, 305 (1942).
(25) J.E. Griffiths, M. Clerc and P.M. Rentzepis, J. Chem. Phys. 60, 3824 (1974); 63, 2262 (1975).
(26) A. Laubereau, J. Chem. Phys. 63, 2260 (1975).
(27) C. Mavroyannis and M.J. Stephen, Molec. Phys. 5, 629 (1962).
(28) P. Claverie in Molecular Interactions: From Diatomics to Biopolymers, B. Pullman, ed., Wiley, New York (1978), Chap. 2.
(29) S.F. Boys and F. Bernardi, Molec. Phys. 19, 553 (1970).
(30) N.S. Ostlund and D.L. Merrifield, Chem. Phys. Letters 39, 612 (1976).
(31) J.S. Dahler and J.O. Hirschfelder, J. Chem. Phys. 25, 986 (1956).

(32) H.C. Longuet-Higgins, Proc. Roy. Soc. A 235, 537 (1956).
(33) A.D. Buckingham in Molecular Interactions: From Diatomics to Biopolymers, B. Pullman, ed., Wiley, New York (1978), Chap. 1.
(34) L. Jansen, Physica 23, 599 (1957).
(35) H. Kreek, Y.H. Pan and W.J. Meath, Molec. Phys. 19, 513 (1970).
(36) A.D. Buckingham in Physical Chemistry. An Advanced Treatise, Vol. 4, H. Eyring, W. Jost and D. Henderson, eds., Academic Press, New York (1970), Chap. 8.
(37) A.D. Buckingham, Int. Rev. Sci., Phys. Chem. Series 1, Vol. 3, D.A. Ramsay, ed., Butterworths, London (1972), Chap. 3.
(38) A.D. Buckingham, Adv. Chem. Phys. 12, 107 (1967).
(39) R.D. Amos, Molec. Phys. 38, 33 (1979).
(40) L.I. Schiff, Quantum Mechanics, 3rd edn., McGraw-Hill, New York (1968), section 35.
(41) A. Dalgarno, Adv. Chem. Phys. 12, 143 (1967).
(42) G. Starkschall and R.G. Gordon, J. Chem. Phys. 54, 663 (1971); 56, 2801 (1972).
(43) P.W. Langhoff, R.G. Gordon and M. Karplus, J. Chem. Phys. 55, 2126 (1971).
(44) K.T. Tang, J.M. Norbeck and P.R. Certain, J. Chem. Phys. 64, 3063 (1976).
(45) F. London, Trans. Faraday Soc. 33, 8 (1937).
(46) L. Pauling and J.Y. Beach, Phys. Rev. 47, 686 (1935).
(47) C.C.J. Roothaan, Rev. Mod. Phys. 23, 69 (1951).
(48) D.R. McLaughlin and H.F. Schaefer, Chem. Phys. Lett. 12, 244 (1971).
(49) P.J. Bertoncini and A.C. Wahl, J. Chem. Phys. 58, 1259 (1973).
(50) B. Liu and A.D. McLean, J. Chem. Phys. 59, 4557 (1973).
(51) W.J. Stevens, A.C. Wahl, M.A. Gardner and A.M. Karo, J. Chem. Phys. 60, 2195 (1974).
(52) R.R. Teachout and R.T. Pack, Atomic Data 3, 195 (1971).
(53) J.O. Hirschfelder, Chem. Phys. Lett. 1, 325 (1967).
(54) P.R. Certain and L.W. Bruch, International Review of Science, Phys. Chem. Series 1, Vol. 1, Chap. 4, W. Byers Brown, ed., Butterworths, London (1972).
(55) J. Gerratt, Proc. Roy. Soc. A 350, 363 (1976).
(56) A. Conway and J.N. Murrell, Molec. Phys. 23, 1143 (1972).
(57) J.N. Murrell and G. Shaw, J. Chem. Phys. 46, 1768 (1967).
(58) J.N. Murrell in Rare Gas Solids, M.L. Klein and J.A. Venables, eds., Academic Press, New York (1976), Chap. 3.
(59) P. Flubacher, A.J. Leadbetter and J.A. Morrison, Proc. Phys. Soc. 78, 1449 (1961). See also T.G. Gibbons, M.L. Klein and R.D. Murphy, Chem. Phys. Lett. 18, 325 (1973).
(60) B.M. Axilrod and E. Teller, J. Chem. Phys. 11, 299 (1943).
(61) A.D. McLachlan, Discussions Faraday Soc. 40, 239 (1965).

(62) O. Sinanoğlu in The World of Quantum Chemistry, R. Daudel and B. Pullman, eds., Reidel, Dordrecht (1974), p. 265.
(63) L. Onsager, J. Amer. Chem. Soc. 58, 1486 (1936).
(64) B. Linder, Adv. Chem. Phys. 12, 225 (1967).
(65) O. Tapia and O. Goscinski, Molec. Phys. 29, 1653 (1975).
(66) B. Pullman, A. Pullman and H. Berthod, Int. J. Quantum Chem. Quantum Biology Symposium 5, 79 (1978).
(67) The Hydrogen Bond. Recent developments in theory and experiments. Vols I, II, III, P. Schuster, G. Zundel and C. Sandorfy, eds., North-Holland, Amsterdam (1976).
(68) G.J. Brealey and M. Kasha, J. Amer. Chem. Soc. 77, 4462 (1955).
(69) T.R. Dyke, B.J. Howard and W. Klemperer, J. Chem. Phys. 56, 2442 (1972).
(70) T.R. Dyke and J.S. Muenter, Int. Rev. Science, Phys. Chem. Series 2, Vol. 2, A.D. Buckingham, ed., Butterworths, London (1975), Chap. 2.
(71) T.R. Dyke and J.S. Muenter, J. Chem. Phys. 60, 2929 (1974).
(72) P.A. Kollman and L.C. Allen, J. Chem. Phys. 51, 3286 (1969). G.H.F. Diercksen, Theor. Chem. Acta, 21, 335 (1971).
(73) H. Popkie, H. Kistenmacher and E. Clementi, J. Chem. Phys. 59, 1325 (1973).
(74) K. Morokuma, Acc. Chem. Res. 10, 294 (1977). P.A. Kollman, Acc. Chem. Res. 10, 365 (1977).
(75) J. Del Bene and J.A. Pople, J. Chem. Phys. 52, 4858 (1970); 58, 3605 (1973).
(76) D. Hankins, J.W. Moskowitz and F.H. Stillinger, J. Chem. Phys. 53, 4544 (1970).
(77) A.D. Buckingham, Trans. Faraday Soc. 56, 753 (1960).

INFRARED AND RAMAN STUDY OF

VIBRATIONAL RELAXATION IN LIQUIDS

S. Bratos

Laboratoire de Physique Théorique des Liquides*
Université Pierre et Marie Curie
4, Place Jussieu, 75005 Paris (France)

1 INTRODUCTION

The purpose of the present lecture course is to review infrared and spontaneous Raman studies of vibrational relaxation in liquids. This process is relative to internal degrees of freedom of a molecular liquid and is generated by a large variety of intermolecular forces, the dipole-dipole, dipole-induced dipole, dispersion, Pauli repulsion forces and hydrogen bonding. The relaxation due to the centrifugal or Coriolis forces is generally, although arbitrarily, separated from the present context and is studied in the theory of rotational relaxation. Vibrational relaxation originates either in de-excitation or in dephasing of molecular vibrations ; these processes are called T_1 and T_2 processes, respectively, in agreement with the well known terminology of nuclear magnetic resonance. The remarkable analogy of relaxation processes in these two fields is due to the quantum-mechanical character of degrees of freedom involved. For earlier reviews of this field, see Bratos, Guissani and Leicknam (1), Bailey (2), Diestler (3) and Oxtoby (4).

The basic formulas of the theory relate the infrared absorption coefficient $\alpha(\omega)$ and the Raman differential scattering cross-section $\partial^2\sigma / \partial\omega\partial\Omega$ to appropriate correlation functions. If $\vec{M}$ denotes the dipole moment of the liquid sample S, $\vec{\vec{\alpha}}$ its macroscopic polarizability tensor, ω the frequency of the incident infrared radiation, $\Delta\omega$ the frequency difference from incident and $\vec{k}_D$ the wave vector of the scattered light one finds

$$\alpha(\omega) = \frac{2\pi}{3c\hbar nV}\,\omega(1-e^{-\frac{\hbar\omega}{KT}})\int_{-\infty}^{\infty} dt\; e^{-i\omega t}\langle\vec{M}(o)\vec{M}(t)\rangle \qquad (1\text{-}1)$$

* Laboratoire associé au C.N.R.S.

$$\frac{\partial^2\sigma}{\partial\omega\partial\Omega} = \frac{1}{2\pi} k_D^4 \int_{-\infty}^{\infty} dt\, e^{-i\Delta\omega t} \langle [\vec{\epsilon}_I \vec{\vec{\alpha}}(o)\vec{\epsilon}_D]\, [\vec{\epsilon}_I \vec{\vec{\alpha}}(t)\vec{\epsilon}_D] \rangle \qquad (1\text{-}2)$$

These expressions due to Gordon (5-7) only apply if (i) the wave vector of the incident infrared radiation and the scattering wave vector are vanishingly small and (ii) the difference between the external and the internal Maxwellian electric field in S can be neglected. The latter assumption is heavily restrictive for polar liquids.

The information contained in infrared and Raman spectra of liquids is relative, in the lowest order approximation, to vibrational and rotational but not to translational motions. This can be seen (i) by writing the dipole moment vector M and the macroscopic polarizability tensor $\vec{\vec{\alpha}}$ of S as a sum of monomolecular quantities, $\vec{M} = \sum \vec{M}_i$, $\vec{\vec{\alpha}} = \sum \vec{\vec{\alpha}}_i$ and (ii) by noticing the invariance of $\vec{M}_i$'s and $\vec{\vec{\alpha}}_i$'s under translations. Moreover, the vibrational and rotational relaxation effects can be disentangled by recording the isotropic and anisotropic Raman spectra ; this can be done by building the scattering experiment in the VV and VH geometries and by measuring the corresponding differential scattering cross-sections $(\partial^2\sigma/\partial\omega\partial\Omega)_{VV}$, $(\partial^2\sigma/\partial\omega\partial\Omega)_{VH}$. Then, denoting by α_i the mean polarisability of the molecule i and by $\vec{\vec{\beta}}_i = \vec{\vec{\alpha}}_i - \alpha_i \vec{\vec{I}}$ the anisotropic component of $\vec{\vec{\alpha}}_i$, the following formulas apply (8)

$$\left(\frac{\partial^2\sigma}{\partial\omega\partial\Omega}\right)_{is} \equiv \frac{1}{2\pi} k_D^4 \int_{-\infty}^{\infty} dt\, e^{-i\Delta\omega t} \sum_{ij} \langle \alpha_i(o)\alpha_j(t) \rangle = \left(\frac{\partial^2\sigma}{\partial\omega\partial\Omega}\right)_{VV} - \frac{4}{3}\left(\frac{\partial^2\sigma}{\partial\omega\partial\Omega}\right)_{VH} \qquad (1\text{-}3)$$

$$\left(\frac{\partial^2\sigma}{\partial\omega\partial\Omega}\right)_{ani} = \frac{1}{2\pi} k_D^4 \int_{-\infty}^{\infty} dt\, e^{-i\Delta\omega t}\, \mathrm{Tr} \sum_{ij} \langle \vec{\vec{\beta}}_i(o)\vec{\vec{\beta}}_j(t) \rangle = 10 \left(\frac{\partial^2\sigma}{\partial\omega\partial\Omega}\right)_{VH} \qquad (1\text{-}4)$$

These equations contain the answer to the present problem ; the trace of a tensor being invariant with respect to translations and rotations of the coordinate frame, the isotropic Raman spectra only depend on vibrational relaxation mechanisms, contrary to the infrared and anisotropic Raman spectra in which rotational relaxation mechanisms are also involved. The spontaneous isotropic Raman spectroscopy thus represents, combined with the stimulated Raman spectroscopy, a particularly powerful technique in studying vibrational relaxation of liquids. The recognition of these facts represents one of earliest results in this field (9,10).

This simple analysis no longer remains valid if the collision induced effects are considered. Then, the dipole moment vector $\vec{M}$ contains, not only single particle contributions $\vec{M}_i$, but also two-, three- etc particle terms which involve vibrational, rotational and translational coordinates ; this also applies to the polarizability tensor $\overleftrightarrow{\alpha}$. The correlation functions of Eqns (1-1 to 1-4) then depend on these three types of coordinates and the corresponding spectral effects can no more be disentangled. The collision induced effects are often considered to be small, but this assumption is not always well controlled.

2 VIBRATIONAL RELAXATION IN VAN DER WAALS LIQUIDS. DILUTED SOLUTIONS

2.1 Generalities

The preliminary work on this simplest problem of the theory of vibrational relaxation in liquids originated from Valiev in early sixty's (11,12). Systematic investigations were undertaken ten years later and concerned nearly exclusively the relaxation a singly excited, non-degenerate normal mode of a diatomic or polyatomic molecule dissolved in an inert solvent. The basic theory is described in papers by Bratož, Rios and Guissani (13), Bratos and Maréchal (9), Bartoli and Litowitz (10) and Nafie and Peticolas (14). See also Morawitz and Eisenthal (15), Brindeau, Bratos and Leicknam (16), Van Konynenburg and Steele (17), Diestler (18), Paul and Fueller (19), Oxtoby and Rice (20) and Leicknam, Guissani and Bratos (21).

2.2 Theory

2.2.1. Basic assumptions

The theory is based on the following assumption. (i) The active molecule executes quantum mechanical vibrations represented by the normal coordinates $n = (n_1 n_2 \ldots n_N)$; they obey the semi-classical Hamiltonian

$$H(n,t) = \left[\frac{1}{2}\sum_i p_i^2 + \frac{1}{2}\sum_i \lambda_i n_i^2 + \frac{1}{6}\sum_{ijk} \lambda_{ijk} n_i n_j n_k + \ldots\right] + \left[V(n,t)\right] \tag{2-1}$$

where $V(n,t)$ is a stochastic potential of the solvent-solute interaction. (ii) The active molecule executes stochastic reorientations. (iii) Vibrations and rotations are statistically independent. (iv) The collision induced processes are absent. It may be noticed that not all of these assumptions are essential. The theory has also been presented in a way avoiding stochastic formulation (4) ; moreover, the effect of vibrations and rotations can easily be included. On the contrary, neglecting collision induced and dielectric effects represents a real limitation of the theory.

2.2.2. Method of calculation

The general results of the theory resting on these premises may be stated as follows. (i) The correlation functions which enter into Eqns (1-1 to 1-4) may be given a simple general form. Designating by n the singly excited normal mode, noting by 0 and 1 the quantum numbers of the vibrational states involved and writting $\partial\vec{M}/\partial n = \frac{\partial M}{\partial n}\vec{u}$ gives

$$\langle\vec{M}(o)\vec{M}(t)\rangle = \left(\frac{\partial M}{\partial n}\right)^2 \langle n_{01}(o)n_{10}(t)\rangle_s \ \langle\vec{u}(o)\vec{u}(t)\rangle_s \qquad (2\text{-}2a)$$

$$\langle\alpha(o)\alpha(t)\rangle = \left(\frac{\partial\alpha}{\partial n}\right)^2 \langle n_{01}(o)n_{10}(t)\rangle_s \qquad (2\text{-}2b)$$

$$\mathrm{Tr}\langle\vec{\vec{\beta}}(o)\vec{\vec{\beta}}(t)\rangle = \langle n_{01}(o)n_{10}(t)\rangle_s \ \mathrm{Tr}\langle\frac{\partial\vec{\vec{\beta}}}{\partial n}(o)\frac{\partial\vec{\vec{\beta}}}{\partial n}(t)\rangle_s \qquad (2\text{-}2c)$$

The vibrational correlation function G(t) defined by the ratio $\langle n_{01}(o)n_{10}(t)\rangle_s/\langle n_{01}(o)n_{10}(o)\rangle_s$ is thus directly obtainable from the normalized isotropic Raman spectra. Alternatively, G(t) may be extracted from infrared or anisotropic Raman spectra if either rotational correlation functions are available from an independent source (9,10) or if the rotational relaxation times are very large as compared with the vibrational relaxation times. (ii) The vibrational correlation function G(t) may conveniently be expressed in terms of the solvent induced frequency increment $\hbar\omega_{10}(t) = V_{11}(t)-V_{00}(t)$. Designating by ω_o the non perturbed frequency of the mode n one can write :

$$G(t) = \exp\left[i\omega_o t\right] \langle\exp\left[i\int_o^t dt'\omega_{10}(t')\right]\rangle_s \qquad (2\text{-}3)$$

It results from this equation that G(t) reaches vanishingly small values for times t such that the dispersion in the argument of the exponential is of the order of unity ; vibrational relaxation is thus a pure dephasing. It is due to the fluctuations of the vibrational frequency of the active molecule ; in turn, these fluctuations are generated by fluctuations in the molecular environment. Population changes are too slow to be detectable spectroscopically, except if there are quasi-resonant vibrational states in the molecule under consideration. (iii) The correlation function G(t) has two simple limits, the slow and the fast modulation limit ; they are defined by inequalities $\omega\tau \gg 1$ and $\omega\tau \ll 1$ where ω indicates the order of magnitude and τ the correlation time of $\omega_{10}(t)$. Then, if $\alpha = \langle\omega\rangle$, $\beta = \langle\omega^2\rangle - \alpha^2$, $\gamma = \beta\tau$, one can write

$$G(t) = \exp\left[i\omega_o t\right] \exp\left[i\alpha t - \frac{1}{2}\beta t^2 + \ldots\right], \ \omega\tau \gg 1 \ ;$$

$$G(t) = \exp\left[i\omega_o t\right] \exp\left[i\alpha t - \gamma t\right], \ \omega\tau \ll 1 \qquad (2\text{-}4)$$

The corresponding spectral density is either a disymmetric distorted Gaussian ($\omega\tau \gg 1$) or a Lorentzian ($\omega\tau \ll 1$).

2.2.3. Models.

Further development of the theory rests on the models proposed by Fischer and Laubereau (22), Rothschild (23,24), Levant (25), Dijkman and Van der Maas (26), Diestler and Manz (27), Metiu, Oxtoby and Freed (28), Cohen and Wilde (29), Oxtoby (30), Arndt and Mac Clung (31) and by Wilde and Cohen (32) ; these models either determine the complete expression for G(t) or describe only its correlation time. (i) In the model due to Rothschild the averaged exponential of Eqn (2-3) is developed into a cumulant expansion series up to terms of the second order and the highest order cumulant is given an exponential form. These assumptions may be rationalized by postulating $\omega_{10}(t)$ to be a Markovian-Gaussian process and by applying the Doob theorem. (iii) The Wilde-Cohen model is basically similar to the Rothschild model. G(t) is calculated starting from a reference-system correlation function called frozen-lattice correlation function. This function is considered as the highest order memory function in the hierarchy of generalized Langevin equations. (iii) Finally, the Levant, Dijkman and Van der Maas models contain the postulate of slow modulation ; moreover, a quasi crystalline local structure of liquids is assumed by the latter two authors. Eqn (2-3) is then evaluated by building in appropriate intermolecular potentials. This is for models of the first group ; the models of the second group may be described as follows. (i) In the Diestler-Manz model the relaxation is ascribed to the vibrational-translational coupling and is examined in the frame of the cell model. (ii) The Fischer and Laubereau model consists to replace the liquid sample S by a linear three-body system formed by a diatomic molecule and an atom. The interaction potential has the form of an exponential. Vibrational motions are treated quantum-mechanically whereas translational motions are described classically. Both, the dephasing and the energy relaxation times are calculated in this way ; the latter is found to be considerably longer than the former. (iii) Finally, Oxtoby, Metiu and Freed proposed a hydrodynamic model in which the relaxation process is controlled by a viscous force ; two versions of this model exist. In its classical version, the decay of the vibrational amplitude of a molecule in solution is calculated by representing it by a cylinder closed at each by a hemi-sphere, immersed in the fluid and vibrating ; the problem then reduces to that of macroscopic hydrodynamics (28). In its semi-classical version, the quantum mechanical expression for $\omega_{10}(t)$ is used and its correlation time is calculated by calculating the correlation times of fluctuating forces which enter into $\omega_{10}(t)$; a hydrodynamic model similar to that just mentioned is applied for that purpose (30). The correlation time furnished by this model in its classical version is the energy relaxation time ; the dephasing time is obtained in its semi-classical version. These models, in their ensemble, provide a qualitative or semi-quantitative description of vibrational relaxation. Unfortunately, the computer simulation of G(t) in diluted solutions is difficult and no results are as yet available.

2.3 Discussion

This Section may be concluded by briefly discussing the important question of the speed of modulation of $\omega_{10}(t)$. In early work the modulation was generally assumed to be slow and the opposite statement is frequent in recent work. Unfortunately, this question can not simply be answered by comparing the observed profiles to those corresponding to Eqns (2-4) ; theoretical and, in part, experimental uncertainties are far too large to allow a meaningful discussion of fine differences between them. The major source of theoretical uncertainty comes from neglect of collision induced absorption and scattering and from neglect of dielectric field effects. Comparing integrated intensities of free and dissolved molecules shows the importance of these two mechanisms.

This crucial problem has as far been analysed along the following lines. (i) The vapor solution frequency shift $\Delta\omega$ is compared to the vibrational half-width $\Delta\omega$ 1/2. In the slow modulation limit $\Delta\omega \sim \Delta\omega$ 1/2 and in the fast modulation limit $\Delta\omega \gg \Delta\omega$ 1/2, a property due to the motional narrowing. The former statement has been found experimentally to apply if $\Delta\omega$ 1/2 > 10 cm^{-1} and the latter if $\Delta\omega$ 1/2 $\lesssim 1$ cm^{-1}. (ii) Vibrational half width $\Delta\omega$ 1/2 (0→v) of successive harmonics is studied as a function of the quantum number v of the upper vibrational state. In the slow modulation limit $\Delta\omega$ 1/2 (0→v)$\sim$v $\Delta\omega$ 1/2 (0→1) and in the fast modulation limit $\Delta\omega$ 1/2 (0→ v)$\sim v^2 \Delta\omega$ 1/2 (0→1) ; a transition between the fast and the slow modulation regime is thus expected to occur in this sequence if the modulation is fast in the case of the fundamental. This effect has recently been observed by Battaglia and Madden in the resonance Raman spectrum of dissolved iodine (33) ; the transition occurs for v such that $\Delta\omega$ 1/2 (0-v)$\sim$10cm^{-1}. (iii) Isotropic Raman half-width $\Delta\omega$ 1/2 is investigated in the critical region. The amplitude of density fluctuations and their correlation time increase when this point is approached. Therefore, if the modulation is fast in normal conditions, one might expect to pass from the fast to the slow modulation regime and to observe critical line anomalies ; no such effect should occur for the slow modulation regime. This problem has been theoretically investigated by Hills and Madden (34) and by Hills (35).

3 VIBRATIONAL RELAXATION IN VAN DER WAALS LIQUIDS. PURE LIQUIDS

3.1 Generalities

The complexity of theoretical analysis increases for an order of magnitude when going from dilute solutions to pure liquids. The problem is to study the relaxation of a system of N vibrators coupled by molecular interaction. Early work in this field is due to Valiev (11), Döge (36,37) and Tokuhiro and Rothschild (38). More recently, a number of papers have been published covering this field ; the theory is described in papers by Van Woerkom, de Bleyser, de Zwart and Leyte (39), Madden and Lynden-Bell (40), Wang (41), Lynden-Bell (42,43), Oxtoby, Levesque and Weis (44), Wertheimer (45,46), Knauss (47) and Oxtoby (4) ; see also Hills (48,49) and Hills and Madden (50). At the present time the theory is firmly established only in its fast modulation limit, in contrast to the situation in dilute solutions where the analysis is available for all modulation speeds. In the discussion below, these papers are grouped according to the theoretical method which was employed.

3.2 Theory

3.2.1. Cumulant expansion method

The theories proposed by Van Woerkom, de Bleyser, de Zwart and Leyte and by Oxtoby are based on use of the generalized cumulant expansion theorem for non-commuting quantum mechanical operators (51-53) ; they involve the following assumptions. (i) The Hamiltonian of the liquid sample S is written $H = H_V+H_B+D+R+V$ where H_V is the Hamiltonian for vibrational degrees of freedom of isolated molecules, H_B is the bath Hamiltonian for rotational and translational degrees of freedom, D describes the environmental fluctuations of vibrational frequency, R the resonant transfer of vibrational energy and V vibrational population changes ; the Hamiltonian matrix is built on eigenfunctions of H_o. (ii) The operators M, α corresponding to the infrared and isotropic Raman spectra and designated by a common symbol Q are written as a sum of monomolecular contributions. (iii) D,R and V are small with respect to H_V+H_B. Several interaction representations are then introduced in which V, or V+R, or else V+R+D are considered as perturbation. (iv) The averaged ordered exponential is developped into a trunkated cumulant expansion series following the equation (continued on following page).

$$\langle\langle \exp_o\{\frac{i}{\hbar}\int_o^t dt'\overset{-x}{H_1}(t')\}\rangle\rangle = \exp\{\frac{i}{\hbar}\int_o^t dt'\langle\langle\overset{-x}{H_1}(t')\rangle\rangle_c +$$

$$+\frac{i^2}{\hbar^2}\int_o^t dt'\int_o^{t'} dt''\langle\langle\overset{-x}{H_1}(t')\overset{-x}{H_1}(t'')\rangle\rangle_c\} \qquad (3\text{-}1)$$

$$\overset{-x}{H_1} = \exp\{\frac{it}{\hbar}\overset{x}{H_o}\}H_1 \;;\; \langle\langle A\rangle\rangle_c = \langle AQ(t)Q(o)\rangle_c\langle Q(t)Q(o)\rangle^{-1}$$

where H_o, H_1 realize a partition of the complete Hamiltonian into a non-perturbed and perturbing parts. Note the generalized averaging prescription $\langle\langle A\rangle\rangle$ and the fact that the ordering prescription in suppressed in Eqn (3-1). (v) The difference between external and internal Maxwellian electric fields is negligibly small.

The validity of theories of this group is restricted by assumptions (ii,iv,v). In particular, the assumption (iv) only applies, such as it stands, in the fast modulation limit $H_1\tau \ll 1$ where H_1 designates the order of magnitude and τ the correlation time of $H_1(t)$; the demonstration involves the use of the Redfield equations. Furthermore, assumptions (ii,v) amount to neglect the collision-induced and internal field effects. Once they have been introduced, the calculation of correlation functions reduces to averaging over rotational and translational degrees of freedom ; computer (44) or stochastic (39) simulation have been used for that purpose.

3.2.2. Redfield method

The second group of theories due to Madden and Lynden-Bell and to Lynden-Bell either employ techniques familiar in nuclear magnetic resonance or use methods derived from them. These theories are semi-classical and are based, except (43), on the following assumptions. (i) The Hamiltonian is written as a sum of a vibrational Hamiltonian H_o and of a stochastic Hamiltonian $H_1(t)$ describing the coupling between the vibrations and the bath. The dipole-dipole and dispersion forces contribute to relaxation. (ii) The operators $Q = \alpha, \vec{M}, \beta$ obey the Redfield equation.

$$\frac{d}{dt}Q^{*}_{\alpha\alpha'} = -\sum_{\beta\beta'} R_{\alpha\alpha'\beta\beta'}\,Q^{*}_{\beta\beta'} \qquad (3\text{-}2)$$

where $Q^{*}_{\alpha\alpha'}$ are matrix elements of Q^{*} over vibrational states, and $R_{\alpha\alpha'\beta\beta'}$ the relaxation matrix ; the asterisk indicates the interaction representation under H_o. (iii) Collision-induced and internal field effects are absent. (iv) Translational motions are assimilated to a translational diffusion and rotational motions are supposed to be very slow.

The theory described above was recently generalized by Lynden-Bell (43). The assumptions (i,iii) remain valid but the Redfield method is abandoned ; other important points are as follows. (i) The action of the perturbing Liouvillian $iL_1 = i/\hbar \left[\ ,H_1 \right]$ on Q generates a number of bimolecular coordinates P_i coupled to Q. For example, if the coupling is due to the dipole-dipole interaction, P_i will be a collective variable depending on pairs of $D^{(1)}$ functions of the orientations of molecules and on their relative positions. (ii) The evolution of P_i's, coupled to that of Q, is exponential in zero order. It is governed by monomolecular rotational diffusion processes. (iii) The system of coupled equations associated with Q and P_i is solved in the fast motion limit in which the off-diagonal elements of L_1 are small compared to the difference of the diagonal elements.

The validity of theories of this group is limited, in both versions, by neglect of collision induced and internal field effects. Moreover, the use of the Redfield equations eliminates from consideration all slow and moderately fast band shaping processes ; this limitation of the theory is attenuated in its last version. Finally, the assumption (iv) amounts to suppress rotational relaxation effects. It is then possible to use infrared, isotropic Raman and anisotropic Raman spectroscopies to study the manifestations of pure vibrational relaxation but the description is somewhat hypothetical.

3.2.3. Zwanzig-Mori method

The last group of theories by Wang, Wertheimer and Knauss are based on the Zwanzig-Mori theory of Brownian motion. These theories are quantum-mechanical in principle although, in practice, the non-vibrational degrees of freedom may be taken to be classical. The basic assumptions of these theories are as follows. (i) The Hamiltonian is written $H = H_o+H_B+H_1$ where H_o is the Hamiltonian of a system of non-interacting vibrators, H_B the bath Hamiltonian and H_1 collects all coupling terms. (ii) The normalized correlation function $G(t) = \langle Q(o)Q(t)\rangle / \langle Q(o)Q(o)\rangle$ obeys the generalized Langevin equation

$$\frac{dG(t)}{dt} = i\,\Omega\, G(t) - \int_o^t dt'k(t-t')G(t') \qquad (3\text{-}3)$$

where Ω is the frequency and k(t) the memory function. Note that the definition of the scalar product varies when going from one author to another. (iii) Ω and k(t) are calculated either by using perturbation type developments (41, 47) or by suppressing H_1 in the expressions for the density matrix and projected propagator (45,46). (iv) Collision-induced and internal field effects are absent. Once these assumptions are introduced, the averages over rotational-translational degrees of freedom are calculated either by applying classical equations for rotational and translational diffusion (41) or by using special models.

Although the Zwanzig-Mori theory by no means is a weak coupling theory, the assumption (iii) combined with methods used to calculate averages over the bath coordinates produces fast modulation type results (41). This is particularly restrictive in (47), but less so in (45,46). Finally, the consequences of the assumption (iv) may be analysed along similar lines as in preceeding Sections.

3.2.4. Some results

The main conclusions reached by these theories are as follows. (i) Vibrational relaxation in pure liquids is due to three mechanisms, the environmental fluctuations of vibrational frequency, the resonant transfer of vibrational energy and the population changes. The former two are dominating if the bands are isolated whereas all of them are operative in the opposite case (39,41-50) ; see also (54). (ii) The efficiency of different mechanisms in creating a dephasing varies according to the type of molecular interaction. Dipole-dipole forces produce a resonant transfer only for infrared active and dispersion forces only for Raman active normal modes. On the contrary, all types of interaction may generate environmental fluctuations of vibrational frequency (42). (iii) The pair correlations affect the spectral density of an isotropic Raman band although they do not affect its integrated intensity. The Raman scattering process in thus only partially incoherent, contrary to what is generally believed (41). (iv) The spectral effect of environmental fluctuations does not depend on molecular concentration whereas the effect of resonant transfer is proportional to it (46,47). (v) The self and distinct pair correlation functions may be observed separately by building appropriate isotropic Raman experiments for liquid isotope mixtures (46).

These conclusions have been tested by a recent molecular dynamics calculation due to Oxtoby, Levesque and Weis (44). The interaction potential has been assumed to be a superposition of atom-atom potentials : The correlation functions have then been calculated by appropriate molecular dynamics runs. This calculation does not only confirm the results cited above but shows in addition that the environmental fluctuations and the resonant transfer are correlated. See also the paper by Riehl and Distler (55).

3.3 Discussion

This Section may be concluded by briefly discussing the problem of separability of vibrational and rotational relaxation processes. In diluted solutions, the infrared or anisotropic Raman correlation functions are generally presented as a product of a vibrational and a rotational correlation function (13). This approximation certainly is an useful zero order approximation ; it can often be improved. The effect of the Coriolis interaction was investigated by Müller, Etique and Kneubühl (56), Müller and Kneubühl (57), Müller (58) and Lynden-Bell (59). Moreover, the effect of the centrifugal rotational-vibrational coupling was studied by Bratos and Chestier (60). Although definitely measurable, these effects turn out to be comparatively small.

The separation of vibrational and rotational relaxation processes is much more questionable in pure liquids. In that case, the infrared and anisotropic Raman correlation functions are not expected, a-priori, to have a product form, even not if vibrational and rotational motions are statistically independent. Then

$$\langle \vec{M}(o)\vec{M}(t)\rangle = \sum_{ij} \langle \vec{M}_i(o)\vec{M}_j(t)\rangle = \left(\frac{\partial M}{\partial n}\right) \sum_{ij} \langle n_i(o)n_j(t)\rangle_s \langle \vec{u}_i(o)\vec{u}_j(t)\rangle_s \qquad (3\text{-}4a)$$

$$\mathrm{Tr}\,\langle \overleftrightarrow{\beta}(o)\overleftrightarrow{\beta}(t)\rangle = \mathrm{Tr}\sum_{ij} \langle \overleftrightarrow{\beta}_i(o)\overleftrightarrow{\beta}_j(t)\rangle =$$

$$= \sum_{ij} \langle n_i(o)n_j(t)\rangle_s \langle \mathrm{Tr}\left(\frac{\partial \overleftrightarrow{\beta}}{\partial n_i}(o)\frac{\partial \overleftrightarrow{\beta}}{\partial n_j}(t)\right\rangle_s \qquad (3\text{-}4b)$$

The correlation functions thus appear as a sum of products of vibrational and rotational factors rather than a product of a single vibrational and a single rotational correlation function. If the rotational motion is very slow the total correlation function may still be termed as purely vibrational but will be different in infrared, isotropic and anisotropic Raman as

$$\langle \vec{u}_i(o)\vec{u}_j(o)\rangle_s \neq \langle \mathrm{Tr}\ \overleftrightarrow{\beta}_i(o)\overleftrightarrow{\beta}_j(o)\rangle_s \qquad (3\text{-}5)$$

However, the notion of vibrational correlation function is ambigous in these circumstances and it seems safer to abandon it altogether.

4 VIBRATIONAL RELAXATION IN HYDROGEN BONDED LIQUIDS

4.1 Generalities

Vibrational relaxation in hydrogen bonded liquids is the most spectacular, complex and extensively studied case of vibrational relaxation. The early work in this field is due to Landsberg and Baryshanskaia (61), Stepanov (62), Bratož and Hadži (63), Sheppard (64), Blinc and Hadži (65), Sandorfy (66), Maréchal and Witkowski (67) and Witkowski and Wojczik (68). These papers offer a qualitative, or semi-quantitative, description of infrared bands due to νAH stretching motions ; infrared bands associated with the corresponding δAH and γAH bending motions have been much less investigated. The interpretation of spectral effects in the frame of the theory vibrational relaxation is of recent date. This step was accomplished by Bratos (69), Yarwood and Robertson (70,71), Robertson and Yarwood (72) and by Maréchal and Bratos (73). See also Rösch and Ratner (74), Weidemann and Hayd (75) and Romanowski and Sobczyk (76). The theory actually is elaborated only for diluted solutions. It is important to notice that Eqn (1-1) is not an appropriate starting point for studying infrared absorption in pure hydrogen bonded liquids ; this is due to the presence of strong dielectric field effects in the fluid. No proposal exists to solve this difficult problem.

4.2 Basic Assumption

The theory is relative to a single H-bonded system AH...B dissolved in an inert, spectrally inactive solution; the two halves of the complex contain N_A, N_B atoms, respectively. It rests on the following assumptions. (i) Internal vibrations are described by normal coordinates $n=(n_1 n_2 .. n_{N_c})$ of the complex where $N_c=3N_A+3N_B-12$; they obey the laws of quantum mechanics. (ii) External vibrations are described by stochastic functions $N(t)=(N_1(t),N_2(t)...N_6(t))$; they obey the laws of classical mechanics. (iii) The Hamiltonian for internal vibrations is that of a system of harmonic vibrators perturbed by a potential $V(n,t)$ anharmonic in $n,N(t)$. $H(n,t)$, $V(n,t)$ and the dipole moment $M(n,t)$ of the complex are all stochastic through $N(t)$ and through the angular coordinates $\Omega(t)$ of the complex :

$$H(n,t) \equiv H(n,N(t)) = \left[\frac{1}{2}\sum_{i=1}^{N_c}(p_i^2+\lambda_i n_i^2)\right] + \left[\sum_{i=1}^{N_c} V_i(t)n_i + \frac{1}{2}\sum_{ij=1}^{N_c} V_{ij}(t)n_i n_j + \frac{1}{6}\sum_{ijk=1}^{N_c} V_{ijk}(t)n_i n_j n_k + ..\right] \tag{4-1a}$$

$$\vec{M}(n,t) \equiv \vec{M}(n,N(t),\Omega(t)) = \vec{M}_o(t) + \sum_{i=1}^{N_c} \vec{M}_i(t)n_i + \dots \tag{4-1b}$$

(iv) The behavior of external modes is described on different levels of approximation. In simplest approximation the time dependence of $N(t)$ is wholly neglected ; this may be justified by considering the time scale $\tau \sim 1/\Delta\omega\ 1/2$ involved (69). Alternatively, the νAH...B mode $N_1(t)$ is assumed to obey the simple Langevin equation containing a harmonic force (70-72). These assumptions seem reasonable as a whole. In particular, the stochastic formulation can be avoided, if desired. The dipole moment is that of the complex ; thus the difficulties which arise from neglect of collision induced effects do not occur here.

4.3 Discussion

4.3.1. Weak hydrogen bonds

Vibrational relaxation in solutions containing weak hydrogen bonds is studied first. Unfortunately, the quantities $V_i(t)$, $V_{ij}(t)$, $V_{ijk}(t)$ and $M_i(t)$ of Eqns (4-1a,b) are not available, a-priori, and experimental data must be used. It can then be shown that, for systems dealt with here, the dominating relaxation mechanism is the anharmonic coupling between the high frequency νAH mode n_1 and the bath controled νAH..B mode N_1. This information may be built into the theory by retaining, in Eqn (4-1a), only $V_1(t)$ and $V_{11}(t)$; similarly, only $\vec{M}_1(t)$ is kept in Eqn (4-1b). The following conclusions may then be reached.

The 1νAH band profile of weak hydrogen bond is due to a pure dephasing. It is generated by the fluctuations in the geometry of hydrogen bonded complexes, a sort of environmental fluctuations. The T_1-type de-excitation, a process termed in the theory of hydrogen bonding as predissociation, has recently been shown to have an entirely negligible role (77). A large variety of band profiles are theoretically predicted. They may be structureless asymmetric distorted Gaussians or have a structure resulting from strongly damped νAH..B stretching vibrations. The first prediction is generally realized in liquids ; the progression in the N_1-mode has also been observed, but only in gases.

4.3.2. Medium strong hydrogen bonds

Two band shaping mechanisms are operative in the case of medium-strong hydrogen bonds. The first still is the anharmonic coupling between the high frequency νAH mode n_1 and the low frequency external modes N whereas the second is the Fermi resonance between states involving n_1 and some other internal modes n_i, n_j etc. perturbed by hydrogen bonding ; they are both bath controlled. This information may be built into the theory by retaining, in Eqn (4-1a), the terms $V_1(t)$, $V_{11}(t)$ and $V_{1ij}(t)$; the former two realize the νAH νAH..B anharmonic interaction and the latter the Fermi-type interaction between the normal modes n_1, n_i and n_j. Moreover, only $\vec{M}_1(t)$ is retained in Eqn (4-1b).

The theory based on these premises essentially describes the interaction between sharp internal quantum states of the molecule and the νAH continuum generated by the bath. Vibrational relaxation process is here a complex multimode relaxation process involving not only the νAH mode n_1 but also several other internal modes n_i, n_j. It depends on, both, dephasing type mechanisms and population changes ; these mechanisms are generated by fluctuations in the geometry of complexes as well as by non-resonant energy transfers ; they are strongly correlated to each other. The theory predicts a large variety of complicated band profiles. Basically, the profile is a distorted Gaussian. The Fermi resonance generates one, or several, transmission windows called Evans holes ; larger are combination or overtone bands more the Evans hole is blurred. Narrow combination bands and weakly interacting modes produce deep and narrow transmission windows ; broad combination bands and strongly interacting normal modes generate flat and broad transmission windows. All these predictions have been tested on a vast body of experimental material.

4.3.3. Strong asymmetric hydrogen bonds

The last problem treated here is vibrational relaxation in solutions containing strong asymmetric hydrogen bonds. Extensive experimental investigations and, more specifically, the study of spectral effects due to the deuteration of the bond proton, have shown that three band shaping mechanisms are operative in this case. The first of them still is the anharmonic coupling between the νAH mode n_1 and the external modes N and the second is the Fermi resonance between n_1, n_2, n_3, the normal modes associated with the νAH, δAH and γAH motions. The third relevant band shaping mechanism is the anharmonicity of the νAH mode n_1 itself. This information can be incorporated into the theory by retaining, in Eqn (4-1a), the terms $V_1(t)$, $V_{11}(t)$, $V_{111}(t)$, $V_{1111}(t)$, $V_{122}(t)$, $V_{123}(t)$ and $V_{133}(t)$. Moreover, the term $\vec{M}_1(t)$ is conserved in Eqn (4-1b) ; compare with (78).

There results a highly complex multimode vibrational relaxation process depending in dephasing-type mechanisms and on population changes. The theory predicts an extremely broad profile with a structure on its high frequency side ; the components of this structure are called A,B,C bands. The deuteration produces little variation of the band position but may modify its fine structure. This theory explains in a satisfactory way all recent experimental findings in this area.

APPENDIX. REVIEW OF BASIC EXPERIMENTAL DATA

A.1 Generalities

The purpose of this Appendix is to review shortly the experimental material underlying the theory of vibrational relaxation in liquids ; the main references on this subject are given in (2, 4, 79, 80). The data for Van der Waals liquids and solutions are extracted from isotropic Raman spectra whereas the infrared spectra are used for hydrogen bonded liquids. In the latter case, rotational relaxation processes are much slower than their vibrational analogues and have a negligible spectral activity.

A.2 Vapor-Liquid Frequency Shifts and Vibrational Half-Widths

The vapor-liquid frequency shifts $\Delta\omega$ and the vibrational half-widths $\Delta\omega$ 1/2 are discussed first. These quantities vary in wide limits when going from Van der Waals to hydrogen bonded liquids ; $\Delta\omega$ is comprised between 0 and 2000cm^{-1} and $\Delta\omega$ 1/2 between 0,05cm^{-1} and 2000cm^{-1}. A particularly well investigated system is liquid nitrogen where $\Delta\omega$ = 5cm^{-1} and $\Delta\omega$ 1/2 = 0,05cm^{-1} ; no system is known showing a weaker intermolecular coupling. $\Delta\omega$ and $\Delta\omega$ 1/2 increase progressively with increasing force of molecular interaction and attain values of the order of 100cm^{-1} for liquids containing weak hydrogen bonds. The upper limit correspond to strong asymmetric hydrogen bonds in systems like KH_2PO_4 and chromous acid. It is important to notice that the variation of $\Delta\omega$, $\Delta\omega$ 1/2 is continous over the whole domain of variation.

The vapor-liquid frequency shifts and vibrational half-widths vary with T, ρ ; they vary also with v, the quantum number of the vibrationnally excited upper state. Most often, $|\Delta\omega|$ decreases with increasing temperature and the band approaches the position it has in a gas ; often this variation only reflects the variation of the density with T. The temperature dependence of ω 1/2 is more complex ; in weakly bonded Van der Waals liquids this quantity decreases with increasing temperature whereas the opposite behavior is found for 1νAH bands of hydrogen bonded systems. There seems to be no unique mechanism producing this variation. The experimental investigation of the v-dependence of $\Delta\omega$ 1/2 (o→v) is unfortunately a difficult task. Overtone bands are weak and are not easily observable in Raman ; if infrared techniques are employed one is faced with the problem of disentangling of vibrational and rotational spectral effects. Recent attempts to use resonance Raman techniques merit attention. Additional problems arise in the case of polyatomic fluids. The normal mode under investigation must be dynamically separated from other modes ; if it is not, an irregular $\Delta\omega$ 1/2(o→v) sequence is obtained, useless for the present analysis. For all these reasons, the problem still remains debatable ; nevertheless, it seems that $\Delta\omega$ 1/2(o→v) $\sim v^2\Delta\omega$ 1/2(o→1) for very narrow and $\Delta\omega$ 1/2(o→v) $\sim v\Delta\omega$ 1/2(o→1) for moderately broad and broad bands. The limits between these two types of behavior may be settled, tentatively, to about 10cm^{-1}.

A.3 Profiles

There exists a large variety of vibrational profiles. As a rule, the band shape is Lorentzian in weakly interacting Van der Waals liquids ; note that the centrifugal rotational-vibrational coupling may introduce an asymmetry, particularly for fluids at low pressures. Stronger is molecular interaction, larger are deviations from the simple Lorentzian behavior ; the bands are often slightly asymmetrical. This tendency is enhanced in liquids containing weak hydrogen bonds where the $1\nu AH$ bands are distinctly asymmetrical but still remain structureless ; note, however, that a structure due to the $\nu AH...B$ vibration is observed for the Millen-type complexes in the gas phase. Finally, the $1\nu AH$ bands of liquids containing medium-strong and strong hydrogen bonds always exhibit a heavy spectral structure ; sharp and narrow Evans-type transmission windows an obŝerved, in particular, in the case of very strong hydrogen bonds.

An interesting source of information is isotropic dilution. If the observed molecule is isolated in a solvent formed by its isotopically substituted analogues, the vibrational bands narrows, a consequence of the fact that the resonant transfer has been interrupted. One should not conclude, however, that this new band corresponds to the self-correlation function of the pure liquid !

If, on the contrary, the observed molecule is itself isotopically substituted and isolated in a solvent formed by non-substituted molecules, the band position varies and its structure may by deeply modified. This technique is of a general use in the hydrogen-band research ; the isotopic substitution is the deuteration of the bond proton.

A.4 Integrated Intensity

The integrated intensity of a vibrational band varies considerably when going from vapor to liquid. As a rule, the changes are considerably larger in infrared than in Raman spectra. In Van der Waals liquids, this change is ascribed to the dielectric field effects ; the ratio A/A_g, where A designates the integrated intensity in the liquid and A_g the corresponding quantity in the gas, may easily attain the value of 2. The corresponding value of A/A_g in hydrogen bonded liquids is, typically, of the order of 10-20. It is generally attributed to the change of molecular dipole moment on hydrogen bond formation. These results show the importance of dielectric field effects ; the role of collision induced effects should neither be neglected.

REFERENCES

(1) S. Bratos, Y. Guissani, J.C. Leicknam, in Mol. Mot. in Liquids, ed. J. Lascombe, Reidel, Dordrecht, p. 187 (1974).
(2) R.T. Bailey, Mol. Spectrosc. 2, 173 (1974).
(3) D.J. Diestler, Top. in Appl. Phys. 15, 169 (1976).
(4) D.W. Oxtoby, Adv. Chem. Phys., to be published.
(5) R.G. Gordon, J. Chem. Phys. 43, 1307 (1965).
(6) R.G. Gordon, J. Chem. Phys. 42, 3658 (1965).

(7) R.G. Gordon, Adv. Mag. Reson. 3, 1 (1968).
(8) L.D. Landau, E.M. Lifschitz, Electrodynamics of Continous Media, Pergamon, Oxford, p.377 (1963).
(9) S. Bratos, E. Maréchal, Phys. Rev. A 4, 1078 (1971).
(10) F.J. Bartoli, T.A. Litowitz, J. Chem. Phys. 56, 404 (1972).
(11) K.A. Valiev, Opt. Spectrosc. 11, 253 (1960).
(12) K.A. Valiev, Soviet Phys. JETP 13, 1287 (1961).
(13) S. Bratož, J. Rios, Y. Guissani, J. Chem. Phys. 52, 439 (1970).
(14) L.A. Nafie, W.L. Peticolas, J. Chem. Phys. 57, 3145 (1972).
(15) H. Morawitz, K. Eisenthal, J. Chem. Phys. 55, 887 (1971).
(16) E. Brindeau, S. Bratos, J.C. Leicknam, Phys. Rev. A 6, 2007 (1972).
(17) P. Van Konynenburg, W.A. Steele, J. Chem. Phys. 56, 4776 (1972).
(18) D.J. Diestler, Chem. Phys. Lett. 39, 39 (1976).
(19) R. Paul, G.G. Fuller, J. Chem. Phys. 64, 3809 (1976).
(20) D.W. Oxtoby, S.A. Rice, Chem. Phys. Lett. 42, 1 (1976).
(21) J.C. Leicknam, Y. Guissani, S. Bratos, J. Chem. Phys. 68, 3380 (1978).
(22) S.F. Fischer, A. Laubereau, Chem. Phys. Lett. 35, 6 (1975).
(23) W.A. Rothschild, J. Chem. Phys. 65, 455 (1976).
(24) W.A. Rothschild, J. Chem. Phys. 65, 2958 (1976).
(25) R. Levant, Mol. Phys. 34, 629 (1977).
(26) F.G. Dijkman, J.H. Van der Maas, J. Chem. Phys. 66, 3871 (1977).
(27) D.J. Diestler, J. Manz, Mol. Phys. 33, 227 (1977).
(28) H. Metiu, D.W. Oxtoby, K.F. Freed, Phys. Rev. A 15, 361 (1977).
(29) S.S. Cohen, R.E. Wilde, J. Chem. Phys. 68, 1138 (1978).
(30) D.W. Oxtoby, J. Chem. Phys. 70, 2605 (1978).
(31) R. Arndt, R.E.D. Mac Clung, J. Chem. Phys. 69, 4280 (1978).
(32) R.E. Wilde, S.S. Cohen, J. Chem. Phys. 70, 4557 (1979).
(33) M.R. Battaglia, P.A. Madden, Mol. Phys. 36, 1601 (1978).
(34) B.P. Hills, P.A. Madden, Mol. Phys. 37, 937 (1979).
(35) B.P. Hills, Mol. Phys. 37, 949 (1979).
(36) G. Döge, Z. Naturforsch. 28a, 919 (1973).
(37) G. Döge, in Mol. Mot. in Liquids, ed. J. Lascombe, Reidel, Dordrecht, p. 225 (1974).
(38) T. Tokuhiro, W.G. Rothschild, J. Chem. Phys. 62, 2150 (1975).
(39) P.C.M. Van Woerkom, J. de Beyser, M. de Zwart, J.C. Leyte, Chem. Phys. 4, 236 (1974).
(40) P.A. Madden, R.M. Lynden-Bell, Chem. Phys. Lett. 38, 163 (1976).
(41) C.H. Wang, Mol. Phys. 33, 207 (1977).
(42) R.M. Lynden-Bell, Mol. Phys. 33, 907 (1977).
(43) R.M. Lynden-Bell, Mol. Phys. 36, 1529 (1978).
(44) D.W. Oxtoby, D. Levesque, J.J. Weis, J. Chem. Phys. 68, 5528 (1978).
(45) R.K. Wertheimer, Mol. Phys. 35, 257 (1978).
(46) R.K. Wertheimer, Mol. Phys. 36, 1631 (1978).
(47) D.C. Knauss, Mol. Phys. 36, 413 (1978).
(48) B.P. Hills, Mol. Phys. 35, 793 (1978).
(49) B.P. Hills, P.A. Madden, Mol. Phys. 35, 807 (1978).
(50) B.P. Hills, P.A. Madden, Mol. Phys. 35, 1471 (1978).

(51) R. Kubo, J. Phys. Soc. Jap. 17, 1100 (1962).
(52) N.G. Van Kampen, Physica (Utrecht) 74, 215 (1974).
(53) N.G. Van Kampen, Physica (Utrecht) 74, 239 (1974).
(54) C.B. Harris, R.M. Shelby, P.A. Cornelius, Phys. Rev. Letters 38, 1415 (1977).
(55) J.P. Riehl, D.J. Diestler, J. Chem. Phys. 64, 2593 (1976).
(56) K. Müller, P. Etique, F. Kneubühl, in Mol. Mot. in Liquids, ed. J. Lascombe, Reidel, Dordrecht, p. 265 (1974).
(57) K. Müller, F. Kneubühl, Chem. Phys. 8, 468 (1975).
(58) K. Müller, Chem. Phys. Lett. 40, 508 (1976).
(59) R.M. Lynden-Bell, Mol. Phys. 31, 1653 (1976).
(60) S. Bratos, J.P. Chestier, Phys. Rev. A 9, 2136 (1974).
(61) G.S. Landsberg, F.S. Baryshanskaia, Izvestia 10, 509 (1946).
(62) B.I. Stepanov, Zhur. Fiz. Khim. 19, 507 (1945).
(63) S. Bratož, D. Hadži, J. Chem. Phys. 27, 991 (1957).
(64) N. Sheppard, in Hydrogen Bonding, ed. D. Hadži, Pergamon, Oxford, p. 85 (1958).
(65) R. Blinc, D. Hadži, Mol. Phys. 1, 391 (1958).
(66) C. Sandorfy, in The Hydrogen Bond, ed. P. Schuster, G. Zundel, C. Sandorfy, North Holland, Amsterdam, p. 613 (1976).
(67) Y. Maréchal, A. Witkowski, J. Chem. Phys. 48, 3697 (1968).
(68) A. Witkowski, M. Wojcik, Chem. Phys. 1, 916 (1973).
(69) S. Bratos, J. Chem. Phys. 63, 3499 (1975).
(70) J. Yarwood, G.N. Robertson, Nature 257, 41 (1975).
(71) J. Yarwood, R. Ackroyd and G.N. Robertson, Chemical Phys. 32, 283 (1978).
(72) G.N. Robertson, J. Yarwood, Chem. Phys. 32, 267 (1978).
(73) E. Maréchal, S. Bratos, J. Chem. Phys. 68, 1825 (1978).
(74) N. Rösch, M.A. Ratner, J. Chem. Phys. 61, 3344 (1974).
(75) E.G. Weidemann, A. Hayd, J. Chem. Phys. 67, 3713 (1977).
(76) H. Romanowski, L. Sobczyk, Chem. Phys. 19, 361 (1977).
(77) C.A. Coulson, G.N. Robertson, Proc. Roy. Soc. A 337, 167 (1974) (1974).
(78) S. Bratos, H. Ratajczak, to be published.
(79) J. Vincent, this volume.
(80) G.C. Pimentel, A.L. Mc Cllelan, The Hydrogen Bond, Freeman, San Francisco, p. 67 (1960).

ROTATIONAL SPECTRAL BAND SHAPES IN DENSE FLUIDS

W.A. Steele

Department of Chemistry
152 Davey Laboratory
The Pennsylvania State University
University Park, Pennsylvania 16802
U.S.A.

1. INTRODUCTION

Although the detailed structure of infrared or Raman vibration-rotation spectra for low pressure gases is ordinarily interpreted in terms of energy levels, one can also view measurements of the rotational transitions as a method of determining quantized values of the angular momentum of an isolated molecule. In this way, one sees that spectral experiments probe the time evolution of molecular orientations even at low density. As the density increases, the rotational lines broaden and eventually merge into a continuous spectral band. At this point, it is no longer useful to describe the molecules as being in well-defined energy or angular momentum states. A more rewarding approach to the understanding of these band shapes can be found by extracting time-correlation functions from the data by Fourier transforming the band shapes. In general, a time-correlation function contains information about the evolution of some relevant molecular variable, averaged over an equilibrium ensemble of molecules; for rotational spectra, the relevant variables are functions of orientation angles. It should be emphasized that a Fourier transform of a spectral intensity distribution does not carry one very far toward solving the problem, which is to extract information about vibrational and rotational motions of molecules in dense phases from the data. It is only when one actually begins to model, either by a physical theory or by computer simulation, that the power of the time-correlation function approach becomes apparent.

In this paper, we will not rederive the general expressions

relating infrared and Raman band-shapes to time-correlation functions. However, starting from these general results (1,2,3) we will discuss some of the problems and pitfalls that must be dealt with if one is to extract rotational information from a band whose shape is partly determined by various non-rotational factors. In addition to instrumental broadening, these factors can include pure vibrational intensity, both from the fundamental transition in a vibration-rotation band and from hot bands; vibration-rotation interactions, especially those due to Coriolis coupling; and collision-induced spectral intensity. Some of these topics are dealt with by other authors in these proceedings (4,5); we will concentrate on vibration-rotation coupling effects on liquid phase spectra, primarily on the grounds that they are not dealt with elsewhere.

The presence of these complicating factors often gives rise to serious uncertainties in the "rotational" information that is finally obtained. Nevertheless, limited information such as rotational spectral moments, correlation times and/or relaxation times can still be calculated from vibration-rotation spectra even when a full-scale analysis is not justified. We will discuss this approach to the problem and will show how such quantities can be related to orientational memory functions. Finally, we will briefly describe a few of the most popular models for reorientation in dense fluids and will attempt to indicate the improvements that are needed to obtain a generally successful theory.

We begin by writing down the relationship between $A(\omega)$, the fraction of the radiant energy absorbed at (angular) frequency ω per unit path length in the sample, and the time-correlation function. For an isolated band one can write (6)

$$A(\omega) = \frac{4\pi^2\omega}{3hcV}\,[1-\exp(-\,h\omega/kT)]\,\cdot$$

$$\frac{1}{2\pi}\int e^{-i\omega t} < \underset{\sim}{m}(0)\cdot\underset{\sim}{m}(t) > dt \qquad (1.1)$$

where $\underset{\sim}{m}(t)$ is the total dipole moment in volume V of an isotropic fluid at time t. Eq. (1.1) is already in an approximate form because the rigorous expression involves the dipole moment due to all sources, and yields the spectral intensity at all frequencies. However, as long as one can identify a particular band with a particular contribution to $\underset{\sim}{m}(t)$ (notably, with the part of $\underset{\sim}{m}(t)$ arising from a specific normal mode of vibration), and as long as overlapping bands can be separated, the approximation is a good one.

We note that theoretical treatments of time-correlation functions such as $< \underset{\sim}{m}(0)\cdot\underset{\sim}{m}(t) >$ are almost invariably based on classical statistical mechanics. In this case, both the intensity and the correlation function are even functions of their arguments. The quantum definition of the orientational part of the correlation function in eq. (1.1) can conveniently be converted to a symmetrized form in which quantum effects are minimized by applying the so-called "detailed balance" correction to the spectral intensity (7). This amounts to multiplying $A(\omega)$ by $\exp[-h\omega'/2kT]$, where $\omega' = \omega-\omega_c$ with ω_c = frequency at the band center.

The time-correlation function expression for a Raman band shape is quite similar to that for dipolar absorption - the principal complication arises from the fact that the polarization vectors $\underset{\sim}{\epsilon}_i$ and $\underset{\sim}{\epsilon}_s$ of the incident and scattered radiation are controlled in modern experiments. If the intensity of incident, monochromatic (with frequency ω_i), plane-polarized light is $\mathcal{J}(\omega_i)$ (in units of watts/cm^2, for example) and the intensity of the scattered light in unit frequency interval at $\omega_f' = \omega_f - \omega_i$ is $I(\omega_f)$ (in units of watts/(sterradian sec^{-1})) one has (2,8)

$$\frac{I(\omega_f')}{\mathcal{J}(\omega_i)} = \frac{\pi^2}{\lambda^4\epsilon_0^2}\,\frac{1}{2\pi}\int e^{-i\omega_f' t} \langle \underset{\sim}{\epsilon}_i\cdot\underset{\approx}{\alpha}(0)\cdot\underset{\sim}{\epsilon}_f\ \underset{\sim}{\epsilon}_i\cdot\underset{\approx}{\alpha}(t)\cdot\underset{\sim}{\epsilon}_f) \rangle\, dt \qquad (1.2)$$

where λ_f = wavelength of the scattered light, ϵ_0 = dielectric constant of the fluid, and $\underset{\approx}{\alpha}$ = polarizability of the fluid in the scattering volume.

We should emphasize that both eqs. (1.1) and (1.2) have been simplified by omitting the factors that account for internal field effects. One reason for this omission is that these factors are a matter of some controversy and are still actively under study (9,10). Of course, absolute measurements of spectral intensity are difficult in any case and thus only band shapes are reported in the majority of experiments. The usual approach is to divide the intensity observed at frequency ω by the integrated intensity (in arbitrary units). It is easy to show that the correlation functions relevant to such data are those given in eqs. (1.1) and (1.2) but normalized to unity at time = 0.

We now focus our attention on the dipole and polarizability time-correlation functions and begin to introduce a molecular description. In the first place, note that the product $\underset{\sim}{\epsilon}_i\cdot\underset{\approx}{\alpha}(t)\cdot\underset{\sim}{\epsilon}_f$ has the effect of extracting one element from the space-fixed second rank tensor; we can denote this by $\alpha_{if}(t)$. This quantity and the dipole moment can now be written as a sum of molecular contributions:

$$\underset{\sim}{m}(t) = \sum_{j=1}^{N} \underset{\sim}{\mu}_j(t) \qquad (1.3)$$

$$\alpha_{if}(t) = \left(\sum_{j=1}^{N} \alpha_j(t)\right)_{if} \tag{1.4}$$

It is easy to enumerate the principal sources of the time dependence of these quantities. First, we note that both the dipole moment and the polarizability are altered when a molecule is placed in a dense fluid (5). For example, the presence of permanent electrostatic moments on neighboring molecules will induce a dipole moment on a polarizable molecule that depends upon the separation distances and the orientations of all the molecules involved. These contributions to the total $\underset{\sim}{\mu}_j(t)$ vary in time as the distances and orientations change, and the part of the spectral intensity that results is ordinarily referred to as "collision-induced" or "translational." The other two important sources of time-dependence arise from the reorientation and vibration of molecule n in the fluid; we write

$$\underset{\sim}{\mu}_j(t) = \mu\, \underset{\sim}{n}_j(t) + \sum_{\ell} \underset{\sim}{v}_{\ell j}(t)\; Q_{\ell j}(t) \tag{1.5}$$

where $\underset{\sim}{v}_{\ell j}(t) = (\partial \underset{\sim}{\mu}_j(t)/\partial Q_\ell)$. Of course, $\underset{\sim}{n}_j$ and $\underset{\sim}{v}_{\ell j}$ are fixed in the molecular axes, but vary in the space-fixed frame as the molecule rotates. In addition, the time-dependence of the ℓ'th vibrational normal coordinate $Q_{\ell j}$ for molecule j is basically harmonic with frequency ω_ℓ. The dipole correlation function now splits into two parts:

$$< \underset{\sim}{m}(0)\cdot \underset{\sim}{m}(t) > = \sum_{j,k} \mu^2 \Big\{< \underset{\sim}{n}_j(0)\cdot \underset{\sim}{n}_k(t) >$$

$$+ \sum_{\ell,\ell'} < \underset{\sim}{v}_{j\ell}(0)\cdot \underset{\sim}{v}_{k\ell'}(t)\; Q_{j\ell}(0)\; Q_{k\ell'}(t) >\Big\} \tag{1.6}$$

In writing eq. (1.6), it has been assumed that there is no correlation between molecular orientation and the vibrational normal coordinates, but that the orientations and/or normal coordinates of neighboring molecules can possibly be correlated. The time-dependence of the first summation in eq. (1.6) is due only to rotational motion and its Fourier transform yields the pure rotational spectral band shapes observed in the microwave or far infrared (or, for extremely viscous liquids, in the radio-frequency regime). Vibration-rotation band shapes are dependent upon the second correlation function in eq. (1.6) and occur at infrared frequencies because of the time-dependence of the normal coordinate $Q_{\ell j}(t) \simeq Q_{\ell j}(0)\exp(i\omega_\ell t)$. Anharmonicity in $Q_{\ell j}(t)$ makes its appearance as overtone bands, and thus gives rise to correlations that are easily separable from those that oscillate at the fundamental frequency. However, vibrational relaxation has a more serious effect upon the simple harmonic time-dependence of $Q_{\ell j}(t)$. The current state of understanding of vibrational relaxation is discussed elsewhere in these proceedings (4), so we give only a schematic treatment here. If we suppose that the time-dependence of a relaxing normal coordinate can be written as $Q_{\ell j}(t) = Q_{\ell j}(0) \exp(i\omega_\ell t) V_\ell(t)$, we now see that this formalism leads to fundamental vibration-rotation bands centered at frequency $\nu_\ell = \omega_\ell/2\pi$ with band shape given by the Fourier transform of

$$C_\ell^{VR}(t) = \sum_{j,k} < \underset{\sim}{v}_{j\ell}(0) \cdot \underset{\sim}{v}_{k\ell}(t) Q_{j\ell}(0) Q_{k\ell}(0) V_\ell(t) > \qquad (1.7)$$

where coupling between different normal modes has been neglected. If the permanent dipole moments in the fluid are not too large (11), the coupling between normal coordinates on different molecules can also be neglected and one can simplify eq. (1.7) to

$$C_\ell^{VR}(t) = N < \underset{\sim}{v}_\ell(0) \cdot \underset{\sim}{v}_\ell(t) Q_\ell^{\,2}(0) V_\ell(t) > \qquad (1.8)$$

where the subscripts denoting molecule j have been deleted on the grounds that all N molecules behave the same, on average.

In the course of this argument, we have made approximations to the general expression in the order of decreasing generality. An additional simplification is needed if one is to arrive at a time-correlation function formalism for the rotational side-bands associated with the ℓ'th vibrational normal mode; namely, that there is no vibration-rotation coupling. Although we will also consider cases where this coupling is important, adoption of this approximation here allows one to write eq. (1.8) as

$$C_\ell^{VR}(t) = N < \underset{\sim}{v}_\ell(0) \cdot \underset{\sim}{v}_\ell(t) > < Q_\ell^{\,2}(0) V_\ell(t) > \qquad (1.9)$$

The crucial rotational factor in eq. (1.9) is $< \underset{\sim}{v}_\ell(0) \cdot \underset{\sim}{v}_\ell(t) >$. We will return to this correlation function after giving a parallel discussion of Raman vibration-rotation spectra.

In introducing a molecular description of the Raman scattering process, one applies essentially the same set of approximations to the polarizability tensor as were made for the dipole moment in the case of infrared spectra. Thus, one begins by ignoring collision-induced changes in $\underset{\approx}{\alpha}_j(t)$, the polarizability of molecule j, so that its time-dependence is due only to rotation and vibration. One then writes

$$\underset{\approx}{\alpha}_j(t) = \underset{\approx}{\alpha}_j^{(0)}(t) + \sum_\ell \frac{\partial \underset{\approx}{\alpha}_j^{(0)}}{\partial Q_\ell} \, Q_\ell \qquad (1.10)$$

where $\alpha_j^{(0)}(t)$ is the polarizability of the molecule in its equilibrium nuclear configuration. Again assuming that the vibrational displacements are uncorrelated with other modes in the same molecule or with modes in other molecules, one finds

$$< \alpha_{if}(0)\alpha_{if}(t) > = \sum_{j,k} < [\alpha_j^{(0)}(0)]_{if}[\alpha_k^{(0)}(t)]_{if} >$$

$$+ N < [t_\ell(0)]_{if}[t_\ell(t)]_{if}Q_\ell(0)Q_\ell(t) > \qquad (1.11)$$

where the tensor $[t_\ell(t)]_{if}$ is defined as $(\partial\alpha_{if}^{(0)}/\partial Q_\ell)$ (for molecule j).

Band shapes of pure rotational Raman spectra are related to the time-correlation function given by the first summation in eq. (1.11); in the liquid, these bands are also referred to as frequency-dependent light scattering (12). The second summation gives spectral bands at larger frequency shifts (from the incident) due to the relatively rapid time variation of the normal coordinate Q_ℓ and thus can be written as the vibration-rotation correlation function for the ℓ'th mode. Again denoting this by $C_\ell^{VR}(t)$, one now has

$$C_\ell^{VR}(t) = N < [t_\ell(0)]_{if}[t_\ell(t)]_{if} > < Q_\ell^2(0)V_\ell(t) > \qquad (1.12)$$

where we have extracted the harmonic time dependence of Q_ℓ and in addition, have assumed that the vibrational and rotational motions of a molecule in the fluid are uncorrelated. From this Raman band shape expression, the shapes of the rotational wings are evidently to be calculated from the correlation function $< [t_\ell(0)]_{if}[t_\ell(t)]_{if} >$ which changes in time as the molecule-fixed tensor elements rotate.

To summarize this section, we have shown how approximations to the general expressions lead to correlation function formulations of infrared and Raman band shapes. When vibrational and rotational motions are uncorrelated, band shapes can be calculated by taking the Fourier transform of the product of a vibrational and a rotational correlation function. If the rotational functions for infrared and Raman bands are denoted by $C_\ell^{IR}(t)$ and $C_\ell^{R,if}(t)$, we can give these a similar appearance by writing

$$C_\ell^{IR}(t) = \sum_m < [v_m(0)]_\ell[v_m(t)]_\ell > \qquad (1.13)$$

where $[v_m]_\ell$ is the m'th component of the vector $\mathbf{v}_\ell$, and

$$C_{\ell}^{R,if} = < [t_{\ell}(0)]_{if}[t_{\ell}(t)]_{if} > \tag{1.14}$$

For the remainder of this discussion, we will be concerned with the evaluation of these functions by experimental and theoretical means. Before proceeding, we note that the part of eq. (1.11) that yields an expression for the pure rotational Raman band shape is lacking the phase factor that accounts for interference between the radiation scattered from two different points in the fluid. These factors are an essential part of the theory of Rayleigh scattering and of scattering from macromolecules (12). They do not appear in the results for vibration-rotation spectra, which are essentially single-molecule or, at the very least, are affected only by correlations of range short compared to the wave-length of the incident radiation.

2. ORIENTATIONAL VARIABLES

Both the vector $v_{\ell}(t)$ and the tensor $t_{\ell}(t)$ are constants in the molecule-fixed frame. In most cases, one does not attempt to evaluate these quantities explicitly, but rather uses group theory to determine which elements are non-zero. This is usually done in cartesian coordinates, and indeed, Nafie and Peticolas (13) have listed these non-zero elements for a large number of normal modes in molecules conforming to the various point groups. However, the cartesian variables are a clumsy formulation of time-correlation functions for reorientation. Thus, we will here rewrite the results of Nafie and Peticolas in a more useful form that directly involves the Eulerian angles α, β, γ that denote the orientation of a molecule. It is convenient to let α, β be the azimuthal and polar angles of the principal symmetry axis, and γ, the angle that describes the orientation of a symmetry plane containing that axis (if such a plane exists; otherwise, the origin of γ is arbitrary). The complete set of orthogonal functions that spans the space of α, β, γ is made up of the Wigner D-functions; we here follow Rose's notation (14) in writing

$$D_{k,m}{}^{j}(\Omega) = e^{-ik\alpha} d_{km}{}^{j}(\beta) e^{-im\gamma} \tag{2.1}$$

For linear molecules, the angle γ is superfluous and one can utilize a subset of the D-functions which is, of course, the spherical harmonics $Y_{j,k}(\beta,\alpha)$.

We introduce these coordinates and functions primarily because rotational equations of motion, for either the molecule or the reorientational distribution function, are conveniently written down (and solved) using these variables. For example, if $W(\delta\Omega;t)$ is the probability that a molecule in an isotropic fluid has undergone a reorientation $\delta\Omega = \delta\alpha, \delta\beta, \delta\gamma$ in time t, one can write quite generally that

$$W(\delta\Omega;t) = \sum_{j,m,m'} \frac{2j+1}{8\pi^2} D^{*}_{m,m'}{}^{j}(\delta\Omega) C_{j,m,m'}(t) \qquad (2.2)$$

where $C_{j,m,m'}(t)$ is an orientational correlation function given by

$$C_{j,m,m'}(t) = < D_{m,m'}{}^{j}(\delta\Omega) > \qquad (2.3)$$

Thus, we wish to express the infrared and Raman orientational correlation functions in terms of the $D_{m,m'}{}^{j}(\delta\Omega)$ so that one can relate Fourier transforms of the spectral bands to the $C_{j,m,m'}(t)$. (We will subsequently discuss aspects of the modelling of these correlation functions).

The most direct way of switching from molecule-fixed cartesian vector and tensor components to Euler angle space is to work in the spherical basis. Thus, the vector v is written as (14):

$$v = \begin{vmatrix} v_1 \\ v_0 \\ v_{-1} \end{vmatrix} = \begin{vmatrix} -\frac{1}{\sqrt{2}}(v_x+iv_y) \\ v_z \\ \frac{1}{\sqrt{2}}(v_x-iv_y) \end{vmatrix} \qquad (2.4)$$

The reason for this is that the time variation of the m'th component of the body-fixed vector v is related to the space-fixed component by

$$v_m]_{space} = \sum_{m'} v_{m'} D^{*}_{m,m'}{}^{1}(\Omega) \qquad (2.5)$$

Now the D-function for the molecular orientation $\Omega(t)$ is related to those for the reorientation $\delta\Omega$ in time t by:

$$D_{m,m'}{}^{j}(\Omega(t)) = \sum_{r} D_{m,r}{}^{j}(\Omega(0))\, D_{r,m'}{}^{j}(\delta\Omega) \qquad (2.6)$$

Since these functions are complete and orthogonal for fixed j, one also has

$$\sum_{m} D_{m,r}{}^{j}(\Omega(0))\, D^{*}_{m,r'}{}^{j}(\Omega(0)) = \delta_{r,r'} \qquad (2.7)$$

Consequently, it can be shown that

$$< \underset{\sim}{v}_\ell(0)\cdot\underset{\sim}{v}_\ell(t) > = \sum_{m,m'} [v_m]_\ell [v^*_{m'}]_\ell < D_{m,m'}{}^1(\delta\Omega) > \tag{2.8}$$

It is now evident that infrared band shapes are obtainable by evaluating Fourier transforms of a simple sum of orientational correlation functions; the contributing terms can be quickly determined for a vibrational mode of given symmetry by ascertaining which spherical components of $\underset{\sim}{v}_\ell$ are non-zero. To take a simple example, vibration-rotation absorption due to a vibrational dipole moment which is parallel to the molecular symmetry axis gives rise to parallel bands; in this case, only v_0 is non-zero and

$$< \underset{\sim}{v}_\ell(0)\cdot\underset{\sim}{v}_\ell(t) >_{\text{parallel}} = v_0 v_0^* < D_{0,0}{}^1(\delta\Omega) > \tag{2.9}$$

$$= v_0 v_0^* < \cos\delta\beta > \tag{2.10}$$

We will now treat the Raman problem in a similar fashion. The tensor $\underset{\sim}{t}_\ell(t)$ is written in the irreducible spherical basis, which means that one defines:

$$\bar{t} = \frac{1}{3}(t_{xx}+t_{yy}+t_{zz}) \tag{2.11}$$

$$t_0 = \sqrt{\frac{2}{3}}\left\{t_{zz} - \frac{1}{2}(t_{xx}+t_{yy})\right\} \tag{2.12}$$

$$t_{\pm 1} = \pm(t_{xz} \pm it_{yz}) \tag{2.13}$$

$$t_{\pm 2} = \frac{1}{2}(t_{xx} - t_{yy} \pm 2it_{xy}) \tag{2.14}$$

where the subscript ℓ has been omitted for simplicity. It turns out that these irreducible space fixed tensor components are related to the molecule-fixed values by:

$$t_m]_{\text{space}} = \sum_{m'} t_{m'} D^*_{m,m'}{}^2(\Omega) \tag{2.15}$$

and that $\bar{t}$ is invariant. After using eq. (2.6) and the orthogonality of the D-functions, an average over $\Omega(0)$ gives the Raman correlation function for the ℓ'th vibrational normal mode as:

$$< [t_\ell(t)]_{if}[t_\ell(0)]_{if} > = [\bar{a}]_{if}^2 \bar{t}^2 +$$

$$\frac{1}{5} \sum_\mu [a_\mu]_{if}^2 \sum_{m,m'} [t_m]_\ell [t^*_{m'}]_\ell < D_{m,m'}{}^2(\delta\Omega) > \tag{2.16}$$

where the if cartesian component of the space-fixed polarizability tensor has been expressed in the irreducible spherical basis by writing

$$t_{if} = [\bar{a}]_{if}\bar{t} + \sum_\mu [a_\mu]_{if} t_\mu]_{space} \tag{2.17}$$

In fact, only two experimental geometries are needed to obtain all the useful polarized and depolarized spectral data (if one is far from resonance), and these are ordinarily chosen to extract a diagonal polarizability element denoted by t_{zz} by setting the incident and scattered beam polarizations to be parallel, and an off-diagonal element denoted by t_{xz} by taking the two polarizations to be perpendicular. Denoting these geometries by VV and VH, respectively, one has non-zero space-fixed spherical tensor elements that are given by

$$t_{VV} = \bar{t} + \sqrt{\frac{2}{3}}\, t_0]_{space} \tag{2.18}$$

$$t_{VH} = \frac{1}{2}\,(t_{-1} - t_1)]_{space} \tag{2.19}$$

Since the values of the a's are defined by eqs. (2.18) and (2.19), (2.16) can now be written:

$$< t_\ell(t)t_\ell(0) >_{VH} = \frac{1}{10} \sum_{m,m'} [t_m]_\ell [t^*_{m'}]_\ell \cdot < D_{m,m'}{}^2(\delta\Omega) > \tag{2.20}$$

$$< t_\ell(t)t_\ell(0) >_{VV} = \bar{t}_\ell^{\,2} + \frac{4}{3} < t_\ell(t)t_\ell(0) >_{VH} \tag{2.21}$$

The remaining problem is to identify which elements of the molecule-fixed irreducible spherical tensor t_ℓ (of rank 2) are non-zero for a given vibrational normal mode. This question, together with the analogous one for infrared bands (the $[v_m]_\ell$ are actually elements of an irreducible spherical tensor of rank 1), can be easily solved by reference to those character tables that show the symmetries of cartesian vector and tensor elements for the various point groups.

The simplest way to demonstrate the power and utility of this approach is to give a couple of examples. First, we note that the symmetry of the $\langle D_{m,m'}^{j}(\delta\Omega) \rangle$ under a sign change in $\delta\Omega$ means that (15)

$$\langle D_{m,m'}^{j}(\delta\Omega) \rangle = (-1)^{m+m'} \langle D_{-m',-m}^{j}(\delta\Omega) \rangle \tag{2.22}$$

If the molecule possesses a mirror plane passing through the symmetry axis, its properties will be symmetric to a switch from right-handed to left-handed rotations, which implies

$$\langle D_{m,m'}^{j}(\delta\Omega) \rangle = \langle D_{-m,-m'}^{j}(\delta\Omega) \rangle \tag{2.23}$$

We will now work out the orientational correlation function expressions that give the vibration-rotation band shapes for two examples: ethylene, which belongs to the D_{2h} point group and has 12 normal modes of vibration; and chloroform, which belongs to C_{3v} and has 9 normal modes. The relevant information is given in Tables I and II, where the symmetries of the irreducible representations are listed, together with the symmetries of both the cartesian and the spherical vector and tensor components. From these tables, one can evaluate eqs. (2.8) and (2.20) or (2.21) for each of the vibrational

Table I

D_{2h}	Vector		Tensor	
A_g			x^2, y^2, z^2	$\bar{t}$, t_0, $t_2 = t_{-2}$
B_{1g}			xy	$t_2 = -t_{-2}$
B_{2g}			xz	$t_1 = -t_{-1}$
B_{3g}			yz	$t_1 = t_{-1}$
A_u				
B_{1u}	z	v_0		
B_{2u}	x	$v_1 = -v_{-1}$		
B_{3u}	y	$v_1 = v_{-1}$		

Table II

C_{3v}	Vector		Tensor	
A_1	z	v_0	x^2+y^2, z^2	$\bar{t}$, t_0
A_2				
E	x, y	v_1, v_{-1}	(x^2-y^2, xy);	(t_2, t_{-2});
			(xz, yz)	(t_1, t_{-1})

modes. For our two examples, one finds that the active modes and their accompanying orientational correlation functions are:

Ethylene:

VH Raman

3 A_g modes: $$C(t) = \delta_0 < D^2_{0,0}(\delta\Omega) > +$$

$$\delta_2 \{ < D^2_{2,2}(\delta\Omega > + < D^2_{2,-2}(\delta\Omega) > \}$$

2 B_{1g} modes: $$C(t) = < D^2_{2,2}(\delta\Omega) > - < D^2_{2,-2}(\delta\Omega) >$$

1 B_{2g} mode: $$C(t) = < D^2_{1,1}(\delta\Omega) > - < D^2_{1,-1}(\delta\Omega) >$$

VV Raman: add a constant term $\bar{\delta}$ to C(t) for the A_g bands; all else unchanged.

Infrared

1 B_{1u} mode: $$< D^1_{0,0}(\delta\Omega) >$$

2 B_{2u} modes: $$< D^1_{1,1}(\delta\Omega) > - < D^1_{1,-1}(\delta\Omega) >$$

2 B_{3u} modes: $$< D^1_{1,1}(\delta\Omega) > + < D^1_{1,-1}(\delta\Omega) >$$

The δ's indicate that the various correlation functions have different weights (due to the differing magnitudes of the polarizability derivatives that contribute to a given band).

Chloroform:

VH Raman

3 A_1 modes: $C(t) = < D_{0,0}^{2}(\delta\Omega) >$

3 E modes: $C(t) = \delta_1' < D_{1,1}^{2}(\delta\Omega) > + \delta_2' < D_{2,2}^{2}(\delta\Omega) >$

VV Raman: add a constant term $\overline{\delta}'$ to C(t) for A_1 band; E band is unchanged.

Infrared

3 A_1 modes: $C(t) = < D_{0,0}^{1}(\delta\Omega) >$

3 E modes: $C(t) = < D_{1,1}^{1}(\delta\Omega) >$

(In the case of the C_{3v} molecule, it has been assumed that the three-fold symmetry axis gives rise to symmetry in the equations of motion that cause $< D_{m,m'}^{j}(\delta\Omega) > \equiv 0$ for $m \neq m'$, $j = 1$ or 2). For C_{3v} symmetry, the A_1 bands are parallel and the E bands are perpendicular, thus indicating the direction of the transition moment relative to the symmetry axes of these molecules.

The argument is readily extended to the normal modes for other point groups. In principle, one can extract a great deal of information about dynamics of reorientation in dense fluids by using this approach to analyze the measured band shapes for all the (hopefully non-overlapping) active modes of a given molecule. Additional information can be obtained from overtone bands in the Raman and infrared, from hyperRaman (16) and from non-linear spectroscopies such as CARS (17). Before undertaking such an analysis, it is wise to be aware of the complications that can arise in real systems. In the first place, there are a variety of factors that can affect the vibrational part of a vibration-rotation band. (We assume that corrections have been made for purely instrumental effects such as slit-width broadening or variations in detector sensitivity with wave length). Isotopic splitting of the central frequency can usually be handled, but other effects such as Fermi resonance and vibrational relaxation are harder to cope with. In addition, a variety of vibration-rotation coupling terms are present in real molecules, and at least one of these can have a major effect on spectral band shapes for degenerate modes; we now discuss this problem.

3. VIBRATION-ROTATION COUPLING

We first list the major kinds of vibration-rotation coupling for gaseous molecules (18) and consider which of these is likely to be significant in the liquid. As a preliminary, we note that dynamics and thus band shapes in liquids are determined not solely by molecular inertial constants and temperature, but are strongly affected by the torques due to intermolecular interactions. With this in mind, we list several factors that can couple vibrational and rotational motions:

1) Both vibrational anharmonicity and harmonic vibrational averaging cause the effective moments of inertia to change with vibrational quantum state; in addition, it is likely that intermolecular potentials will also change with vibrational state because of changes in the electronic wave function.

2) Centrifugal stretching alters the molecular dimensions and thus also affects both the effective moments of inertia and the intermolecular potential function.

For the continuous bands observed in liquids, both of these effects should be unimportant, at least for relatively rigid molecules at moderately low temperatures.

3) Coriolis coupling (19): When a nucleus in a molecule rotating with velocity $\underset{\sim}{\omega}$ also has a velocity $\underset{\sim}{v}$ due to its vibrational motion, there is a force $\underset{\sim}{F}_{cor}$ on it which is given by

$$\underset{\sim}{F}_{cor} = -2m\, \underset{\sim}{v} \times \underset{\sim}{\omega}$$

where m is the nuclear mass. This force appears as a perturbation term in the vibrational equation of motion. When this equation of motion is split into equations for the various normal modes, it emerges that the effect of the Coriolis force upon a given mode is strongly dependent on molecular symmetry. There is a so-called second-order Coriolis effect which can mix two non-degenerate vibrational modes. The amount of mixing depends inversely upon the difference between the two unperturbed frequencies as well as linearly upon the angular velocity. Unless the frequency difference is quite small, such effects can be neglected for liquid state spectra. However, a first-order Coriolis effect appears when one has a degenerate mode of the correct symmetry. The formal theory for a molecule in the liquid does not differ significantly from that for the gas, so we will merely quote the results here. Let us consider a specific case: the E mode for a C_{3v} molecule that is rotating about its own symmetry axis with velocity ω_z. Schematically, one has:

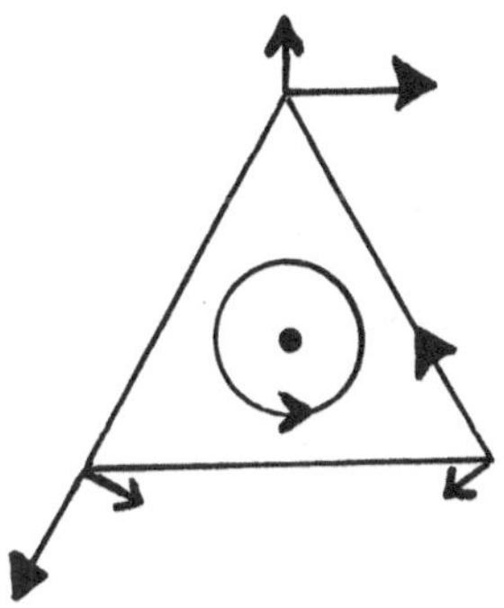

The long arrows indicate vibrational motion in one component of the unperturbed E mode; the short arrows indicate the directions of the Coriolis force exerted on the atoms at the corners of the triangle when the molecule is rotating as shown by the circular arrow. However, the short arrows are also parallel to the vibrational displacements in the second component of the E mode. The result is that the vibrational transition moment changes from one component to the other by rotating around the molecular symmetry axis with the velocity $-\zeta\,\omega_z$, where ζ is the Coriolis coupling constant. This constant is a function only of the vibrational displacements in the two normal modes that are being coupled, and thus is expected to be essentially the same in the liquid and gas. After a time interval t, the vibrational transition moment will have rotated internally by an angle $-\zeta t\overline{\omega}_z$, where $\overline{\omega}_z$ is the average velocity for rotational around the symmetry axis (20,21,22). This should be added to the change in orientation of the transition moment that is due to the rotation of the molecular frame. Consequently, one writes the correlation functions for such an E type band as $< \exp[-i\ \zeta t\overline{\omega}_z] D_{m,m'}{}^{j}(\delta\Omega) >$. The theoretical range of ζ is from +1 to -1, and the observed gas phase values of this constant for a degenerate mode are often quite close to the limit for heavy molecules. Chloroform (21) and nitrogen trifluoride (22) are two examples, having $\zeta = 0.97$ and -0.90 for their degenerate modes, respectively. The inclusion of this coupling has a large effect on the correlation function of the free rotor, as indicated in Fig. 1 for $CHCl_3$, and is likely to have an equally large effect in the liquid state. However, one must introduce a specific model for reorientation if one is to proceed further in evaluating the correlation function, with or without the Coriolis coupling factor.

4. SPECTRAL MOMENTS AND CORRELATION TIMES

As a simple illustration of "typical" data, the polarized and depolarized Raman spectra are shown in Fig. 2 for an A_{1g} band of benzene (23). The "sharp" peak in the VV spectrum is due to pure vibration, with broadening due primarily to relaxation and instrumental effects. The featureless (and noisy) VH spectrum is a rotational spectrum convoluted with all the factors that give rise to width in the VV spectrum. Since these spectra also contain collision-induced intensity, a limited amount of reliable orientational information can be extracted. Realistically, only a more limited procedure than Fourier transforming the band is justified.

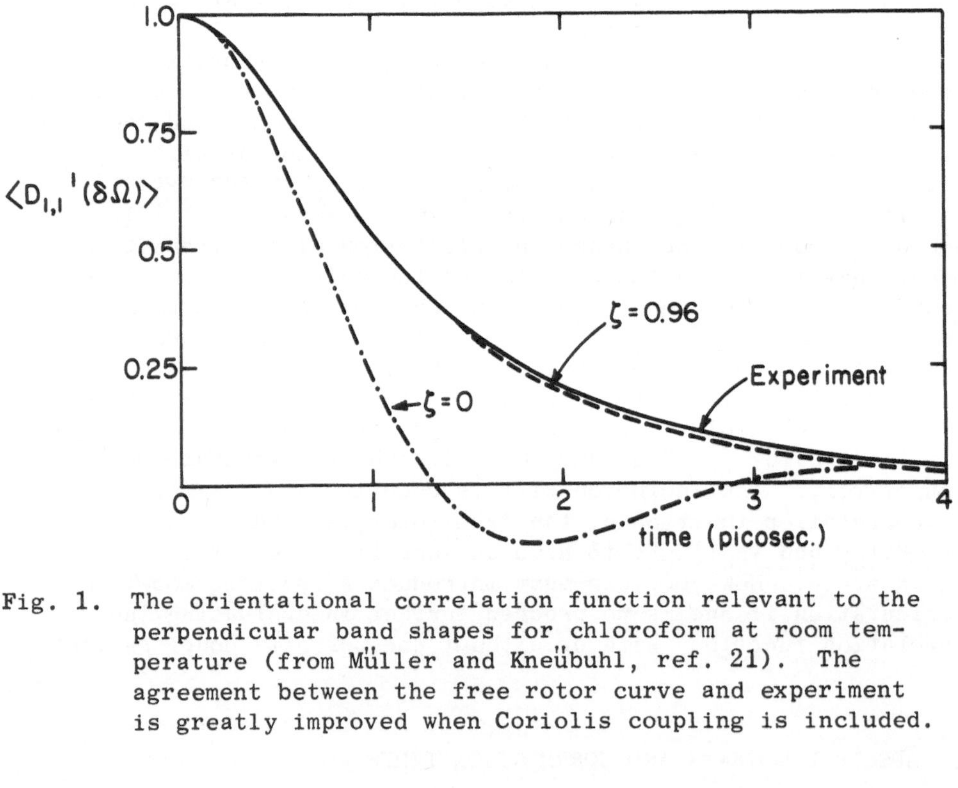

Fig. 1. The orientational correlation function relevant to the perpendicular band shapes for chloroform at room temperature (from Müller and Kneübuhl, ref. 21). The agreement between the free rotor curve and experiment is greatly improved when Coriolis coupling is included.

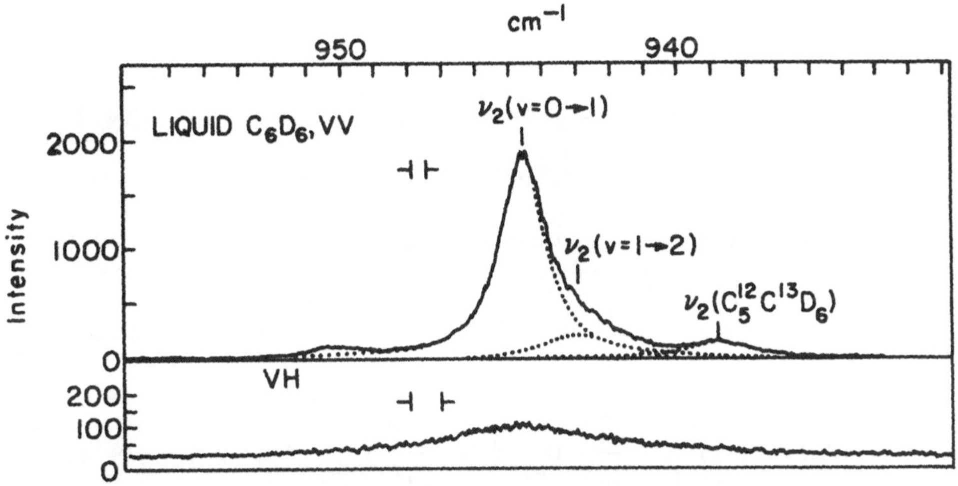

Fig. 2. Experimental Raman spectra for benzene at room temperature (from Gillen and Griffiths, ref. 23). Instrumental broadenings for the polarized and depolarized spectra are indicated; also, the polarized vibration-rotation band is approximately decomposed into the fundamental, a hot band, a C^{13} band and an impurity band that is probably due to C_6HD_5. All these bands contribute to the depolarized spectrum, but cannot be identified in the experimental curve.

Most often, this amounts to a calculation of a few spectral moments and the correlation time. In particular, the n'th moment (M(n) is defined as:

$$M(n) = \frac{\int_{band} I(\omega')\omega'^{n} d\omega'}{\int_{band} I(\omega') d\omega'} \qquad (4.1)$$

where ω' is the frequency relative to the band center and $I(\omega')$ is the observed intensity after correction for detailed balance. Furthermore, the correlation time τ_c is defined as:

$$\tau_c = \int_0^\infty C(t)\,dt \tag{4.2}$$

which is thus

$$\tau_c = \frac{\pi I(0)}{\int_{band} I(\omega')\,d\omega'} \tag{4.3}$$

Of course, there will be a number of possible values both for M(n) and τ_c for a given molecule under given conditions, depending upon the indices on the $< D_{m,m'}{}^j(\delta\Omega) >$ that characterize the rotational spectrum. For the special case where m = m', it is straightforward to show that

$$M_{j,m}(2n) = (-1)^n \sum_k < \frac{d^n}{dt^n} D_{k,m}{}^j(\Omega)\ \frac{d^n}{dt^n} D_{-k,m}{}^j(\Omega) > \tag{4.4}$$

for the contribution that $< D_{m,m}{}^j(\delta\Omega) >$ makes to the spectral band, if orientational correlations are unimportant. When correlations are present, it is possible to write an expression for the second moment $M_{j,m}{}^{(c)}(2)$ which is

$$M_{j,m}{}^{(c)}(2) = \frac{\sum_n \sum_k < \underset{\sim}{\omega}(1)\cdot\frac{d}{d\underset{\sim}{\psi}_1} D_{k,m}{}^j(\Omega_1)\ \underset{\sim}{\omega}(n)\cdot\frac{d}{d\underset{\sim}{\psi}_n} D_{-k,m}{}^j(\Omega_n) >}{\sum_n \sum_k < D_{k,m}{}^j(\Omega_1)\ D_{-k,m}{}^j(\Omega_n) >} \tag{4.5}$$

where $\underset{\sim}{\psi}$ denotes the angles of twist around the three principal axes of the molecule. In taking the ensemble averages in eq. (4.5) one only need remember that the angular velocities $\underset{\sim}{\omega}(1)\ldots.\underset{\sim}{\omega}(n)$ of different molecules are uncorrelated in an equilibrium ensemble. We also define the equilibrium ensemble average of $D_{m,m}{}^j(\Omega_{1n})$ to be $f_{j,m}$; here Ω_{1n} denotes the orientation of molecule 1 relative to that of molecule n, with $1 \neq n$. It then emerges that

$$M_{j,m}{}^{(c)}(2) = \frac{M_{j,m}(2)}{1 + N f_{j,m}} \tag{4.6}$$

We will not discuss the case of correlated intermolecular orientations further here, since it is not particularly relevant to the analysis of vibration-rotation band shapes. By taking the derivatives indicated in eq. (4.4) for the uncorrelated case, one finds (24)

$$M_{j,m}(2) = [j(j+1)-m^2] \frac{<\omega_x^2 + \omega_y^2>}{2} + m^2 <\omega_z^2> \left(1 - \frac{\zeta}{m}\right)^2 \quad (4.7)$$

where ζ is set equal to zero except for a degenerate perpendicular band, where it is the Coriolis coupling constant. Note that

$$<\omega_i^2> = \frac{kT}{I_i} \quad (4.8)$$

where I_i is the i'th principal moment of inertia. Eq. (4.8) is a consequence of the well-known fact that average molecular velocities are independent of density and molecular interactions in classical statistical mechanics. Consequently, the rotational second moment is an invariant whose calculation from experimental data gives a stringent test of the accuracy of the procedure used to extract orientational information from the raw spectral data. Theoretical expressions for the rotational fourth moment (24) show that it depends upon the mean square torque exerted on a molecule in the fluid. In favorable cases, it has proved possible to obtain useful estimates of this quantity from spectra. However, the higher moments are increasingly sensitive to the intensity in the far wings of the band, which are significantly affected by overlap with other bands, by collision-induced effects and by the poor signal/noise ratio inherent in weakly absorbing or scattering spectral regions.

In the absence of the complete correlation function, it is the correlation times that contain the most useful and accessible information about hindrance to free rotation in dense fluids. These times can also be obtained from other measurements such as nuclear spin relaxation; for molecules with complex shapes, it is usually necessary to combine data from several techniques to get a reasonably complete description of the anisotropic reorientation that occurs (3). To give just one relatively simple example, we denote the correlation time for $< D_{m,m}^j(\delta\Omega) >$ by $\tau_{j,m}$ and consider the experimental results for liquid benzene at room temperature. Data from a variety of experiments on this symmetric rotor (including that shown in Fig. 2) indicate that

$$\begin{aligned} \tau_{2,0} &= 2.8 \pm 0.3 \text{ picosec} \\ \tau_{2,2} &= 0.8 \pm 0.3 \text{ picosec.} \end{aligned} \quad (4.9)$$

Thus we see that the decay of $< D_{0,0}^{2}(\delta\Omega)>$, which is sensitive only to the reorientation of the symmetry axis, is considerably slower than that of $< D_{2,2}^{2}(\delta\Omega) >$, which is sensitive both to in-plane and out-of-plane reorientations. Qualitatively, this result is not unexpected, because the high axial symmetry implies that the torques that hinder in-plane motion should be much smaller than those that hinder reorientation of the plane of the molecule. The real problem is to deduce a model which can give a quantitative description of such anisotropic rotation.

We will now give a brief description of a couple of currently popular models for reorientation in dense fluids, but will emphasize the characteristic behavior of the memory functions for the models rather than the equations of motion themselves. One can define a memory function (25,26) $K_{j,m}(t)$ (for simplicity, we consider only the correlation functions with m=m' and denote these by $C_{j,m}(t)$) by writing

$$\frac{d}{dt} C_{j,m}(t) = - \int_0^t K_{j,m}(t-\tau)\, C_{j,m}(\tau)d\tau \qquad (4.10)$$

This equation can be solved numerically (25) or by Laplace transform, so that an assumption for $K_{j,m}(t)$ will generate $C_{j,m}(t)$ in a straightforward way.

We note two general features of the memory functions: first, the correlation time $\tau_{j,m}$ is given by

$$\tau_{j,m}^{-1} = \int_0^\infty K_{j,m}(t)dt \qquad (4.11)$$

Secondly, if one considers times large compared to the decay time of $K_{j,m}(t)$, and a correlation function $C_{j,m}(t)$ which is non-zero but decaying slowly, one can solve eq. (4.10) to give

$$C_{j,m}(t) \propto \exp(- t/\tau^{*}_{j,m}) \qquad (4.12)$$

where the relaxation time $\tau^{*}_{j,m}$ is related to the correlation time by

$$\tau^{*}_{j,m}/\tau_{j,m} = 1 - \int_0^\infty t\, K_{j,m}(t)dt \qquad (4.13)$$

5. MODELS FOR REORIENTATION

Turning now to specific models, we begin with one of the few cases that is capable of predicting anisotropic reorientation. This is the rotational diffusion model. Physically, one describes the motion as due to a series of random reorientational steps (which may be of different mean magnitudes for motion around different axes). If the average length and duration of these steps are small compared to the total orientation change and time interval, an equation of motion for $W(\delta\Omega)$, the probability distribution for reorientations $\delta\Omega$ in time t, can be derived by standard methods of Brownian motion theory. The result is

$$\frac{\partial W}{\partial t} = \underset{\approx}{\mathcal{R}} : \frac{\partial^2 W}{\partial \underset{\sim}{\psi} \partial \underset{\sim}{\psi}} \tag{5.1}$$

where $\underset{\approx}{\mathcal{R}}$ is a rotational diffusion constant tensor. In the context of this model, it can also be shown that

$$\mathcal{R}_{ij} = \frac{1}{2} \int_0^\infty < \omega_i(0)\ \omega_j(t) > dt \tag{5.2}$$

The implicit assumption made in eq. (5.2) is that the angular velocity correlation function shown there decays extremely rapidly. This implies that the magnitudes and directions of successive steps in angle space are uncorrelated, as noted above. As far as the memory function for such a process is concerned, one deduces that it must decay extremely rapidly so that

$$\int_0^\infty tK_{j,m}(t)dt \simeq 0 \tag{5.3}$$

Consequently, the decay of diffusional correlation functions (for symmetric diffusers) is exponential at all times, with $\tau^*_{j,m} = \tau_{j,m}$. In order to obtain explicit solutions, one goes back to the equation of motion (5.1) and assumes that the molecular axes are chosen to diagonalize the diffusion tensor $\underset{\approx}{\mathcal{R}}$. Then

$$\frac{\partial W}{\partial t} = \sum_i \mathcal{R}_{ii} \frac{\partial^2 W}{\partial \psi_i^2} \tag{5.4}$$

This equation has the same form as the Schroedinger equation for a symmetric top and thus has the same eigenfunctions ($D_{mm'}{}^j(\delta\Omega)$, for a symmetric diffusor) and eigenvalues. Thus if $\mathcal{R}_{xx} = \mathcal{R}_{yy} = \mathcal{R} \neq \mathcal{R}_{zz}$, one finds

$$\tau_{j,m}^{-1} = j(j+1)\mathcal{R} + m^2(\mathcal{R}_{zz} - \mathcal{R}) \qquad (5.5)$$

Asymmetric diffusor solutions are considerably more complex, but are known for j=1 or 2 (3,12).

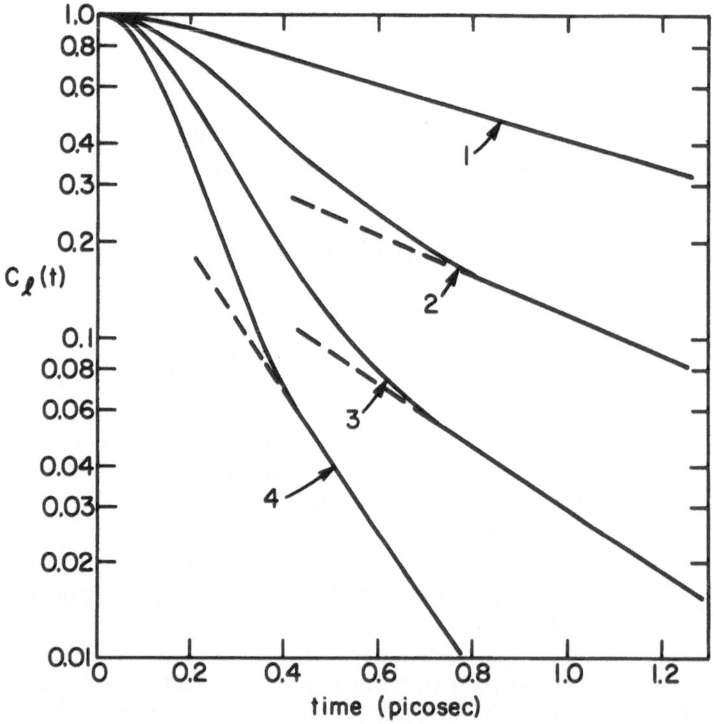

Fig. 3. Orientational correlation functions from a computer simulation of a dense fluid of diatomic molecules. Interaction potentials were assumed to be Lennard-Jones site-site interactions, with a molecular length/thickness ratio characteristic of Br_2. Masses and moments of inertia were taken to be those of Br_2. (see ref. 27 for details).

Unfortunately, thos model does not agree very well with experiment (3) or with molecular dynamics computer simulations (27), both of which show that the decay is not simply exponential and that neither the correlation times or the relaxation times vary with j in the way predicted by eq. (5.5). As an example, correlation functions are shown in Fig. 3 for a computer simulation model of the linear molecule Br_2 at a density approximately corresponding to the liquid, at a temperature above the normal boiling point but below the critical point. We emphasize that these curves are typical of those observed for small molecules whose motion is strongly hindered. They are obviously not exponential except at long times. In fact, the initial curvatures are due to a remnant of free rotation, and are nicely accounted for if one writes

$$\lim C_{\ell}(t) = 1 - \frac{1}{2} M(2)\, t^2 + \ldots.. \tag{5.6}$$

$$= 1 - \frac{\ell(\ell+1)}{2} \frac{kT}{I}\, t^2 + \ldots.. \tag{5.7}$$

where we omit the subscript m on $C_{\ell,m}(t)$ because it is always zero for linear molecule correlation functions. If it is assumed that this initial inertial motion is quickly damped out and that the reorientation at long times can be described as the result of a large number of uncorrelated steps, eq. (5.5) would lead one to expect exponential decays with slopes proportional to $\ell(\ell+1)$. Unfortunately, slopes of the exponential portions of the curves shown in Fig. 3 do not conform to this proportionality. In fact, memory functions calculated numerically from these correlation functions are shown in Fig. 4; it is evident that they do not decay extremely rapidly on the time scale of the decay of the correlation functions. Consequently, one should not be surprised to find that a random walk description of this system is incorrect, either for short or for long times.

This analysis illustrates one feature that is often quite significant in studies of reorientation; namely, that the torques which determine the hindrance to free rotation are not necessarily large enough to allow one to neglect inertial effects even in the liquid. This is particularly important for highly symmetric molecules (28) like methane (29), but can be significant in other systems as well. For example, it is quite likely that the reorientation of benzene around its own symmetry axis is dominated by inertial effects even though the tumbling of the plane of the molecule is highly hindered. Thus, the rotational diffusion model might give a good account of the $\tau_{\ell,0}$ correlation times of this molecule while failing to describe the $\tau_{\ell,m}$, $m \neq 0$.

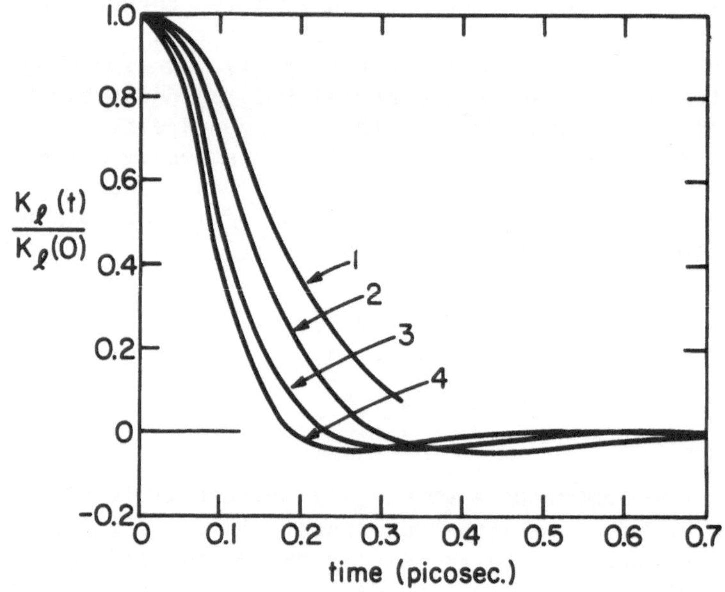

Fig. 4. Orientational memory functions calculated by numerically solving eq. (4.10), using the correlation functions of Fig. 3 as input.

It is possible to include inertial effects upon isotropic reorientation by introducing the so-called extended diffusion model (30). However, this theory has not yet been applied to the problem of anisotropic reorientation, primarily because of the severe computational problems involved. The model has several alternative forms; we limit the discussion here to J-diffusion. In this approximation, a molecule is assumed to undergo free rotation in the intervals between a series of infinitely short collisional events which randomize the molecular angular momentum without affecting the orientation. In effect, this collision destroys the orientational memory. Since the fraction of the molecules left uncollided in time t is $\exp(-\beta t)$, where β is the collision frequency, the memory functions are written as (31)

$$K_{j,m}(t) = \exp[-\beta t]\; K_{j,m}^{(0)}(t) \qquad (5.8)$$

where $K_{j,m}^{(0)}(t)$ is the free rotor memory function. When m is equal to zero, the J-diffusion model reduces to the rotational diffusion for $\beta \gg (kT/I)^{1/2}$, the characteristic frequency for free rotation. Consequently, it suffers from the same errors as the diffusion model in that limit. In fact, the extended diffusion model is most realistic in the gas where the molecules actually are undergoing collision-interrupted free rotation, but does not appear to be adequate to describe those liquids where the hindrance to rotation is important. It is evident that the model goes over smoothly into free rotor motion as $\beta \to 0$. One also obtains the correct initial curvatures of $C_\ell(t)$ for finite β; this is a (rigorous) consequence of using a memory function $K_\ell(t)$ that has the correct value at t = 0. However, since both $C_{\ell,m}(t)$ and $K_{\ell,m}(t)$ have Taylor series expansions containing only even powers of time (for a classical ensemble), the J-diffusion memory function of eq. (5.8) is incorrect at all times <u>except</u> t = 0.

The weight of the available evidence indicates that neither of the physical models discussed here is capable of quantitatively describing reorientation of simple molecules in liquids. Consequently, many workers have been exploring other approaches to this dynamical problem; one of the most promising lines of attack is to devise approximations to the memory functions. For the most part the theories have been limited to memory functions for reorientation of the molecular symmetry axis (i.e., $K_{\ell,0}(t) \equiv K_\ell(t)$). M. W. Evans (26) has reviewed many of the approximate forms used to date. We note here that there are at least three rather different ways of dealing with the problem. One is to make some kind of ansatz for the memory function $K_\ell(t)$ which would (hopefully) eliminate the defects of eq. (5.8). A second approach has been called the Mori continued fraction theory and is based on the definition of a hierarchy of memory functions for $K_\ell(t)$ by means of:

$$-\frac{dK_\ell^{(n-1)}(t)}{dt} = \int_0^t K_\ell^{(n)}(t-\tau)\, K_\ell^{(n-1)}(\tau)\,d\tau \qquad (5.9)$$

Thus, eq. (4.10) can be viewed as the n=1 case of eq. (5.9). The Mori continued fraction result is a consequence of the assumption that there is no memory in one of the higher $K_\ell^{(n)}(t)$; i.e. that

$$K_\ell^{(n)}(t) = \alpha_\ell \delta(t) \qquad (5.10)$$

for some n > 1. This closure of the hierarchy of equations generates solutions for all the lower $K_\ell^{(n)}(t)$. In particular,

$$K_\ell^{(n-1)}(t) = K_\ell^{(n-1)}(0)\ e^{-\alpha_\ell t} \tag{5.11}$$

$$K_\ell^{(n-2)}(t) = K_\ell^{(n-2)}(0)\ e^{-\beta_\ell t}[\cos(\beta_\ell t/Y_\ell) + Y_\ell \sin(\beta_\ell t/Y_\ell)] \tag{5.12}$$

etc. where $\beta_\ell = \alpha_\ell/2$ and $Y_\ell = \alpha_\ell/(4K_\ell^{(n-1)}(0) - \alpha_\ell^2)^{1/2}$. The Laplace transforms of these memory functions have particularly simple forms and yield a continued fraction for the transform of $C_\ell(t)$.

Although eq. (5.10) is incorrect in principle, it can yield useful approximations for $C_\ell(t)$, especially if the value of n is large so that the truncation of the hierarchy occurs at a point far removed from the experimentally relevant functions. A number of adjustable parameters is generated by this theory, which are α_ℓ plus the initial time values of each of the memory functions of order n-1 to 1. (Since the zero order memory function is the correlation function, its initial value can be set equal to unity by experimental normalization). Thus, the fact that good fits between theory and experiment are obtained with n=2 or 3 may well be a consequence of the increasing number of fitted constants rather than a physical realization of eq. (5.10). On the other hand, rigorous expressions for the initial values of several of the lower memory functions are known. For example, in the case of linear molecules,

$$K_\ell^{(1)}(0) = \ell(\ell+1)\frac{kT}{I} \tag{5.13}$$

$$K_\ell^{(2)}(0) = \frac{kT}{I}\left(2\ell(\ell+1) - 2 + \frac{\langle N^2 \rangle}{2(kT)^2}\right) \tag{5.14}$$

where $\langle N^2 \rangle$ is the mean square of the torque that resists reorientation of the symmetry axis. If the "free" parameters in the continued fraction approach are constrained to conform to these equations, the calculated rotational spectra will automatically possess correct second moments (from $K_\ell^{(1)}(0)$), fourth moments (from $K_\ell^{(2)}(0)$), etc. In addition, we see that intermolecular forces now appear explicitly via the mean square torque. Thus, one may begin to consider how alterations in molecular shape and/or polarity will affect reorientation in dense fluids. Ordinarily, it is quite difficult to extract information concerning these questions from values of parameters obtained by fitting spectral data to either of the diffusion models discussed above.

A third form of memory function theory that may be useful in describing reorientation is one in which the time dependence of a correlation function is not given by a single integral equation but by a set of coupled equations:

$$-\frac{d}{dt}\underset{\sim}{Y}(t) = \int_0^t \underset{\approx}{M}(t-\tau\cdot\underset{\sim}{Y}(\tau)d\tau \qquad (5.15)$$

where $\underset{\sim}{Y}(t)$ represents a 1 x n array of correlation functions and $\underset{\approx}{M}(t)$ is an n x n array of memory functions. Eq. (5.15) is put to practical use by assuming that one can guess what variables give the entire set of slowing decaying, coupled (by symmetry) correlation functions. If $\underset{\sim}{Y}$ is constructed from the correlation functions for these variables, the matrix $\underset{\approx}{M}$ will contain only rapidly decaying memory functions and one can write

$$-\frac{d}{dt}\underset{\sim}{Y}(t) = \underset{\approx}{\Gamma}\cdot\underset{\sim}{Y}(t) \qquad (5.16)$$

where

$$\underset{\approx}{\Gamma} = \int_0^\infty \underset{\approx}{M}(t)dt \qquad (5.17)$$

Eq. (5.16) is now soluble by standard methods and gives the $Y_i(t)$ as sums of simple exponential decays.

A limited number of computer simulations (27,32-34) have been reported that might help decide whether the variables relevant to reorientation can be split into slow and fast categories as required by eq. (5.17). The tentative conclusion is that the separation is not possible, at least for small molecules rotating in the liquid. In fact, the most successful application of this coupled-variable approach to reorientation has not been to the single-molecule problem but rather to those cases where dynamical and static intermolecular angular correlations are important. The theory is due primarily to Kivelson and coworkers (35); it appears to give good results when applied to the calculation of correlation functions such as those given by the first terms on the left hand sides of eqs. (1.6) and (1.11). However, the computer simulations have not yet progressed to the point of providing a critical test of this theory, mainly because of difficulties in determining the long range part of the angular correlations between molecules with non-spherical interactions.

We will not attempt to go further into the long list of models that have been proposed over the years for molecular reorientation in dense phases. However, it is useful to note the major areas of improvement needed to go from the models described here to a more successful theory. In the first place, it appears that memory effects in orientation space are important in many systems of physical interest; thus one requires memory functions with finite decay times (or their equivalent). Secondly, a molecule in a dense fluid that interacts with smoothly varying potential functions is continually under the influence of a fluctuating torque, so that the idea of the extended diffusion models, which is that reorientation is interrupted free rotation, is only adequate (at liquid densities) for very special molecules such as rough hard spheres. (However, simulations (36) indicate that extended diffusion is not adequate to model even these systems, at high density). It seems that descriptions which explicitly introduce the torque would represent a considerable step forward. Librational models are cases in point. The hydrodynamic descriptions that are currently in vogue also utilize the idea that reorientation is controlled by the torque, but attempt to relate this torque to hydrodynamic properties rather than directly to the intermolecular interaction law. Thirdly, much more work dealing with the problem of anisotropic reorientation is needed, particularly for systems where either (or both) inertial and memory effects are present. For workers interested in understanding the shapes of vibration-rotation spectral bands, this last question is certainly the most important of those listed here. In any case, much remains to be done in this field, both on the theoretical and the experimental sides.

REFERENCES

(1) R.G. Gordon, Adv.Mag.Res., 3, 1 (1968)
(2) R.T. Bailey, in Spec.Per.Reports, Molecular Spectroscopy, Vol.2, Chemical Soc., London (1972)
(3) W.A. Steele, Adv.Chem.Phys., 34, 1 (1976)
(4) D. Oxtoby, this volume
(5) G. Birnbaum, this volume
(6) R. Kubo, in "Lectures in Theoretical Physics", Vol.I, Ed. W.E. Britten and L.G. Dunham, Interscience (1959) p.151
(7) J.T. Hynes, Ph.D. Thesis, Princeton University (1965)
(8) D.A. Long, "Raman Spectroscopy", McGraw-Hill (1977) Chap.3
(9) C. Brot, in Spec.Per.Reports, Dielectric and Related Molecular Processes, Vol.2 (1973) p.1; S. Adelman and J.M. Deutch, Adv.Chem.Phys., 31 (1975)
(10) See, for example, B.M. Ladanyi and T. Keyes, Molec.Phys., 33, 1067, 1247 (1977); T. Keyes and B.M. Ladanyi, Molec.Phys., 33, 1099, 1271 (1977)
(11) P. Mirone and G. Fini, J.Chem.Phys., 71, 2241 (1979) and references contained therein
(12) B.J. Berne and R.J. Pecora, "Dynamic Light Scattering", Wiley (1976)

(13) L.A. Nafie and W.L. Peticolas, J.Chem.Phys., 57, 3145 (1972). See also F.J. Bartoli and T.A. Litovitz, J.Chem.Phys., 56, 404, 413 (1972)
(14) M.E. Rose, "Elementary Theory of Angular Momentum", Wiley (1957)
(15) W.A. Steele, Molec.Phys., in press
(16) P.D. Maker, Phys.Rev., A 1, 923 (1970)
(17) J. Bjarnason, B.S. Hudson and H.C. Andersen, J.Chem.Phys., 70, 4130 (1979)
(18) H.H. Nielsen, Rev.Mod.Phys., 23, 90 (1951)
(19) J.H. Meal and S.R. Polo, J.Chem.Phys., 24, 1119 (1956)
(20) R.M. Lynden-Bell, Molec.Phys., 31, 1653 (1976)
(21) K. Müller and F. Kneubühl, Chem.Phys., 8, 468 (1975)
(22) M. Gilbert, P. Nectoux and M. Drifford, J.Chem.Phys., 68, 679 (1978)
(23) K.T. Gillen and J.E. Griffiths, Chem.Phys., Letters, 17, 359 (1972)
(24) R.G. Gordon, J.Chem.Phys., 38, 2788 (1963); 40, 1973 (1964); 41, 1819 (1964)
(25) B.J. Berne and G.D. Harp, Adv.Chem.Phys., 17, 63 (1970)
(26) M.W. Evans, Spec.Per.Reports, "Dielectric and Related Molecular Processes", Vol.3 (1977) p.1
(27) W.A. Steele and W.B. Street, Molec.Phys., in press
(28) A.A. Maryott, M.S. Malmberg and K.T. Gillen, Chem.Phys., Letters, 25, 169 (1974)
(29) G. Levi and M. Chalaye, Chem.Phys., Letters, 19, 263 (1973); J.P. Marsault, F. Marsault-Herail and G. Levi, Molec.Phys., 33, 735 (1977); J. Vincent-Geisse, J. Soussen-Jacob, T. Nguyen Tan and R.E.D. McClung, Can.J.Phys., 57, 564 (1979)
(30) R.G. Gordon, J.Chem.Phys., 44, 1830 (1966)
(31) F. Bliot, C. Abbar and E. Constant, Molec.Phys., 24, 241 (1972); F. Bliot and E. Constant, Chem.Phys., Letters, 18, 253 (1973); 29, 618 (1974)
(32) K. Singer, J.V.L. Singer an d A.J. Taylor, Molec.Phys., 37, 1239 (1979)
(33) B. Quentrec and C. Brot, Phys.Rev., A 12, 272 (1975)
(34) G.H. Wegdam, G.J. Evans and M. Evans, Adv.Molec.Relax. and Interaction Processes, 11, 295 (1977)
(35) D. Kivelson and P. Madden, Molec.Phys., 30, 1749 (1975); T. Keyes and D. Kivelson, J.Chem.Phys., 56, 1057 (1972); B. Guillot and S. Bratos, Molec.Phys., 37, 991 (1979)
(36) J. O'Dell and B.J. Berne, J.Chem.Phys., 63, 2376 (1975)

COMPUTER CALCULATION OF VIBRATIONAL BAND PROFILES IN LIQUIDS

David W. Oxtoby

James Franck Institute
University of Chicago
Chicago, Illinois 60637 USA

Research on vibrational band profiles in condensed phases has as its goal an understanding of the way in which microscopic structure and dynamics manifest themselves in experimentally measured line shapes. Computer simulations can give insight into the interpretation of experimental line shapes and can test the assumptions and predictions of theories. Computer studies have two fundamental advantages: first, they are calculations on known systems so that uncertainties about the nature of intermolecular potentials in real liquids are avoided; second, they have great flexibility in that questions can be answered that are not directly accessible to experiment. Examples of the latter are: What effect does vibrational anharmonicity have? and: What is the relative importance of repulsive forces compared with attractive forces? Such questions can be answered through simulations simply by changing parameters, but would be much more difficult to answer unambiguously through experiment.

The method of molecular dynamics (1) is based on the direct numerical solution of the classical equations of motion for N particles in a box. The usual practice is to employ periodic boundary conditions, in which the central box is surrounded by replicas of itself; the results are essentially independent of N when N is chosen to be several hundred to a thousand. The quantities simulated are correlation functions, so it is necessary to express the spectral line shape in terms of appropriate time correlation functions; the results are well known (2). Raman scattering depends on the molecular polarizability tensors $\underset{\approx}{\alpha}_i$, which can be separated into a mean polarizability α_i

$$\alpha_i \equiv \frac{1}{3} \mathrm{Tr}\, \underset{\approx}{\alpha}_i \qquad (1)$$

and a polarizability anisotropy $\underset{\approx}{\beta}_i$

$$\underset{\approx}{\beta}_i = \underset{\approx}{\alpha}_i - \alpha_i \underset{\approx}{I} \,. \qquad (2)$$

If plane-polarized light is used, the scattered light detected can be polarized either parallel to the incident beam (giving $I_{||}(w)$) or perpendicular to it (giving $I_{\perp}(w)$). We can define the isotropic and anisotropic line shapes as

$$I_{iso}(w) = I_{||}(w) - \frac{4}{3} I_{\perp}(w) \qquad (3)$$

$$I_{aniso}(w) = I_{\perp}(w) \,. \qquad (4)$$

Then (2)

$$I_{iso}(w) \propto \int_{-\infty}^{\infty} dt\, e^{iwt} \sum_{i,j=1}^{N} \langle \alpha_i(t)\alpha_j(0) \rangle \qquad (5)$$

$$I_{aniso}(w) \propto \int_{-\infty}^{\infty} dt\, e^{iwt} \sum_{i,j=1}^{N} \langle \mathrm{Tr}\underset{\approx}{\beta}_i(t) \cdot \underset{\approx}{\beta}_j(0) \rangle \,. \qquad (6)$$

Vibrational Raman spectra arise from changes in the polarizability tensor due to vibrational motion, so that

$$I_{iso}(w) \propto \left| \frac{\partial \alpha}{\partial Q} \right|^2 \int_{-\infty}^{\infty} dt\, e^{iwt} \sum_{i,j=1}^{N} \langle Q_i(t)\, Q_j(0) \rangle \qquad (7)$$

$$I_{aniso}(w) \propto \int_{-\infty}^{\infty} dt\, e^{iwt} \langle \mathrm{Tr} \sum_{i,j=1}^{N} \frac{\partial \underset{\approx}{\beta}_i(t)}{\partial Q_i} \cdot \frac{\partial \underset{\approx}{\beta}_j(0)}{\partial Q_j} \times$$

$$Q_i(t) Q_j(0) \rangle \qquad (8)$$

where Q_i is the vibrational normal mode of interest on molecule i. It is frequently assumed (although, as we shall see, not generally true) that Q_i and Q_j are uncorrelated for $i \neq j$. In this case,

$$I_{iso}(w) \propto \int_{-\infty}^{\infty} dt\, e^{iwt} \langle Q_i(t)\, Q_i(0) \rangle \tag{9}$$

If we make a further assumption that rotations and vibrations are uncorrelated, then for a molecule with a threefold or higher axis in the direction $\underset{\sim}{u}_i$

$$I_{aniso}(w) \propto \int_{-\infty}^{\infty} dt\, e^{iwt} \sum_{i,j=1}^{N} \langle P_2(\underset{\sim}{u}_i(t) \cdot \underset{\sim}{u}_j(0) \times$$

$$Q_i(t) Q_j(0) \rangle \tag{10}$$

$$\approx \int_{-\infty}^{\infty} dt\, e^{iwt} \langle P_2(\underset{\sim}{u}_i(t) \cdot \underset{\sim}{u}_i(0) \rangle \langle Q_i(t)\, Q_i(0) \rangle \tag{11}$$

where $P_2(x)$ is the second Legendre polynomial. In a similar fashion the infrared absorption line shape can be shown to be

$$I_{inf}(w) \propto \int_{-\infty}^{\infty} dt\, e^{iwt} \sum_{i,j=1}^{N} \langle \underset{\sim}{u}_i(t) \cdot \underset{\sim}{u}_j(0)\, Q_i(t) Q_j(0) \rangle \tag{12}$$

$$\approx \int_{-\infty}^{\infty} dt\, e^{iwt} \langle \underset{\sim}{u}_i(t) \cdot \underset{\sim}{u}_i(0) \rangle \langle Q_i(t)\, Q_i(0) \rangle . \tag{13}$$

All computer simulations to date have evaluated either the separate rotational correlation functions $\langle \underset{\sim}{u}_i(t) \cdot \underset{\sim}{u}_i(0) \rangle$ and $\langle P_2(\underset{\sim}{u}_i(t) \cdot \underset{\sim}{u}_i(0) \rangle$ or the vibrational correlation function $\sum_{j=1}^{N} \langle Q_i(t)\, Q_j(0) \rangle$. No tests have been made of the assumption of separability of rotational and vibrational contributions, an assumption which has recently been challenged on theoretical (3) and experimental (4) grounds.

A large number of rotational correlation functions have been simulated by molecular dynamics; we mention only a few here. Quentrec and coworkers (5) simulated liquid nitrogen with an atom-additive Lennard-Jones potential. They calculated not only the rotational correlation function but also other related correlation functions (not measurable by experiment) to study such questions as the effect on rotational relaxation of local density and temperature fluctuations. Cheung and Powles (6) included a quadrupolar interaction in addition to an atom-atom Lennard-Jones potential for nitrogen and showed that it had little effect on the short time behavior of the rotational correlation function but improved the agreement with experiment at long times. Van Gunsteren et al. (7) studied the rotational relaxation of liquid water and showed that inclusion of a reaction field significantly shortened the rotational correlation time.

The vibrational correlation function $\sum_i \langle Q_i(t)\, Q_i(0) \rangle$ is more difficult to calculate directly for two reasons. First of all, while it is reasonable to treat translations and rotations classically in a liquid, the quantum mechanical nature of the vibrations may be important, so that errors may be introduced in a classical simulation. Second, and more important, a direct simulation of vibrating molecules is difficult because for moderate to high frequencies the vibrational motion is fast compared to motion of other degrees of freedom. As a result, the time step for the numerical integration must be very short, resulting in lengthy calculations and increasing the numerical errors. In addition, relaxation times for vibrational correlations are frequently 10-100 psec, while simulations can only be carried out for periods of a few picoseconds both because of the expense and because spurious boundary effects begin to appear after a few picoseconds. As a result, the only direct simulation of a vibrational correlation function is by Riehl and Diestler (8), who studied a one-dimensional chain of oscillators with an artificially small (10 cm^{-1}) vibrational frequency. Shugard et al. (9) simulated vibrational relaxation of a low frequency diatomic impurity in a one-dimensional solid matrix through an approach (10) in which only the nearest neighbors are directly simulated and the effect of other atoms is replaced by a stochastic force whose properties are determined from the phonon spectrum of the solid. While it may be possible to do similar stochastic simulations in liquids, the method, like other direct simulations, will probably only be useful for low frequency vibrational modes.

The difficulty with direct simulations is that vibrational motion is generally much faster than rotational and translational; this same feature can in fact be used to advantage in an indirect simulation method introduced by Oxtoby et al. (11, 12). The separation of time scales allows one to make an adiabatic approximation, in which changes in the interaction of a molecule with its surroundings are slow enough that vibrational population changes do not take place while the molecular energy levels fluctuate. This approximation neglects the contribution to line broadening from vibrational population relaxation, an effect which is usually small for small molecules. Suppose we write the full Hamiltonian as

$$\mathcal{H} = \mathcal{H}_o + \mathcal{H}_B + V \tag{14}$$

where $\mathcal{H}_o$ is the Hamiltonian for the vibrational degrees of freedom of the isolated molecules, $\mathcal{H}_B$ is that for the bath degrees of freedom (rotations and translations) which will be treated classically, and V the coupling between the two. Let us first consider the "self" contribution from Eq. (7) with j = i. In the interaction representation V becomes a function of time due to bath motion, and fluctuations in its diagonal matrix elements give rise, in the adiabatic approximation, to energy level fluctuations for the vibrational levels. The oscillator frequency becomes a time dependent function w(t), with

$$\hbar w(t) = \hbar w_0 + V_{11}(t) - V_{00}(t) \tag{15}$$

for the fundamental 0 - 1 band, for example. The oscillator phase at time t, relative to its phase at time 0, is

$$\varphi(t) - \varphi(0) = \int_0^t dt' \; w(t'). \tag{16}$$

Thus the vibrational correlation function is simply proportional to (2, 12)

$$\langle Q_i(t)Q_i(0) \rangle \propto \langle \exp \left[i \int_0^t dt' \; \Delta w(t') \right] \rangle e^{iw_0 t} . \tag{17}$$

While this expression can be simulated directly on the computer, it is frequently true that the fluctuations $\Delta w(t)$ are small enough that Eq. (17) can be expanded in cumulants and truncated at second order. In this case (weak coupling limit)

$$\langle Q_i(t)Q_i(0) \rangle \propto \exp \left[i (w_0 + \langle \Delta w \rangle) t \right] \exp(-t/\tau_v) \tag{18}$$

where $\langle \Delta w \rangle$ is the Raman line shift and τ_v is the vibrational correlation time, given by

$$\tau_v^{-1} = \int_0^\infty dt \; \langle \Delta w(t) \Delta w(0) \rangle_c \tag{19}$$

$$\equiv \langle \Delta w(0)^2 \rangle_c \; \tau_c \; . \tag{20}$$

Here, the cumulant average is defined by

$$\langle A^2 \rangle_c = \langle A^2 \rangle - \langle A \rangle^2 . \tag{21}$$

$\langle \Delta w(0)^2 \rangle_c$ is the mean square frequency fluctuation, and τ_c is the correlation time for the frequency fluctuations. Equations (18) and (19) can be derived from Eq. (17) under the condition

$$\langle \Delta w(0)^2 \rangle_c^{1/2} \tau_c << 1 . \tag{22}$$

If this condition holds, the line shape will be Lorentzian and the coupling weak. If the line is non-Lorentzian, then the coupling is strong and the full expression (17) must be simulated. So far we have considered only the single-molecule terms in the vibrational correlation; terms with $j \neq i$ will contribute if resonant transfer of the vibrational quantum to an identical neighbor is possible. In the weak coupling limit, the full vibrational correlation function is given (12) by the same expression as Eq. (18)

$$\sum_j \langle Q_i(t)\, Q_j(0) \rangle \propto \exp [\, i\, (w_0 + \langle \Delta w \rangle)\, t\,]\, \exp (\, -t/\tau_v); \tag{23}$$

the only difference is that $\Delta w(t)$ includes an additional contribution from resonant transfer:

$$\Delta w(t) \rightarrow \Delta w(t) + \Delta w_{RT}(t). \tag{24}$$

To proceed with the simulation, it is necessary to specify the nature of the coupling V. Coupling occurs in two ways: first, the potential between two molecules depends on the positions of the atoms and therefore on the vibrational coordinates, and second, the rotational Hamiltonian involves the moments of inertia which in turn depend on the vibrational coordinates (vibration-rotation coupling). V can be approximated very accurately by an expansion to second order in the vibrational coordinates (higher order terms could be included if necessary):

$$V(t) \approx \frac{\partial V(t)}{\partial Q_i}\Big|_{\tilde{Q}=0} Q_i + \frac{1}{2} \frac{\partial^2 V(t)}{\partial Q_i^2}\Big|_{\tilde{Q}=0} Q_i^2 + \sum_{j \neq i} \frac{\partial^2 V}{\partial Q_i\, \partial Q_j} Q_i Q_j \tag{25}$$

$$\equiv F_1^{\,i}(t)\, Q_i + F_2^{\,i}(t)\, Q_i^2 + \sum_j F_3^{\,ij}(t)\, Q_i Q_j \,. \tag{26}$$

The third term here gives rise to resonant transfer. For the 0 - 1 transition,

$$\hbar \Delta w(t) = F_1^{\,i}(t)\, [Q_{11} - Q_{00}] + F_2^{\,i}(t)\, [(Q^2)_{11} - (Q^2)_{00}\,] + \sum_j F_3^{\,ij}(t)\, [\,Q_{10}]^2 \,. \tag{27}$$

The matrix elements of Q and Q^2 can be evaluated from the full (anharmonic) potential energy surface and are simply put in as constants. F_1, F_2, and F_3 are all derivatives of V evaluated at the equilibrium separation $\underset{\sim}{Q} = 0$ and depend only on translational and rotational degrees of freedom. It is here that the tremendous simplification of the indirect simulation approach can be seen: simulations of the vibrational correlation function can be carried out on rigid, non-vibrating, classical molecules. Since the F's have a short correlation time (a fraction of a picosecond for nitrogen) the simulations are straightforward.

Two simulations of liquid nitrogen near its boiling point have been carried out, using two different potentials. The first simulation (11) used an atom-atom Lennard-Jones potential with parameters chosen to fit thermodynamic data. Vibration-rotation coupling was not included. The τ_v calculated was 125 psec, in good agreement with the experimental value (13) of 150 psec. Of more interest than the numerical value are a number of qualitative conclusions that were reached. First, the calculation gives

$$\langle \Delta w(0)^2 \rangle_c^{\;1/2}\, \tau_c \approx 0.035 << 1 \tag{28}$$

so that the coupling is weak. τ_c was found to be 0.15 psec, indicating that $\Delta w(t)$ loses correlation on the time scale of single collisions. A calculation of the linewidth was carried out with and without the resonant transfer term and it was found that the line was narrower when resonant transfer was present. This is contrary to the intuitive expectation that resonant transfer, which in rough terms gives rise to an exciton band, should give a broader line. The reason that this argument does not hold is clear from Eqs. (19) and (24); because the cross terms between $\Delta w(t)$ and $\Delta w_{RT}(t)$ can be either positive or negative depending on the potential energy surface, resonant transfer can give rise to either a narrowing or a broadening of the isotropic Raman line. Finally, a calculation was carried out in which the

N_2 molecules were assumed to be harmonic oscillators; the resulting τ_v was thirty times longer than experiment. This demonstrates the tremendous importance of vibrational anharmonicity and shows that model calculations which take the vibrations to be harmonic cannot give qualitatively accurate results.

A second simulation was carried out (14) using a multiparameter potential fit to various solid phase data by Raich and Gillis (15). It includes an atom-atom repulsive part, a central dispersion interaction, and quadrupolar interaction. Vibration-rotation coupling was included in this calculation. The results showed that both the quadrupole terms and the vibration-rotation coupling had only a relatively small (5-10%) effect on the linewidth. The short range and dispersion contributions were both important, and in fact there was substantial cancellation between the two so that the Raman line was significantly narrower when both were included than when only one was present. Because of this near-cancellation the results were sensitive to the dependence of the dispersion coefficient C_6 on vibrational coordinate; for reasonable assumptions on $C_6(Q)$, τ_v varied from 50 to 150 psec. These results show that both repulsive and attractive contributions must be included in a complete theory. They are discouraging in that they suggest that the potential must be known quite accurately in order to theoretically predict Raman line shapes; on the other hand one could take a more positive point of view and welcome the fact that line shapes can be sensitive probes of molecular interactions.

Nitrogen is a representative example of a weakly coupled liquid. As a result, Eq. (17) is very difficult to calculate, but the approximations leading to Eq. (18) are well satisfied and a calculation of $\langle \Delta w(t) \Delta w(0) \rangle$ is straightforward. In the opposite limit of strongly coupled liquids, Eq. (17) relaxes rapidly and is directly calculable. Simulations have in fact been carried out (16) on a nitrogen-like liquid in which the coupling was artificially increased, and direct simulations of Eq. (17) have been shown to be feasible. The most interesting applications would of course be to dipolar and hydrogen-bonded liquids in which the coupling is strong and the vibrational lines non-Lorentzian. It would also be possible and interesting to simulate the full vibration-rotation correlation functions of Eqs. (10) and (12) for the anisotropic Raman and infrared line shapes, to determine whether the usual approximation of separability of vibrational and rotational degrees of freedom is valid. Finally, it would be straightforward to apply these same methods to the computer simulation of vibrational line shapes in moderate to high temperature solids.

Acknowledgment: I would like to thank Drs. D. Levesque and J.-J. Weis for the extended collaboration which has made this work possible. Acknowledgment is made to the donors of the Petroleum Research Fund of the American Chemical Society (Grant No. 10047-AC6) for partial support of this research.

REFERENCES

(1) See, for example, A. Rahman, Phys. Rev. 136, A405 (1964); L. Verlet, Phys. Rev. 159, 98 (1967).
(2) S. Bratos, J. Rios, and Y. Guissani, J. Chem. Phys. 52, 439 (1970); S. Bratos and E. Marechal, Phys. Rev. A4, 1078 (1971).
(3) P. C. M. van Woerkom, J. de Bleyser, M. de Zwart, and J. C. Leyte, Chem. Phys. 4, 236 (1974); R, Lynden-Bell, Mol. Phys. 33, 907 (1977).
(4) A. M. Amorim da Costa, M. A. Norman, and J. H. R. Clarke, Mol. Phys. 29, 191 (1975).
(5) J. Barojas, D. Levesque, and B. Quentrec, Phys. Rev. A 7, 1092 (1973); B. Quentrec and C. Brot, Phys. Rev. A 12, 272 (1975).
(6) P. S. Y. Cheung and J. G. Powles, Mol. Phys. 30, 921 (1975); 32, 1383 (1976).
(7) W. F. van Gunsteren, H. J. C. Berendsen, and J. A. C. Rullmann, Disc. Farad. Soc. 66 (1978).
(8) J. P. Riehl and D. J. Diestler, J. Chem. Phys. 64, 2593 (1976).
(9) M. Shugard, J. C. Tully, and A. Nitzan, J. Chem. Phys. 69, 336 (1978); 69,2525 (1978).
(10) S. A. Adelman and J. D. Doll, J. Chem. Phys. 64, 2375 (1976).
(11) D. W. Oxtoby, D. Levesque, and J.-J. Weis, J. Chem. Phys. 68, 5528 (1978).
(12) D. W. Oxtoby, Adv. Chem. Phys. 40, 1 (1979).
(13) W. Clements and B. P. Stoicheff, Appl. Phys. Lett. 12, 246 (1968); A. Laubereau, Chem. Phys. Lett. 27, 600 (1974).
(14) D. Levesque, J.-J. Weis, and D. W. Oxtoby, J. Chem. Phys., to be published.
(15) J. C. Raich and N. S. Gillis, J. Chem. Phys. 66, 846 (1977).
(16) D. Levesque, J.-J. Weis, and D. W. Oxtoby, unpublished work.

VIBRATIONAL RELAXATION OF HYDROGEN-BONDED SPECIES IN SOLUTION: FINE STRUCTURE OF THE ν_s(XH) BAND

G.N. Robertson

Department of Physics, University of Cape Town

7700 Rondebosch (South Africa)

1 INTRODUCTION

I shall represent the hydrogen-bonded species formed between a proton donor XH and a base Y as X-H ... Y, and shall regard it for simplicity as a linear triatomic species. There are then two vibrational modes : the XH stretch ν_s(XH) with a frequency $\overline{\nu}_1$ typically in the region of 3000 cm^{-1} , and the hydrogen bond stretch ν_σ(XH ... Y) with a frequency $\overline{\nu}_3$ typically of about 150 cm^{-1}. The normal coordinates r_1 and r_3 can very roughly be identified with the XH and X ... Y interatomic distances.

The characteristic feature of the infra-red spectra of these species in dilute solution in an inert solvent such as carbon tetrachloride is that the ν_s(XH) absorption band is broad, with a width at half height of at least 100 cm^{-1}, and featureless. The half-width decreases as the temperature is reduced, and is roughly proportional to $T^{\frac{1}{2}}$ except at low temperatures (1). On deuteration the half-width decreases, by a factor of 1.5 or 1.6, which is definitely intermediate between $\sqrt{2}$ and 2, though the factor tends to $\sqrt{2}$ at lower temperatures. For comparison, the frequency ratio $\nu_1(H)/\nu_1(D)$ is usually about 1.35 (being somewhat less than $\sqrt{2}$ because of anharmonicity).

If one analyses experimental ν_s(XH) bandshapes in detail, one finds that they cannot be satisfactorily represented either in terms of a Gaussian or in terms of a Lorentzian function. Furthermore, the band-shapes of the protium and of the corresponding deuterium species differ : the ν_s(XD) profile has a perceptibly narrower centre and more extended wings than the ν_s(XH) profile.

Sometimes a fine structure is observed. Usually this takes the form of one or more relatively narrow transmission windows within the broad ν_s(XH) band (2), (3).

The first progress towards an understanding of these phenomena was made by Bratos in 1975, who argued that the broadening of the ν_s(XH) band was entirely due to vibrational phase relaxation and that the mechanism responsible for the very rapid relaxation involved the coupling of the ν_s(XH) and ν_σ(XH ... Y) modes (4). Bratos made certain simplifying approximations in his treatment of the dynamics of the relaxation process, which are now known to be insufficiently accurate. In my talk today I shall show that by treating the dynamics of the ν_σ(XH ... Y) mode rather more precisely than Bratos did one can obtain a complete understanding of the ν_s(XH) lineshapes of several typical hydrogen-bonded species in solution. The results of the analysis fully vindicate Bratos' ideas.

The basic ideas are as follows. In solution the ν_σ(XH ... Y) mode will be in close energetic contact with the brownian motion of the solvent molecules, since its vibrational quantum is comparable with kT at room temperature. The amplitude of this mode thus fluctuates rapidly. However, it is known both from bandshape analysis of gas-phase spectra and by comparison of crystallographic with spectroscopic data that the ν_s(XH) and ν_σ(XH ... Y) modes are strongly coupled together, and that to a first approximation the ν_s(XH) force constant is a linear function of the X ... Y separation. Since the ν_σ(XH ... Y) mode in solution has a randomly fluctuating amplitude, it follows that the ν_s(XH) vibration suffers a random frequency modulation as a result of the coupling between the modes. This will produce a broadening of the absorption band. No transitions are involved : the ν_s(XH) mode will follow the frequency fluctuations adiabatically and will lose its phase coherence while preserving its amplitude. Relaxation of the excited state population should thus be slow in comparison with the phase relaxation.

From the breadth of the featureless ν_s(XH) absorption band one can estimate the timescale of this T_2 relaxation process. According to the fluctuation-dissipation theorem, the normalised infrared absorption profile is given by

$$I(\omega) = \frac{\omega}{2\pi}\int_{-\infty}^{\infty} e^{i\omega t}\,\phi(t)\,dt, \tag{1.1}$$

where

$$\phi(t) = \langle\langle \mu_{01}(t)\,\mu_{01}(0) \rangle\rangle \langle\langle \mu_{01}(0)^2 \rangle\rangle^{-1} \tag{1.2}$$

is the autocorrelation function of the dipole transition moment. Inverse Fourier transformation of experimental absorption bands shows that the dipole transition moment relaxes in 0.2 picoseconds or less. It follows that the rotation of the molecule has a negligible effect on the bandshape : the moments of inertia of hydrogen-bonded complexes are substantial, and a complex cannot change its orientation appreciably within 0.2 ps. Hence one is justified in ignoring the rotation motion completely.

2 A MODEL FOR ν_s(XH) VIBRATIONAL PHASE RELAXATION (5)

In order to make these ideas quantitative it is necessary to adopt some sort of stochastic model for the ν_σ(XH ... Y) mode. The simplest model for describing a harmonic oscillator subjected to random collisions is the classical Ornstein-Uhlenbeck stochastic process, which obeys a Langevin equation

$$\ddot{r}_3 + \gamma\dot{r}_3 + \omega_3^2 r_3 = m_3^{-1}F(t) \tag{2.1}$$

where the random force F(t) is assumed to have a Gaussian marginal distribution and an infinitesimal correlation time. The autocorrelation function $\phi(t)$ is then given by

$$\psi(t) = \langle\langle r_3(t)\, r_3(0) \rangle\rangle \langle\langle r_3^2 \rangle\rangle^{-1} \tag{2.2}$$

$$= \exp(-\tfrac{1}{2}\gamma t)\ \{\cos\tilde{\omega}t + (\gamma/2\tilde{\omega})\sin\tilde{\omega}t\}$$

where $$\tilde{\omega}^2 = \omega_3^2 - \gamma^2/4 \tag{2.3}$$

in the periodic case. Strictly speaking one should treat the ν_σ(XH ... Y) mode quantum-mechanically, since $\hbar\omega_3$ is comparable with kT. Although several quantum-mechanical generalisations of the Ornstein-Uhlenbeck stochastic process have been proposed, the autocorrelation functions all have the same classical limiting form at long times $t \gg \hbar/kT$, and for the present purpose it is sufficient to describe the long time behaviour of $\psi(t)$ correctly.

We first consider the case where the XH molecule has only one vibrational mode in the 3000 cm^{-1} region so that no Fermi resonances are possible. The hamiltonian governing the ν_s(XH) vibration is

$$\hat{H}(t) = \hat{p}_1^2/2m_1 + \tfrac{1}{2}m_1\omega_1^2 r_1^2 + K_{113}r_1^2 r_3(t) \tag{2.4}$$

This form of cubic coupling term has been widely used in the analysis of gas phase spectra. Since the hydrogen-bond stretching coordinate $r_3(t)$ is a stochastic process, the hamiltonian $\hat{H}(t)$ governing the ν_s(XH) vibration is a stochastic operator. It may be written as

$$\hat{H}(t) = \hat{p}_1^2/2m_1 + \tfrac{1}{2}m_1\omega^2(t)\, r_1^2 \quad , \tag{2.5}$$

where

$$\omega(t) = \omega_1[1 + (2K_{113}/m_1\omega_1^2)r_3(t)]^{\frac{1}{2}} , \tag{2.6}$$

showing that the $\nu_s(XH)$ frequency suffers a random modulation because of the coupling between the modes.

The time-dependence of the dipole transition movement $\mu_{01}(t)$ can be calculated by applying the Heisenberg equations of motion. One finds that

$$\dot{\mu}_{01}(t) = -i\omega(t)\mu_{01}(t) \quad , \tag{2.7}$$

where $$\omega(t) = \omega_1 + \left(\frac{K_{113}}{m_1\omega_1}\right) r_3(t) \tag{2.8}$$

to a good approximation. By hypothesis $r_3(t)$ is a Gaussian random variable, since it is governed by the Ornstein-Uhlenbeck stochastic differential equation. Thus the fluctuating part of $\omega(t)$ is Gaussian. This equation of motion for $\mu_{01}(t)$ is identical to the basic equation of Kubo's theory of an oscillator frequency modulated by a Gaussian random process (6). The only complication is that here it is an <u>oscillatory</u> stochastic process which modulates the oscillator. The theory of Kubo carries over without difficulty, and one finds that the dipole transition movement autocorrelation function $\phi(t)$ is given by

$$\phi(t) = \langle\langle \mu_{01}(t)\mu_{01}(0) \rangle\rangle / \mu_{01}^2 = \exp[-i\omega_1 t - \Delta^2 \int_0^t (t-x) \ \mathrm{Re}\{\psi(x)\}dx] \tag{2.9}$$

where $$\Delta = K_{113} \langle\langle r_3^2 \rangle\rangle^{\frac{1}{2}}/m_1\omega_1 \tag{2.10}$$

and $\psi(t)$ is the autocorrelation function of the Ornstein-Uhlenbeck stochastic process. Δ is the parameter which Kubo calls the "amplitude of modulation". That only the real part of $\psi(t)$ enters equation (2.9) provides added justification for representing $r_3(t)$ by a classical stochastic process in this context.

By analogy with Kubo's theory one can define a characteristic time

$$\tau_c = \int_0^\infty \psi(t) \ dt \quad , \tag{2.11}$$

and for this particular model it turns out that

$$\tau_c = \gamma/\omega_3^2 \quad . \tag{2.12}$$

Kubo calls τ_c the correlation time of the modulation, which is appropriate if the spectral density of $\psi(t)$ has a single maximum. In the overdamped and critically damped cases $(\gamma/2\omega_3 \gtrsim 1)$ $\psi(t)$ is a monotonically decreasing function and Kubo's theory is applicable without modification, but in the underdamped case it must be applied with caution. Nevertheless, τ_c as defined above remains a useful parameter for characterising the speed of modulation.

A few words about the concept of "speed of modulation" would be in order at this point. The relaxation function $|\phi(t)|$ of the randomly modulated oscillator decays in a time τ_r, the relaxation time; the behaviour of $|\phi(t)|$ is qualitatively different according as the ratio τ_c/τ_r is large or small, i.e. according as the modulation process is slow in comparison with the relaxation of the oscillator or vice versa.

It is easily shown that

$$\tau_c \tau_r = (\tau_c \Delta)^2 \quad , \tag{2.13}$$

so that one can consider $\tau_c\Delta$ to be a parameter which characterises the speed of modulation. The modulation is slow if $\tau_c\Delta \gg 1$, so that $\tau_c \gg \tau_r$ and the ν_s(XH) transition dipole moment relaxes much more rapidly than the hydrogen bond vibration. In these circumstances the ν_s(XH) bandshape will be Gaussian, reflecting only the probability distribution but not the dynamics of the modulation process. As the Kubo parameter $\tau_c\Delta$ decreases the speed of modulation increases, and the kurtosis of the spectrum becomes more and more pronounced. This is interpreted as a narrowing of the line.

However, in the present model τ_c is related to the damping constant γ (cf. equation 2.12), so that it is impossible to discuss the speed of modulation without reference to the damping of the ν_σ(XH ... Y) mode. Indeed, the Kubo parameter $\tau_c\Delta$ can be written as

$$\tau_c\Delta = \Delta(\gamma/\omega_3^2) = (2\Delta/\omega_3)(\gamma/2\omega_3) \quad . \tag{2.14}$$

For given amplitude of modulation and ν_σ(XH ... Y) frequency, the Kubo parameter is proportional to the damping constant. This means that the speed of modulation will depend both on the strength of the coupling between the modes and on the strength of the interaction between the hydrogen bond and the solvent. Illustrations are given in reference (5).

3 EXPERIMENTAL TESTS OF THE THEORY

The most direct way of testing the theory is to record infrared spectra digitally and to fit the theoretical intensity expression

$$I(\omega) = \frac{C\omega}{2\pi}\int_{-\infty}^{\infty} e^{i\omega t}\phi(t;\underline{P})\, dt \tag{3.1}$$

to the experimental results. The dipole moment autocorrelation function $\phi(t;\underline{P})$ depends on four parameters $\underline{P}$, viz. ω_1, ω_3, Δ and τ_c; these are determined by a non-linear least squares method.

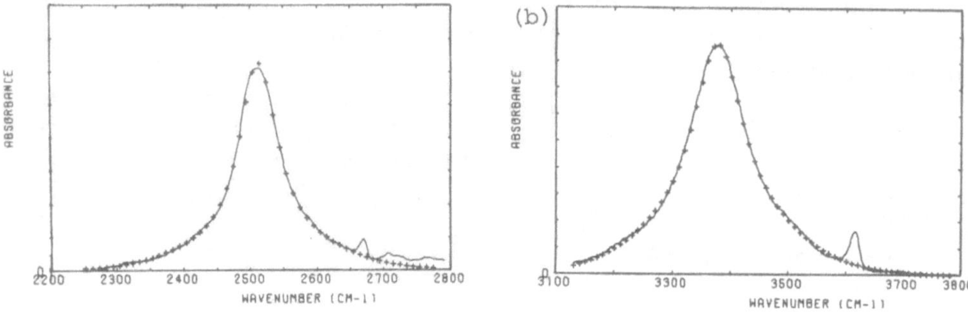

Fig. 1. Observed and fitted spectra of (a) phenol (OD)-dioxan and (b) phenol (OH)-dioxan in $CC\ell_4$ at 313 K.

Figures 1 to 3 show the experimental and fitted spectra of phenol (OH) - and phenol (OD)-dioxan in $CC\ell_4$; phenol (OH) - and phenol (OD) - acetonitrile in $CC\ell_4$; and phenol (OH) - acetonitrile in $CDC\ell_3$. The experiments were performed by Drs. J. Yarwood and R. Ackroyd (7). The fits seem reasonably satisfactory to the eye, so it is worthwhile to perform proper goodness-of-fit tests.

A simple test is to estimate the (external) variance of each data point by repeating the experiment ten times. The internal variance can be estimated from the sum of squares of the residuals between the observed and fitted spectra. The null hypothesis is that the discrepancy between the experimental and fitted curves is entirely due to errors of observation, so that the two estimates of the variance are the same (within sampling error).

In the case of PhOD-dioxan we found a variance ratio of 1.0, indicating that the theory gives an essentially complete account of the infrared spectral profile. Since the theory only takes

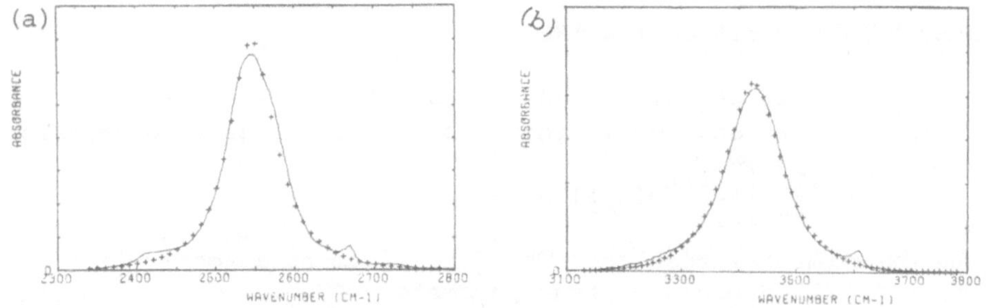

Fig. 2. Observed and fitted spectra of (a) phenol (OD) - acetonitrile and (b) phenol (OH) - acetonitrile in $CC\ell_4$ at 313 K.

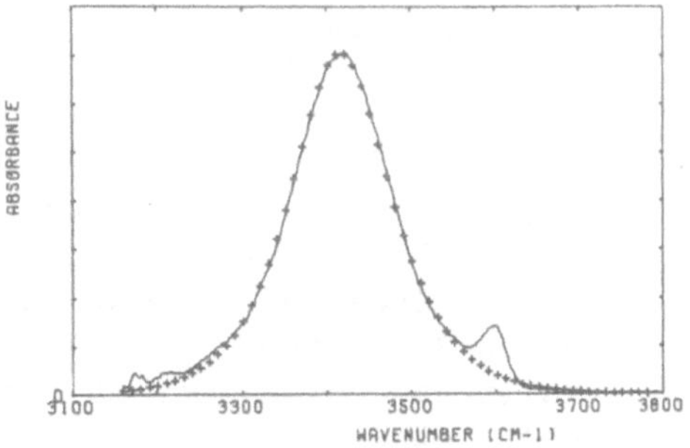

Fig. 3. Observed and fitted spectra of phenol (OH) - acetonitrile in $CHCl_3$ at 313 K.

vibrational phase relaxation into consideration, this shows indirectly that neither population relaxation nor vibration-rotation coupling influences the spectrum, which confirms Bratos' qualitative argument (4).

Other cases are less favourable; but in several cases where the experimental baseline could be fairly accurately determined we obtained variance ratios in the region of 5. Where there were difficulties with the background this increased to between 10 and 20.

The theory predicts that on deuteration the ν_s(OH) amplitude of modulation should decrease in the same ratio as the frequency, i.e.

$$\Delta(H)/\Delta(D) = \omega_1(H)/\omega_1(D) \quad . \qquad (3.2)$$

On the other hand, γ and τ_c should be unaffected by deuteration. The frequency ratio is estimated to be 1.34 for both the phenol-dioxan and the phenol-acetonitrile species, while the amplitude of modulation ratios are estimated to be 1.29 ± 0.01 and 1.36 ± 0.01. This is fairly satisfactory. For comparison, the ratio of the half-widths of the spectra are 1.61 and 1.55 respectively. This is easily explained in terms of the Kubo parameter $\tau_c\Delta$, which is smaller for the deuterium than for the corresponding protium species because of the smaller Δ. The estimates of τ_c agree within experimental error in each case, as they should. Evidently motional narrowing is somewhat more pronounced for the deuterated species, which is observed in each case to have a narrower profile.

We can also understand the effect on the spectrum of changing the solvent in terms of the speed of modulation. The ν_s(OH) half-width of the phenol-acetonitrile complex in CCl_4 is 116 cm^{-1}, while that of the same species in $CDCl_3$ is 148 cm^{-1}. The amplitude of modulation Δ is very similar in both cases. (This is a parameter which depends only on the structure of the complex and not on its environment.) However, τ_c is smaller for the complex in CCl_4 than in $CDCl_3$ (0.05 and 0.07 ps respectively) so that the Kubo parameter $\tau_c\Delta$ is smaller. The speed of modulation is greater in CCl_4 than in $CDCl_3$, so that the absorption band has a greater kurtosis and appears narrower in the centre. The mechanism responsible for the differences is easy to understand : in the invert solvent CCl_4 the random forces acting on the ν_σ(OH ... Y) coordinate are smaller than in the polar solvent $CDCl_3$ and thus the damping constant γ is smaller. Since $\tau_c = \gamma/\omega_3^2$ (cf. equation 2.12) the speed of modulation is greater in the inert than in the polar solvent. There is little evidence that the structure of the complex is affected by a change of solvent.

The most detailed prediction of the theory is that the amplitude of modulation Δ is proportional to $\ll r_3^2 \gg^{\frac{1}{2}}$, so that it varies with temperature as $[\coth(\hbar\omega_3/2kT)]^{\frac{1}{2}}$. In order to test this relation Dr. J. Bournay recorded the infrared spectra of self-associated CD_3OH in freon 11 ($CFCl_3$) at 10 K intervals between 280 K and 160 K, and we analysed the ν_s(OH) absorption band so as to find Δ and $\tau_c\Delta$ at each temperature (8). Figure 4 shows a typical absorption spectrum with its theoretical reconstruction superimposed. Plotting Δ as a function of $[\coth(\hbar\omega_3/2kT)]^{\frac{1}{2}}$ gives a very satisfactory straight line, with a correlation coefficient of 0.99. Furthermore, the intercept is zero to within experimental error. This provides strong support for the theory.

A very interesting phenomenon occurs when the solution freezes, which takes place somewhere between 160 and 150 K. At 170 K (liquid) the halfwidth of the ν_s(OH) band is 175 cm^{-1}; at 150 K (solid) the halfwidth is 210 cm^{-1}, and the bandshape is nearly gaussian. However the amplitude of modulation Δ actually decreases slightly, from 95 cm^{-1} to 93 cm^{-1} across the freezing point. There is no discontinuous change; the $\Delta = b[\coth(\hbar\omega_3/2kT)]^{\frac{1}{2}}$ relationship is

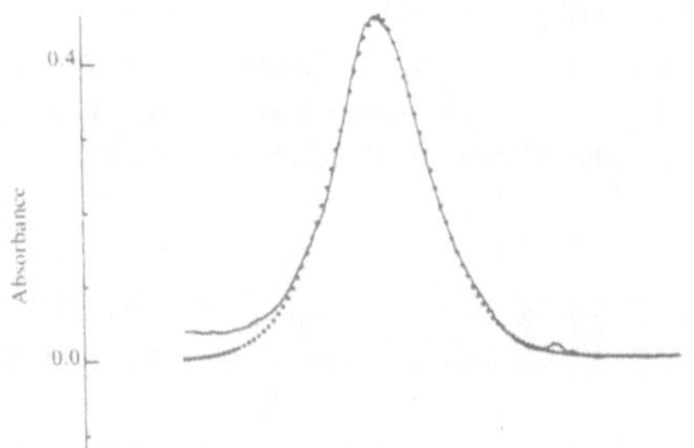

Fig. 4. Observed and fitted spectra of self-associated $CD_3OH/CFCl_3$ at 240 K.

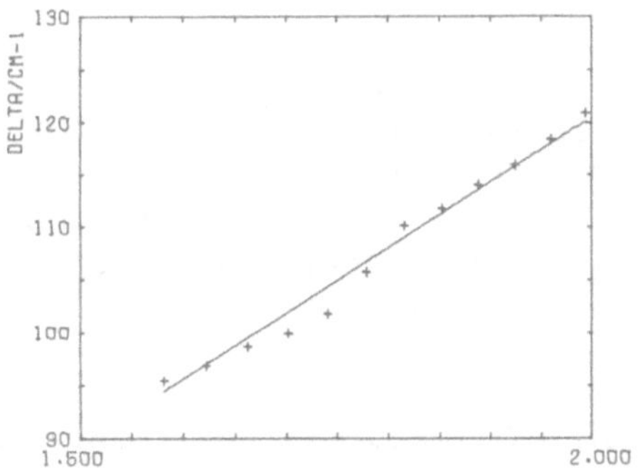

Fig. 5. Regression of amplitude of modulation Δ on $[\coth(\hbar\omega_3/2kT)]^{\frac{1}{2}}$ for self-associated CD_3OH.

still followed. What does change is the speed of modulation. For the liquid at 170 K the Kubo parameter τ_c = 0.83, indicating that there is appreciable motional narrowing. For the solid at 150 K τ_c = 7.0, which shows that the system is close to the slow modulation limit. On freezing the characteristic time τ_c of the modulation increases by nearly an order of magnitude, from 0.05 ps to 0.4 ps. This is perfectly understandable, since $\tau_c = \gamma/\omega_3^2$. It simply means that the viscous damping of the hydrogen bond stretching vibration ν_σ(XH ... Y) is an order of magnitude greater in solid than in liquid solution. In the liquid we estimate $\gamma/2\omega_3$ = 0.4, so that the vibration is underdamped, though not far from the critical damping point. In the solid $\gamma/2\omega_3$ = 4.0, indicating that the ν_σ(XH ... Y) mode is heavily overdamped.

On further cooling of the solid solution to liquid helium temperature the band profile remains Gaussian and Δ decreases from 93 to 80 cm^{-1} (the halfwidth decreases from 210 to 190 cm^{-1}). Although this is significantly greater than the 95% confidence estimate (58 ± 6) cm^{-1} obtained by extrapolating to zero temperature the relationship

$$\Delta = a + b \coth(\hbar\omega_3/2kT)^{\frac{1}{2}} \tag{3.3}$$

with parameters a = (-4 ± 10) cm^{-1} and b = (62 ± 6) cm^{-1} determined from measurements on the liquid between 160 K and 280 K, at least part of the discrepancy may be due to error in the literature value of ω_3 , which we think is too small (8).

We conclude that at 6 K the phase relaxation of the ν_s(OH) mode is produced by the zero-point oscillation of the hydrogen bond. This is much the most plausible way of accounting for the large residual breadth and the Gaussian profile of the ν_s(OH) band of the solid solution at 6 K. The ν_s(OH) band reflects the probability distribution of the r_3 coordinate, which is Gaussian with variance $\hbar/2m_3\omega_3$.

An unusual feature is that the ν_σ(OH ... O) mode is heavily overdamped at 6 K; we estimate that $\gamma/2\omega_3$ is roughly 5.0. When we examine the ν_s(OH) band of self-associated CD_3OH at liquid helium temperature we are indirectly observing the zero-point motion of an overdamped stochastic oscillator.

4 ν_s(XH) PHASE RELAXATION COMPLICATED BY FERMI RESONANCE

Let us turn now to the situation where the XH molecule is polyatomic and has an internal mode (or combination of modes) with a similar frequency to the ν_s(XH) mode. I shall denote this mode by ν_2 and the coordinate by r_2 .

The hamiltonian governing the coupled motion of the ν_1 and ν_2 modes can be written

$$\hat{H}(t) = (\hat{p}_1^2/2m_1 + \tfrac{1}{2}m_1\omega_1^2 r_1^2) + (\hat{p}_2^2/2m_2 + \tfrac{1}{2}m_2\omega_2^2 r_2^2) + V(r_1,r_2) + K_{113}r_1^2 r_3(t) + K_{223}r_2^2 r_3(t) \quad , \tag{4.1}$$

where it has been assumed that the ν_2 mode is also coupled to the hydrogen bond stretching mode and suffers a random frequency modulation. It is convenient to represent the quantum states $\upsilon_1 = 0$, $\upsilon_2 = 0$ by $|0\rangle$; $\upsilon_1 = 1$, $\upsilon_2 = 0$ by $|1\rangle$; and $\upsilon_1 = 0$, $\upsilon_2 = 1$ by $|2\rangle$. After a straightforward but fairly lengthy calculation one finds that the μ_{01} transition dipole autocorrelation function $\phi(t)$ is given by

$$\phi(t) = \tfrac{1}{2}[1 + (1 + V_{12}^2/\delta^2)^{-\frac{1}{2}}] \exp[x + (\delta^2+V_{12}^2)^{\frac{1}{2}}] + \tfrac{1}{2}[1 - (1 + V_{12}^2/\delta^2)^{-\frac{1}{2}}] \exp[x - (\delta^2+V_{12}^2)^{\frac{1}{2}}] \tag{4.2}$$

$$\text{where } x(t) = -\frac{i}{2}(\omega_1 + \omega_2)t - \tfrac{1}{2}(\Delta_1^2 + \Delta_2^2)\, g(t) \quad , \tag{4.3}$$

$$\delta(t) = -\frac{i}{2}(\omega_1 - \omega_2)t - \tfrac{1}{2}(\Delta_1^2 - \Delta_2^2)\, g(t) \tag{4.4}$$

$$V_{12}(t) = -i \langle 1|V(r_1,r_2)|2\rangle\, t/\hbar \quad , \tag{4.5}$$

$$g(t) = \int_0^t (t - x)\, \psi(x)\, dx \quad , \tag{4.6}$$

$$\Delta_1 = K_{113} \ll r_3^2 \gg^{\frac{1}{2}}/m_1\omega_1 \quad , \tag{4.7}$$

$$\Delta_2 = K_{223} \ll r_3^2 \gg^{\frac{1}{2}}/m_2\omega_2 \quad , \tag{4.8}$$

and $\psi(t)$ is the autocorrelation function of $r_3(t)$. It has been assumed that $\mu_{Q2} \ll \mu_{Q1}$, so that transitions to the interfering mode are effectively forbidden. However, the beating of the two modes profoundly affects the way in which the allowed transition dipole moment relaxes.

If the ν_1 and ν_2 modes are only weakly coupled together, and $V_{12}(t)/\delta(t)$ is small for all times t, the expression for $\phi(t)$ simplifies considerably, and can be written approximately as

$$\phi(t) \approx (1 - V_{12}^2/4\delta^2) \quad \exp(-i\omega_1 t - \Delta_1^2 g(t) + \tfrac{1}{2}V_{12}^2/\delta) + (V_{12}^2/4\delta^2) \quad \exp(-i\omega_2 t - \Delta_2^2 g(t) - \tfrac{1}{2}V_{12}^2/\delta) \tag{4.9}$$

The effect of the Fermi resonance between the ν_1 and ν_2 modes is easily understood. In the absence of any phase relaxation (i.e. if both Δ_1 and Δ_2 are zero), the ratio $V_{12}(t)/\delta(t)$ is constant, and is given by

$$V_{12}(t)/\delta(t) = 2 < 1|V|2 > /\hbar(\omega_1 - \omega_2) \quad . \tag{4.10}$$

This indicates that there is a transfer of intensity from the strongly allowed $|0> \rightarrow |1>$ transition to the forbidden or weakly allowed $|0> \rightarrow |2>$ transition. However, if the ν_1 and ν_2 modes are each coupled to the rapidly relaxing ν_σ(XH ... Y) hydrogen bond stretching mode, each will suffer a random frequency modulation, and the extent of the coupling between the ν_1 and ν_2 modes will be subject to fluctuations and hence time-dependent. The quantity

$$V_{12}(t)/\delta(t) = \frac{-i < 1|V|2 > t\hbar^{-1}}{-\frac{i}{2}(\omega_1-\omega_2)t - \frac{1}{2}(\Delta_1^2-\Delta_2^2)\ g(t)} \tag{4.11}$$

has the asymptotic form

$$V_{12}(t)/\delta(t) \sim \frac{2\hbar^{-1} < 1|V|2 >}{(\omega_1-\omega_2) - i(\Delta_1^2-\Delta_2^2)\ \tau_c} \tag{4.12}$$

which is complex.

An interesting phenomenon occurs when the real part of V_{12}/δ is much smaller than the imaginary part, so that the real part of $(V_{12}/\delta)^2$ is negative. Reference to equation (4.9) shows that the absorption feature at ω_2 is then losing intensity to the ω_1 feature. The condition for this to be possible is that

$$(\Delta_1^2 - \Delta_2^2)\tau_c \gg |\omega_1 - \omega_2| \tag{4.13}$$

showing that the strongly allowed ω_1 absorption feature must be appreciably broader than the ω_2 feature, and that the separation between the frequencies must be small. In these circumstances the narrow feature at ω_2 will be overlapped by the broad feature centred at ω_1. The transfer of intensity away from ω_2 will result in a narrow transmission window at ω_2 within the broad band centred at ω_1. This transmission window is known as an "Evans hole", (3) after its discoverer. The connexion between vibrational relaxation and Evans hole formation was first discussed by Bratos (4), and the above treatment represents a slight extension of Bratos' work.

In order to test the theory, it is desirably to study a system where the frequency mismatch $\omega_1-\omega_2$ can be varied continuously, so that the evolution of the Evans hole can be followed through resonance. Figure 6 shows the infrared spectrum of self-associated CH_3OD in solution in freon 11 at 20 K intervals between 160 K and 300 K (9). The $\nu_s(OD)$ absorption band is broad, but it has a doublet structure as a result of the transmission window at about 2460 cm^{-1}. At room temperature the more intense component is on the high frequency side; at 260 K the doublet is nearly symmetrical; below this temperature the more intense component is on the low frequency side. For every 20 K reduction in temperature the frequency ν_1 of the broad band decreases by about 10 cm^{-1}, while the frequency of the interfering mode remains constant. These frequency shifts are a consequence of the strong anharmonic coupling between the $\nu_s(OH)$ and $\nu_\sigma(OH\ ...\ O)$ modes (10).

These absorption bands can be analysed in much the same way as before. We exploit the fact that τ_c and Δ_1 are known for the CD_3OH complex at each temperature, assume that τ_c is unchanged and Δ reduced by a factor of 1.345 (i.e. of $\omega_1(H)/\omega_1(D)$) for the CH_3OD complex, and try to find Δ_2 and V_{12} by least-squares curve fitting.

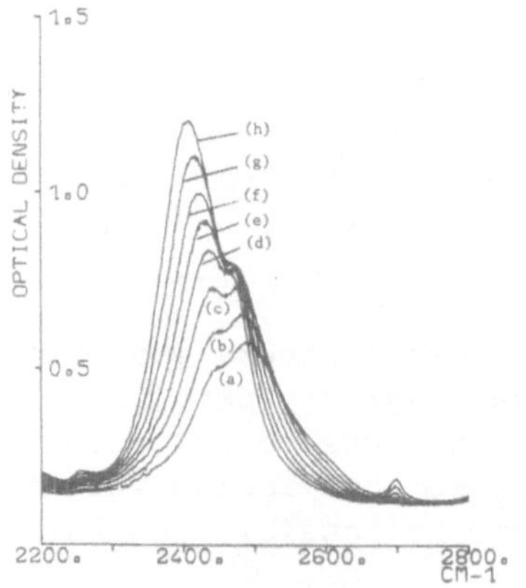

Fig. 6. Absorption spectrum of self-associated CH_3OD at 20 K intervals between 280 K and 160 K.

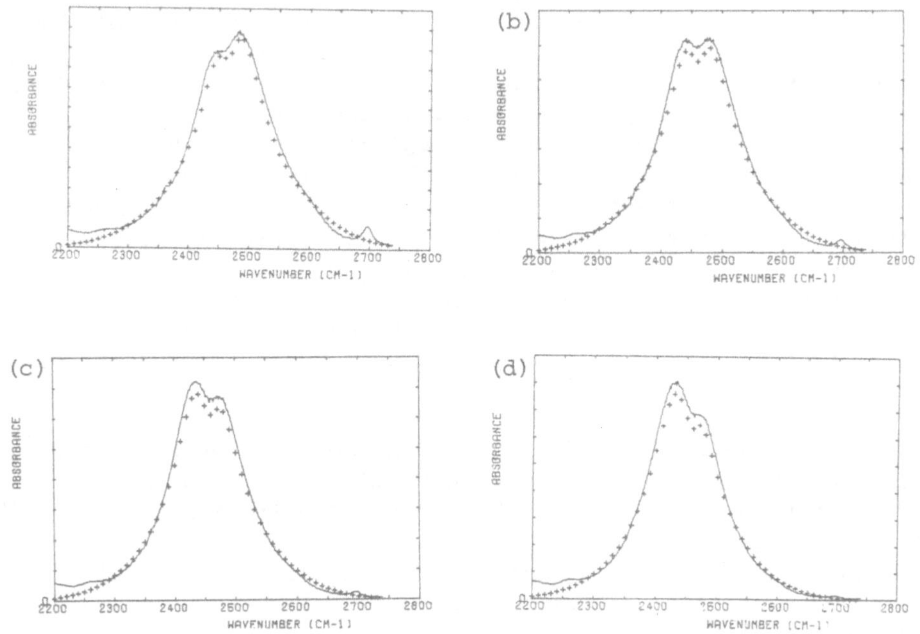

Fig. 7. Observed and fitted spectra of self-associated CH_3OD at (a) 280 K (b) 260 K (c) 240 K (d) 220 K.

Examples of the experimental and fitted spectra are shown in fig.7.

The results of the analysis are that $V_{12}/hc = (15.1 \pm 1.1)\ cm^{-1}$ at the 95% confidence level; there is no evidence of any systematic variation of V_{12} with temperature. This is in complete agreement with theory. However, the dependence of Δ_2 on $[\coth(\hbar\omega_3/2kT)]^{\frac{1}{2}}$ does not seem to be quite linear. If a linear relationship is nevertheless assumed, the correlation coefficient is found to be 0.96 and the intercept $(-27 \pm 13)\ cm^{-1}$. Since this estimate of the intercept does differ significantly from zero, we cannot claim to have shown that Δ_2 is proportional to the root-mean-square amplitude of the ν(XH ... Y) coordinate r_3, as required by equation (4.8). Nevertheless, the agreement between theory and experiment is fairly satisfactory. Error in the assumed value of ω_3 may be responsible for some of the discrepancies.

5 CONCLUSIONS

We conclude that the ν_s(XH) bandshapes of simple hydrogen-bonded species in solution in an inert solvent are largely determined by the phase relaxation of the ν_s(XH) mode, which takes place on a sub-picosecond timescale. The low-frequency hydrogen bond stretching mode can exchange energy rapidly with the solvent and is well modelled by a classical Brownian oscillator. The rapid dephasing of the ν_s(XH) mode is a consequence of its random frequency modulation by the stochastic ν_σ(XH ... Y) oscillator, to which it is strongly coupled. Analysis of the ν_s(XH) bandshape allows estimates of the ν_σ(XH ... Y) frequency and damping constant to be made. The observed effects on the spectra of deuteration, change of solvent, and change of temperature are all correctly predicted by the model. The model can also explain the spectral changes observed when the solvent freezes, and can account qualitatively for the large residual width of the ν_s(XH) band of a hydrogen-bonded species in dilute solid solution near the absolute zero of temperature.

The model can be extended very simply so as to describe the situation where the ν_s(XH) mode is in Fermi resonance with an infrared inactive overtone or combination. A criterion can be found for the existence of a narrow transmission window within the broad ν_s(XH) band. A good qualitative account is obtained of the evolution of the Evans hole as the frequency mismatch between the two interfering modes is varied. However, there seems to be some deviation from the expected law $\Delta_2 \propto [\coth(\hbar\omega_3/2kT)]^{\frac{1}{2}}$ governing the temperature dependence of the amplitude of modulation of the interfering infrared-inactive mode. It is not possible at this stage to say whether the discrepancy is due to experimental error, or whether the simple model hamiltonian of (4.1) is inadequate.

ACKNOWLEDGMENTS

I wish to thank Drs. R. Ackroyd, J. Bournay and J. Yarwood for their collaboration. I am indebted to Professor S. Bratos for making available unpublished calculations and for many helpful discussions.

REFERENCES

(1) M. Asselin and C. Sandorfy, Chem. Phys. Lett. 8, 601 (1971); C. Sandorfy, in The Hydrogen Bond : Recent Developments, P. Schuster, G. Zundel and C. Sandorfy, eds., North-Holland, Amsterdam (1976).
(2) M.F. Claydon and N. Sheppard, Chem. Commun. (1969) 1431.
(3) J.C. Evans, Spectrochim. Acta 16, 994 (1960).
(4) S. Bratos, J. Chem. Phys. 63, 3499 (1975).
(5) G.N. Robertson and J. Yarwood, Chem. Phys. 32, 267 (1978).

(6) R. Kubo, in Fluctuation, Relaxation and Resonance in Magnetic Systems, D. ter Haar, ed., Oliver and Boyd, Edinburgh (1961).
(7) J. Yarwood, R. Ackroyd and G.N. Robertson, Chem. Phys. 32, 283 (1978).
(8) J. Bournay and G.N. Robertson, Nature 275, 46 (1978).
(9) J. Bournay and G.N. Robertson, Chem. Phys. Lett. 60, 286 (1979).
(10) J. Bournay and G.N. Robertson, Molec. Phys. (in the press).

EXPERIMENTAL STUDY OF ROTATIONAL AND VIBRATIONAL RELAXATION IN LIQUIDS FROM INVESTIGATION OF THE IR AND RAMAN VIBRATIONAL PROFILES

J. Vincent-Geisse

Service de Spectroscopie Moléculaire en Milieu Condensé
Laboratoire de Recherches Physiques[x], Université P. et M. Curie, 4 place Jussieu, Tour 22, 75230 PARIS CEDEX 05

The purpose of this talk is to give a general survey of the experimental studies on vibrational and rotational relaxation of liquids carried out by Raman and infrared vibrational spectroscopy, in the absence of specific interactions.

We shall first briefly recall the theoretical bases and the assumptions that they imply, we then follow with the experimentally determined quantities and the various phenomena investigated. Their practical application will be shown on different molecular systems. We shall conclude with the present state of research and the problems still left to elucidate.

1 - THEORETICAL BASES

They are by now wellknown (1 to 4) and we shall chiefly point out the underlying hypotheses and their applicability. The experimental data are, on one hand, obtained by IR absorption and Raman scattering band profiles and their Fourier transforms, or from the bandwidths, the spectral moments and the relaxation times on the other. The basic formulae are :

$$\mathrm{FT}\ G_{IR} = \Phi_{IR} = \Phi_V\ \Phi_{1R} \qquad [1]$$

$$\mathrm{FT}\ I_{iso} = \Phi_V \qquad [2]$$

$$\mathrm{FT}\ I_{aniso} = \Phi_V\ \Phi_{2R} \qquad [3]$$

[x]Laboratoire Associé au C.N.R.S.

As they are widely used, we shall not define each quantity, but only make a few remarks. The correlation functions are single-particle functions. The rotational IR correlation function, Φ_{IR}, has a simple expression,

$$\Phi_{IR} = < \vec{u}(o)\ \vec{u}(t) > \qquad [4]$$

where $\vec{u}$ is the unit vector of the transition dipole moment. The Raman correlation function, on the contrary, is only simply expressed, and therefore simply interpreted, for certain molecular vibrations of some symmetry species. For example :

$$\Phi_{2R} = \frac{1}{2} < 3 \left[\vec{u}_z(o)\ .\ \vec{u}_z(t)\right]^2 - 1 > \qquad [5]$$

in the case of parallel vibrations of a linear molecule or symmetric rotor. As far as a low-symmetry molecule is concerned, Φ_{2R} is a function of different $< \vec{u}_i(o)\ .\ \vec{u}_j(t) >$, where i and j stand for x, y or z. Knowing Φ_{2R} for one vibration yields no practically utilizable information.

On the other hand, this separation of Φ into the product of rotational and vibrational functions with a same Φ_V for the IR, anisotropic and isotropic Raman scattering is only rigorous under precise conditions. For a dilute solution, the solvent must be inert (no collision induced processes). The coupling between vibration and rotation must be negligible. The case of pure liquids is more complex and will be taken up later on. The above formulae, however, are still valid when there are no cross correlations between molecules.

For the following, we shall assume, first, that these conditions are realized, since this is the generally adopted hypothesis, and we shall, later on, note a few cases where they do not appear to be verified.

As regards the practical applicability of the formulae, if we recall that

$$I_{iso} = I_{VV} - \frac{4}{3} I_{VH} \qquad [6]$$

$$I_{aniso} = I_{VH} \qquad [7]$$

I_{iso} can be determined only when I_{VH}/I_{VV} is $\neq 3/4$, that is, for practical purposes, only for totally symmetric vibrations. If ρ=0, the experiments are simpler (one single Raman measurement), but Φ_{2R} cannot be determined. For ρ between 0 and 3/4, two measurements are necessary, and under the best conditions, one may therefore obtain Φ_V, Φ_{1R}, Φ_{2R}.

This outline we are giving here concerns itself with the basic theories used to transform the experimental data. With respect to the interpretation of these data, we call upon very general theories, such as those by Gordon on the spectral moments, the theories on memory functions by Berne and Harp, and also on models. The literature on all this is very ample and its discussion would exceed the frame of our talk. We therefore assume that the theoretical concepts are known from other sources and will only give the corresponding references for particular cases.

2 - PHENOMENA UNDER INVESTIGATION. METHODS

We discuss investigations of vibrational and rotational relaxation. The vibrational relaxation is determined from the isotropic Raman spectrum, or, very simply, through the I_{VV} spectrum if $\rho = 0$. The studies on this subject are relatively recent, and we shall discuss them at the end and examine in the first place, the very numerous works on rotational.relaxation. We encounter there all intermediate cases between very slow modulation (nearly free rotation and profiles close to that of a gas) and fast modulation (rotational diffusion by very small angular steps and a lorentzian profile). I shall practically omit old studies, where for lack of any better hypothesis, the IR profile has been assumed to be completely rotational, that is, that the vibrational relaxation was negligible. In some cases, however, it has been shown that this assumption was acceptable and some of these results are still valid.

Which experimental techniques are utilized ? The more accurate is the study of both isotropic Raman scattering and infrared absorption, or of anisotropic and isotropic Raman scattering, or of the three together. This however, is not always possible since the vibration under study must be at the same time, sufficiently intense and both Raman and IR active (which excludes, for example, molecules with a centre of symmetry) and since Φ_V can be determined only for totally symmetric vibrations. Supposing one wishes to compare the rotations around the various axes of the molecule, which is an interesting question, this method will then only be applicable to one direction of the transition dipole moment; with regard to the others, we must resort to a different technique. I shall therefore briefly recall Rakov's method (5), which still remains very useful in some cases, provided one is conscious of its limitations and applies it correctly. To this effect, we start from relations [1] to [3]. If the various Φ are nearly exponential, a condition very often realized, these relations may be expressed by equivalent ones, involving bandwidths. For example :

$$(\Delta\nu_{1/2})_{IR} = (\Delta\nu_{1/2})_V + (\Delta\nu_{1/2})_{1R} \qquad [8]$$

with $\Delta\nu_{1/2} = 1/C\,\Pi\,\tau$ [9]

and $\Phi = \exp(-t/\tau)$ [10]

Rakov's method consists in assuming that the temperature dependence of $(\Delta\nu_{1/2})_{IR}$ obeys and Arrhenius law and that $(\Delta\nu_{1/2})_V$ is independent of temperature.

$$(\Delta\nu_{1/2})_{IR} = \delta + A\, e^{-U/kT} \quad [11]$$

A study of $(\Delta\nu_{1/2})_{IR}$ as a function of T then permits us to determine the preexponential factor A, the activation energy U and the vibrational width δ. A systematic application of this relation may prove disappointing. Actually, while the first assumption is practically always satisfied, the second ($(\Delta\nu_{1/2})_V$ constant) is generally not. It has been theoretically and experimentally shown, and this point will be taken up further on, that generally $(\Delta\nu_{1/2})_V$ slightly decreases if T increases. This being the case, one may show that [11] furnishes the higher limit of $(\Delta\nu_{1/2})_V$ for δ, when the lowest temperature is reached. If δ=0, quite a frequent case for a light molecule diluted in an inert solvent, one is then certain that $(\Delta\nu_{1/2})_V$ may be considered zero and the vibrational relaxation negligible, whichever the temperature. When δ is small ($<\frac{1}{10}(\Delta\nu_{1/2})_{IR}$ at ambiant temperature), the variation with T may be neglected to a first approximation and one obtains an acceptable $(\Delta\nu_{1/2})_{IR}$. Beyond, one must be very circonspect and only utilize the result (higher limit of $(\Delta\nu_{1/2})_V$) as one among other data. Finally, one must recall that this method cannot be applied when the profile appreciably differs from a lorentzian curve and approaches that of the free rotor. This case is realized if the rotational relaxation is preponderant; it is then possible to neglect the vibrational relaxation to a first approximation. To conclude, in spite of the above mentioned restrictions, Rakov's method is liable to be very helpful for the determination of rotational relaxation when the latter is relatively important.

We shall point out finally that, in the case of a perpendicular band of a symmetric top, Rakov's method separately applies to the diffusion constants $D_\perp$ and $D_{//}$ and not always to the bandwidth itself.

Thus recalling the theoretical and experimental techniques, we note that quite a large uncertainty remains with respect to the determination of the experimental data on which further interpretation will be based. Fortunately, these relaxation phenomena may be brought to light and measured by means of very numerous techniques, each offering a definite range of applicability, limitations and uncertainties. Eventually, it is the comparison of the results obtained by different techniques which will enable to assure us of the accuracy of the assumptions we have advanced and of the conclusions we have drawn.

3 - ROTATIONAL RELAXATION

Let us now give a survey of the different molecules which have been investigated within the preceding frame. The number of publications to this date is very large and we shall not attempt to mention them all. Particularly we shall omit all works in which the vibrational relaxation was ignored or assumed incorrectly to be negligible. These studies were interesting at the time of their publication, when they offered a first step towards investigation of these phenomena and provided approximate results which directly showed the extremely wide range of phenomena met with, and, in some cases, the existence of a nearly free rotation in the liquid phase. Among them, we shall only retain those which presently give still utilizable results. The molecules will be classified according to growing complexity. This order roughly corresponds to a more and more hindered rotation and to more and more diversified purposes.

3.1 - Diatomic Molecules

This is the simplest case since there is only one vibrational band, which moreover is generally not perturbed by any hot or adjacent bands. This is also one of the first cases investigated. A distinction must be made at the start between heteronuclear molecules, to which one may apply both IR and Raman techniques, and homonuclear molecules, the vibrations of which are only Raman active.

Among heteronuclear molecules, HCl and its analogues have been the most frequently investigated. The existence of rotational wings in solution has been pointed out as early as 1958, and been correctly interpreted as a persistence of rotation in the liquid phase. As concerns bibliography of the early works, refer to (7). A manifest confirmation of the possibility of significant rotation for hydrogen halides in the liquid phase is obtained by observation of the fine structure lines of HF, DF and HCl in SF_6 (6) (fig.1). After the publication of the above mentioned theoretical studies (1 to 4), very accurate measurements have been carried out, by Raman scattering as well as by infrared absorption (7,8,9), of the fundamentals and the first harmonics. These studies have yielded unexpected results : the HCl spectrum in CCl_4, for example, presents a very intense central peak for the $0\to1$ band, an intensity which decreases for the $0\to2$ band and completely disappears for the $0\to3$ band (fig.2). Neither a consideration of the relatively weak vibrational relaxation, nor of the rotation-vibration interaction (10) can explain this phenomenon. Recalling that the intensity variations for these harmonics are very different for gas and solution, it has been suggested that the induced dipole moment which had not been taken into account by any one of these theories, plays an important part. This result is to be compared to that of fig.1, which shows, besides the rotation lines similar to that of the gas, an intense central

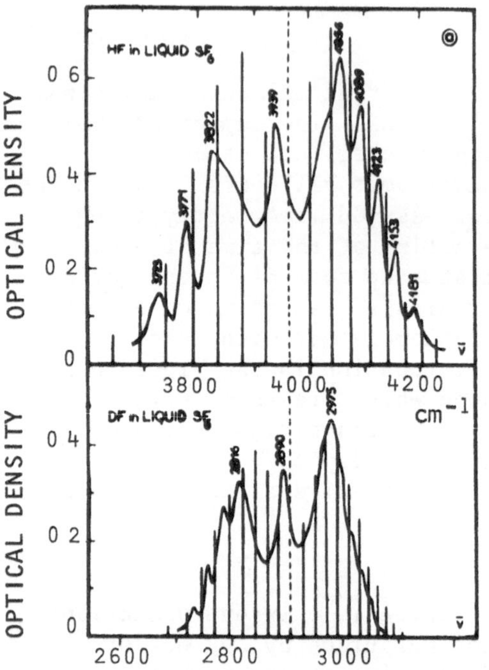

Fig.1 Infrared spectra of HF and DF in liquid SF_6. Ref. 6

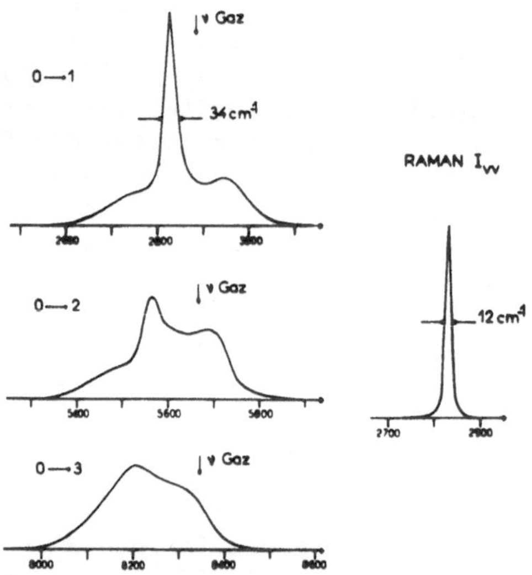

Fig.2 Infrared and I_{VV} Raman spectra of HCl in CCl_4. Ref. 11

peak, forbidden by the selection rules in the free molecule. Such a peak may be observed for gaseous HCl compressed by inert gases, while it does not appear for some other molecules, as we shall see further along.

It must be pointed out that a more detailed interpretation, for the fundamental band of HCl in CCl_4 (12) has been proposed on the basis of a quantum theory. For the harmonics, a similar experimental study had been made later. It remains to be verified whether such a theory may be extended to this case.

Furthermore, for HCl, the center of mass is not a center of action for forces, and a translation-rotation coupling may occur. It seems that such an effect might explain (13) some anomalies in the Raman spectrum (9).

The Raman scattering of pure liquid HCl has also been investigated (14,15). The results are not in agreement with N.M.R. data, and one concludes that there exists a coupling between rotational and vibrational effects due to hydrogen bonding in the liquid.

The case of HCl is therefore exceptional in many respects and further developments of the theory seem necessary for a complete interpretation of the numerous experimental data.

Carbon monoxide, pure or in solution, has been the matter of several studies by Raman scattering or IR absorption. Most of them, however, have been carried out at low temperature for the pure liquid (16,17) as well as for solutions in liquefied gases (18), and these are not the best conditions to bring out rotational effects. However, a study has been performed on CO diluted in CCl_4, C_2Cl_4 and SO_2 (19). The profiles of the transitions $0 \rightarrow 1$ and $0 \rightarrow 2$ are nearly similar in the IR, and the Raman I_{VV} spectrum presents a very weak vibrational width; the results are easily interpreted in terms of the above mentioned theories and allow us to compute the mean free rotation angle, (near 50°), the spectral moments and the mean squared torque.

We shall omit the IR studies on induced bands of homonuclear molecules as they are not within the scope of this lecture. The most interesting molecule is hydrogen, which presents rotational fine structure for all phases. H_2 and D_2 in SF_6 have been investigated by Raman scattering (20). The observed spectrum is identical to that of the gas but simply with broadened lines. Hydrogen illustrates the extreme case of a nearly free rotation in the liquid phase. The Raman spectra of O_2 and N_2 have also been observed, either pure at low temperature (21) or diluted in liquid SF_6 (22). In the latter case, the anisotropic spectrum essentially reproduces the envelope of the spectrum in the gas phase.

Chlorine has also been studied by Raman scattering at various temperatures (23,24). The effect of the chlorine isotopes, which give two distinct lines in the isotropic spectrum, must be taken into account. The results obtained for the correlation time τ_{2R} are in satisfactory agreement with NMR data.

Other diatomic molecules have been investigated in liquid cryosystems. A comprehensive bibliography on the subject will be found in reference (25).

3.2 - Linear Molecules

Here again, a distinction must be made between asymmetric molecules which can be accurately studied by both Raman scattering and IR absorption, and symmetric molecules, in which one given vibration is either Raman or IR active.

The most precise and accurate study to this date deals with N_2O. The first evaluations of correlation functions were achieved for hexane solutions of this molecule, using the ν_1, ν_2 and ν_3 bands in the IR (26). As early at this time, one had demonstrated the characteristic behavior of correlation functions and determined the connection between the functions Φ_{1R} for a parallel and a perpendicular band through a relation which has been proved later rigorously. In a second stage, the parallel bands ν_1 and ν_3, in various solvents and at variable temperature have been investigated by both Raman and IR (27). Hot bands were substracted. At ambiant temperature, the vibrational relaxation is practically negligible in SF_6 and remains small in CCl_4 (fig.3). The $\Phi_R(t)$ curves, traced on a semi-log scale show a characteristic aspect which is found in practically all cases of hindered rotation. After a curved section at short times, more important for SF_6 than for CCl_4, one finds a linear section corresponding to a complete loss of the original memory on account of the numerous intervening collisions. An important point, which we may attempt to clear up in this particular case, is the identity of Φ_{1R} (and Φ_{2R}) for ν_1 and ν_3. With respect to Raman scattering and infrared in SF_6 solutions, the rotational widths are very large, the eventual differences being of the order of the experimental uncertainty. For CCl_4 solutions in the IR, a systematic and constant difference of 1.5 cm^{-1} appears between these widths. This difference, which is larger than the experimental uncertainty, becomes relatively important at low temperatures and shows the limit of the applied theories. The correlation functions obtained, Φ_{1R} and Φ_{2R}, are compared to those derived from various models. Gordon's M and J models (27) are not applicable. Fixman Rider's friction model fits none the better. One of Fauquembergue's models (M or J model with an anisotropic reorientation of the angular momentum) furnishes a simultaneous fitting of Φ_{1R} and Φ_{2R} in CCl_4, but is far less satisfactory in SF_6. Finally, another Fixman Rider's model, intermediate between Gordon's M and J models, permits us to reproduce with fair precision either the Φ_{1R} or the

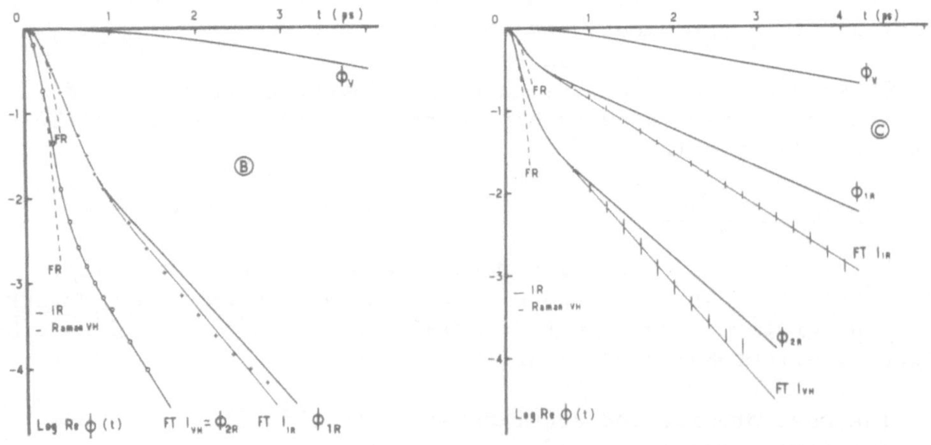

B in SF_6 C in CCl_4

Fig.3 Correlation function for ν_3 band of N_2O at 23°C Ref.27a

Φ_{2R} curve, but never both (28). This last point shows the necessity of testing a given model on several experimental results and the more strictly so as the more adjustable parameters are included in the model.

Furthermore, the Raman spectra of pure liquid N_2O were measured along the coexistence curve between the critical and the triple point (29). Vibrational relaxation appears to be important, particularly at low temperatures. The rotational correlation function Φ_{2R}, which is found to be identical for ν_1 and ν_3, is interpreted by the microscopic version of the J model according to Chandler. Agreement is good near the critical point. The mean-squared torque acting on the N_2O molecule was estimated with the help of the memory formalism and compared with that computed from an intermolecular potential.

OCS is another molecule which has been well investigated. A first paper has dealt with ν_1 in isopentane and dichloromethane for the Raman and for the IR (30). For isopentane, a very inert solvent like all hydrocarbons, the study of the widths as function of temperature (fig.4) shows that the vibrational width is negligible at ambiant temperature and, to the contrary, preponderant at low temperature. The band contour is quite ordinary; the observed profiles are lorentzian and the Φ_{1R} are linear on a semi-logarithmic scale and pass close the origin, with a curved section at very short times. These are very favourable conditions for testing Rakov's method and comparing his results with those derived from a rigorous separation technique. The residual width δ obtained is in effect the common limiting value to which all the three curves in fig.4 tend. A study of ν_1 and ν_3 bands in hydrocarbons is now in progress (31) and will complement the preceding results.

In the case of symmetric molecules, as complete a study as this is no longer possible, but one may still reach interesting conclusions by examining, on one hand, the IR spectra, and on the other, the Raman spectra, but for different vibrations, and by using Rakov's method in the IR with all the above mentioned precautions.

As regards CO_2, the ν_1 band for the pure liquid has been investigated both by Rayleigh and Raman scattering in the pressure range of 100 to 3000 atmosphere (32). The vibrational relaxation is negligible and the anisotropic spectrum directly provides Φ_{2R}. This function is independent of pressure and it follows Gordon's J model with good approximation. The comparison with the Rayleigh spectrum is interesting since the latter is, in contrast, pressure dependent and we come to the conclusion that the collective aspects of molecular dynamics are becoming important at high pressures. This band has also been studied in another laboratory (33) and a correction for collision induced scattering has been applied. In this case, the effect is weak and here again Gordon's J model describes well the experimental results. Application of Rakov's method in the IR

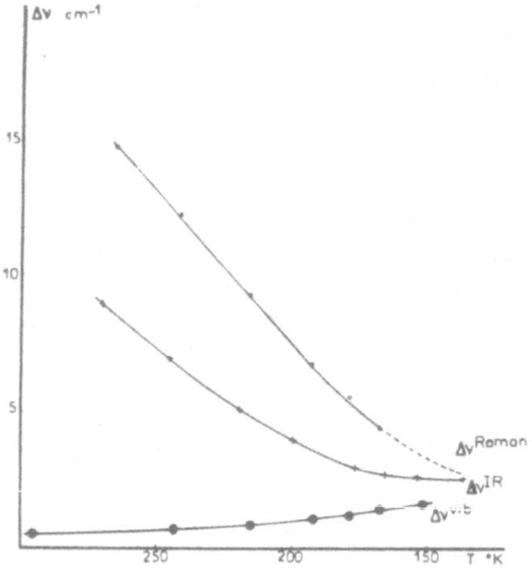

Fig.4 Bandwidths for ν_1 of OCS in isopentane • Ra_{iso} o Ra_{aniso} + IR Ref.30

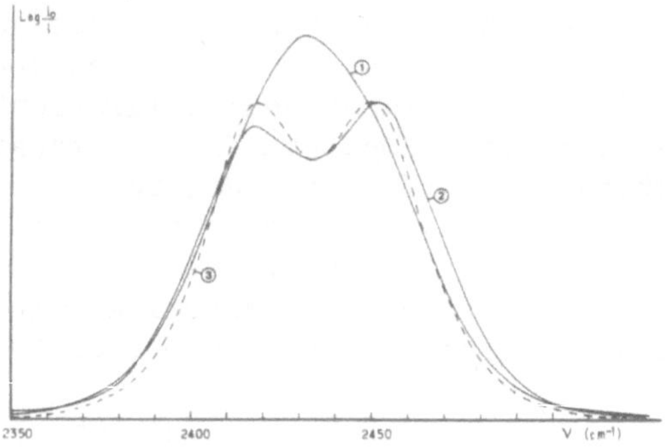

Fig.5 Profile of the ν_3 band of C_2D_2 in SF_6 1) exp.curve at T=5°C 2) exp.curve at T=40°C 3) J model Ref.35

study of ν_3 of CO_2 in hydrocarbons shows that the vibrational width is negligible at ambiant temperature; the same is true for ν_1 in Raman (31). Consequently, we can interprete the profiles as purely rotational. The ν_3 band of CO_2 in SF_6 departs largely from a lorentzian shape, showing a rotation by large angles.

The same may be noted for C_2H_2 and C_2D_2 in inert solvents. The IR study of ν_3 for these molecules in SF_6 (34) shows a very rapid variation of the profile with temperature. It is nearly lorentzian at low temperature, subsequently develops a plateau and then separates into P and R branches at +40°C (fig.5). It must be pointed out that this spectrum is very different from that of HCl and never shows a central peak with shoulders. The case of these molecules in SF_6 solution is therefore particularly informative with respect to the testing of rotational models since all intermediate cases between relatively free rotation (mean free rotational angle of 70° in SF_6 at 40°C) and rotational diffusion are encountered. The agreement is really not very good, whichever the model, but Gordon's J model gives a general account of the spectral shape at all temperatures.

3.3 - Spherical Rotors

The most interesting molecule is methane and its deuterated analogue. It is also one of the first molecules where the persistence of a significant rotation in the liquid phase has been observed, in 1952 for pure liquid, and in 1953 for solutions. Other strictly qualitative studies followed (for a bibliography concerning these older works, see, for example (38)). The development of theories of the liquid state gave a fresh impetus to these experimental studies. A study performed in the IR on CH_4 and CD_4 as pure liquids or diluted in liquid rare gases (36), has been interpreted by McClung (37) by means of Gordon's M and J models extended to spherical rotors. CH_4 has recently been studied in solution in SF_6 (38). The profile varies strongly with temperature and, at 40°C, one observes a structure caused by the intervention of the nuclear spin in the population distribution of the rotational levels (fig.6). In this case, the usual rotational models dealing with classical mechanics are not sufficient and one must resort to Eagles and McClung's semi-classical models established to account for the CH_4 spectra in the compressed gas phase. The M and J models describe fairly well the obtained profiles and a free rotation angle which attains 165° at 40°C is derived. On the other hand, the spectral moments were measured, unfortunately with some uncertainty due to the presence of a very weak combination band located on the wing of the P branch at 200 cm^{-1} off the central peak. The mean-squared torque is deduced and, within the limits of the experimental error, is found to be proportional to the density of the solvent (fig.7). This last result shows that the behavior of CH_4 in liquid SF_6 is the same as in an inert compressed gas. In fact, the spectra are actually very similar.

For CH_4 in cryosystems, see also (25). One also finds all the experimental results about CH_4 and their given theoretical interpretations in a review article by Steele (39)

Other studies have been carried out on SF_6 and CF_4 by Raman scattering (40,41). Results obtained for several bands of different symmetry species have been interpreted approximately by Gordon -

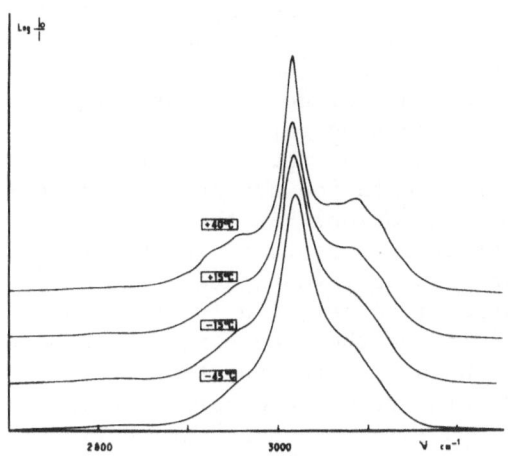

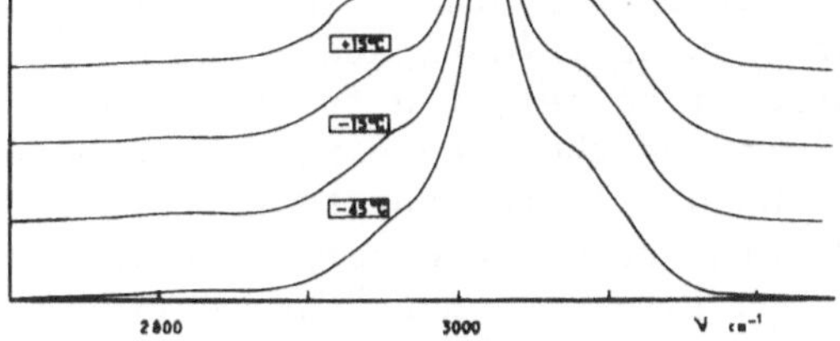

Fig.6 Profile of the ν_3 band of CH_4 in SF_6 at various temperature. Ref.38

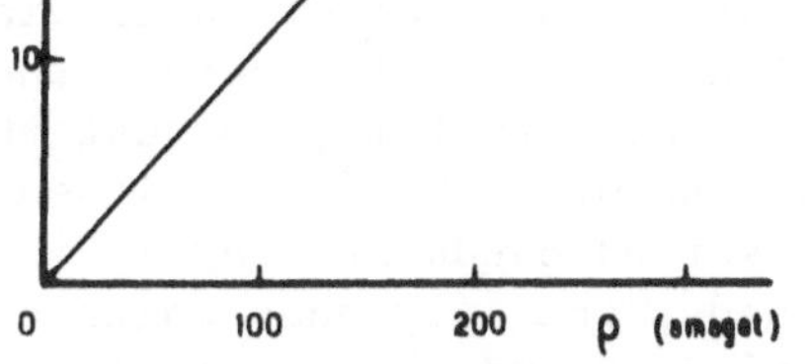

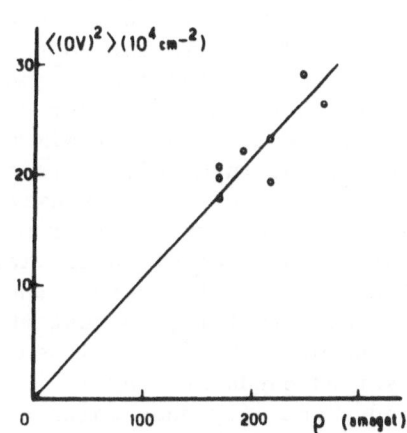

Fig.7 ν_3 band of CH_4 in SF_6. Mean squared torque versus density. Ref.38

McClung's J model, taking into account the Coriolis coupling for modes of species F. The small discrepancies observed can be explained by the existence of vibrational relaxation which is probably not negligible at low temperature and which cannot be rigorously measured for non-totally symmetric modes. One may also mention a study on tetramethylsilane (42) : this molecule has 19 vibrational modes, 14 of which are Raman active; the selection of bands utilizable for purposes of investigation leads to 5, two of which finally provide coherently interpretable results in agreement with N.M.R. data and with the J model.

From all these studies on spherical rotors one can derive a few general conclusions in contrast to what has been noted for linear molecules, where each one presented a particular case. The rotation in this case is always important even for pure liquids and heavy molecules. This obviously, is to be attributed to their globular shapes. Besides, at temperatures below the liquid phase, these molecules present a reorientational plastic phase. The observed profiles are in acceptable agreement with those obtained by means of Gordon-McClung's J model. Finally, the vibrational relaxation is generally weak or negligible, if one excludes very intense IR bands for pure liquids, where transition dipole-transition dipole coupling determines some particular phenomena which will be examined further on.

3.4 - Symmetric Rotors

From the point of view of rotational motion, the symmetric rotors are relatively simple to handle and the studies on the subject are extremely numerous. A distinction must be made at the start between parallel bands (transition dipole moment parallel to the axis of the rotor) and perpendicular degenerate bands. As concerns the formers, the simple theories applied (1 to 4) are still valid and both IR and Raman studies, similar to those described for linear molecules, are still possible and lend themselves to fruitful comparisons with the results derived from other techniques, such as far-IR, microwaves, Rayleigh scattering, or N.M.R.. Unfortunately, the information thereby obtained only bears on the reorientation around axes perpendicular to the molecular axis (tumbling motion). As these molecules generally possess a larger number of atoms than those in the preceding systems, several of their vibrations can be investigated, and one may attempt to verify whether the rotational correlation functions are the same, that is, to test the validity of the theoretical hypotheses. On the other hand, the measurements become less accurate since the observed bands are rarely isolated. With respect to perpendicular bands, the only ones which can give us information on the reorientation around the molecular axis (spinning motion) we encounter problems which are still more arduous. The vibrational profile is not directly measurable; one must work either in the IR or in the

Raman, eventually using Rakov's method and, for interpretational purposes, take the Coriolis coupling into account. Comparison with N.M.R. results are still possible.

To the difficulties inherent in the interpretation of the results, others, purely experimental, are added. For solutions, the solvent must be suitable for both the Raman and the IR and, as far as possible, be transparent in a wide range covering several bands of the solute. If one deals with weak Raman bands, the investigation becomes difficult and, for the isotropic Raman, the distortions due to the apparatus function may be important. In contrast, the base line may be known with certitude (pure solvent). For the pure liquid, besides the additional assumptions to be met by the theories, which have been previously recalled, the experimental errors get more important. The base line, in particular, is ill-defined for wide bands since it is impossible to go far into the wings without encountering other bands. Under such conditions, computing the correlation function, and even more so, the spectral moments, may become illusive. The only data accessible without too much error is then the bandwidth from which the correlation time τ may be derived by means of formula [9], and its inverse τ^{-1}, which is a diffusion coefficient, assuming a lorentzian profile. It is also possible to obtain a direct comparison of the observed profile with that derived from a model. Finally, still for pure liquids, the relevant quantity for very intense bands in the infrared is no longer the absorption coefficient but the imaginary part of the local susceptibility; so the refractive index must also be known. With respect to all these considerations concerning the precision of the measurements, base line interference and corrections applied to correlation functions and moments see ref. (19, 43 to 46).

To conclude, the difficulties in this type of investigation are very important and this accounts for the large spread, or even the contradictions of the results. We finally shall show that, prior to any interpretation, it is absolutely necessary to take many experimental results into account. Even barring the experimental uncertainty, and we have noted how important it is, the interpretation of isolated data may lead to erroneous conclusions. As an example, let us only quote the application of Gordon's J model, extended by McClung to symmetric rotors. This model utilizes one single parameter τ_J, the mean time between two collisions, and one collision interrupts simultaneously the rotation around each axis. Then it has been shown that, for $\tau_J \to 0$, the J model tends towards Debye's diffusion model. That is to say, the J model may represent any reorientational situation between free rotation and true rotational diffusion. This explains its success. In the case of a symmetric rotor, any individual band may generally be interpreted by means of the J model, but the only true test will be that which ascertains whether the model is applicable to different types of vibrations with the same τ_J (a discussion on these models will be found in ref. (47) for example).

Let us recall here some useful formulae for the rotational diffusion limit (3,47). With the notations $D_{xx} = D_{yy} = D_\perp$ and $D_{zz} = D_{//}$ we have (3) :

$$\Phi_{1R} = \exp(-2D_\perp t) \qquad [12]$$

$$\Phi_{2R} = \exp(-6D_\perp t) \qquad [13]$$

for parallel bands and

$$\Phi_{1R} = \exp\left[-(D_\perp + D_{//})t\right] \qquad [14]$$

$$\Phi_{2R} = x\exp\left[-(5D_\perp + D_{//})t\right] + (1-x)\exp\left[-(2D_\perp + 4D_{//})t\right] \qquad [15]$$

for perpendicular bands,
where x ($0 \leqslant x \leqslant 1$) depends on the relative values of the elements $\alpha_{2,\pm 1}$ and $\alpha_{2,\pm 2}$ of the polarizability tensor associated to the transition, in spherical coordinates.

We are in this limiting case if $D_{ii} << (kT/I_i)^{1/2}$. There is no a priori relation between $D_\perp$ and $D_{//}$.

In the diffusion limit of the Gordon McClung's J model (τ_J going to zero) we have a relation between $D_{//}$ and $D_\perp$ (47).

$$D_{ii} = \frac{kT}{I_i}\tau_J \qquad [16]$$

which gives $$\frac{D_{//}}{D_\perp} = \frac{I_x}{I_z} \qquad [17]$$

In order to have the J model verified at all temperatures the ratio $D_{//}/D_\perp$ must be temperature independent; in other terms $D_{//}$ and $D_\perp$ must have an equal activation energy. For an isotropic rotation we should have $D_{//} = D_\perp$. The J model corresponds, on the contrary, to an anisotropic D tensor.

Griffiths has given an extensive survey of the studies on molecular relaxation for symmetric rotors in the liquid phase (48), recalling the theoretical bases of all the spectroscopic techniques and reviewing the various molecules investigated.

With respect to rotational motion, the most interesting molecules are the compounds CH_3X or CD_3X, where X stands for D or H,F,Cl, Br,I or CN. These compounds have been extensively studied. Firstly, CH_3D and CD_3H are the simplest since vibrational relaxation is negligible while rotation is important. These molecules have been investigated in the IR, in solution in rare gases, nitrogen and oxygen, for the whole range of the liquid phase (49). The J model gives a good account for the profiles observed for a parallel band and for two perpendicular bands at low densities but fits far less well at high densities near the triple point. The authors extended the application of the memory function, which, to first order, provides the J model, to second order, which allows them to calculate the components of the mean-squared torque. They are the same for the J model. Figure 8 shows the anisotropy of the torques, calculated in this way, at high densities.

Like preceding molecules the methyl halides have six vibrational modes. Of these, 3 are of symmetry species A1 (parallel bands) and 3 of symmetry species E (perpendicular bands). All are IR and Raman active. The relative simplicity of their spectra, together with a large difference between their moments of inertia, make these molecules very good candidates for an extensive study. Very early, their spectra were examined in solution and it was noticed that the parallel bands were narrow and the widths of the perpendicular bands varied very much. This width was often very large and the contour generally followed the band envelope of the vapour phase. From this, it was concluded that the tumbling motion approached true rotational diffusion and that the spinning motion was nearly free. On the bases of sound theories, these studies were taken up again. It was the methyl iodide CH_3I on which the first determination of Φ_V and Φ_{2R} was performed (50). Subsequently numerous studies were done, essentially on the parallel bands of methyl iodide (51 to 57),(see (39) and (48) for a more complete bibliography, which also deals with the various techniques). Unfortunately, no overall conclusion can be drawn from these measurements. The only band which has been really studied is the ν_3 made for the pure liquid. On may consider that the correlation time τ_{2R} and the diffusion coefficient $D_\perp$ are well known, $D_{//}$ being determined by N.M.R. techniques. With respect to the determination

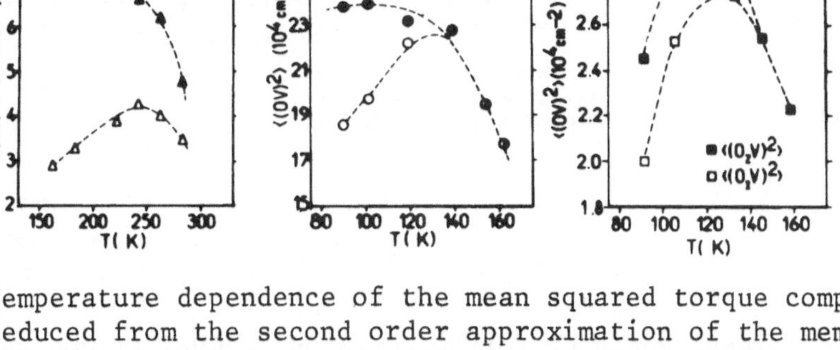

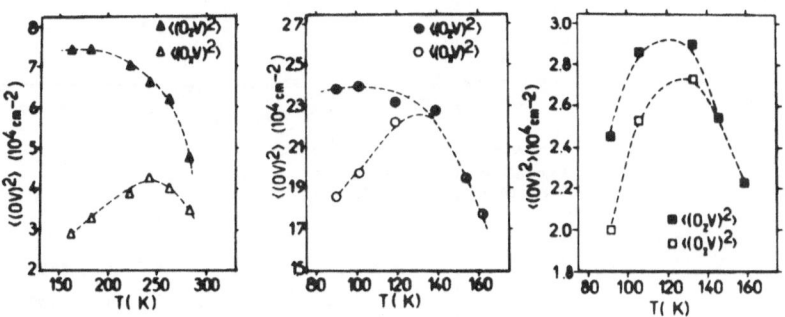

Fig.8 Temperature dependence of the mean squared torque components, deduced from the second order approximation of the memory function for CD_3H dissolved in liquid Xenon (left), argon (middle), oxygen (right). Ref.49

of $D_{//}$, it should be noted that the diffusion coefficient is defined as the inverse of the correlation time. This definition agrees with the theory only in case of Debye diffusion, but is very different for nearly free rotation. Similarly, calculating an activation energy in the last case is open to question. The most interesting work seems to be that of Jonas et al. (56) who studied the ν_3 band of the pure liquid at various pressures and temperatures. In the usual temperature dependence studies one deals with two simultaneously variable parameters, density and temperature. Measurements at constant density permit a discussion of the effects separately due to each one of the two parameters (see fig.9). This rapid survey of the literature on CH_3I shows that a great deal remains to be done. Concerning the other methyl halides, no study of importance has been performed.

An analogous conclusion can be drawn from an examination of the studies on acetonitrile, in spite of the very large number of publications relative to this molecule (51, 57 to 61) (see (48) for a more complete bibliography). This molecule possesses 4 vibrational modes of species A_1 and 4 modes of species E. The perpendicular bands seem to be specially interesting, the Coriolis coupling constant varying between +0.9 and -0.4 and passing through zero for ν_5. An examination of the obtained data on the parallel bands show extremely dispersed results. This can be explained, in part, by the overlapping of several bands, the presence of hot bands, and also by the uncertainty of the base line for the pure liquid. In spite of these uncertainties, it seems that these bands show anomalies and that the rotational widths are not found equal for the four bands, even in solutions (60). This questions the applied theories in the case of acetonitrile. Confirmation of this fact would be desirable. The study of the perpendicular bands is much less frequently attempted. It is difficult to apprize the importance of the vibrational relaxation. Moreover, the ν_5 band seems to present, as does the ν_1 band, an anomaly in the width which is presently not explained (both are CH bands) (61). In all cases, these profiles are far removed from a Lorentzian shape.

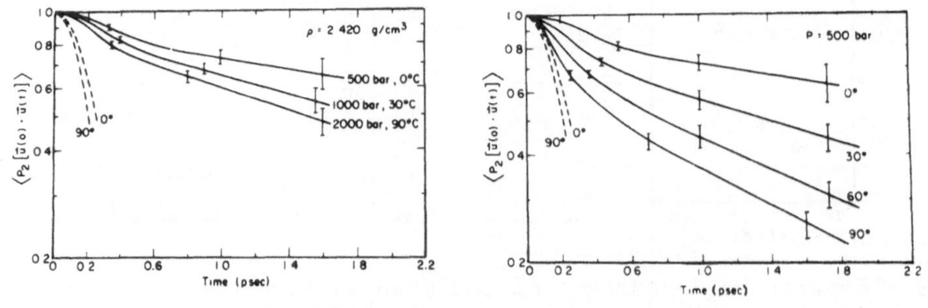

Fig.9 Rotational Raman CF for the ν_3 band of CH_3I at constant density (left), at constant pressure (right). The dashed lines are the free rotor CF. Ref.56

Another molecule which is relatively simple and has been extensively studied is chloroform and its analogues (62 to 70). The ν_{CH} band in this compound is well isolated and has, naturally, been the subject of numerous studies. Unfortunately, this band presents important anomalies in the IR (62). These anomalies may possibly be explained by the fact that its intensity is near zero in the vapour phase, (the first harmonic is much more intense than the fundamental) and much larger and very variable in solution. These anomalies manifest themselves by a very marked assymmetry of the band in certain solvents and by a variation of the profile and the width, which cannot be explained by the usual theories, upon examination of the successive harmonics. The other bands do not show such anomalies, but their study is more difficult. Present studies concern themselves mainly with Raman spectra of $CHCl_3$ and $CDCl_3$, which permit an easier interpretation and whose results can be compared with those of N.M.R.. A study of fluoroform has also been carried out (70). In these compounds, vibrational relaxation becomes important and the rotational relaxation is close to diffusion by small angular steps. The majority of the studies deal with the pure liquid, some at variable pressures,(63,67,69) other in solution (68). One observes a rotational second moment, in studies at constant temperature and variable density, which greatly exceeds its theoretical value and decreases with increasing density (69). This result contradicts the theoretical predictions of Gordon and suggests, in the opinion of the authors, the existence of collision-induced scattering. An analogous process could explain the anomalies in the IR spectrum (62).

Another symmetric top molecule which has been studied in Raman and in the IR is allene C_3H_4 (71,72). The ratio of the moments of inertia in this molecule is approximately the same as that for acetonitrile, but the spectra are very different. The widths of the perpendicular bands are much narrower for allene and have a profile not far from a lorentzian one. That suggests that the spinning motion is much less free than that of the CH_3 group. A comparison study involving a parallel and a perpendicular band permits us, in each case, to calculate the diffusion coefficients $D_\perp$ and $D_{//}$. The rotation is anisotropic and does not correspond to the predictions of the J model.

The ammonia molecule, which is one of the simplest and the most interesting with respect to rotation, has curiously hardly been studied. This probably can be explained by the ease of hydrogen bond formation. A study in inert solvents might turn out interesting. In contrast, the PH_3 molecule has been studied in the Raman (73). There are anomalies : the peak position of the polarized and depolarized components do not coincide and the profile of the depolarized component is very asymmetric. We shall briefly mention some other works by Raman scattering : - a study of liquid trimethylamine $(CH_3)_3N$ at variable temperature and pressure (74). The hypothesis of isotropic rotation, which is suggested by N.M.R. measurements is compatible with the observed profiles, - a study of liquid ClO_3F (75) which shows that the J diffusion limit of the extended rotational diffusion model is a good approximation, - two papers on cyclohexane (76,77), the conclusions of which are somewhat different among themselves, - a study on NF_3 (78), to which we shall return later on, - a study on t-butyl chloride and t-butyl bromide (79).

To conclude this review of symmetric rotors, we consider Benzene which has a different shape than the preceding molecules and which is one of the most extensively studied (see ref. (39) and (48) for a more complete bibliography which also deals with various techniques). Because of its centre of symmetry no band is simultaneously IR and Raman active. Vibrational relaxation is certainly not negligible for such a heavy molecule. The only precise method to study the rotation is therefore a Raman investigation in comparison with the N.M.R.. That is effectively what has been done (80 to 83). The obtained results show without ambiguity very easy rotation around the C_6 axis. One has to be very careful of the more detailed conclusions that have been drawn from these studies and which call upon hypotheses and models whose validity is far from being proved.

In conclusion, the results obtained for symmetric rotors are quite disappointing, except those which deal with the molecules CH_3D and CD_3H whose spectral profiles can be interpreted by pure rotational relaxation. For the others, the correlation functions are often inaccessible and we must content ourselves with data on bandwidths and correlation times. But even the definition of the last poses a problem, since the correlation function cannot be calculated and all other definitions varying from author to author, call on hypotheses which are often not proven. There is no dearth of data but many others would be necessary and a large effort to combine them remains to be done.

3.5 - Asymmetric Rotors

There are relatively few studies on these molecules owing to the difficulties met in their interpretation. The simplest are the triatomic molecules such as H_2O, D_2O, SO_2. They present three modes of vibration, ν_1 and ν_2 of A_1 species, ν_3 of B_1 species. In the infrared, water in an inert solvent such as CCl_4, shows extremely broad bands and the ν_1 and ν_3 bands, very near one another, widely overlap, thus rendering a precise determination of the correlation functions (84) very difficult; one may however discern without ambiguity a nearly free rotation. On the contrary, a mixture of H_2O with D_2O permits to observe the broad and disymmetric Raman and IR profiles of the HOD molecule and to calculate their correlation functions (85), nearly identical in both techniques, but, in this case, it is no longer the rotation alone that is observed.

The SO_2 molecule has been the matter of a quite elaborate study (86). An IR study of ν_1 and ν_3 in solution in CCl_4 by Rakov's method gives a residual width zero (86). This has been confirmed by a study of ν_1 in Raman scattering which yielded a nearly constant vibrational width of 0.5 cm^{-1} (87). The very wide profile obtained for ν_1 can therefore be interpreted in purely rotational terms. The obtained correlation functions are given in fig.10; for hexane, it is very near that of the free rotor. The interpretation of the results may be carried further on, by considering SO_2 as a symmetric rotor to a first approximation. The rotation around the axis of smallest moment of inertia is very important, the free rotation angle reaching 50° in hexane. It must be pointed out that this rotation around the axis of smallest moment of inertia involves that of the dipole moment which thus appears to play no part in these rotational motions in an inert solvent.

Other molecules have led to partial studies but the encountered difficulties greatly increase and the interpretation get very arduous. These studies have been performed either on the comparison between Raman and Rayleigh scattering, or the comparison between Raman and Infrared. In the first group, we may mention a study on CH_2Br_2 (88) and another one on 1,2,5 thiadiazole (89). In the case of CH_2Br_2 the Raman and Rayleigh spectra were studied in solution

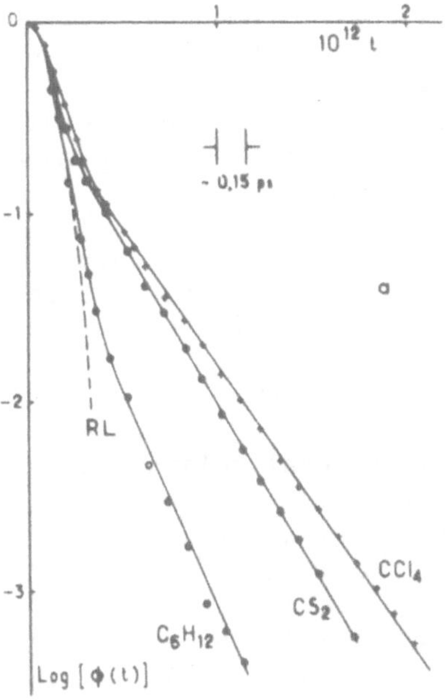

Fig.10 Rotational infrared CF of the ν_1 band of SO_2 in various solvents. Ref.86b

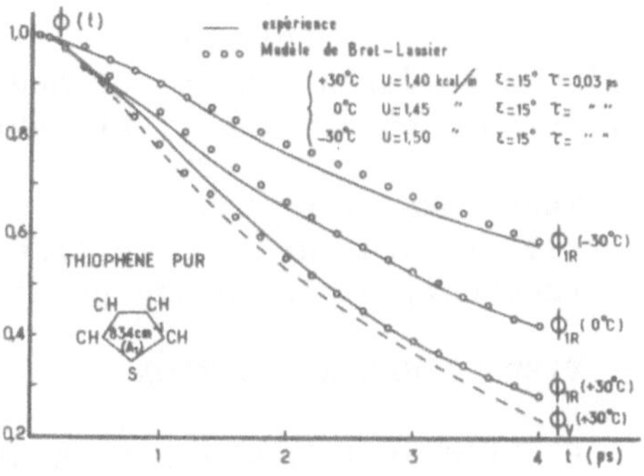

Fig.11 CF for a A_1 band of Thiophene. Comparison with the Brot-Lassier's model. U is the activation energy for reorientation, 2ξ the angle between two neighbouring potential wells and τ the time between two soft collisions in a potential well. Ref.92

in terms of concentration and temperature. The rotational correlation functions obtained for 4 Raman bands of totally symmetric species and for Rayleigh scattering are not identical and an interpretation of the differences is given. In particular, it is shown that the vibrational correlation times appearing in Φ_{iso} and Φ_{aniso} are not equal for a vibrational mode associated to the motion of a proton, this motion, of large amplitude, provoking an interaction with the rotation. This interpretation might perhaps account for the above mentioned anomalies observed in the case of several CH bands in symmetric rotors. Similar conclusions appear in the study on 1,2,5 thiadiazole. An analogous study has been carried out on 1-4 liquid dioxane (90). The pair orientational correlation was found to be negligible and the experimentally determined reorientation times compared with those predicted by hydrodynamic models.

As regards parallel Raman and IR studies, a word has dealt on furan and thiophene (91). The width of several infrared bands of different symmetry species have been investigated as a function of temperature for the liquid and plastic phases. Totally polarized A_1 bands have been studies in Raman and infrared and Φ_V and Φ_{IR} are derived. The Φ_{IR} function is successfully interpreted, whichever the temperature, by means of Brot Lassier's model (hindered libration inside a well followed by a large angle jump into another well (92) (see fig.11)).Furan has also been studied in solution (93).

On another hand, a comparative study of the Raman and IR spectra of quinoline (94) provides interesting results : the obtained correlation functions are identical to experimental uncertainties. It is concluded that relaxation if purely vibrational, a predictible result, considering the shape and the number of atoms of the molecule but, in this case, clearly brought out to light.

These examples show that, in the case of asymmetric rotors, one may still obtain interesting results provided several techniques are simultaneously employed and particular cases investigated.

4 - VIBRATIONAL RELAXATION

Experimental investigation of the vibrational relaxation has only recently been initiated; the choice techniques in this case are isotropic Raman scattering and picosecond laser spectroscopy. This relaxation can have several causes :

a) a damping of the vibration due to intra or intermolecular energy transfers : though such effects may eventually be brought out (99, 107), they correspond to particular cases, and for all cases of diluted solutions, in which we are specially interested, they can be neglected. However, for pure liquids and very intense IR vibrational bands, an energy transfer occurs by resonance between neighbouring molecules (transition dipole - transition dipole coupling). This effect can be very important and has been the matter of many studies.

b) a dephasing of the vibration in course of time, or, in other words, a vibrational frequency modulation due to variable environment of the molecule. This process is generally the most important, specially for diluted solutions, and we shall examine it in the first place.

4.1 - Vibrational Dephasing

It is equivalent, but more suggestive, to evoke this process in terms of frequency fluctuations. The vibrational frequency, slightly shifted from that of the free molecule, is dependent on the intermolecular potential to which the active molecule is submitted and two limiting cases can be distinguished : in the first one (slow modulation), the variation of the intermolecular potential is sufficiently slow to let the molecule react to each value and the observed profile is that of the frequency distribution, assumed to be gaussian. The vibrational width $(\Delta\nu_{1/2})_V$ and the frequency shift between gas and solution are approximately equal. In the second case (fast modulation) the variation of the potential is too rapid to let the active molecule follow it and the mean vibrational frequency is observed. The vibrational width is small and the profile is lorentzian. It is the motional narrowing phenomenon (95) otherwise wellknown, principally in compressed gases. Recent theories are based on one or the other of these hypotheses and we shall take up this question further on.

Experimental investigations bearing exclusively on vibrational relaxation by isotropic Raman scattering are very recent, and very scarce, since, precedently, the interest was principally centered on rotational relaxation and vibrational relaxation only intervened as an awkward phenomenon, usually very weak in the selected cases, but which had to be evaluated in order to correct infrared data, for example. In these conditions, computing the correlation functions was illusive, and fruitless, and the first works on this subject have dealt with vibrational widths and their variations with temperature. In fact, Bratos et al. had theoretically predicted (97) that $(\Delta\nu_{1/2})_V$ decreases while temperature increases, which had been verified on CH_3I in solution (50). This phenomenon had elsewhere been brought out on thiophene (98) but no interpretation had been proposed at the time. Further works have shown that the variation followed practically always the predicted way (27b,30,60,86,87,91b). However, in a recent work (99) the vibrational correlation functions of all the parallel bands of CH_3I have been determined and an anomalous comportment of the ν_1 CH stretching vibration, $(\Delta\nu_{1/2})_V$ increasing very slightly with T, has been observed. We see for example, on fig.12 the vibrational correlation function for the isotropic Raman band of ν_2 of CH_3I and its variation with temperature.

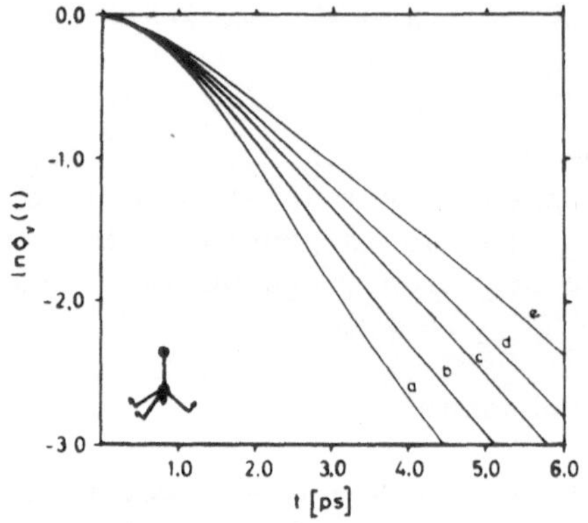

Fig.12 Vibrational CF for isotropic Raman band of ν_2 of CH_3I
a = -60°C, b = -40°C, c = -20°C, d = 0°C, e = 20°C ref.99

A similar study has been carried out on the variation of Φ_V and τ_V with respect to pressure and to temperature at constant density (100). Such measurements are particularly interesting as a test of theory. Unfortunately, they deal with the same ν_1 band of CH_3I and CD_3I whose anomalous comportment has been pointed out. These measurements come to confirm the preceding but it would be desirable to carry out similar studies in a more favourable case.

As an overall work on vibrational relaxation, we shall cite Dijkman's study (101) on the CH stretching vibrations in compounds containing the CH_3 or the C=C-H group.

Parallelously to these experimental investigations, the interpretation of Φ_V functions and correlations times τ_V have been attempted by means of different theories in two directions. Firstly, Rothschild (102) takes up a formula proposed by Kubo, valid for vibrational as well as for rotational relaxation, for slow or fast modulation, and including two parameters : the spectral second moment M_2 (expressed in S^{-2}) and the correlation time τ_c of the frequency fluctuation ω_1. This formula is written :

$$\Phi_V(t) = \exp \left\{- M_2[\tau_c t + \tau_c^2 (e^{-t/\tau_c} - 1)]\right\} \qquad [20]$$

with $\langle\omega_1\rangle = 0$ $\qquad \langle\omega_1^2\rangle = M_2$

$$\langle \omega_1(0) \ \omega_1(t) \rangle = M_2 \exp \left(- \frac{t}{\tau_c}\right) \qquad [21]$$

τ_c is the correlation time and $\sqrt{M_2}$ the amplitude of the frequency fluctuation. $\Phi_V(t)$ presents the two limiting forms of a gaussian for $t \gg \tau_c$ and of an exponential for $t \ll \tau_c$. Furthermore, the modulation is slow if $\tau_c\sqrt{M_2} \gg 1$ and fast if $\tau_c\sqrt{M_2} \ll 1$. This formula has often been applied by taking the measured moment for M_2 and adjusting τ_c to experiment. The agreement is generally good (see for example ref.27b,67,103) and in such a case an argument in favour of the slow or fast character of the modulation is obtained by computing $\tau_c \sqrt{M_2}$. For N_2O, for example, we have $\tau_c \sqrt{M_2} \simeq 0.5$.

Another very successful theory is that of Fisher and Laubereau (104). These authors carry out an a priori calculation of τ_V taking into account the binary head on collision of a diatomic molecule with an atom, with a repulsive potential. This calculation may be extended to vibrations of polyatomic molecules with various approximations but then rather provide an order of magnitude. In consideration of these difficulties, the comparison of measured and calculated τ_V may be considered satisfactory. For application of this formula, see for example (105).

A recent article by Oxtoby gives a review of the theoretical and experimental works on dephasing of molecular vibrations in liquids (114).

Finally, we shall discuss the question of the slow or fast character of the modulation. This point is the matter of many discussions owing to the fact that the adopted criterion is nearly always unique and varies according to authors. From the above result, obtained for N_2O for example, one might conclude to the intermediate character of the modulation. Several authors have based their studies on this criterion ($\tau_c \sqrt{M_2}$ derived from Kubo formula) and concluded to a quite rapid modulation. Some of them have utilized other theoretical considerations. Comparing width $(\Delta\nu_{1/2})_V$ to displacement $\Delta\nu$, the modulation may be considered as rapid for SO_2 and CO_2. The other cases investigated are less clear. Lastly, the question has been approached (106) from a comparative study of the vibrational widths of the 1st harmonic and fundamental of a given vibration. Actually,the ratio of these two widths must equal 2 for a slow modulation and 4 for a fast modulation; the ratios observed on several bands of various compounds vary from 1.3 to 3.2. These results, moreover fraught with large uncertainty, are therefore not conclusive. The only case in which we reach a definite conclusion is nitrogen, for which several concording results have been obtained by picosecond pulsing technique, molecular dynamic calculations and Raman scattering measurements (96) and show an important motional narrowing together with a fast modulation. On the whole, the question remains therefore hitherto unsolved and complementary measurements would be necessary to decide on the problem.

The bulk of the data show that this field of researches is now in full progress but that the experimental results are still too few to offer a comprehensive view from which a general conclusion might be reached, as is actually the case for rotational relaxation.

4.2 - Transition Dipole - Transition Dipole Coupling

This regards the resonant excitation transfer between two neighbouring molecules in the pure liquid. This phenomenon had been predicted as early as 1961 in a theoretical study by Valiev (108), practically unnoticed at the time, owing to the fact that it dealt simultaneously with several causes of vibrational broadening and the comparison with the few data available at the time was not very convincing. This broadening of the bands has been clearly shown in 1968 (109). It appeared that a given IR band of the pure liquid was always wider than the corresponding band of the solution and that this overwidth followed the intensity of the band. In these conditions, the interpretation became obvious and a classical approached calculation of the coupling between neighbouring oscillators showed a broadening proportional to intensity. Precise measurements of this phenomenon are delicate in the infrared since, for very intense bands, the refractive index inside the band varies and the absorption coefficient must be replaced by the imaginary part of local susceptibility. An experimental study, performed later on, took both effects into account (110). In spite of the imprecision of the measurements (the rotational width is supposed to be identical for all these solvents and the vibrational width due to dephasing, to be a linear function of the frequency shift), the phenomenon is sufficiently important for intense bands to be shown without any ambiguity. To Döge (111) we owe the investigation of this same effect in Raman scattering, where experiments can be more accurate, and to have proposed a thorough interpretation from a parallel study of liquid and solid in terms of vibrational excitons. A similar work was nearly simultaneously carried out in the IR (112,113) by isotopic dilution method and the influence of temperature was examined. The characteristic aspect of the curves showing the preponderant effect of vibrational relaxation at low temperature and the apparition of rotational relaxation at high temperature are seen on fig.13. In this case, Φ_V and Φ_{IR} cannot be separated owing to the influence of rotation on this intermolecular coupling.

Later experimental studies on this matter have nearly exclusively dealt with isotropic Raman scattering in which the isotopic dilution method permits a precise quantitative study. Among them, we shall cite works on CH_3I (99), on CH_3CN and CD_3CN (100).

Furthermore, in the case of extremely intense IR bands, this transition dipole - transition dipole interaction may present a particular and unexpected aspect in the liquid phase. Precedently mentioned Raman studies on CF_4 (41b) and NF_3 (78) show that at low temperature, the ν_3 band of CF_4 and ν_1 and ν_3 of NF_3 present a structure nearly similar to that of the plastic phase which get less visible as temperature increases (fig.14). This continuous variation with temperature shows that the same phenomenon is involved

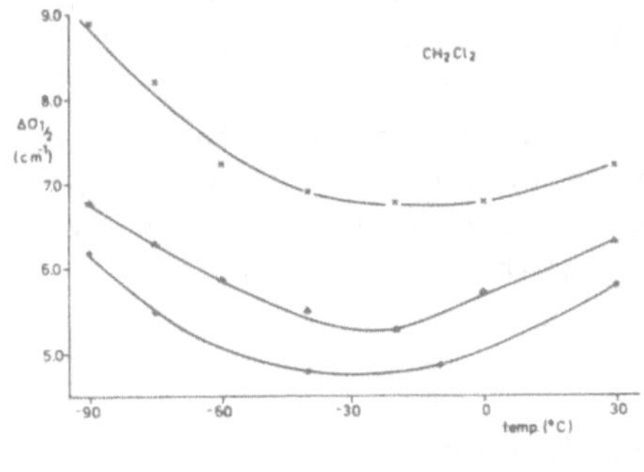

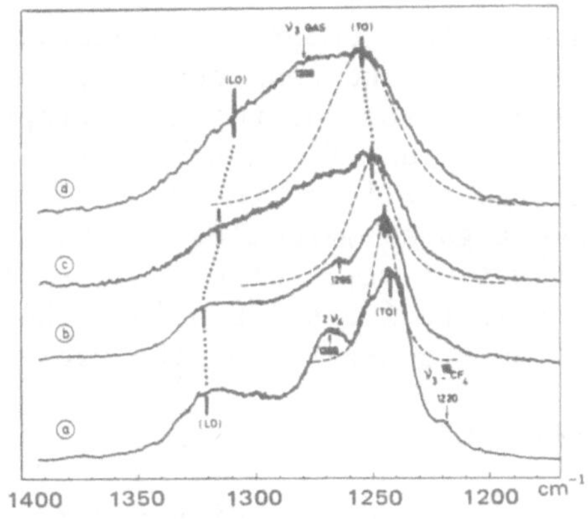

Fig.13 Temperature dependence of the IR band width of CH_2Cl_2 ($1265\ cm^{-1}$) in CD_2Cl_2 at several concentrations. molfr o 0.10, Δ 0.39, × 1 Ref.113

Fig.14 Raman profile of the strong IR line ν_3 of CF_4 a) plastic crystal b)c)d) liquid at 85 K, 135 K, 185 K respectively Solid line, experimental profiles. Dotted line, pure rotational spectra in the J model. Ref.41b

in solid and liquid. This effect is wellknown for cubic crystals where a triply degenerate mode is decomposed into an optical longitudinal mode and an optical transverse mode with a splitting proportional to the intensity of the IR band. Here, the same interpretation is therefore adopted for the liquid.

This transition dipole - transition dipole coupling is thus now well investigated and understood. Owing to considerable variations in the intensity of the IR bands, this effect can be either very important or totally negligible, but it always exists theoretically and must eventually be taken into account each time a pure liquid is concerned.

5 - CONCLUSION

From this review of the experimental works performed on vibrational and rotational relaxation in the liquid phase, we shall derive a few conclusions. First of all, as regards the precision of the obtained results, there are two very favourable cases : pure rotational relaxation as it can be seen in the infrared for very light molecules diluted in an inert solvent such as SF_6 or hydrocarbons, and pure vibrational relaxation which can be observed in isotropic Raman scattering. Great difficulties are met in all intermediate cases. Deconvolution of spectra are always delicate and are sometimes replaced by mere subtraction of bandwidths; when the molecules present numerous vibrational modes, overlapping of the bands often renders very arduous the determination of the correlation functions, and consequently, that of the correct correlation times. A problem such as the anisotropy of the various tensors describing the rotation of a non spherical or non linear molecule has not been satisfactorily solved to this day, even for as simple and widely investigated molecules as CH_3I or CH_3CN. Finally, vibrational relaxation still offers a vast field of researches before one may be able to foresee a priori the effects to be expected and the slow or rapid character of the vibrational modulation.

REFERENCES

(1) S. Bratos, J. Rios and Y. Guissani, J. Chem. Phys. 52,439(1970)
(2) S. Bratos and E. Marechal, Phys. Rev. A4, 1078 (1971)
(3) F.J. Bartoli and T.A. Litovitz, J. Chem. Phys. 56, 413 (1972)
(4) L.A. Nafie and W.L. Peticolas, J. Chem. Phys. 57, 3145 (1972)
(5) A.V. Rakov, Research in molecular spectroscopy, ed. by D.V. Skobeltzyn, Consultant Bureau, 27, 109 (1965)
(6) P.V. Huong, M. Couzi and M. Perrot, Chem. Phys. Lett. 7, 189 (1970)
(7) M. Perrot, P.V. Huong and J.Lascombe, J. Chim. Phys. 614 (1971)
(8) M. Perrot and J. Lascombe, J. Chim. Phys. 5 (1973)
(9) J.P. Perchard, W.F. Murphy and M.J. Bernstein, Mol. Phys. 23 499 and 519 (1972)
(10) M. Perrot, P.B. Caloine and J. Lascombe, C.R. Acad. Sc. Paris 274, 104 (1972)
(11) P. Perrot and J. Lascombe, J. Chim. Phys. 1486 (1973)
(12) L. Galatry, D. Robert, P.V. Huong, J. Lascombe and M. Perrot Spectrochim. Acta, 25A, 1693 (1969)
(13) G. Turrell, J. Mol. Spectrosc. 69, 383 (1978)
(14) C.H. Wang and P.A. Fleury, J. Chem. Phys. 53, 2243 (1970)
(15) C.H. Wang and R.B. Wright, Mol. Phys., 27, 345 (1974)
(16) G.E. Ewing, J. Chem. Phys. 37, 2250 (1962)
(17) J. Bruining and J.H.R. Clarke, Mol. Phys. 31, 1425 (1976)
(18) J.P. Marsault, F. Marsault-Herail and G. Levi, J. Chem. Phys. 62, 893 (1975)
(19) a) B. Caloine, Thesis of Docteur Ingénieur, Bordeaux 1972
b) Y. Guissani and J.C. Leicknam, Can. J. Phys. 51, 938 (1973)
(20) Y. Le Duff and W.Holzer, Chem. Phys. Lett. 24, 212 (1974)
(21) M. Scotto, J. Chem. Phys. 49, 5362 (1968)
(22) Y. Le Duff,J. Chem. Phys. 59, 1984 (1973)
(23) G. Viossat, R. Cavagnat and J. Lascombe, J. Raman Spectrosc. 8, 299 (1977)
(24) E.B. Gill and D. Steele, Mol. Phys. 34, 231 (1977)
(25) M.O. Bulanin, J. Mol. Struct. 19, 59 (1973)
(26) a) J. Vincent-Geisse, J. Soussen-Jacob and Nguyen-Tan Tai, C.R. Acad. Sc. Paris, 268 , 1020 (1969)
b) J. Vincent-Geisse, J. Soussen-Jacob, Nguyen-Tan Tai and D. Descout, Can. J. Chem. 48, 3918 (1970)

(27) a) C. Breuillard and R. Ouillon, Mol. Phys. 33, 747 (1977)
b) R. Ouillon, C. Breuillard, J. Soussen-Jacob and J. Tsakiris J. Mol. Struct. 46, 447 (1978)
(28) C. Dreyfus, C. Breuillard, Nguyen-Tan Tai and R. Ouillon Chem. Phys. Lett. 62, 246 (1979)
(29) M.E. Costines, J.P. Marsault, G. Levi and F. Marsault-Herail Mol. Phys. (to be published)
(30) D. Legay, J.P. Perchard, A.M. Goulay-Bize and J. Soussen-Jacob J. Chim. Phys. 863 (1975)
(31) A.M. Goulay, E. Dervil and J. Vincent-Geisse (to be published)
(32) M. Perrot, J. Devaure and J. Lascombe, Mol. Phys. 30, 97 (1975) and 36, 921 (1978)
(33) P.V. Konynenburg and W.A. Steele, J. Chem. Phys. 62, 2301 (1975)
(34) J. Soussen-Jacob, J. Bessiere, J. Tsakiris and J.Vincent-Geisse Spectrochim. Acta, 33A, 805 (1977)
(35) J. Vincent-Geisse and J. Soussen-Jacob, Organic Liquids, ed. by J. Wiley and Sons, chap. 4, 71 (1978)
(36) A. Cabana, R. Bardoux and A. Chamberland, Can. J. Chem. 47, 2915 (1969)
(37) R.E.D. McClung, J. Chem. Phys. 55, 3459 (1971)
(38) J. Vincent-Geisse, J. Soussen-Jacob, T. Nguyen-Tan and R.E.D. McClung, Can. J. Phys. 57, 564 (1979)
(39) W.A. Steele, The rotation of molecules in dense phases. Adv. Chem. Phys. vol. 34 (1976) ed. by Prigogine and Rice. Wiley.
(40) F. Marsault-Herail, J.P. Marsault, G. Michoud and G. Levi Chem. Phys. Lett. 31, 335 (1975)
(41) M. Gilbert and M. Drifford, J. Chem. Phys. a) 65, 923 (1976) b) 66, 3205 (1977)
(42) S. Sunder and R.E.D. McClung, Can. J. Phys. 54, 211 (1976)
(43) T. Fujiyama and B. Crawford, J. Phys. Chem. 73, 4040 (1969)
(44) P. Van Konynenburg and W.A. Steele, J. Chem. Phys. 56, 4776 (1972)
(45) J. Vincent-Geisse and E. Dayan, Mol. Phys. 31, 1233 (1976)
(46) J. Vincent-Geisse, J. Soussen-Jacob, C. Breuillard, J.C. Briquet and T. Nguyen-Tan, Mol. Phys. 33, 145 (1977)
(47) R.E.D. McClung, Adv. Mol. Rel. Int. Proc. 10, 83 (1977)
(48) J.E. Griffiths, Vibrational spectra and structure, vol. 6 273 (1977) ed. by R. Durig, Elsevier
(49) J.P. Marsault, F. Marsault-Herail and G. Levi, Mol. Phys. 33 735 (1977)
(50) M. Constant, M. Delhaye and R. Fauquembergue, C.R. Acad. Sc. Paris, 271, 1177 (1970)
(51) F.J. Bartoli and T.A.Litowitz, J. Chem. Phys. 56, 404 (1972)
(52) M. Constant, R. Fauquembergue, Adv. Raman Spectrosc. vol.1 413 (1972) Heyden
(53) J.E. Griffiths, Chem. Phys. Lett. 21, 354 (1973)
(54) R.B. Wright, M. Schwartz and C.H. Wang, J. Chem. Phys. 58, 5125 (1973)

(55) R. Arndt, R. Moormann and A. Schaffer, Mol. Motions in Liquids ed. by J. Lascombe, Reidel (1974) 217
(56) J.H. Campbell, J.F. Fisher and J. Jonas, J. Chem. Phys. 61 346 (1974)
(57) D.R. Jones, H.C. Andersen and R. Pecora, Chem. Phys. 9, 339 (1975)
(58) J.E. Griffiths, J. Chem. Phys. 59, 751 (1973)
(59) C. Breuillard-Alliot and J. Soussen-Jacob, Mol. Phys. 28, 905 (1974)
(60) J. Yarwood, R. Arndt and G. Doge, Chem. Phys. 25, 387 (1977)
(61) C. Dreyfus, C. Breuillard, T. Nguyen-Tan, A. Grosjean and J. Vincent-Geisse, J. Mol. Struct. 47, 41 (1978)
(62) J. Soussen-Jacob, E. Dervil and J. Vincent-Geisse, Mol. Phys. 28, 935 (1974)
(63) J.H. Campbell and J. Jonas, Chem. Phys. Lett. 18, 441 (1973)
(64) A.E. Boldeskul, S.S. Esman and V.E. Pogorelov, Opt. Spectr. 37, 521 (1974)
(65) W.G. Rotschild, G.J. Rosasco and R.C. Livingston, J. Chem. Phys. 62, 1253 (1975)
(66) C. Brodbeck, I. Rossi, Nguyen-Van-Thanh and A. Ruoff, Mol. Phys. 32, 71 (1976)
(67) J. Schroeder, V.H. Schiemann and J. Jonas, Mol. Phys. 34, 1501 (1977)
(68) R. Arndt and R.E.D. McClung, J. Chem. Phys. 69, 4280 (1978)
(69) J. Schroeder, V.H. Schiemann and J. Jonas, J. Chem. Phys. 69 5479 (1978)
(70) J. DeZwaan, D.W. Hess and C.S. Johnson,J.Chem.Phys.63,422(1975)
(71) M. Bouachir and J. Lascombe, J. Ram. Spectr. 7, 271 (1978)
(72) R. Seloudoux, J. Soussen-Jacob and J. Vincent-Geisse, Chem. Phys. (to be published)
(73) M. Schwartz and C.H. Wang, Chem. Phys. Lett. 25, 26 (1974)
(74) M. Besnard, J. Devaure and J. Lascombe, J. Chim. Phys. 72, 453 (1975)
(75) S. Sunder, K.E. Hallin and R.E.D. McClung, J. Chem. Phys. 61, 2920 (1974)
(76) J. Schulz, Z. Naturforsch. 29a, 1636 (1974)
(77) M.L. Bansal and A.P. Roy, Chem. Phys. Lett. 50, 513 (1077)
(78) M. Gilbert, P. Nectoux and M. Drifford, J. Chem. Phys. 68 679 (1978)
(79) M. Constant, R. Fauquembergue and P. Descheerder, J. Chem. Phys. 64, 667 (1976)
(80) K.I. Gillen and J.E. Griffiths, Chem. Phys. Lett. 17,359(1972)
(81) M.N. Neuman and G.C. Tabisz, Chem. Phys. 15, 195 (1976)
(82) K. Tanabe and J. Jonas, J. Chem. Phys. 67, 4222 (1977)
(83) K. Tanabe, Chem. Phys. 31, 319 (1978)
(84) a) J. Jacob, J. Leclerc and J. Vincent-Geisse, J. Chim. Phys. 66, 970 (1969) b) J.G. David,Spectrochim.Acta, 28A,977(1972)
(85) T.H. Wall, J. Chem. Phys. 51, 113 (1969) and 52, 2792 (1970)
(86) a) J.G. David and H.E. Hallam, J. Mol. Struct. 5, 31 (1970) b) J.C. Briquet and J.Soussen-Jacob, Mol. Phys. 28, 921 (1974)

(87) R. Ouillon and Y. Le Duff, Adv. Raman Spectr. vol.1 (Heyden) 428 (1972)
(88) C.H. Wang, D.R. Jones and D.H. Christensen, J. Chem. Phys. 64, 2820 (1976)
(89) D.R. Jones, C.H. Wang, D.H. Christensen and O.F. Nielson, J. Chem. Phys. 64, 4475 (1976)
(90) S.K. Satija and C.H. Wang, Chem. Phys. Lett. 46, 352 (1977)
(91) a) J. Loisel, J.P. Pinan-Lucarre and J. Vincent-Geisse, Adv. Mol. Rel. Proc. 6, 201 (1974)
b) J.P. Pinan-Lucarre, L. Colombo, J. Loisel, M. Le Postollec and T. Nguyen-Tan, Adv. Mol. Rel. Proc. 10, 1 (1977)
(92) J.P. Pinan-Lucarre, (to be published)
(93) P. Dorval and P. Saumagne, Mol. Motions in liquids, ed. by J. Lascombe, Reidel, 319 (1974)
(94) W.G. Rotschild, Mol. Motions in liquids, ed. by J. Lascombe Reidel, 247 (1974)
(95) R. Ouillon, Chem. Phys. Lett. 35, 63 (1975)
(96) Y. Le Duff, J. Chem. Phys. 59, 1984 (1973)
(97) R. Levant and S. Bratos, C.R. Acad. Sc. Paris, B276,603 (1973)
(98) G. Doge, Z. f. Natur. 23a, 1130 (1968)
(99) G. Doge, R. Arndt and A. Khuen, Chem. Phys. 21, 53 (1977)
(100) J. Schroeder, V.H. Schiemann, P.T. Sharko and J. Jonas, J. Chem. Phys. 66, 3215 (1977)
(101) F.G. Dijkman, Thesis, Utrecht, 1978
(102) W.G. Rotschild, J. Chem. Phys. 65, 455 (1976)
(103) A. Moradi-Araghi and M. Schwartz, J. Chem. Phys. 68, 5548 (1978)
(104) S.F. Fisher and A Laubereau, Chem. Phys. Lett. 35, 6 (1975)
(105) K. Tanabe and J. Jonas, Chem. Phys. Lett. 53, 278 (1978)
(106) R. Arndt and Y. Yarwood, Chem. Phys. Lett. 45, 155 (1977)
(107) J.E. Griffiths, M. Clerc and P.M. Rentzepis, J. Chem. Phys. 60, 3824 (1974)
(108) K.A. Valiev, Opt. Spectr. 11, 253 (1961) and Soviet Physics, JETP, 13, 1287 (1961)
(109) J. Vincent-Geisse, Spectrochim. Acta, 24A, 1 (1968)
(110) M. Lafaix, and J. Vincent-Geisse, Spectrochim. Acta, 29A 177 (1973
(111) G. Doge, Z. f. Natur. 28a, 919 (1973)
(112) P.C.M. Van Woerkom, J. de Bleijser, M. de Zwart and J.C.Leyte Chem. Phys. 4, 236 (1974)
(113) P.C.M. Van Woerkom, J. de Bleijser, M. de Zwart, M.P.J. Burgess and J.C. Leyte, Ber. d. Buns. Ges. 78, 1303 (1974)
(114) D.W. Oxtoby, Adv. Chem. Phys. 40, 1 (1979)

COLLISION-INDUCED VIBRATIONAL SPECTROSCOPY IN LIQUIDS

George Birnbaum
National Bureau of Standards
Washington, D.C. 20234

1 INTRODUCTION

Pressure or collision-induced absorption arises from transient dipoles produced by distortion of the electronic distribution of molecules in binary, ternary and higher order interactions. The induced dipole is modulated by the vibration and rotation of the colliding molecules and, because of its strong dependence on the intermolecular separation, by their relative translational motion. Thus a variety of pressure-induced spectra are known ranging from pure translation and rotation in the microwave and far infrared regions to fundamental and overtone rotation-vibration spectra in the infrared region. Such spectra have been observed in pure substances and in mixtures in the gaseous, liquid and solid phases. Collisions also induce anisotropic polarizability in atoms and spherical molecules producing depolarized Rayleigh spectra. More recently, vibrational Raman bands have been studied which are forbidden by symmetry in the isolated molecule but are produced by intermolecular fields.

The aim here is to discuss collision-induced vibrational spectra in liquids observed in infrared absorption and Raman scattering. Although there has been little progress in the theoretical analysis of such spectra, studies of gases have been important in unraveling the induced interactions in the liquid state. Since the interpretation of the effects are undoubtedly easier for simple molecules, we shall consider N_2, O_2, Br_2, CS_2 and CO_2 in absorption and CS_2, CO_2 and SF_6 in Raman scattering, in some instances

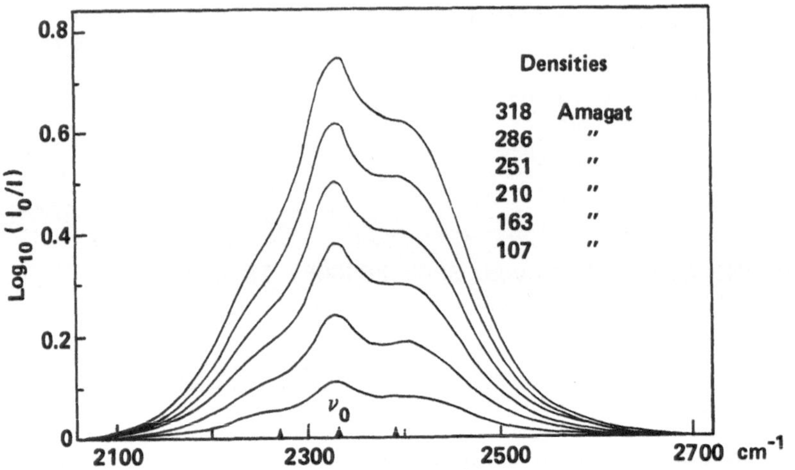

Figure 1. The fundamental absorption band of gaseous N_2 at 298K at various densities. The triangular marks on the frequency axis, on the left and right sides of ν_o, represent O_{max} and S_{max}, respectively (from (8)).

in the compressed gas and the liquid. These molecules have a center of symmetry and transitions that are allowed in absorption are forbidden in the Raman spectrum and conversely. We also discuss infrared absorption in the polar molecule HCl.

Reviews have appeared which deal with a number of the areas discussed here and provide useful background information. These include a treatment of band shapes and molecular dynamics in liquids (1), pressure effects in molecular spectra (2), infrared spectra of liquified gases in inert solvents (3), pressure-induced far-infrared absorption (4) and collision-induced light scattering (5). The very extensive work on pressure-induced absorption in H_2, which is outside the scope of this work, has been reviewed in detail (6, 7).

2 COLLISION-INDUCED VIBRATIONAL ABSORPTION BANDS

2.1 General Features

At sufficiently low densities where collisions are bimolecular, the absorption coefficient in cm^{-1},

$$\alpha(\nu) = L^{-1}\ln\left[I_o(\nu)/I(\nu)\right] \tag{2.1}$$

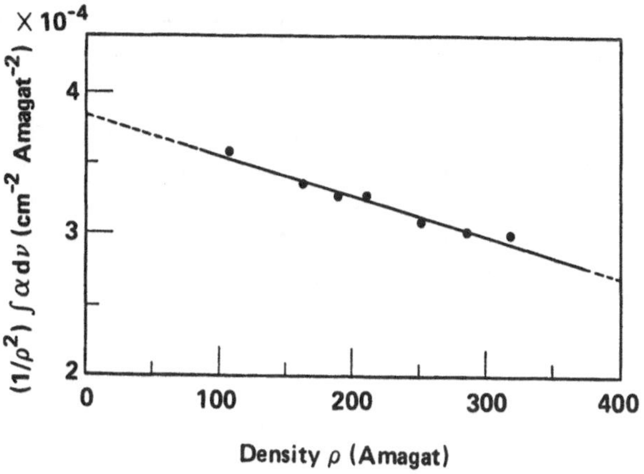

Figure 2. Relation between the integrated absorption coefficient of the fundamental band of N_2 at 298K and density (from (8)).

is proportional to ρ^2, where ρ is the density. Here ν is in cm^{-1}, $I_o(\nu)$ is the intensity of radiation transmitted by the empty cell of length L and $I(\nu)$ is that transmitted by the cell filled with fluid. Figure 1 (8) shows the fundamental band of gaseous N_2 whose shape appears to be preserved with increasing density. As shown in Fig. 2, however, $\rho^{-2}\int\alpha(\nu)d\nu$ decreases with increasing ρ. This is due to three-body interactions which make a negative contribution to $\int\alpha(\nu)d\nu$, the absorption coefficient integrated over a given band, because of a partial cancellation of the induced dipole (9).

At not too high densities, in a binary mixture of gas at the partial densities ρ_1 and ρ_2, $\int\alpha(\nu)d\nu$ of a given band can be expanded in a power series in the form

$$\int\alpha(\nu)d\nu = a_o\rho_1 + a_1\rho_1{}^2 + a_2\rho_1{}^3 \ldots$$

$$+ b_1\rho_1\rho_2 + b_2\rho_1\rho_2{}^2 + b_2'\rho_1{}^2\rho_2 + \ldots \qquad (2.2)$$

where the coefficients a_1, a_2. . . are temperature dependent. The term a_o is a property of the individual molecule and is equal to zero if the transition is forbidden. The coefficients a_1 and a_2 arise from collisions of molecules of type 1, whereas those involving b_1 and b_2 constitutes the enhancement of the absorption

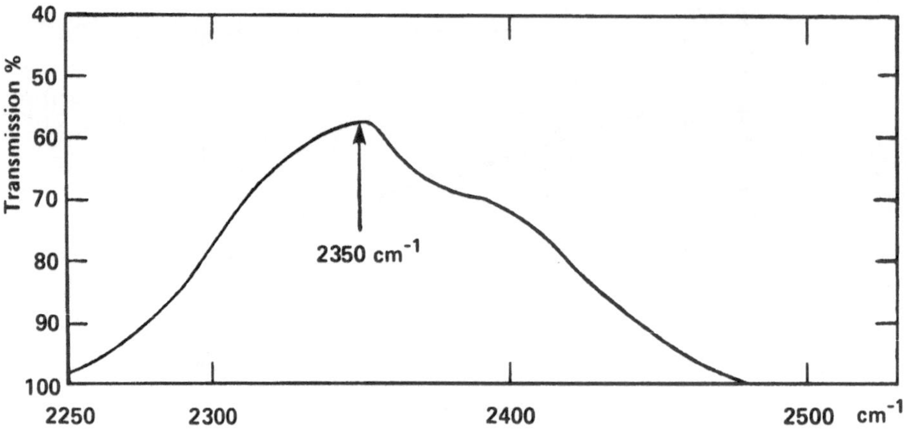

Figure 3. Transmission in the fundamental band of liquid N_2 at 69K; path length 0.75 cm (from (10)).

by collisions between the two types of molecules. In single component fluids, the terms involving the coefficients b are equal to zero. The ternary coefficients are negative and with the high degree of local symmetry possible in liquids can produce a large cancellation of the induced absorption.

The infrared absorption spectrum in liquid N_2 shown in Fig. 3 has a shape which is rather similar to that of the compressed gas in Fig. 1. A rough estimate of a_1 in the liquid shows that it is considerably smaller than that in the gas, although an accurate comparison cannot be made because of the difference in temperatures.

The width of pressure-induced transitions is determined by the time, τ, during which the interacting molecules are within each other's range. For bimolecular collisions one has

$$\Delta\nu = (2\pi c\tau)^{-1} = (2\pi c)^{-1}(\bar{v}/\Delta\bar{R}), \qquad (2.3)$$

where $\Delta\nu$ is the line width in cm^{-1}, $\bar{v}$ is the average relative velocity and $\Delta\bar{R}$ is the effective range of the square of the induced dipole moment. For the molecules and temperatures considered here, $\Delta\nu \sim 10cm^{-1}$ which is much greater than the spacing between the rotational lines comprising the band. Thus no line structure is seen in Figs. 1 and 3. The general features of collision-induced Raman bands are similar to those considered here.

2.2 Theoretical Aspects (11)

As the molecular interaction may to a good approximation be considered to have a very small influence upon the vibrational energy levels, the vibrational wave functions of the collision pair may be expressed as

$$\psi_{12} = \psi_1(\xi_1 \ldots)\psi_2(\xi_2 \ldots) \tag{2.4}$$

where $\xi_1 \ldots$ are the normal coordinates of molecule 1 and $\xi_2 \ldots$ are those of molecule 2.* The induced dipole expressed as a function of the nuclear coordinates

$$\underline{\mu} = \underline{\mu}(\xi_1 \ldots \xi_2 \ldots), \tag{2.5}$$

may be expanded in a Taylor series

$$\underline{\mu} = \underline{\mu}_0 + \left(\frac{\partial \underline{\mu}}{\partial \xi_1}\right)_0 \xi_1 + \left(\frac{\partial \underline{\mu}}{\partial \xi_2}\right)_0 \xi_2 + \ldots$$

$$+ \frac{1}{2}\left(\frac{\partial^2 \underline{\mu}}{\partial \xi_1^2}\right)_0 \xi_1^2 + \frac{1}{2}\left(\frac{\partial^2 \underline{\mu}}{\partial \xi_2^2}\right)_0 \xi_2^2 + \left(\frac{\partial^2 \underline{\mu}}{\partial \xi_1 \partial \xi_2}\right)_0 \xi_1 \xi_2$$

$$+ \ldots \tag{2.6}$$

where for simplicity the dependence of $\underline{\mu}$ on nuclear coordinates is shown only for ξ_1 and ξ_2 and $\underline{\mu}_0$, which is independent of ξ_1 and ξ_2, appears in the description of the rotational-translational spectrum.

The transition moment, $\int \psi_1 \psi_2 \underline{\mu} \psi_1' \psi_2' d\tau$, where $\underline{\mu}$ is given by (2.6), contains the following terms:

$$\left(\frac{\partial \underline{\mu}}{\partial \xi_1}\right)_0 \int \psi_1 \psi_2 \xi_1 \psi_1' \psi_2' d\tau = \left(\frac{\partial \underline{\mu}}{\partial \xi_1}\right)_0 \int \psi_1 \xi_1 \psi_1' d\tau \tag{2.7}$$

$$\left(\frac{\partial^2 \underline{\mu}}{\partial \xi_1^2}\right)_0 \int \psi_1 \psi_2 \xi_1^2 \psi_1' \psi_2' d\tau = \left(\frac{\partial^2 \underline{\mu}}{\partial \xi_1^2}\right)_0 \int \psi_1 \xi_1^2 \psi_1' d\tau \tag{2.8}$$

*The subscripts 1 and 2 may refer to molecules of the same or different species.

$$\left(\frac{\partial^2 \underline{\mu}}{\partial\xi_1\partial\xi_2}\right)_0 \int\psi_1\psi_2\xi_1\xi_2\psi_1{}'\psi_2{}'d\tau = \left(\frac{\partial^2 \underline{\mu}}{\partial\xi_1\partial\xi_2}\right)_0 \int\psi_1\xi_1\psi_1{}'d\tau$$

$$\times \int\psi_2\xi_2\psi_2{}'d\tau \qquad (2.9)$$

The vibrational moments (2.7) to (2.9) describe, respectively, fundamentals, overtones and simultaneous transitions in which a single quantum is absorbed by two molecules.

For the total dipole induced in a pair of molecules by the action of molecular fields, we write

$$\underline{\mu} = \alpha_1 \underline{F_1} + \alpha_2 \underline{F_2} \qquad (2.10)$$

where $\underline{F_1}$ is the field at molecule 1 due to molecule 2 and α_1 is the polarizability of molecule 1. It is sufficient here to consider α to be isotropic. The field may be represented by $\underline{F} \sim \underline{\underline{Q}}_\ell \nabla^{\ell+1}(R_{12})^{-1}$, where $\underline{\underline{Q}}_\ell$ is a tensor representing the molecular multipole and R_{12} is the intermolecular separation. The magnitude of $\underline{\underline{Q}}_2$, for example, is proportional to Θ, the quadrupole moment. The terms in (2.6) that depend on ξ_1 and ξ_2 are

$$\left(\frac{\partial\underline{\mu}}{\partial\xi_1}\right)_0 \xi_1 + \left(\frac{\partial\underline{\mu}}{\partial\xi_2}\right)_0 \xi_2 = \left[\underline{F_1}\left(\frac{\partial\alpha_1}{\partial\xi_1}\right)_0 + \alpha_2\left(\frac{\partial\underline{F_2}}{\partial\xi_1}\right)_0\right]\xi_1$$

$$+ \left[\underline{F_2}\left(\frac{\partial\alpha_2}{\partial\xi_2}\right)_0 + \alpha_1\left(\frac{\partial\underline{F_1}}{\partial\xi_2}\right)_0\right]\xi_2 \qquad (2.11)$$

The field due to the vibrating multipole is given by $(\partial\underline{F}/\partial\xi)\xi$, which is proportional to $\partial|\underline{\underline{Q}}_\ell|/\partial\xi$. Expressions similar to (2.11) are obtained for the terms that depend on $\xi_1{}^2$ and $\xi_2{}^2$.

For two-component mixtures there are two kinds of induced absorption. The first is of the type considered in (2.11) and involves the enhancement of a given band of molecule 1 by collisions with molecule 2 and vice versa. If the second component is a rare gas, then only the enhancement spectrum of 1 is possible. The second kind arises from the terms in (2.6) that depend on $\xi_1\xi_2$ and produce simultaneous transitions at the sum and difference of the vibrational frequencies of molecules 1 and 2. The simultaneous transition moment from (2.6) and (2.10) is given by

$$\left(\frac{\partial\underline{\mu}}{\partial\xi_1\partial\xi_1}\right)_0 \xi_1\xi_2 = \left[\left(\frac{\partial\underline{F_2}}{\partial\xi_1}\right)_0\left(\frac{\partial\alpha_2}{\partial\xi_2}\right)_0 + \left(\frac{\partial\underline{F_1}}{\partial\xi_2}\right)_0\left(\frac{\partial\alpha_1}{\partial\xi_1}\right)_0\right]\xi_1\xi_2 \qquad (2.12)$$

If the vibration in a centro-symmetric molecule is antisymmetrical with respect to the center of symmetry, then for that vibration $\partial\alpha/\partial\xi = 0$ and only one term remains in (2.12). Simultaneous transitions in two molecules of the same type are, of course, also possible.

An important quantity is the integrated absorption coefficient, (2.2). However, it is useful to define a closely related quantity,

$$\int \frac{\alpha(\nu)}{\nu}\, d\nu = \frac{8\pi^3}{3hc}\left(\frac{2-\delta_{12}}{2}\right)\rho_1\rho_2 \int 4\pi R^2 g(R)\,|\langle\psi_1\psi_2|\underline{\mu}|\psi_1'\psi_2'\rangle|^2 dR \quad (2.13)$$

where $\alpha(\nu)/\nu$ has the nature of a transition probability. If molecules 1 and 2 belong to different species, then $(2 - \delta_{12})/2 = 1$. For two-body interactions in the gas, $g(R)$, the pair distribution function, is known and (2.13) may be evaluated. Although $g(R)$ is also known for many simple liquids, a calculation of (2.13) based only on interacting pairs is clearly inadequate. It is nevertheless of interest to make such calculations to assess the effect of many-body interactions. However, since the phases of the vibration on different molecules are random, it is sufficient to consider only three-body interactions in which molecule 1 induces a dipole in molecules 2 and 3.

Thus far we have considered only vibrational transitions, although the molecules must undergo rotational transitions as well. Quadrupolar induction, for example, produces the following transitions: $\Delta J = +2$ (S-branch), $\Delta J = -2$ (O branch) and $\Delta J = 0$ (Q-branch). A detailed account of rotational selection rules are given elsewhere (12). The band positions for the fundamental in N_2 are shown in Fig. 1.

2.3 Single Component Systems

Synthetic profiles in fair agreement with experiment have been calculated for the fundamental bands of gaseous O_2 (300K) and N_2 (77K and 300K) on the basis of quadrupole induction. With the basic induction mechanism established, the quantity $\partial\Theta/\partial\xi$ was evaluated by equating the measured value of $\int(\alpha(\nu)/\nu)d\nu$ with the theoretical expression (2.13) (13, 14).

Consider next the liquid spectrum of O_2 at 55K shown in Fig. 4, with the spectrum of solid γ-O_2 at 44K (15). The high-frequency shoulder (absorption in excess of that given by the stick spectrum) in the liquid spectrum disappeared on dilution with liquid N_2. Since in addition, γ-O_2 is supposed to consist

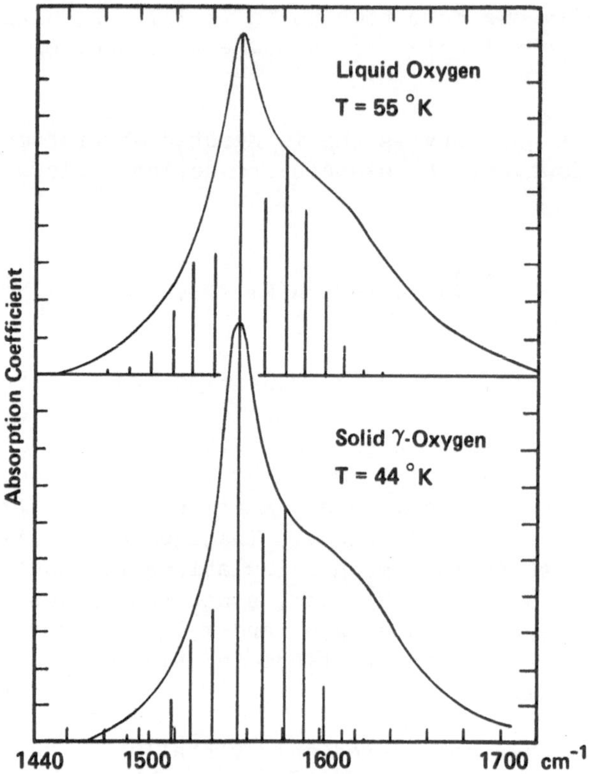

Figure 4. Induced fundamental band of oxygen at low temperature in the liquid and solid γ phase

almost entirely of O_4 molecules, it was suggested that the shoulder is connected with vibrations of the O_2 dimer. Recent low density measurements in the gas phase at 90K has revealed a spectrum attributable to O_2 dimers near the induced fundamental band of O_2 (16).

Induced overtone bands have been observed and are shown for N_2 in Fig. 5. The liquid and solid band shapes are quite similar. For some unknown reason, the overtone band shows less structure than the fundamental band (Figs. 1 and 3).

Collision-induced absorption of the symmetric vibration of CO_2 and CS_2 in liquid CCl_4 has been observed and a theory satisfactorily accounting for the integrated absorption has been presented (17). This theory evaluates the induced dipole on the basis of a quasi-crystalline cell model.

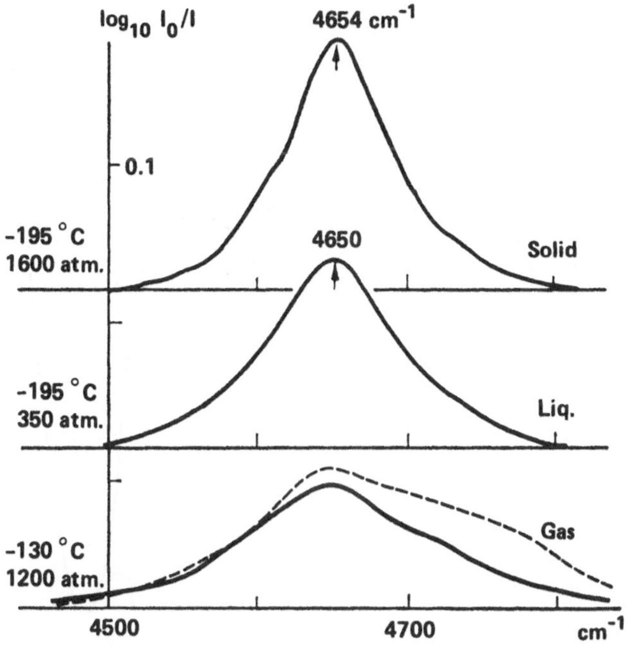

Figure 5. Induced first overtone band of N_2; path length, 49 cm. For comparison, the dashed profile of the gas spectrum was obtained at -195°C, but with a different density (from (2)).

2.4 Simultaneous Vibrational Transitions in Two Component Mixtures

Following the observation of simultaneous vibrational transitions in the gaseous mixtures, $CO_2 + N_2$, $CO_2 + O_2$ and $CO_2 + H_2$ (11), this type of transition was observed in liquid mixtures (18-20). Figure 6 shows bands in a mixture of Br_2 and CS_2 on both sides of the very intense ν_3 band of CS_2 at 1510 cm^{-1} (the value found for dilute solutions of CS_2 in Br_2) not present in the spectra of either of the components. The comparison between observed and calculated frequencies are as follows:

Observed: $Br_2 \pm CS_2$ = 1807, 1204 cm^{-1}
Calculated: 1510 ± 306.1 = 1816, 1204 cm^{-1}

where the vibrational frequency for liquid Br_2 was taken to be 306.1 cm^{-1}. The ratio of the intensity of the difference to the sum band is $I_-/I_+ = \exp(h\nu c/kT)$, where ν is the vibrational frequency of Br_2. The ratio observed near the peak of the absorption band is 0.19, which is close to the calculated value of 0.22 (19).

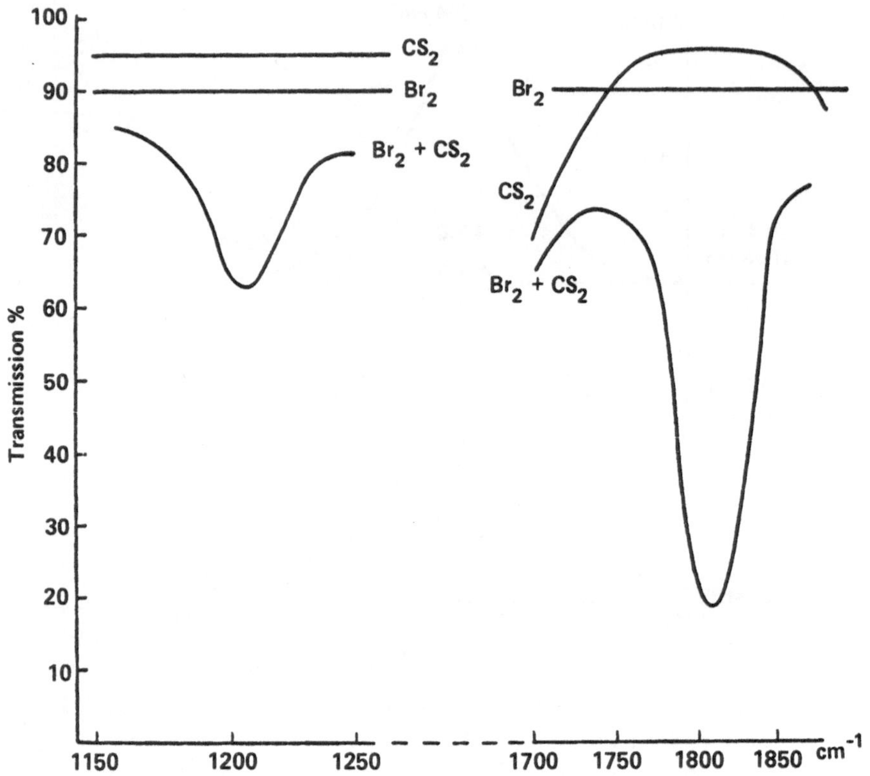

Figure 6. Transmission curves for liquid CS_2, Br_2 and a mixture with 0.54 mole fraction of Br_2; cell length, 5 mm (from (19)).

Simultaneous collision-induced transitions, which are determined by terms such as (2.12) in the expansion of the dipole moment of a pair of molecules, have been observed in numerous mixtures (3, 18-20). However, this kind of spectrum can only be observed in those cases where both components are relatively transparent in relatively thick layers.

Simultaneous induced transitions can also occur in single component systems and have been observed in gaseous and liquid CO_2 (21).

It is interesting to note that the cancellation effect in three body collisions is present for single but not for simultaneous transitions. Consequently, such transitions in the liquid must be due only to molecular pairs, a fact which presents an opportunity to study bimolecular interactions in liquid mixtures, uncomplicated by three-body interactions.

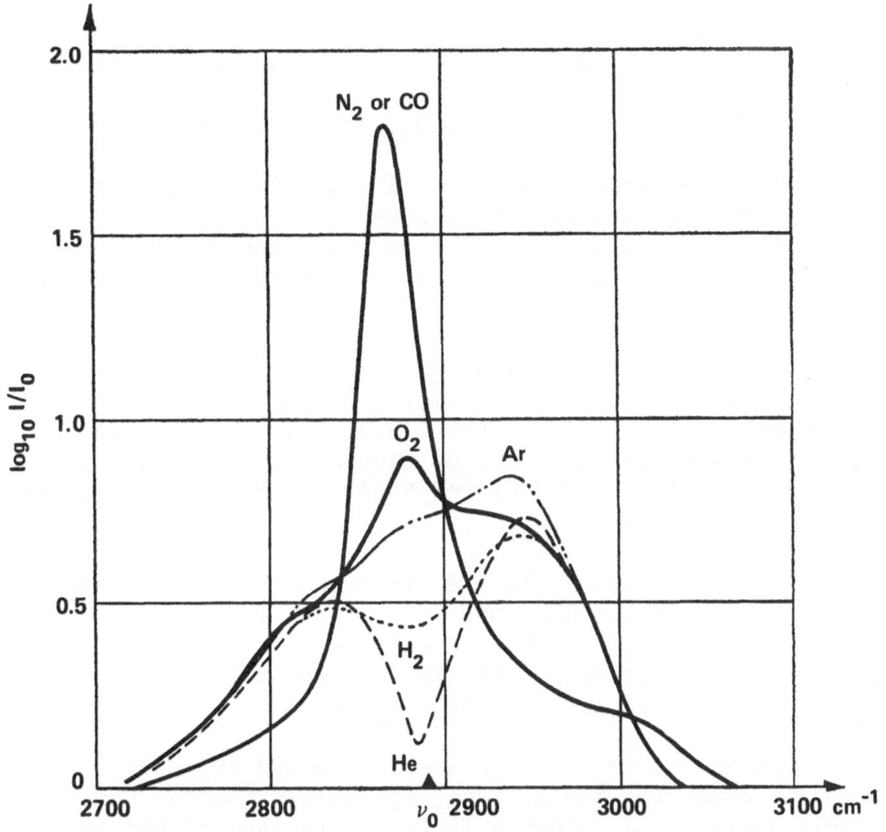

Figure 7. Fundamental band of HCl perturbed by He, H_2, Ar, O_2, N_2 and CO in the gas phase. The band profiles obtained with N_2 and CO are almost identical and for simplicity have been represented by a single profile. Densities: HCl about 5 amagat; perturbers about 400 amagat. Cell length = 50 mm, resolution = 5 cm^{-1}, T = 80°C (from (2)).

3 VIBRATIONAL ABSORPTION IN POLAR MOLECULES

Among the most intensively studied vibrational spectra of polar molecules are those of the hydrogen halides and in mixtures with a wide variety of perturbers (2). Shown in Fig. 7 is the variation of the gaseous band shape of HCl with different perturbers at a density of about 400 amagat. Whereas N_2 or CO produces an intense Q-branch, which is forbidden in the individual

HCl Molecule, H_2 and He are relatively ineffective in producing this feature. The rotational structure of HCl is completely obliterated at this density, although the P and R band shape is clearly delineated for HCl-He. That these results may be attributed to collision induction is in agreement with the observation that $\int \alpha(\nu)d\nu/\rho(HCl)$ shows no variation with density of He but does increase with increasing density of H_2, Ar and N_2 (2).

The peak of the induced Q-branch is seen from Fig. 7 to be shifted to lower frequencies. Frequency shifts of vibrational bands in compressed fluids are well-known and are due largely to a change in the potential energy of the vibrating molecule in the fluid (22).

The spectrum of HCl dissolved in liquid xenon (23) is shown in Fig. 8 and its shape is quite similar to that of gaseous HCl in argon at nearly the same temperature (Fig. 7). The band in Fig. 8 is attributed to the superposition of the induced transitions, O ($\Delta J = -2$), Q ($\Delta J = 0$) and S ($\Delta J = +2$) with the allowed transitions P ($\Delta J = -1$) and R ($\Delta J = +1$).

Mechanisms other than collision induction have been suggested for the appearance of the Q-branch in HCl. At low temperatures and pressures, the idea of orbiting pairs has been advanced (2). The influence of hindering potentials for free rotation of the dissolved molecule (24) has also been proposed.

It is convenient to distinguish between two types of induced absorption in condensed media. One type arises because the radiation field acting on a molecule is due to the sum of the incident field and the dipole fields of all the other molecules induced by the incident field. The measured absorption coefficient, $\alpha(\nu)$, is then related to $\alpha_o(\nu)$, the value for the dilute gas, by

$$\alpha(\nu) = \alpha_o(\nu)(n_o{}^2 + 2)^2/9n_o \qquad (3.1)$$

where n_o is the frequency independent contribution to the refractive index of the fluid coming from the electronic resonances (9).

For HCl dissolved in liquid xenon, an intensity enhancement by a factor 2.8 compared to the dilute gas was obtained, whereas the enchancement due to (3.1) is only 1.23 (25). After taking this effect into account, however, a reaction field model for induction due to the vibrating HCl dipole gave an enhancement factor in good agreement with experiment (25).*

*Note that this differs from the quadrupole-induced dipole model presented in (23).

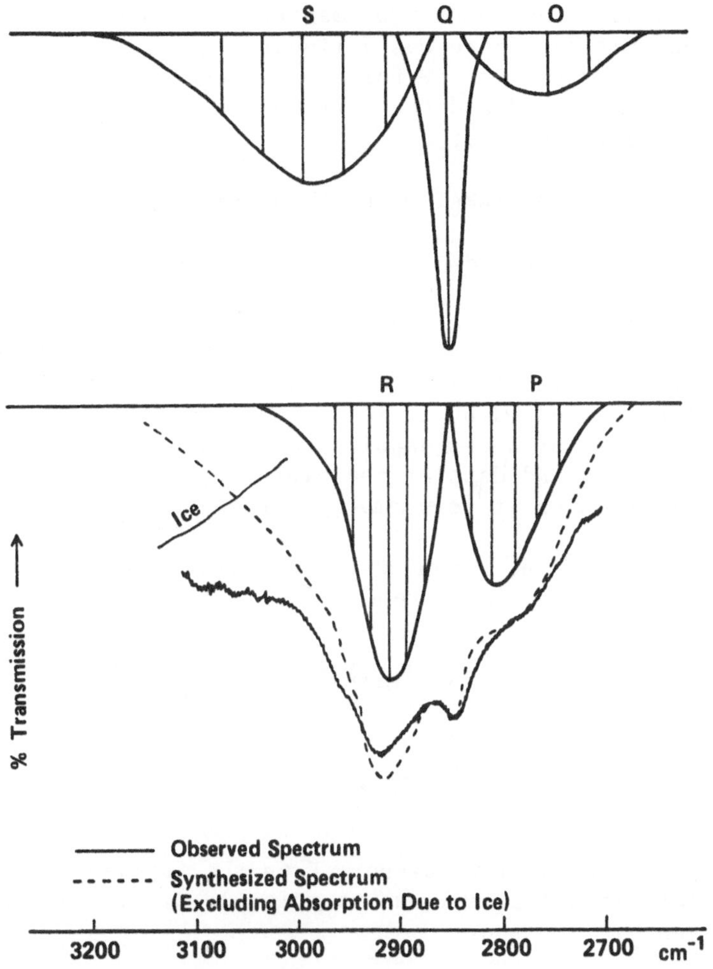

Figure 8. The fundamental band of HCl dissolved in liquid xenon at T = 185K. The synthesized spectrum is a superposition of O, P, Q, R, and S branches. The relative intensities of the P and R branches are those for the infrared spectrum of HCl vapor at T = 185K. The relative intensities of the S, Q, and O branches are those for the Raman spectrum of the vapor at the same temperature. The relative intensity of the Q branch with respect to the R and P branches was chosen so as to best fit the observed spectrum. The intensities are somewhat approximate, since broadening of the branches and intensity perturbations by the solvent were not taken into account (from (23)).

We finally consider the polar molecule NO (26), whose fundamental band was studied at room temperature and in mixtures with He and Ar at densitites up to 500 amagat. Although no enhancement was found for He, there was a linear increase in the integrated absorption with density for Ar. However, the enhancement given by (3.1) dominated and the part due to true induced absorption could not be readily estimated.

4 COLLISION-INDUCED VIBRATIONAL RAMAN SPECTRA

4.1 Theoretical Background

We have mentioned in Section 3 that the absorption coefficient is enhanced in dense media because the radiation field acting on a given molecule is the sum of the incident field and the dipole fields induced by it in neighboring molecules. This enhanced field also produces depolarized light scattering in spherical molecules and depolarized vibrational Raman scattering in molecules for which it is normally forbidden (5).

The induced dipole, expanded to the first order in vibrational coordinates, may be represented by

$$\underline{\mu} = (\underline{\underline{\alpha}}_1 + \frac{\partial \underline{\underline{\alpha}}_1}{\partial \xi_1} \xi_1)(\underline{E} + \underline{F}_1) + (\underline{\underline{\alpha}}_2 + \frac{\partial \underline{\underline{\alpha}}_2}{\partial \xi_2} \xi_2)(\underline{E} + \underline{F}_2)$$

$$+ \underline{\underline{\alpha}}_1 \frac{\partial \underline{F}_1}{\partial \xi_2} \xi_2 + \underline{\underline{\alpha}}_2 \frac{\partial \underline{F}_2}{\partial \xi_1} \xi_1 \qquad (4.1)$$

Here $\underline{\underline{\alpha}}$ is the polarizability tensor, $\underline{E}$ is the applied field and $\underline{F}_1$ is the field at molecule 1 due to molecule 2. The molecular field is given by

$$\underline{F}_1 = \underline{\mu}_2 \, \underline{\underline{T}}_{12} \qquad (4.2)$$

where $\underline{\mu}_2 = \underline{E}\underline{\underline{\alpha}}_2$ is the dipole induced in molecule 2 by the radiation field and $\underline{\underline{T}}_{12} = -\nabla\nabla(R_{12})^{-1}$ is a tensor which describes the dipole-dipole coupling between molecules 1 and 2. The terms in (4.1) which do not depend on the nuclear coordinates describe Rayleigh scattering and those which do describe vibrational Raman scattering. The terms responsible for collision-induced scattering contain the molecular field, which in general is not parallel to the applied field. In the next section, we shall encounter other mechanisms which can produce collision-induced Raman spectra.

4.2 Experimental Results

Depolarized Raman scattering has been observed at the symmetric vibration frequency ν_1 (773.5 cm^{-1}) in gaseous and liquid SF_6 at 300K (27). Since in this case $(\partial\alpha/\partial\xi)$ is isotropic, the depolarized scattering must arise from molecular interaction, namely,

$$\underline{E}\alpha_1(\partial\alpha_2/\partial\xi_2)\xi_2\underline{\underline{T}}_{21} + \underline{E}\alpha_2(\partial\alpha_1/\partial\xi_1)\xi_1\underline{\underline{T}}_{12}$$

It was confirmed that the observed effect is collision-induced since measurements in the gaseous phase showed that the depolarization ratio varied linearly with density. This ratio in the liquid is less by a factor of 5 than that extrapolated from the low density gas. Clearly the dipole-induced dipole (DID) pair approximation is no longer valid at liquid densities where the higher symmetry of interaction with several nearest neighbors of a molecule has to be considered.

Raman scattering in SF_6 at the forbidden ν_3 (940 cm^{-1}) and ν_6 (344 cm^{-1}) transitions has been observed by recording the sum of the I_{VV} and I_{VH} components of the scattered light (28). Since for these bands $\partial\alpha/\partial\xi = 0$, the DID mechanism cannot be responsible for these observations. A possible mechanism may be due to the fact that in the presence of a strong electric field, $\underline{F}_1$, due to the multipoles of a neighboring molecule, the polarizability becomes $\underline{\underline{\alpha}}_1 + \underline{\underline{\beta}}_1\underline{F}_1$, where $\underline{\underline{\beta}}_1$ is the hyperpolarizability tensor and $\partial\underline{\underline{\beta}}/\partial\xi \neq 0$ (29, 30). However, another mechanism which should be considered arises from the induced dipole terms $\underline{\underline{A}}_1\underline{F}_1' + \underline{\underline{A}}_2\underline{F}_2'$ where $\underline{F}_1'$ is the field gradient at molecule 1 due to molecule 2 and $\underline{\underline{A}}$ is the dipole-quadrupole polarizability tensor (31).

Raman scattering from the normally forbidden ν_2 and ν_3 modes in liquid CS_2 (Fig. 9) and liquid CO_2 (32) have been observed. Plots of the logarithm of the high frequency side of the scattered intensity of CS_2 in Fig. 10 demonstrate the exponential shape of the polarized ν_2 and ν_3 bands. Exponential band shapes are frequently observed in collision-induced spectra (4). As shown in Fig. 10, the far-infrared spectrum of liquid CS_2 has practically the same shape, suggesting the possibility of similar dynamical processes.

It happens that Raman scattering at the inactive vibrations, which is completely negligible in the dilute gas in CO_2, is relatively strong for the inactive ν_3 vibration in gaseous SF_6. This could be due to several sources of residual scattering in the case of SF_6 (28): (a) single-molecule hyperpolarizability terms; (b) the presence of allowed hot bands, originating from molecules in excited vibrational states with lower symmetry and (c) the presence of accidental combination bands. These are single-molecule

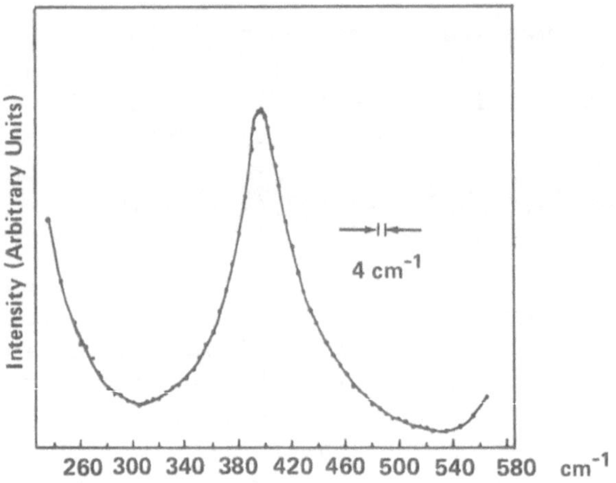

Figure 9. The collision-induced polarized Raman spectrum of ν_2 in liquid CS_2 taken with 4 cm^{-1} slits; T = -85°C (from (30)).

mechanisms which contribute an intensity proportional to the density and may be separated from the induced spectrum by a careful study of the density dependence. The presence of residual linearly varying absorption should also be considered in connection with induced absorption spectra.

5 PROBLEM OF SUBTRACTING COLLISION-INDUCED FROM ALLOWED BANDS

Thus far we have been concerned with vibrational transitions which are primarily cossision-induced. However, such transitions must be considered in investigating allowed transistions in order to subtract the complicating effects of the fromer. A method for doing this is based on the idea that the band shape for a free classical rotator is gaussian and decays at high frequencies much more rapidly than an underlying collision-induced band whose high-frequency wings decay exponentially (34). To subtract the contribution of the collision-induced band, it is convenient to work with the normalized correlation function, C(t), which is proportional to the Fourier transform of the spectral intensity, I(ω). Accordingly, C(t), is written as

$$C(t) = f_r C_r(t) + f_c C_c(t), \tag{5.1}$$

where f_r is the fraction of the intensity due to the rotational spectrum and $f_c = 1 - f_r$. Then the non-vibrational part of the second moment, M(2), becomes

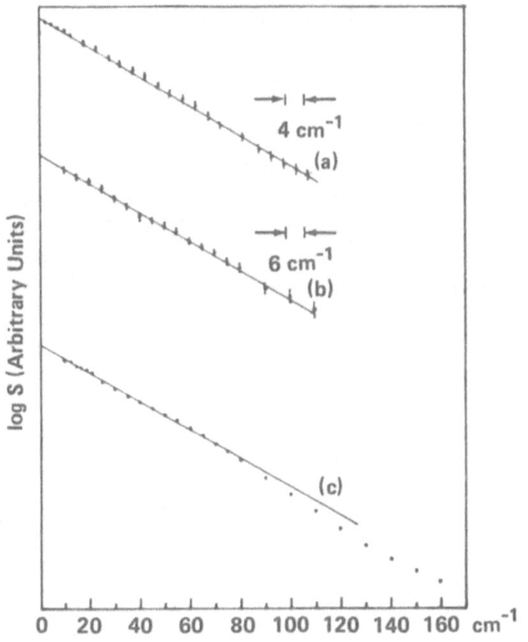

Figure 10. Plots of log S versus frequency for: (a) the high frequency side of the polarized Raman spectrum of ν_2 in liquid CS_2 taken with 4 cm^{-1} slits; (b) the high frequency side of the Raman spectrum of ν_3 in liquid CS_2 taken with 6 cm^{-1} slits; (c) far infrared data taken from (33) (from (30)).

$$M(2) = (1 - f_r)M_c(2) + f_r M_r(2) \qquad (5.2)$$

where the experimental value is defined by

$$M(2) = \int_{-\infty}^{+\infty} \omega^2 I(\omega)d\omega / \int_{-\infty}^{+\infty} I(\omega)d\omega \qquad (5.3)$$

and the angular frequency ω is measured from the vibrational frequency. For classical rotors, $M_r(2)$ is computed from (1)

$$M_r(2) = \ell(\ell + 1)(kT/I) \qquad (5.4)$$

where $\ell = 1$ for infrared bands, $\ell = 2$ for Raman bands and I is the moment of inertia. A cross term, $M_r(2)M_c(2)$, which is not necessarily small, is neglected in this analysis (35).

To proceed further, a line shape or correlation function must be assumed for the collisional contribution, a step which makes this procedure quite uncertain until the character of collision-induced spectra is better understood. With this assumption, however, $M_c(2)$ can be computed and (5.2) solved for f_r. It is then possible to find the experimental $C_r(t)$ from (5.1) and make comparisons with theoretical models for the molecular rotational motion in dense fluids. Despite the uncertainty of this procedure, it does give a reasonable explanation for the deviation of experimental second moments from the predictions of (5.4) for infrared and Raman bands. However, to obtain reliable moments, the baseline of the experimental spectrum must be well established. If this is not the case, the second moment is often used to establish the baseline, in which case the above analysis cannot be used.

6 CONCLUSION

A wide variety of collision-induced infrared and Raman spectra have been discussed here. Except for some scattered results, the emphasis has been largely in the characterization of the spectra and in indentifying the induction mechanisms. The inverse problem of using such spectra as a probe for studying the liquid state has hardly begun.

ACKNOWLEDGMENT

The author thanks Drs. A. Abramowitz and A. Weber for helpful suggestions in the preparation of this paper.

REFERENCES

(1) J.H.R. Clarke, Adv. Infrared Raman Spectrosc. 4, 109 (1978).
(2) B. Vodar and H. Vu, J. Quant. Spectrosc. Radiat. Transfer 3, 397 (1963).
(3) M.O. Bulanin, J. Mol. Spectrosc. 19, 59 (1973).
(4) G. Birnbaum, in Intermolecular Spectroscopy and Dynamical Properties of Dense Systems, J. Van Kranendonk, ed., E. Fermi International School of Physics, Italian Physical Society (1980).
(5) G.C. Tabisz, Specialist Periodical Reports, British Chem. Soc. 29 (1979).
(6) H.L. Welsh, in International Reviews of Science, Physical Chemistry, Vol. 3: Spectroscopy, Butterworths, London, p. 33 (1972).
(7) J. Van Kranendonk, Physica 73, 156 (1974).
(8) S.P. Reddy and C.W. Cho, Can. J. Phys. 43, 2331 (1965).
(9) J. Van Kranendonk, Physica 23, 825 (1957).
(10) A.L. Smith, W.E. Keller and H.L. Johnston, Phys. Rev. 79, 728 (1950).
(11) The discussion in this section follows the treatment given by J. Fahrenfort, Thesis, University of Amsterdam (1955).

(12) B.S. Frost, J. Chem. Soc., Faraday Trans. II, 69, 1142 (1973).
(13) M.M. Shapiro and H.P. Gush, Can. J. Phys. 44, 949 (1966).
(14) De T. Sheng and G.E. Ewing, J. Chem. Phys. 55, 5425 (1971).
(15) A.L. Smith and H. L. Johnston, J. Chem. Phys. 20, 1972 (1952).
(16) C.A. Long and G.E. Ewing, J. Chem. Phys. 58, 4824 (1973).
(17) L. Saighi and D. Robert, Spectrochim. Acta 26A, 1731 (1970).
(18) J.A.A. Ketelaar and F.N. Hooge, J. Chem. Phys. 23, 1549 (1955).
(19) J.A.A. Ketelaar and F.N. Hooge, J. Chem. Phys. 23, 749 (1955).
(20) A theory of simultaneous vibrations has been presented by F.N. Hooge and J.A.A. Ketelaar, Physica 23, 423 (1957).
(21) K. Ozawa, Rev. of Phys. Chem. (Japan) 29, 1 (1959).
(22) A.D. Buckingham, Proc. Roy. Soc. A248, 169 (1958).
(23) J. Kwok and G.W. Robinson, J. Chem. Phys. 36, 3137 (1962).
(24) L. Galatry, Spectrochim. Acta 10, 849 (1959).
(25) J.Chesnoy, D. Richard and C. Flytzanis, Chem. Phys. 42, 337 (1979).
(26) G. Chandraiah and C.W. Cho, J. Mol. Spectrosc. 47, 134 (1973).
(27) W. Holzer and Y. le Duff, Phys. Rev. Lett. 32, 205 (1974).
(28) W. Holzer and R. Ouillon, Chem. Phys. Letters 24, 589 (1974).
(29) R. Samson and A. Ben-Reuven, J. Chem. Phys. 65, 3586 (1976).
(30) T.I. Cox and P.A. Madden, Chem. Phys. Lett. 41, 188 (1976).
(31) A.D. Buckingham, private communication.
(32) W. Holzer and R. Ouillon, Mol. Phys. 36, 817 (1978).
(33) G.J. Davies and J. Chamberlain, J. Chem. Soc. Faraday Trans. II 69, (1973).
(34) P. van Konynenburg and W.A. Steele, J. Chem. Phys. 56, 4776 (1972).
(35) D. Kivelson, J.P. McTague and T. Keyes, J. Chem. Phys. 55, 4096 (1971).

PICOSECOND VIBRATIONAL SPECTROSCOPY

A. Laubereau

Physikalisches Institut der Universität Bayreuth

Bayreuth, W. Germany

1 INTRODUCTION

The advance in laser technology and in understanding of nonlinear optical phenomena has opened up the possibility of generating intense picosecond light pulses in the spectral region extending from the near uv to the infrared. (1,2) These pulses allow direct studies of ultrafast molecular processes with an exceptional time resolution of a fraction of a picosecond. The present chapter discusses applications to the study of vibrational dynamics in liquids. Several recent experiments will be presented to illustrate the state of the art and the potential of our new techniques.

For several decades conventional infrared and Raman spectroscopy have been the main source of information on vibrational dynamics. The analysis of the spectroscopic data faces several important difficulties. The finite instrumental resolution and the problem of overlapping neighboring bands is briefly mentioned here; contributions of overtones and hot bands of other vibrational modes - although not accurately known - have to be taken into account in many practical cases; they represent a serious source of experimental uncertainty. Once a band contour is determined the question arises about its interpretation. In general, many physical processes may contribute to an observed band shape (3). Rotational motion and various kinds of vibrational dephasing and of energy redistribution processes affect a vibrational band. The different contributions to the band contour cannot be separated unambiguously in conventional spectroscopy.

Some of these difficulties are overcome using time resolved techniques. Fig. 1 outlines different methods and various results obtained (4). The excitation of the molecules is achieved by an intense laser pulse via stimulated Raman scattering or by resonant absorption of an infrared pulse. After the passage of the first pulse the excitation process rapidly terminates and free relaxation of the excited mode occurs. A second weak pulse properly delayed with respect to the first one monitors the instantaneous state of the excited system. Three probing processes were used to obtain the dynamical information listed on the right side of Figure 1.

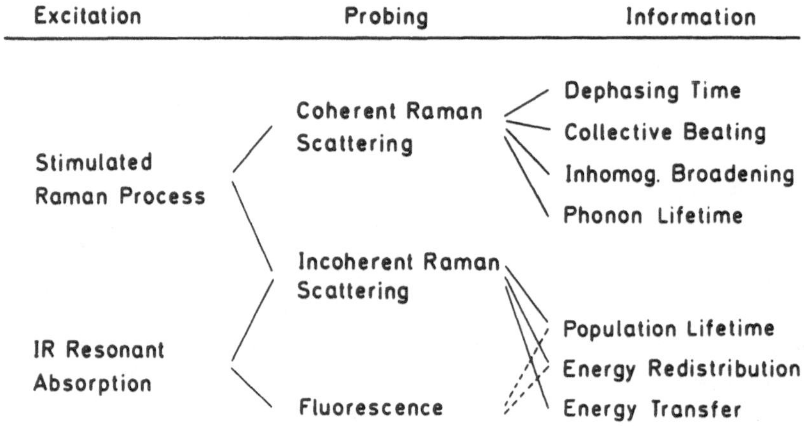

Fig. 1. Physical processes for the excitation and probing of molecular vibrations in liquids. The information is listed on the right.

Comparing spectroscopic and time-resolved measurements, the different physical situations should be noted. The former experiments study the molecular system close to thermal equilibrium while specific non-equilibrium situations are prepared in the latter investigations. In this way new experimental information is obtained which is based on the correct understanding of the excitation and probing processes.

2 EXCITATION PROCESS

Some remarks on transient stimulated Raman scattering are given first. It is convenient to treat the vibrational system as a two-level model; transitions involving higher excited states are negligible on account of the anharmonic frequency shift. In this picture the coherent interaction of a vibrational mode with a non-resonant light pulse via stimulated Raman scattering is quite analoguous to the well-known case of magnetic dipole transitions of a spin system. For our purposes it is advantageous to consider the more general case of an ensemble of two-level systems with a distribution of vibrational transition frequencies. Molecules in a small frequency interval of this distribution are grouped together to vibrational components j with frequency ω_j and number density Nf_j. N denotes the total number density with $\Sigma f_j = 1$. Introducing the set of frequencies ω_j includes the case of isotopic line splitting and inhomogeneous line broadening in the present calculation.

Treating the light field classically, the Hamiltonian if the subensemble j interacting with the light field has the form:

$$H = H^o - \frac{1}{2} q_j \frac{\partial\alpha}{\partial q} E^2 \tag{2.1}$$

q_j denotes the normal mode operator. $\partial\alpha/\partial q$ represents the change of polarizability with vibrational amplitude. We note that the molecular orientation does not explicitly enter Eq. 2.1 when $\partial\alpha/\partial q$ is a scalar. This approximation holds for the experiments to be discussed here. A careful analysis including the anisotropy of the Raman scattering tensor shows that the molecular rotation has a negligible effect on the excitation and probing processes, even for a moderately large depolarization factor $\rho_s \lesssim 0.2$ (4). This point is an important advantage as compared to conventional Raman spectroscopy where deconvolution procedures are required (5). The vibrational excitation of the molecules is described by the expectation values of the normal mode operators, $\langle q_j \rangle$, and the excited state population, n_j. For the two level model mentioned above the equations of motions have the form (4,6,7)

$$\left(\frac{\partial^2}{\partial t^2} + \frac{2}{T_2}\frac{\partial}{\partial t} + \omega_j^{\,2}\right)\langle q\rangle = \frac{1}{2m}\frac{\partial\alpha}{\partial q} E^2 \left[1-2\,n_j\right] \tag{2.2}$$

$$\left(\frac{\partial}{\partial t} + \frac{1}{T_1}\right)(n_j-\bar{n}) = \frac{1}{2\hbar\,\omega_o}\frac{\partial\alpha}{\partial q} E^2 \frac{\partial}{\partial t}\langle q_j\rangle \;. \tag{2.3}$$

$\bar{n}$ is the thermal equilibrium population of the upper levels. The time constants T_1 and T_2, the reduced mass m and the coupling parameter $\partial\alpha/\partial q$ are assumed to be equal for all subensembles j. We note that $\langle q_j\rangle$ follows a differential equation of form similar to a classical oscillator.

Of special interest are the two relaxation times T_1 and T_2 introduced in Eqs. (2.2) and (2.3). T_2 denotes the dephasing time of the vibrational amplitude of a molecular group with transition frequency ω_j. T_1 represents the population lifetime of the first excited vibrational state (4).

Eqs. (2.2) and (2.3) predict exponential decay of $\langle q_j\rangle$ or n_j after the excitation process has terminated. This time dependence does not hold for times shorter than τ_c, where $\tau_c \sim 10^{-13}$ s is a correlation time connected to intermolecular interactions (e.g., rapid translational motion in the liquid). This limitation is not critical in our experiments since experimental time resolution is > τ_c and we observe time constants T_1 and $T_2 >> \tau_c$. The exponential time constants are well supported by experimental results (4).

The physical processes which determine the time constant T_1 and T_2 are different. In fact, in several experiments where T_1 and T_2 were determined for the same mode and molecule, a considerable difference was observed. For instance, a ratio of $T_1/T_2 = 10^{12}$ was found for the fundamental vibrational mode of liquid N_2 (8-10), and T_1/T_2 = 2.6 was observed for a CH_3-stretching mode of CH_3CCl_3 at 300 K (11).

We point to the analogy to a two-level spin system which is governed by the well-known Bloch equations. For optical dipole transitions a pseudo-spin vector has been introduced (12) describing the instantaneous state of the two-level system. The z-component of this vector represents the population of the system, while the x and y components account for the induced dipole moment. An analogous vector may be defined for our Raman transition. The coherent amplitude $\langle q \rangle$ is proportional to the induced dipole moment. For this model the relaxation of $\langle q \rangle$ is termed the free induction decay.

Eqs. (2.2) and (2.3) are supplemented by the classical wave equation for the light field:

$$\Delta E - \frac{1}{c^2}\frac{\partial^2}{\partial t^2}(\mu E) = \frac{4\pi}{c^2}\frac{\partial^2}{\partial t^2} P^{NL} \qquad (2.4)$$

μ denotes the refractive index of the medium; the nonlinear polarization P^{NL} represents the stimulated light scattering off the vibrational mode of interest. The coherent amplitude $\langle q_j \rangle$ corresponds to the induced dipole moment $\partial\alpha/\partial q \langle q_j \rangle E$ of the individual molecule. Summing over the molecular subensembles with number density $f_j N$ yields

$$P^{NL} = N \frac{\partial\alpha}{\partial q} E \sum_j f_j \langle q_j \rangle \qquad (2.5)$$

Eqs. (2.2) to (2.5) represent a complete set describing the generation of an intense Stokes pulse and of a coherent material excitation by stimulated scattering.

We have made a detailed study of Eqs. 2 to 5 using material parameters relevant for the experiments below. Some of the findings are briefly summarized:

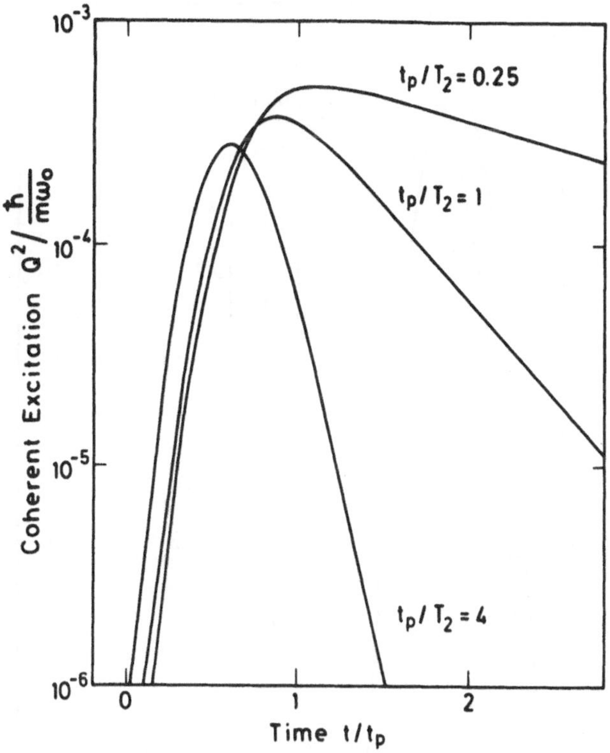

Fig. 2. Coherent vibrational excitation in units of $\hbar/m\omega_0$ versus time for three values of the parameter t_p/T_2 (duration of pump pulse t_p to dephasing time T_2). Note the exponential decay of the freely relaxing system with time constant $T_2/2$ (Stokes conversion of the excitation process 1 %, interaction length 1 cm).

The excitation and subsequent free relaxation of the molecules is illustrated by numerical data depicted in Figs. 2 and 3 for two different physical situations (4). Fig. 2 considers the simple case of a homogeneous line. The exponential decay of $|\langle q_j \rangle|^2$ with time constant $T_2/2$ should be noted.

It is interesting to relate the exponential relaxation with time T_2 to the line shape observed in conventional Raman spectroscopy. For the homogeneous system the exponential decay corresponds to a Lorentzian line with full width (FWHM in units of cm^{-1})

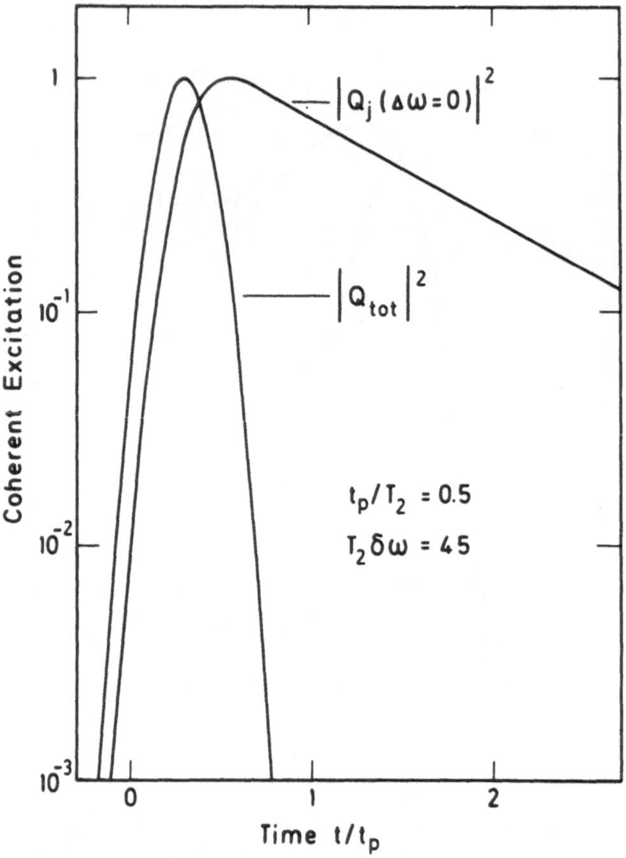

Fig. 3. Calculated coherent excitation of vibrational system with a Gaussian distribution of transition frequencies (bandwidth $\delta\omega$). The solid curve represents the total system. The broken line shows subensemble with negligible spread of transition frequencies, decaying exponentially with time constant $T_2/2$.

Quite different is the result shown in Fig. 3 for a quasi-continuous distribution of transition frequencies of width $\delta\omega\ T_2 = 45$. The total excitation $|Q_{tot}|^2 = |\Sigma f_j\langle q_j\rangle|^2$ disappears very rapidly on account of the broad frequency distribution. The individual components, on the other hand, decay more slowly with time constant $T_2/2$ (broken curve in Fig. 3). The calculations indicate that the Stokes emission builds up rapidly during the stimulated scattering. This amplification process may start from a very low level, e.g. quantum noise, and may extend over many orders of ten. The Stokes emission is highly coherent; i.e. a Stokes field of constant phase is produced. This point is important for the generation of well defined k-vectors of the material.

3 COHERENT RAMAN SCATTERING OF PICOSECOND PULSES

The coherent excitation is monitored by coherent Raman scattering of a delayed probe pulse. The amplitudes $\langle q_j \rangle$ give rise to a macroscopic polarization $P = N \partial\alpha/\partial q \; E \; \Sigma f_j \langle q_j \rangle$, where E denotes the electromagnetic field of the probe pulse (4,13). The polarization generates the scattering emission, which is shifted by the vibrational frequency to larger (anti-Stokes) or smaller (Stokes) frequencies. The molecules vibrate with a defined spatial phase relation described by a wave vector $k_o = k_L - k_S$, prepared in the excitation process by the laser and stimulated Stokes pulse. The subensembles j behave like oscillating three-dimensional phase gratings. Scattering-off these gratings occurs when the k-matching condition for the wave vectors of the incident probe pulse (k_{L2}), the scattered light and the vibrational components is fulfilled (4). For Stokes scattering we have the condition:

$$|\Delta k_j| \Delta \ell = |k_{L2} - k_{S_j} - k_o| \Delta \ell \leq 0.5 \qquad (3.1)$$

k_{Sj} denotes the wave vector of the Stokes component generated by subgroup j; $\Delta\ell$ is the effective interaction length. A similar condition holds for anti-Stokes scattering of the probe pulse. Due to dispersion effects in condensed matter Δk_j in Eq. (3.1) is different for different subgroups j. Those molecules with approximate k-matching, $\Delta k_j \Delta \ell \leq 0.5$, determine the coherent scattering signal. Components with large mismatch $\Delta k_j \Delta \ell >> 0.5$ give negligible contributions, only. The frequency interval of allowed transition frequencies depends on the interaction length $\Delta\ell$. We wish to distinguish two limiting cases (13, 14):

(i) Selective k-matching. For large values of $\Delta\ell$ ($\gtrsim$ 1 cm) a highly selective k-vector geometry can be adjusted. In this way probe scattering of a molecular subensemble with negligible spread of transition frequencies may be observed. This subgroup is expected to decay with a characteristic dephasing time T_2.

(ii) Non-selective k-matching. For a small interaction length ($\Delta\ell <<$ 1 cm), the k-matching is not sensitive to the different transition frequencies. In this case, coherent superposition of various subgroups occurs and fast dephasing by the distribution of frequencies can be observed; i.e., the total excitation of Fig. 3 may be studied.

The different values of $\Delta\ell$ are experimentally adjusted by proper choice of the scattering geometry.

Experimentally we detect scattering signals of several 10^{-12} s which is several orders of ten shorter than the time resolution of conventional photo-detectors. As a result, time integrated scattering signals $S^{coh}(t_D)$ are measured at the appropriate Stokes (or anti-Stokes) position versus time delay between excitation and probing pulse. For the simple case of homogeneous broadening $\omega_j = \omega_o$ and for a probe pulse of amplitude $E_{L2}(t-x/v)$ we find the coherent scattering signal at the end of the sample ($x = \ell$):

$$S^{coh}(t_D) \propto \frac{\kappa_1^2 \Delta\ell^2}{1+(2\Delta k\Delta\ell)^2} \left| \int_{-\infty}^{\infty} dt \, E_{L2}(\ell, t-t_D) \, \langle q \rangle \right|^2 \qquad (3.2)$$

Eq. (3.2) represents a convolution of the coherent excitation (Fig. 2) with the time dependence of the probe intensity.

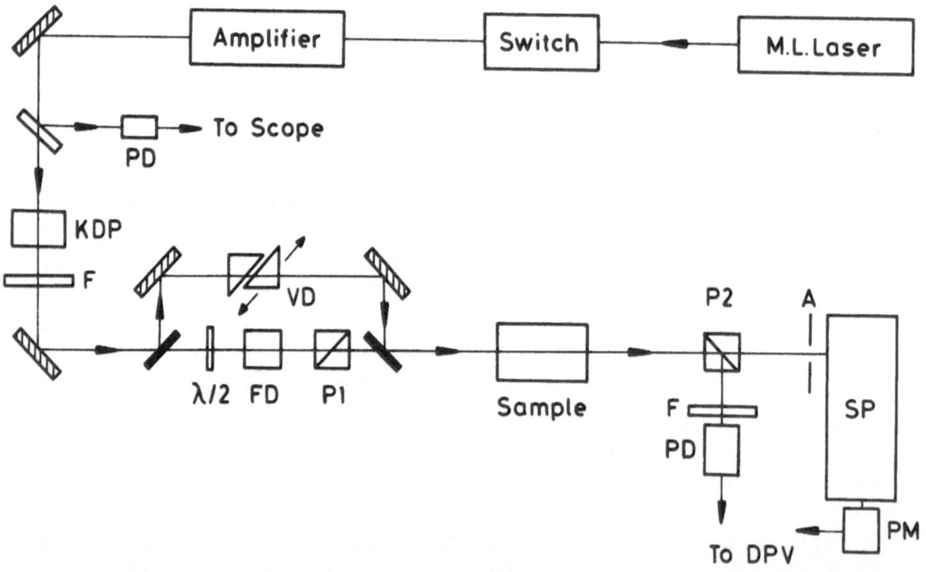

Fig. 4. Experimental set-up to measure time resolved coherent Stokes scattering in a collinear geometry. A beam splitter generates the probe pulse which is properly delayed before travelling collinearly with the exciting pulse. The scattered signals of the two pulses are separated by the two polarizers P1 and P2.

Our experimental system for coherent probe scattering with collinear wave vector geometry is depicted in Fig. 4 (14). For the generation of the ultrashort light pulses we use a Nd:glass laser system consisting of a mode-locked laser oscillator, an electrooptic switch to select a single pulse, and a laser amplifier. We emphasize the need to work with pulses of well defined duration, pulse shape and peak intensity. Bandwidth limited pulses (15) are required to achieve high k-vector resolution. After frequency doubling in a KDP crystal we work with pulses of frequency ν_L = 18,910 cm^{-1}, width 5 cm^{-1}, duration $t_p \simeq$ 3.5 ps and approximately Gaussian shape. The probing pulse is generated by a first beam splitter and delayed in a variable delay system. A second beam splitter directs the probe pulse collinearly with the exciting pulse through the medium. The intensity of the probe pulse is less than 1% of the excitation pulse. The pump pulse passed a $\lambda/2$ plate and polarizer P1 defining the plane of polarization of the excitation process. The stimulated Stokes emission of the pump pulse is effectively blocked by a factor of $\gtrsim 10^5$ with the help of the second polarizer P2. The probe pulse and the coherent Stokes signal pass P2 without significant attenuation. An aperture, A, determines the angle of acceptance γ of the detection system consisting of a spectrometer and a photomultiplier.

An experimental result for the simple case of homogeneous line broadening is shown in Fig. 5. The fundamental vibrational mode of liquid N_2 at 2326 cm^{-1} is investigated (8). From the exponential decay of the signal curve we deduce a dephasing time of T_2 = 150 ps, which corresponds to a dynamic line broadening of 0.070 cm^{-1}. This number is in excellent agreement with the linewidth of the isotropic scattering component observed in spontaneous Raman spectroscopy (16, 17); i.e. the dephasing time T_2 fully accounts for the spectroscopic line broadening.

It is interesting to note that an expectionally long population lifetime T_1 of the fundamental mode of approximately one minute was reported by several authors (9, 10). As a result, population decay does not effect the dephasing process of liquid N_2. Similar results were reported for mixtures of N_2 with liquid Ar. A comparatively small variation of T_2 with concentration and temperature was found (18) in good agreement with spectroscopic findings.

There are vibrational modes where evidence exists of a distribution of transition frequencies. For instance, liquids with strong hydrogen bonding show extended OH-bands which are believed to be inhomogeneously broadened. Spontaneous spectroscopic techniques do not allow a separation of rapid dephasing processes and longer-lived frequency shifts of liquid molecules. Our novel picosecond Raman technique allows for the first time to study this problem.

To demonstrate the potential of the technique we consider vibrational bands with discrete substructure due to isotopic multiplicity. Many liquids are composed of several isotopic species. $SnBr_4$ is discussed here as an example. The two isotopes ^{79}Br and ^{81}Br give rise to five lines of the totally symmetric vibration around 221 cm^{-1}. The isotopic structure is not resolved. The results for coherent probe scattering with a selective k-geometry are depicted in Fig. 6a. The decaying part of the signal curve represents dephasing of a single isotope component with a time constant of T_2 = 6 ps. This time constant corresponds to a line broadening of a single isotope component

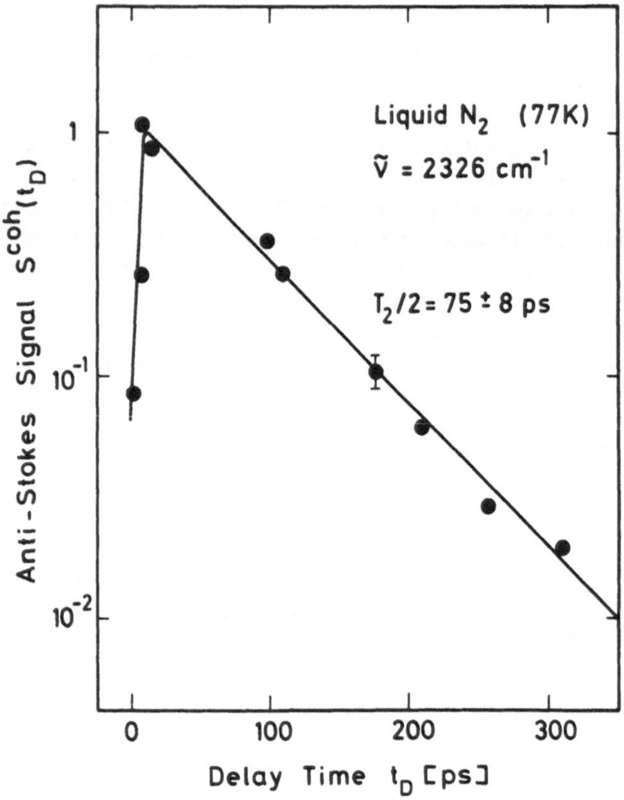

Fig. 5. Coherent probe scattering signal of liquid N_2 versus delay time t_D. The dephasing time of the fundamental mode at $\tilde{\nu}$ = 2326 cm^{-1} is found to be T_2 = 150 ps.

of 1.8 cm^{-1}, which is significantly smaller than the bandwidth of 3.2 cm^{-1} observed in conventional Raman spectroscopy.

Additional information is obtained by coherent probe scattering with non-selective k-matching. In these experiments a coherent superposition of the five isotopic species of $SnBr_4$ is observed. The experimental data are shown in Fig. 6 b. The molecules are first excited with approximately equal phases (maximum part of the signal curve). During the subsequent free relaxation, the different frequencies of the isotopic subgroups give an additional contribution to the loss of phase correlation and an accelerated decay is observed. The solid curve in Fig. 6 b is calculated using the frequency spacing of the isotopic lines as a fitting parameter. Good agreement of the theoretical curve with the data points is obtained for a value of $\Delta\omega$ = 0.7 cm^{-1} in accordance with theoretical estimates. Using the picosecond data of Fig. 6 and the known relative abundance of the $SnBr_4$ species, the total Raman band was evaluated as the sum of

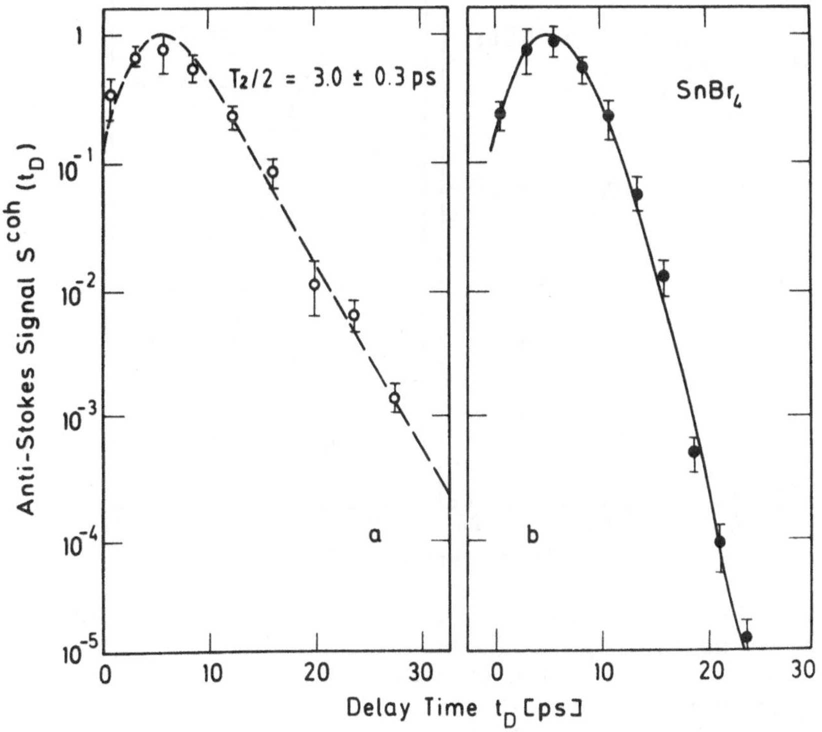

Fig. 6. (a) Coherent probe scattering signal versus delay time of one isotope component of $SnBr_4$; a selective k-matching geometry was applied. (b) Coherent probe scattering signal observed in a non-selective off-axis geometry. The accelerated decay should be noted which is the result of the coherent superposition of five isotopic subcomponents.

five isotope lines of width 1.8 cm^{-1} and spacing 0.7 cm^{-1}; this calculated band contour accounts well for the band shape observed by conventional spectroscopy.

The results clearly demonstrate that the selective k-matching technique allows to distinguish dephasing via rapid processes from the loss of phase correlation by a distribution of vibrational frequencies. Estimates show that the experimental k-resolution is equivalent to a frequency resolution of better than 0.5 cm^{-1}.

We have investigated several liquid systems with our selective k-matching technique. Fig. 7 represents data for a strongly associated liquid. The CH_2-stretching mode at 2935 cm^{-1} of ethylene glycol is studied (14). This mode displays are very broad band in the Raman spectrum of width $\sim$ 60 cm^{-1}.

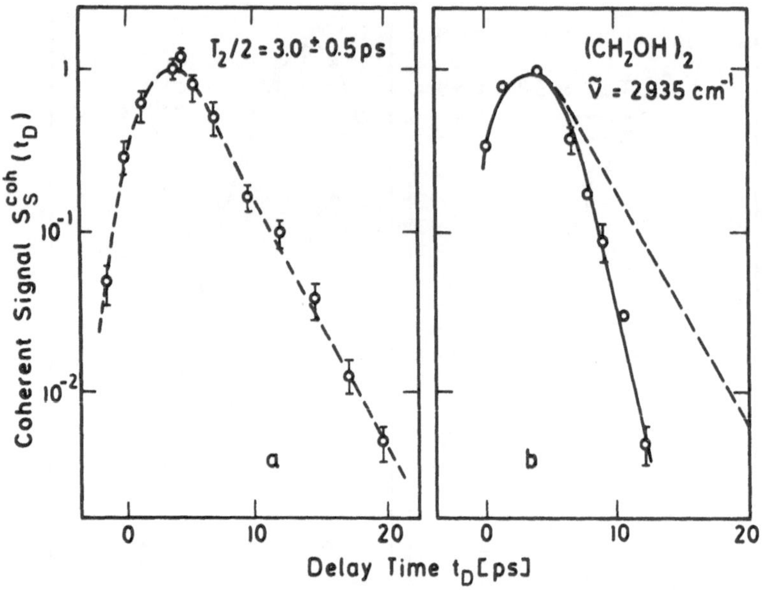

Fig. 7. a) Coherent Stokes probe scattering versus delay time measured in $(CH_2OH)_2$ for a highly selective k-matching geometry. A dephasing time of T_2 = 6 ps is obtained corresponding to a linewidth of 1.8 cm^{-1}. b) Coherent Stokes signal for a less selective k-matching geometry. The signal decays rapidly on account of destructive interference of molecules with different transition frequencies.

In Fig. 7 a, a highly selective k-matching situation is used with a sample length of 10 cm and with a small divergence of the Stokes beam of 3 mrad. The data indicate an exponential decay with a time constant of $T_2 = 6 \pm 1$ ps. The time dependence is different for less selective k-matching (Fig. 7 b) using a shorter cell of 1 cm and a larger Stokes divergence of 10 mrad. Under these experimental conditions the coherent scattering signal disappears rapidly.

The comparison of Fig. 7 a and b provides direct evidence that the Raman band of width 60 cm^{-1} represents a distribution of vibrational frequencies. Our value of T_2 suggests a homogeneous linewidth which is smaller than the Raman band by a factor of approximately 30. Similar results were also obtained for the CH-mode of CH_3OH at 2835 cm^{-1}.

As an example of a small inhomogeneity we discuss the totally symmetric ring vibration at 945 cm^{-1} of deuterated benzene C_6D_6 (19). This mode gives rise to a strong and narrow Raman line of width 1.5 cm^{-1} which has been measured repeatedly by conventional Raman equipment (20). We have investigated this mode with our picosecond technique and found interesting results. The data for coherent probe scattering with selective k-vector geometry are depicted in Fig. 8 a. Monitoring the coherent decay over a factor of 200 we see approximately exponential relaxation with a time constant of T_2 = 10 ps. The broken line is a theoretical curve. The small deviation from exponential decay results from the finite k-resolution of the experimental system with interaction length $\Delta\ell \simeq 0.4$ cm, taken into account in the calculation.

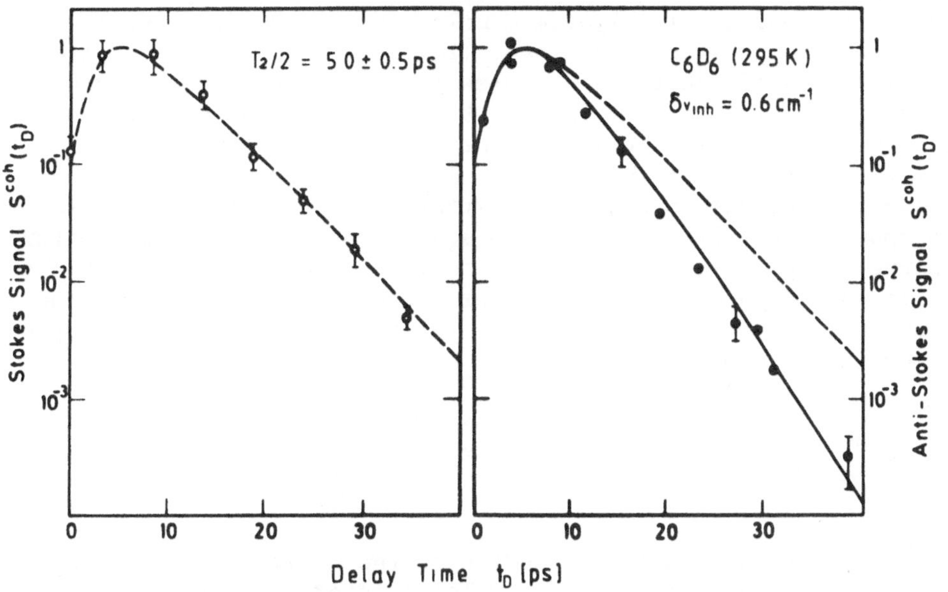

Fig. 8. The symmetric ring mode of C_6D_6 at 945 cm^{-1}. a) Coherent probe scattering versus delay time using a selective k-vector geometry. b) Coherent probe signal under non-selective k-matching conditions. Data suggest a distribution of vibrational frequencies. Curves are calculated for T_2 = 10 ps and $\delta\tilde{\nu}_{inh}$ = 0.6 cm^{-1} taking into account the different k-matching situations.

For non-selective k-matching, Fig. 8 b shows an accelerated decay indicating a distribution of vibrational frequencies. The solid line is a theoretical curve calculated for the value T_2 = 10 ps (from Fig. 8 a) and assuming a Gaussian distribution of width $\delta\tilde{\nu}_{inh}$. Fitting the curve to the data points yields $\delta\tilde{\nu}_{inh} \simeq 0.6$ cm^{-1}.

Additional support to our picosecond results is obtained by a comparison with spectroscopic data. Our value of T_2 = 10 ps corresponds to a Lorentzian line of width 1.0 cm^{-1}. Convolution of the Lorentzian with the (assumed) Gaussian frequency distribution of width 0.6 cm^{-1} leads to a Voigt profile which is fully consistent with the spontaneous Raman band of this mode.

4 INCOHERENT RAMAN SCATTERING OF PICOSECOND PULSES

Spontaneous anti-Stokes probe scattering allows the study of the instantaneous population of a vibrational level. With this technique it was possible, for the first time, to observe population lifetimes, energy transfer and energy redistribution (6,21-25). Time constants between 1 ps and 100 ps were measured for different dynamical processes in a number of polyatomic molecules. The experimental method is related to μs-studies of gases (26). It consists of two steps. First, a powerful short light pulse traverses the sample and excites the vibrational mode of interest. Stimulated Raman scattering or resonant infrared absorption can produce an excess population $n_i(x,t)$ of the excited vibrational state. A second weak pulse of different frequency probes the instantaneous vibrational excitation via spontaneous anti-Stokes Raman scattering. The scattered intensity observed under a large scattering angle is proportional to the instantaneous population of the upper vibrational state and to the incident laser intensity. Experimentally, we observe a time-integrated signal denoted by S_i^{inc}. We have

$$S_i^{inc}(t_D) = \frac{c\mu_A}{8\pi} N \left(\frac{d\sigma}{d\Omega}\right)_i \Delta\Omega \int_{-\Delta x/2}^{+\Delta x/2} dx \int_{\infty}^{\infty} dt\, n_i(x,t) \left|E_{L2}(t-t_D)\right|^2 \qquad (4.1)$$

x denotes the propagation direction of the excitation and probe pulse. Δx represents the observed local interval in the sample. N is the number density of the molecules in the liquid. $(d\sigma/d\Omega)_i$ corresponds to the spontaneous Raman cross section of the observed excited mode i and ΔΩ to the solid angle of acceptance of the probing pulse. Equation (4.1) represents a convolution of the probe pulse of delay time t_D with the vibrational population n_i.

The experimental system for measuring the population lifetime T_1 is presented schematically in Fig. 9 (24). A powerful single picosecond pulse at 9455 cm^{-1} is generated with a mode-locked Nd:glass laser system. In one experiment, the laser pulse traverses the sample excitating the normal mode with largest Raman gain via stimulated Raman scattering. In a second experiment, the laser pulse passes through two $LiNbO_3$ crystals (shown in the Figure) producing the desired infrared pulse of tunable frequency (27). At a frequency of $\tilde{\nu}$ = 3000 cm^{-1}, the pulse duration is ∿3 ps and the pulse energy corresponds to 10^{15} quanta. The infrared pulse with properly adjusted frequency is focused into the sample cell; it is absorbed by the infrared-active molecular vibration and generates the desired excited-state population. A beam splitter in the input beam produces a weak pulse, which is converted to the second-harmonic frequency at 18,910 cm^{-1} in a KDP crystal and serves as a probe pulse with variable delay t_D. The spontaneous anti-Stokes Raman signal $S^{inc}(t_D)$ is observed at a scattering angle of 90° with a monochromator or a set of dielectric filters. With approximately 10^{15} photons in the probe pulse, one estimates for an occupation number $n \sim 10^{-3}$, an anti-Stokes signal of 10^3 photons in a solid angle $\Delta\Omega \simeq 0.2$ sr.

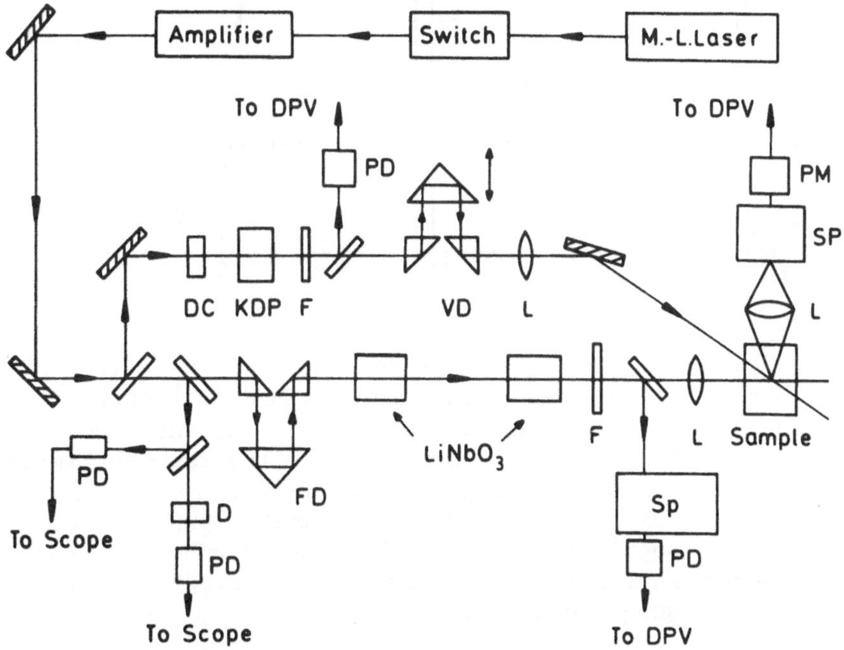

Fig. 9. Schematic of the experimental system to measure population lifetimes. A picosecond pulse of a Nd:glass laser system serves directly as excitation pulse or generates ultrashort infrared pulses in a single pass through two $LiNbO_3$ crystals. Tuning of the infrared frequencies is achieved by adjusting the crystal orientation. Part of the laser pulse is converted to the second harmonic frequency and interrogates the vibrationally excited molecules.

We have investigated the population lifetime and energy relaxation processes in a number of pure liquids and in liquid mixtures. Fig. 10 shows data of a diluted solution of CH_3I (24). The symmetric (ν_1) or asymmetric (ν_4) CH_3-stretching mode is excited (Figs. a and b, respectively), while probe scattering of the ν_1-mode is observed in the two measurements. The results indicate rapid decay of the population of the ν_1-mode with T_1 = 1 ps and energy redistribution

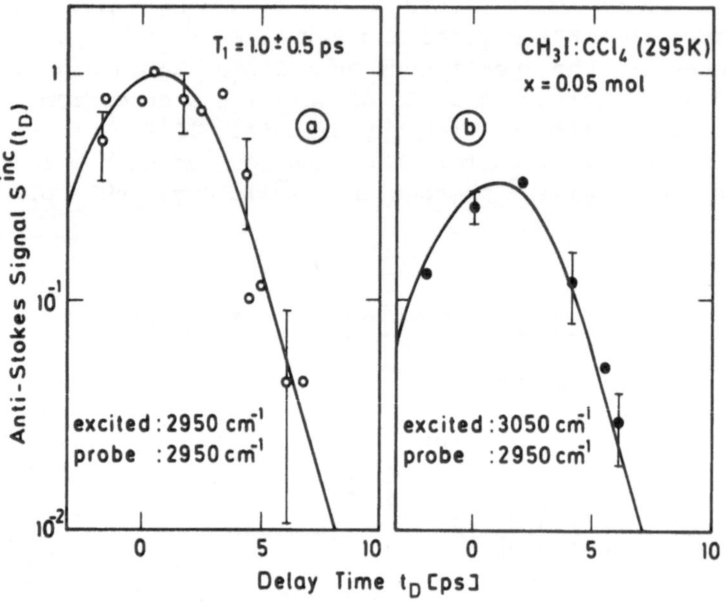

Fig. 10. Incoherent anti-Stokes probe scattering S^{inc} (t_D) of the symmetric CH_3-stretching mode at 2950 cm^{-1} of CH_3I dissolved in CCl_4, mole fraction 0.05. (a) The mode at 2950 cm^{-1} is directly populated by the resonant infrared excitation pulse. (b) The asymmetric CH_3-stretching mode at 3050 cm^{-1} is excited by the tunable pump pulse. The excess population of the vibration at 2950 cm^{-1} is observed indicating rapid energy redistribution between the neighboring modes.

between the two modes with a time constant of $\sim$ 1.5 ps. For an explanation of the fast vibrational depopulation we present in Fig. 11 the energies of the vibrational modes of CH_3I together with several overtones and combination modes around 3000 cm^{-1}. In this molecule rotational coupling, Fermi resonance, and Coriolis coupling are important. Theoretically, rapid equilibrium between the ν_1 and ν_4 modes and a fast rate ($k \sim 5\times10^{12}$ s) for the transition from the ν_1 mode to the overtone $2\nu_5$ is predicted. Rotational coupling is also responsible for the fast decay of the overtone $2\nu_5$ (39).

The importance of intermolecular energy transfer is demonstrated by our last example shown in Fig. 12. The mixture CH_3CCl_3:CD_3OD containing a mole fraction of 60 % CH_3CCl_3 is investigated. The symmetric CH_3-stretching mode at ν_H = 2939 cm^{-1} of CH_3CCl_3 is excited by stimulated Raman scattering of an intense laser pulse. The population decay of this mode was monitored via probe scattering at the corresponding anti-Stokes frequency. The open circles in Fig. 12

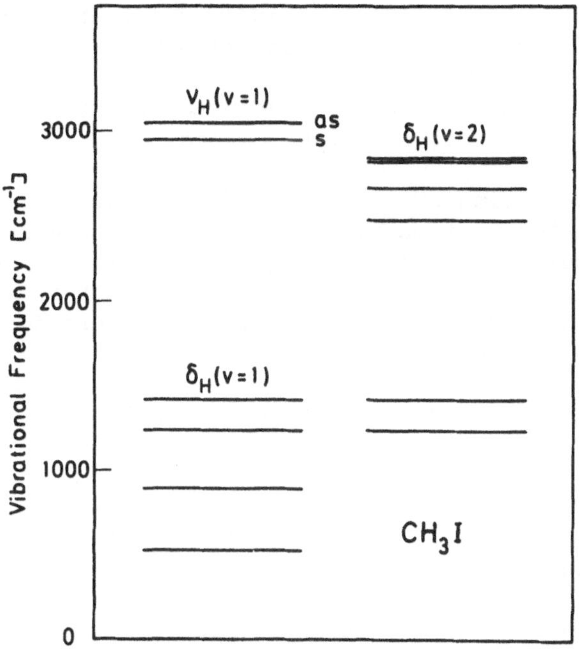

Fig. 11. Energy values of the normal vibrational modes of CH_3I and several overtones and combination modes.

represent the scattering data (23) indicating a population lifetime of T_1 = 6.5 ps. Then the population of the stretching mode of deuterated methanol at ν_D = 2227 cm^{-1} is measured and also shown in the Figure. The full circles clearly show the delayed rise and decay of the population of this lower-frequency mode. Knowing the spontaneous scattering cross-sections of the vibrational modes and taking the ratio of the corresponding maxima from the curves in Fig. 12, we estimate a high efficiency of ∿60 % for the intermolecular transfer process $\nu_H \rightarrow \nu_D$ in our liquid system (23).

5 DISCUSSION OF THE RELAXATION PROCESSES

The picosecond experiments show ultrafast dephasing and energy relaxation occuring in a number of investigated samples. The dephasing data are discussed first.

Exponential decay of the coherent excitation is observed over several orders of magnitude represented by the time constant T_2. To explain this behavior in terms of stochastic theory (28), relaxation in the limiting case of fast modulation has to be considered; i.e., homogeneous broadening with motional narrowing is suggested by the experimental data. We note that a time scale of ∿ 1 ps or less is required for the individual interaction event in the fast modulation limit. The small amplitude rattling motion of the molecules in their liquid cages (time scale 10^{-13} s) and/or the fast molecular rotation are important factors for an exponential dephasing time T_2 (29-31).

The following processes have been suggested to contribute to the measured T_2 values of polyatomic molecules:

(i) "Direct" dephasing by quasi-elastic interaction of nearest neighbors via the repulsive part of the intermolecular potential (29-33). Numerical estimates agree with experimental results within a factor of three. Improved agreement is reported when the coupling with rotational motion and vibrational anharmonicities are included (33).

(ii) Indirect dephasing by anharmonic coupling to low frequency vibrational modes (34, 35).

(iii) Resonant exchange of vibrational quanta via the attractive part of the intermolecular potential of transition dipole-dipole interaction (36, 37).

(iv) Energy transfer to neighboring vibrations (38, 39).

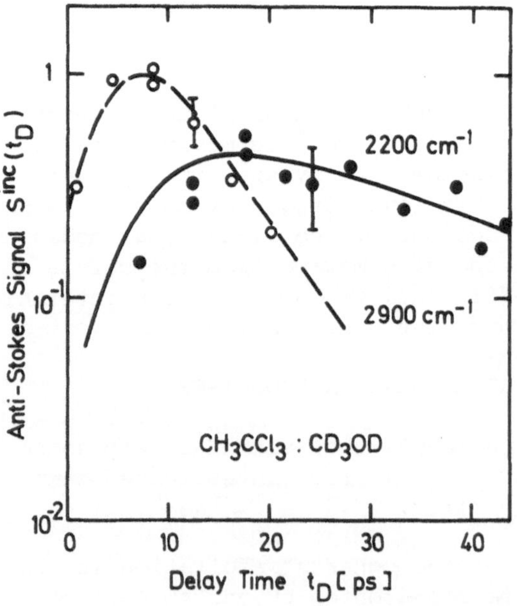

Fig. 12. Incoherent probe signals of CH_3CCl_3:CD_3OD (with mole fraction 0.6 of CH_3CCl_3). The open circles represent the occupation of the primary excited ν_H vibration of CH_3CCl_3. Vibrational energy transfer leads to subsequent occupation of the ν_D vibration of CD_3OD (closed circles).

Processes (i) to (iii) have been termed "pure dephasing". Processes (iv) are distinguished from the mechanisms (i) to (iii) measuring the time evolution of the excited state population (T_1-techniques).

For the symmetric CH-stretching mode of CH_3CCl_3, theoretical estimates suggest process (i) as dominant T_2 mechanism while process (ii) appears to be unimportant (35). The small concentration dependence of the linewidth rules out a significant contribution of (iii). Approximately 20 % of the total line broadening is due to population decay (iv). For many vibrational modes the dominant mechanism of T_2 cannot be stated definitely at the present time.

Different mechanisms have been considered for the energy decay and redistribution process mentioned above. We wish to distinguish collision-induced intramolecular transitions and intermolecular transitions which can be shortrange collision induced or long range resonance transfer processes.

(i) The intramolecular collision-induced vibrational transitions have been interpreted in terms of extended SSH models including mode matching and rotational coupling; i.e. details of the normal mode displacements and of the rotational motion are considered (38, 40). Calculations show that a large amount of the energy mismatch of a vibrational transition is transferred to molecular rotation.

(ii) Fermi resonance and Coriolis coupling. Due to anharmonic coupling normal modes with same symmetry can mix. As a result the transition probability between stretching modes and bending modes of molecular groups such as CH_3- is considerably augmented. Coriolis coupling additionally allows coupling of vibrational states of different symmetry (41).

(iii) Resonance transfer. Concentration studies of systems, e.g. CH_3CCl_3 and C_2H_5OH show that intermolecular transitions are involved. Long range transfer occurs via dipole-dipole coupling (42).

A qualitative understanding of the relative role of mechanisms (i) to (iii) exists for a number of liquid systems at the present time (39).

We turn now to the question of inhomogeneous broadening. For the rapid dephasing processes the fast translational and rotational motion of the molecules in their liquid cages appears to be important. Next-neighbor-interaction is thought to be dominant for these relaxation phenomena. In addition, there are indications of a slow component of the intermolecular interaction leading to an inhomogeneous line broadening. The picosecond data present evidence for a distribution of vibrational frequencies in several strongly and weakly associated liquids. The observed frequency distribution appears to be fairly static on the picosecond time scale, not changing notably for several 10^{-11} s or even longer. The different frequencies of individual molecules may be interpreted as variations of the solvent shift exerted by the specific molecular environment. Existing theories explain the well-known frequency shifts of the band centers (43) and the generation of a frequency distribution (44). The important factor is the slow time scale of the mechanism required to explain the experimental findings. A preliminary physical picture suggests that long range intermolecular interaction including non-nearest neighbors gives rise to a frequency distribution which is governed by slow diffusional motion on the time scale of $\sim 10^{-10}$ s.

6 CONCLUDING REMARKS

This paper discusses several experiments which are devoted to ultrafast vibrational processes and time-resolved Raman spectroscopy in liquids. It is shown that new experimental methods have been developed which benefit from the specific molecular excitation and time resolution achieved with high quality picosecond light pulses. Detailed information on different aspects of molecular dynamics is obtained. Further application of these techniques will advance our understanding of the liquid state and of polyatomic molecules.

REFERENCES

(1) A.J. DeMaria, D.A. Stetser and J. Heyman, Appl. Phys. Lett. 8, 174 (1966); W. Schmidt and F.P. Schäfer, Phys. Lett. A26, 558 (1968); D.J. Bradley, A.J.F. Durant, F.O'Neill and B. Sutherland, Phys. Lett. 30A, 535 (1969); E.P. Ippen, C.V. Shank, and A. Dienes, Appl. Phys. Lett. 21, 348 (1972).

(2) A. Laubereau, L. Greiter, and W. Kaiser, Appl. Phys. Lett. 25, 87 (1974); A.H. Kung, ibid 25, 653 (1974); for recent summaries see "Picosecond Phenomena", eds. C.V. Shank, E.P. Ippen and S.L. Shapiro, Springer (Berlin 1978).

(3) R.T. Bailey, in "Molecular Spectroscopy", Vol. 2 (The Chemical Society, London, 1974)

(4) A. Laubereau and W. Kaiser, Rev. Mod. Phys. 50, 607 (1978).

(5) R.G. Gordon, J. Chem. Phys. 40, 1973 (1964); 42, 3658 (1965); 43, 1302 (1965); S. Bratos and E. Marechal, Phys. Rev. A4, 1078 (1971); L.A. Nafie and W.L. Peticolas, J. Chem. Phys. 57, 3145 (1972).

(6) J.A. Giordmaine, W. Kaiser, Phys. Rev. 144, 676 (1966).

(7) M. Maier, W. Kaiser, and J.A. Giordmaine, Phys. Rev. 177, 580 (1969).

(8) A. Laubereau, Chem. Phys. Lett. 27, 600 (1974).

(9) W.F. Calaway and G.E. Ewing, Chem. Phys. Lett. 30, 485 (1975); J. Chem. Phys. 63, 2842 (1975).
(10) S.R. Brueck, R.M. Osgood, Jr.: Chem. Phys. Lett. 39, 568 (1976).
(11) D. von der Linde, A. Laubereau and W. Kaiser, Phys. Rev. Lett. 26, 955 (1971).
(12) R.P. Feynman, F.C. Vernon, R.W. Hellwarth: J. Appl. Phys. 28, 49 (1957).
(13) A. Laubereau, G. Wochner, and W. Kaiser, Phys. Rev. A13, 2212 (1976).
(14) A. Laubereau, G. Wochner, and W. Kaiser, Chem. Phys. 28, 363 (1978).
(15) W. Zinth, A. Laubereau, W. Kaiser: Opt. Commun. 22, 161 (1977).
(16) W.R.L. Clements, B.P. Stoicheff: Appl. Phys. Lett. 12, 246 (1968).
(17) M. Scotto, J. Chem. Phys. 49, 5362 (1968)
(18) H.M.M. Hesp, J. Langelaar, B. Bebelaar, J.D.W. van Voorst, Phys. Rev. Lett. 39, 1376 (1977); S.A. Akhmanov, N.I. Koroteev, R. Yu. Orlov, I.L. Shumay, JETP Lett. 23, 276 (1976).
(19) A. Laubereau, G. Wochner and W. Kaiser, to be published
(20) M.J. Colles and J.E. Griffiths, J. Chem. Phys. 56, 3384 (1972).
(21) A. Laubereau, D. von der Linde, and W. Kaiser, Phys. Rev. Lett. 28, 1162 (1972).
(22) R.R. Alfano and S.L. Shapiro, Phys. Rev. Lett. 29, 1655 (1972).
(23) A. Laubereau, L. Kirschner and W. Kaiser, Opt. Comm. 19, 182 (1973).
(24) K. Spanner, A. Laubereau, and W. Kaiser, Chem. Phys. Lett. 44, 88 (1976).
(25) A. Laubereau, G. Kehl and W. Kaiser, Optics Commun. 11, 74 (1974); P.R. Monson, L. Patumtevapibal, K.J. Kaufmann, and P.W. Robinson, Chem. Phys. Lett. 28, 312 (1974).
(26) F. DeMartini and J. Ducuing, Phys. Rev. Lett. 17, 117 (1966).
(27) A. Seilmeier, K. Spanner, A. Laubereau, and W. Kaiser, Opt. Commun. 24, 237 (1978).
(28) R. Kubo, in "Function, Relaxation and Resonance in Magnetic Systems", ed. Ther Haar (Oliver and Boyd, Edinburgh, 1962).
(29) S.F. Fischer, A. Laubereau, Chem. Phys. Lett. 35, 6 (1975).
(30) P.A. Madden, R.M. Lynden-Bell, Chem. Phys. Lett. 38, 163 (1976).
(31) W.G. Rothschild, J. Chem. Phys. 65, 2958 (1976).
(32) D.W. Oxtoby, S.A. Rice, Chem. Phys. Lett. 42, 1 (1976).
(33) R.K. Wertheimer, Mol. Phys. 35, 257 (1978).
(34) S. Bratos, J. Chem. Phys. 63, 3499 (1975).
(35) S.F. Fischer, A. Laubereau, Chem. Phys. Lett. 55, 189 (1978).
(36) D. Döge, R. Arndt, A. Khuen, Chem. Phys. 21, 53 (1977).
(37) M. Gilbert, M. Drifford, J. Chem. Phys. 66, 3205 (1977).
(38) A. Miklavc, and S.F. Fischer, Chem. Phys. Lett. 44, 209 (1976).
(39) A. Laubereau, S.F. Fischer, K. Spanner and W. Kaiser, Chem. Phys. 31, 335 (1978).

(40) A. Miklavc and S.F. Fischer, J. Chem. Phys. 69, 281 (1978).
(41) R. Zygan-Maus and S.F. Fischer, Chem. Phys. 41, 319 (1979).
(42) T. Förster, Z. Naturforsch. A4, 321 (1949).
(43) A.D. Buckingham, Proc. Roy. Soc. London A248, 169 (1958).
(44) S. Bratos, J. Rios, Y. Guissany, J. Chem. Phys. 52, 439 (1970).

VIBRATIONAL BANDSHAPES IN VISCOUS LIQUIDS AND GLASSES

C. A. Angell

Department of Chemistry
Purdue University
West Lafayette, IN 47907 USA

INTRODUCTION

Most of the current interest in vibrational bandshapes has been focussed on the spectra of simple molecular liquids such as N_2 and CO_2 which, for reasons associated with the simplicity of the molecular shapes, are only observed in highly fluid states. Any attempt to decrease their fluidity and concomitantly to increase the various associated structural and rotational relaxation times, e.g. by decrease of temperature or increase of pressure, is quickly frustrated by crystallization. There exist, however, liquids of moderate complexity, which have much wider stable liquid ranges (which we will describe in reduced temperature units $(T_b - T_m)/T_m$) than the above, some of which may be extensively supercooled, thus making possible the study of vibrational bandshapes over very wide ranges of fluidity. Such extensions of the temperature range, and accordingly of the relaxation time range, are desirable for the purpose of maximizing the empirical base against which the adequacy of theoretical bandshape interpretations may be judged. Some examples of common molecular liquids and their liquid ranges are given in Table 1.

When reduced liquid ranges exceed unity (group 3 in Table 1) it is commonly found that crystallization does not occur at all during continuous cooling at moderate rates, e.g. 10 K min^{-1}, and in this case it is observed invariably (quantum fluids excepted) that a rigid glassy condition is encountered somewhere in the range $T = (0.25 - 0.50)T_b$.[1,2] Rigidity, of course, implies exceedingly long relaxation times and, as discussed in more detail below, we find that between the so-called "glass transition temperature" T_g and the boiling point T_b, the various measurable relaxation times change by some 13 orders of magnitude.[3,4]

Table 1. Liquid Ranges for Molecular Substances*

	1. Small Liquid Range			2. Medium Liquid Range			3. Wide Liquid Range		
	N_2	HCl	Ar	CCl_4		CH_3CN	S_2Cl_2	CH_3 	2 Me-butane
T_b/K	77.3	188.2	87.4	350	353	355	389	383	306
T_m/K	63.3	158	83.9	250	279	232	193	178	113
$\frac{T_b-T_m}{T_m}$	0.22	0.19	0.04	0.40	0.27	0.53	1.01	1.15	1.70
T_g/K							118	115	69
T_b/T_g							3.30	3.33	4.43

*A more extensive compilation, with glass transition temperature estimated by extrapolation for Group 2 liquids, is given in Ref. 2.

It is the purpose of this contribution to discuss some of the phenomenology of this spectroscopically little-studied viscous (and ultimately rigid) region of liquid state behavior, and to describe some of the spectroscopic observations which have been made within it. We note that the region in question contrasts with the plastic crystal regime. The latter is distinguished by drastic change of order from the liquid without much change in rotational relaxation time, while, with supercooling liquids, we preserve most of the disorder, but dramatically change the relaxation time.

SUPERCOOLING AND THE GLASS TRANSITION

Let us discuss first the vitrification phenomenon which sets a practical if not theoretical limit on studies of the liquid state in those cases where crystallization can be circumvented. Vitrification is commonly understood as the process by which a liquid loses its ability to flow, and thus becomes brittle, during continuous cooling. Since viscosity essentially measures the rate of response of a liquid to the sudden application of a shear stress, the above definition can be made more precise (perhaps at the expense of easy understanding) by asserting that vitrification occurs when the temperature of the liquid falls into the range in which its response time for recovery of the equilibrium (unperturbed) state, after a shearing perturbation, becomes long with respect to the time (order of seconds) that the observer is usually willing to wait for a response. Failure to respond to a shear stress is, in fact, only the most obvious consequence of a <u>general inability</u> of the liquid to regain a state of internal equilibrium, after perturbation, which arises in the same temperature range. Since temperature change itself is a perturbation of a pre-existing equilibrium state, it becomes clear that the "glass transition" seen on continuous cooling is simply the "falling out of internal equilibrium", of the system during cooling. Such a process implies certain changes of the derivative thermodynamic properties (heat capacity, compressibility, etc.) which depend on the occurrence of fluctuations in extensive properties (entropy, volume) during the measurement period. It is, indeed, through the observation of changes in C_p, that the glass transition phenomenon is most meaningfully defined. An example, (ethanol, data of Haida et al[5]) is shown in Fig. 1.

Figure 1 allows us to infer a theoretical limit for the liquid state as defined by, for example, the exhibition of the characteristically high heat capacity observed above T_g in Fig. 1. The theoretical limit is recognized[6] by noting that, since the liquid state has a characteristically higher heat capacity than the crystalline state of the same substance, it loses entropy more rapidly than the crystal when both are cooled over the same temperature interval. Therefore it is inevitable that during continuous cooling of the (internally equilibrated) liquid, a point in temperature would be reached at which the excess entropy of the liquid at the fusion point (ΔS_f) has all vanished, i.e.

$$S \text{ (liquid)} = S \text{ (crystal)} \tag{1}$$

At the latter temperature, T_o(cal), (if not above) it would appear that some internal equilibrium transition in the liquid state would have to occur to avoid an unacceptable inversion of the expected relation between ordered and disordered phase entropies, (or, at lower temperatures, a violation of the third law of thermodynamics).[7] The temperature at which this theoretical limit is reached is usually found to lie less than 50° below the observed glass transition temperature. The inset to Fig. 1 which shows the variation with temperature of the difference between liquid and crystal entropies demonstrates the problem graphically, though the area matching shown in the main figure gives a more accurate estimate of T_o(cal). Several cases for which the necessary thermodynamic data are available have been discussed in Ref. 8.

The question arises: "Is the observed glass transition, which is a kinetic phenomenon, a reflection of the impending thermodynamic transition associated with a relation between the relaxation time and the decreasing number of available configurational states of the system?" The common observation that liquids in the range of viscosity 1 - 10^4 poise (where molecular motions become cooperative and all configurational modes of motion are fully coupled) depart from Arrhenius behavior and instead obey the VTF equation, Eq. (2), for their

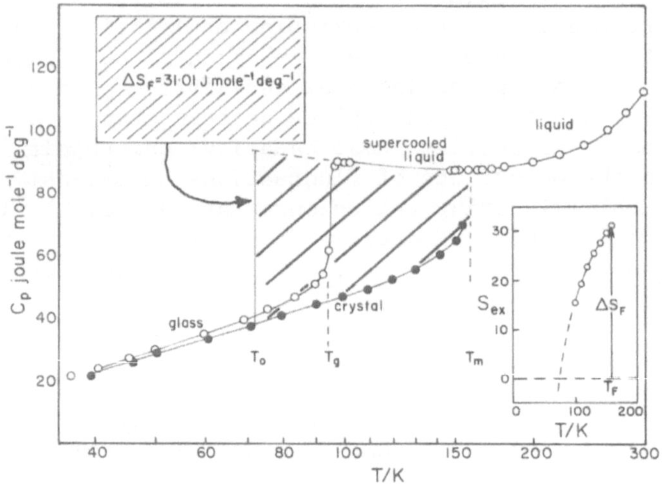

Fig. 1. Heat capacities of liquid, glass and crystalline states of ethanol. Inset: Variation of liquid excess entropy with decreasing temperature showing approach to ideal glass state. (From Ref. 3.)

various relaxation times

$$\tau_i = A_i \exp - B_i/(T-T_o) \qquad (2)$$

suggests the answer is affirmative, since the parameter T_o is usually within a few degrees of the entropy equalization (Eq. (1)) temperature, $T_{o(cal)}$.[8,9] Under the conditions which yield $T_{o(\tau)} = T_{o(cal)}$ it is found that T_o, at which $\tau_i \to \infty$, is the same (again with a few degrees) for each transport process or relaxation time studied.[9,10] Some examples[11] of this type of transport behavior are given in Fig. 2 for one of the few cases for which vibrational spectra, which are diagnostic for reorientations, (to be discussed below), have been consciously measured into the viscous and glassy regions of behavior.[12] The best documented system, unfortunately, contains molecular ions (NO_3^-) rather than simple molecules, which perhaps illustrates the extent of the data gap in this research area. It will be of interest to decide to what extent relaxation times with this characteristic temperature dependence can be extracted from spectroscopic measurements.

For many liquids the simplicity of behavior implied by Eq. (2) and Fig. 3 is lost in the liquid region of very high viscosity ($\eta > 10^6$p) near the glass transition, where the relaxation behavior

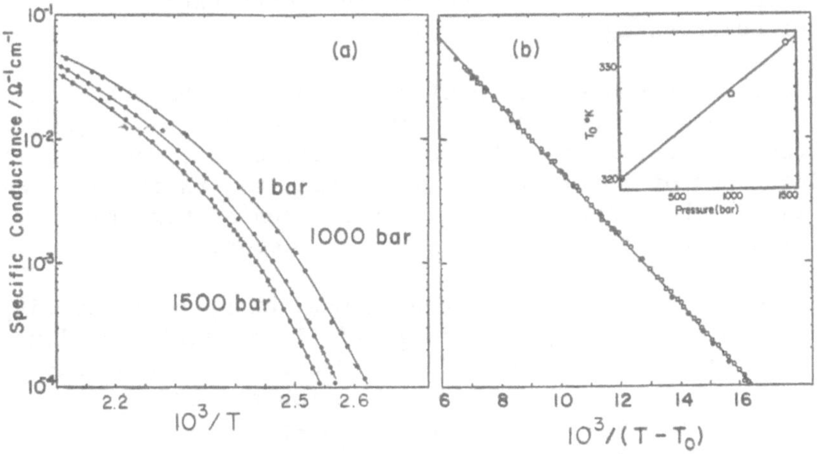

Fig. 2. (a) Arrhenius plots of electrical conductivity of a glass-forming $Ca(NO_3)_2 + KNO_3$ mixture at pressures 1, 1000, and 1500 bar.

(b) Same data linearized and superposed by VTF plot with only T_o varying. Inset: Linear pressure dependence of T_o parameter. (From Ref. 11.)

often tends to return to Arrhenius form.[13,14] The departure from Eq. (2) behavior commences at different temperatures for different relaxation processes as they become increasingly decoupled with decreasing thermal excitation. These complexities are, at the moment, insufficiently investigated for satisfactory interpretations to be possible. We note, however, that since all these changes must be related to subtle changes in structure, i.e. molecular packing, there may be features of the vibrational spectra of the liquid, particularly in the far IR region, which are related to them. Far IR spectroscopic properties in this temperature region are almost uninvestigated.

GLASS TRANSITIONS ON "NORMAL" LIQUID TIME SCALES

Before attempting to discuss the few experimental vibrational spectroscopic investigations which have been made into the low temperature region, it is of interest to enquire into the possibility that the glass transition phenomena we have described (in the context of liquids falling out of internal equilibrium on long time scales) can be observed in the temperature range of ordinary experience, by the study of very short time responses to instantaneous perturbations, or of very fast continuous perturbations, e.g. rapid cooling or compressing.

Two different routes of enquiry convince us that the same general observations can be made, i.e. that structural equilibrium can be arrested independent of vibrational equilibrium, and that associated decreases of thermodynamic properties to fractions of their normal liquid-like values, can be observed even at temperatures where the fluidities are no greater than that of water.

The most direct comparison can be made using heat capacity vs temperature plots for computer-simulated fluids during cooling at rates of the order of 10^{12} deg sec^{-1}. Decreases of heat capacity comparable with those of Fig. 1 have been observed for both an ionic fluid[15] and for Lennard Jones argon.[16] A separate comparison based on laboratory data can be made using microwave or even far IR measurements of the dielectric constants of polar liquids, in which a decrease of ε_o from near-static values to much lower, glass-like, values may be observed.[17] It is notable, however,[18] that in the high temperature range, as a direct consequence of Eq. (2), the "glass transformation range" or region of time-dependent behavior (in which equilibrium is neither completely established or completely lost) is spread out over much wider temperature ranges than at the normal glass transition (where the transformation range may be as small as 5 K). In general we may conclude that spectral features, or changes of features, which we may observe in ordinary time scale experiments passing through the glass transition will have their counterparts at temperatures of ordinary liquid state studies. We now examine the evidence for what these effects may be.

BANDSHAPE VARIATIONS BETWEEN LIQUID AND GLASSY STATES

Internal Modes of Molecular or Molecular Ionic Liquids

Probably the simplest liquid for which vibrational spectra have been measured in the viscous liquid or glassy states is toluene, which is easily vitrified by cooling the optical cell containing it in liquid nitrogen. Clarke and Miller[12] studied the polarized and depolarized Raman spectra of this substance in the glassy state and room temperature liquid state, with results (for the symmetric ring-breathing mode at 1004 cm^{-1}) shown in Fig. 3. Since molecular reorientation affects only the anisotropic component bandshape, the polarized and depolarized spectra should superimpose in the glassy state, from which reorientation should be absent, as is seen in Fig. 4. At higher temperatures where the substance is highly fluid, an additional width assignable to reorientation effects becomes apparent. It amounts to only 1 cm^{-1} out of a total of 5 cm^{-1}. Thus most of the Raman bandwidth must be of heterogeneous origin associated with the spread of vibrational potential well-depths and shapes characteristic of the amorphous phase.

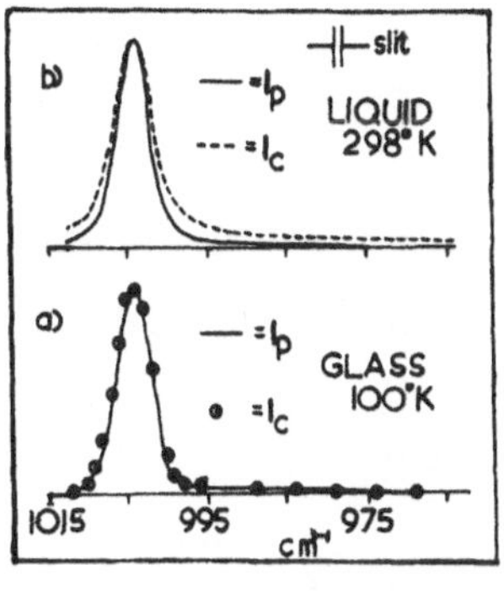

3

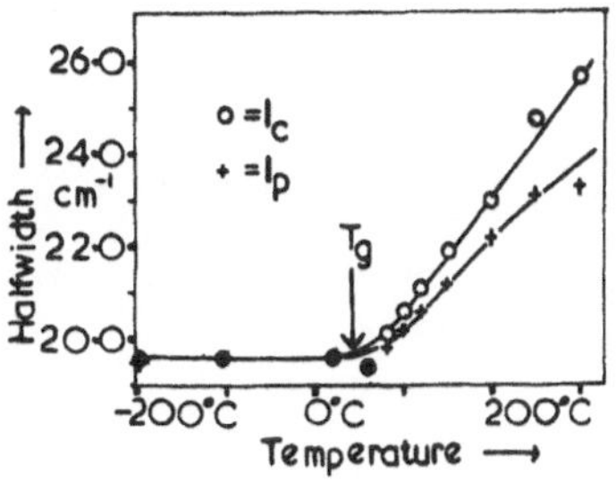

4

Figs. 3 & 4. Fig. 3: Profiles of the 1004 cm^{-1} Raman band of toluene (a) at 100°K and (b) at 298°K. I_p and I_c indicate the scattering with electric vector polarized respectively parallel and perpendicular to that of the incident light. Spectral slit width -1cm^{-1}. (From Ref. 12.) Fig. 4: Variation with temperature of the halfwidth (full width at half height) of the v_1 band of NO_3^- at ≈ 1050 cm^{-1} for glassy and liquid $Ca(NO)_2 . \frac{3}{2}$ KNO_3. T_g is the glass transition temperature. I_p and I_c are defined as for fig. 3. (From Ref. 12.)

In general, the heterogeneous bandwidth should show a temperature dependence associated with the changing structure, though this is not evident in the toluene case. More interesting in this respect is the solution of 40% $Ca(NO_3)_2$ in the system KNO_3 - $Ca(NO_3)_2$ also studied by Clarke and Miller, Fig. 4. The band at 1050 cm^{-1} (depolarized component) is sensitive to rotation about the two fold axis of the NO_3-ion. The plots show additional broadening of this band becoming evident above T_g. The additional width ∿ 2 cm^{-1} at 300°C corresponds to a re-orientation time of ≈ 2.6 x 10^{-12}s, though this contains an inaccuracy due to the inadequacy of a simple subtraction to give the appropriate rotational bandwidth component. The order of magnitude, at least, is compatible with the relaxation time of 6 x 10^{-12}s for the liquid as a whole obtained from the electrical conductance data shown in Fig. 3 for this liquid using the relation[19]

$$\tau_\sigma = \frac{e_o \varepsilon_\infty}{\sigma_o} , \qquad (3)$$

where ε_∞ is the high frequency (vibrational) dielectric constant measured by Howells et al,[20] and e_o is the permittivity of free space.

In the last case the isotropic component broadens markedly with increasing temperature indicating an increasing spread of configurations containing the nitrate ion. In some mixed nitrate sys-

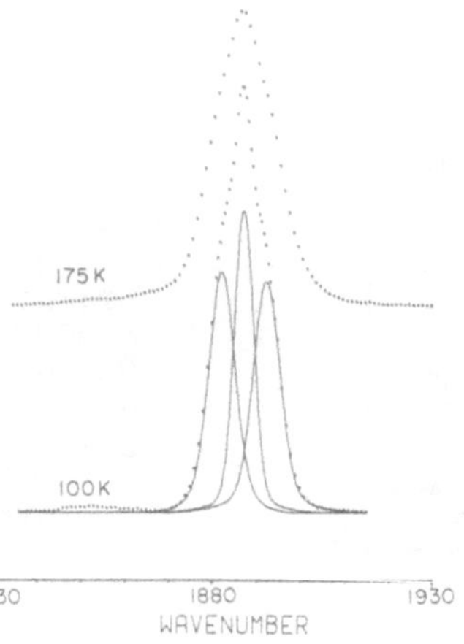

Fig. 5. Spectrum of $Co(CO)_4^-$ in MTHF + oxetane (4:1) at 175 and 100 K showing resolution of single band into a complex of three bands, two due to presence of C_{2v} symmetry species produced by "close ion pairs" and one (central component) from T_d symmetry species. (From Ref. 23.)

tems,[21,22] this spread of configurations becomes resolved, at low temperatures, into two bands representing two distinct NO_3^- ion environments. This latter type of resolution of a high temperature band into distinct structural site bands on approach to T_g is common in the spectroscopy of glassforming solutions. It has been observed recently by Edgell and coworkers[23] for vibrational spectra of the complex ion $Co(CO)_4^-$ dissolved in the glass-forming solvent methyltetrahydrofuran + oxetane (4:1), for which spectra are shown in Fig. 5. These show the single higher temperature band attributed to a site of T_d symmetry becoming resolved into a three band complex of which the outer two components arise from a postulated "close ion pair"[23] of C_{2v} symmetry.

We should note that in none of these cases has a high temperature regime of motional narrowing, which would eliminate the heterogeneous bandwidth, been observed. The latter must lie at still higher reduced temperatures and higher fluidities.

The only other internal mode spectra of simple molecular or molecule-ion species which have been followed into the glassy state are of the hydroxyl ion symmetric stretch which has been studied by Mazzacurati et al[24] as part of a study focussed on the low frequency quasi-lattice modes discussed below. The OH^- band is broadened at all temperatures by the wide distribution of hydrogen bonding interactions in which the molecule-ion participates. Unfortunately there are insufficient data to show any change of behavior at T_g.

Time-Dependence of Internal Mode Bandshapes for Direct Relaxation Time Studies in the Viscous Regime

Where most effort has been devoted to analyzing the time-independent bandshape for dynamic information, it is also possible, for heterogeneously broadened bands in cases where the half-width is a function of temperature or pressure, to obtain kinetic data from measurements of the half-widths or peak heights as functionsof time following perturbations from equilibrium. It is not possible, spacewise, to deal fully here with this application of vibrational (or electronic) spectral band shape analysis, and the reader is referred to original literature for details.[25-27] Measurements of the relaxation times in this longer time range can in principle be made in the time range 10^{-6} to 10^4s though, because vibrational bandshapes are generally not very sensitive to temperature and pressure changes, the short time measurements with their greater signal/noise ratios are more difficult to perform successfully. Published work in this area is limited to analysis of overtone bands of hydrogen bonded liquids near T_g.[27] Also in these measurements[27] unlike their predecessors[26] the measurements were made during temperature ramping rather than after temperature jumps because of the much greater information acquisition rate. An interesting feature of this type of measurement is that it is possible to obtain the structural relaxation kinetics of the micro-region around the probe species, distinct from the relaxation of the whole structure. The two can be quite different in character.[27]

Low Frequency Quasi-Lattice Modes

Low frequency modes essentially characteristic of the liquid quasi-lattice have been studied through the low temperature region into the glassy state for isolated cases of ionic liquids and solutions and very recently some polar molecular liquids and their binary solutions.

Angell, Wegdam and Van der Elsken[28] studied the 50 - 300 cm^{-1} IR spectrum of $ZnCl_2$, a network liquid related to the tetrahedral network glasses BeF_2, SiO_2, GeO_2 and GeS_2 discussed in this volume by Thorpe. They observed the behavior of the $[ZnCl_4]$ tetrahedral unit asymmetric stretch at $\sim$ 260 cm^{-1} and a more complex mode at 100 cm^{-1}, and observed band broadening with increasing temperature above T_g. They analyzed the data by Fourier transformation to obtain damped cosine wave dipole moment correlation functions, and showed that the damping constant b (Fig. 6 inset) decreased with decreasing temperature down to the glass transition temperature. It then stayed constant for the glass (fixed structure) but continued to decrease for the crystal (decreasing anharmonicity). Some data are shown in Fig. 6.

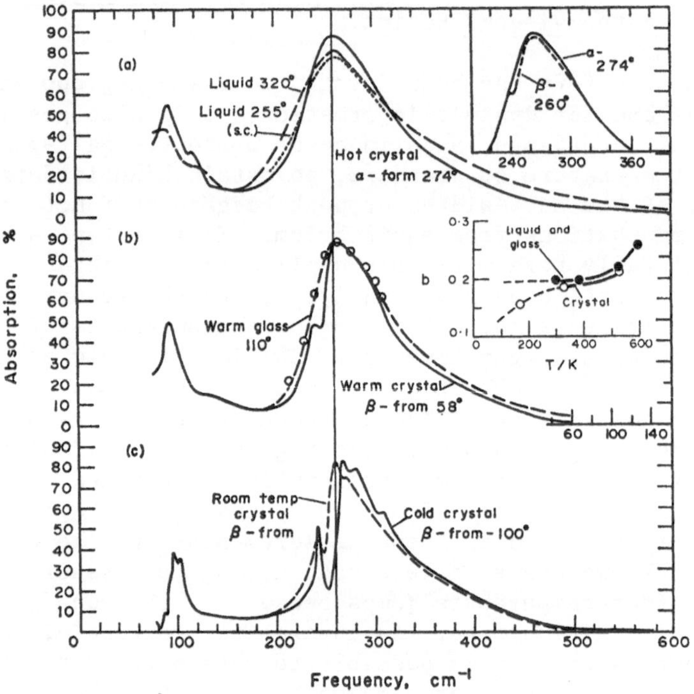

Fig. 6. Low frequency IR spectra of ZnCl$_2$ in liquid, crystal, and
         glass states. Inset: Comparison of temperature dependence
         of damping constant for amorphous and crystal states.
         (From Ref. 28.)

$$\Box = I_0(\omega)\left[\frac{1+n(\omega)}{1+n_0(\omega)}\right]$$

$I_0(\omega)$ is from 25°C spectrum

$n(\omega)$ = Bose factor = $\left[\exp(h\omega/kT)^{-1}\right]^{-1}$

$n_0(\omega)$ = $n(\omega)$ at 25°C

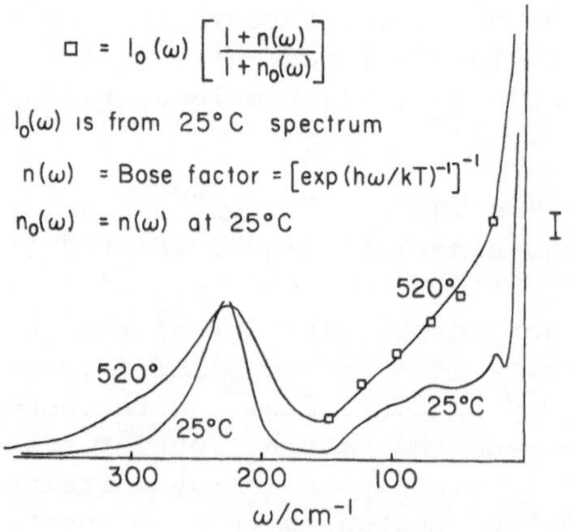

Fig. 7. Raman spectra of ZnCl$_2$ showing Bose factor origin of low
         frequency intensity changes. (From Ref. 29.)

Complimentary Raman spectra were obtained at about the same time by Clarke, Miller and Angell[29] but have to date remained unpublished. The pronounced low frequency band intensity changes, as shown in Fig. 7, are entirely accounted for by the population factor. Changes in bandwidth for the high frequency mode with band maximum at 220 cm^{-1} are consistent with those seen in the 260 cm^{-1} infrared band.

Similar changes of quasi-lattice mode Raman intensities have been observed in the study of NaOH + H_2O solutions by Mazzacurati et al[24] who demonstrated[30] a general relation between anisotropic low frequency Raman spectra and the vibrational density of states for various amorphous materials.

Low frequency vibrational spectra for simple molecular systems in viscous liquid and glassy states remained uninvestigated until very recently when Reid and Evans[31] measured the far IR spectra of solutions of polar molecules in non-polar solvents in an attempt to broaden the empirical base for an itinerant oscillator interpretation of the normal temperature for IR spectra of these solutions. Examples of their findings are shown in Fig. 8. It is clear that, in comparison with the internal modes seen in Figs. 3 and 4, the temperature dependence of these bands, both in peak energies and

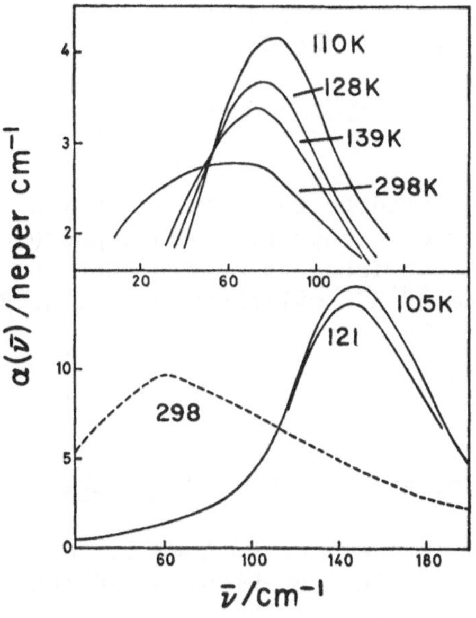

Fig. 8. Low frequency IR spectra of (a) decalin + tetrahydrofuran (b) CH_2Cl_2 in pyridine - toluene solvent. (From Ref. 31.)

half-widths, is very strong in the liquid state. Again the shape becomes essentially "frozen" below T_g, reflecting the structural origin of the band width. Reid and Evans show clearly the contrast between the effect of temperature on this resonance-type absorption (the maximum of which moves to higher frequencies as the structure becomes more compact with decreasing temperature) and that of the low frequency relaxation (dielectric loss) peak which moves rapidly to lower frequencies with a temperature dependence of Eq. (2) form. Plotted as an absorptivity, rather than as a dielectric loss, the dielectric relaxation appears as a weak shoulder on the low frequency edge of the resonance band, and in fact corresponds to overdamped elements of this band. A secondary dielectric loss peak, which is of intermediate frequency and weaker temperature dependence, may be observed at low temperatures. It is completely merged with the broad resonance band at ordinary temperatures (corresponding to high fluidities). The relation of relaxational to resonance absorptions in these cases is discussed for a variety of systems in Ref. 4, Chapter 11.

CONCLUDING REMARKS

The study of vibrational band shapes in liquids with broad liquid ranges and glass-forming ability is a field of considerable interest which has to date received very little attention from spectroscopists or glass scientists. In contrast to the case of liquids of extremely high fluidity, the bandshapes of liquids in the less fluid ranges seem to be dominated by inhomogeneous effects.

ACKNOWLEDGEMENTS

The author is indebted to the National Science Foundation for glass structure and spectroscopy research support under Grant No. DMR 77-04318A1. He also wishes to acknowledge the helpful remarks and valuable criticism of his colleague Dr. J. H. R. Clarke.

REFERENCES

1. D. Turnbull and M. H. Cohen, "Concerning Reconstructive Transformation and Formation of Glass", J. Chem. Phys., 29:1049 (1958).
2. J. M. Sare, E. J. Sare and C. A. Angell, "Glass Transition Temperatures for Simple Molecular Liquids and Their Binary Solutions", J. Phys. Chem., 82:2622 (1978).
3. M. Goldstein, "Viscous Liquids and the Glass Transition: A Potential Energy Barrier Picture", J. Chem. Phys., 51:3728 (1969).

4. J. Wong and C. A. Angell, "Glass: Structure by Spectroscopy", Marcel Dekker, New York, NY (1976) (Chapter 1).
5. O. Haida, H. Suga and S. Seki, "Calorimetric Study of the Glassy State XII. Plural Glass-Transition Phenomena of Ethanol", J. Chem. Thermo., 9:1133 (1977).
6. W. Kauzmann, "The Nature of the Glassy State and the Behavior of Liquids at Low Temperatures", Chem. Rev., 43:219 (1948).
7. J. H. Gibbs, "Modern Aspects of the Vitreous State", Ed. J. D. McKenzie, Butterworth Sci. Publ. Ltd., London, (1960), Chapter 7.
8. C. A. Angell and K. J. Rao, "Configurational Excitations in Condensed Matter and the Bond Lattice Model for the Liquid-Glass Transition", J. Chem. Phys., 57:470 (1972).
9. C. A. Angell and J. C. Tucker, "Heat Capacities and Fusion Entropies of the Tetrahydrates of Calcium Nitrate, Cadmium Nitrate, and Magnesium Acetate. Concordance of Calorimetric and Relaxational 'Ideal' Glass Transition Temperatures", J. Phys. Chem., 78:278 (1974).
10. D. W. Davidson and R. H. Cole, "Dielectric Relaxation in Glycerol, Propylene Glycol, and n-Propanol", J. Chem. Phys., 19:1484 (1951).
11. C. A. Angell, L. J. Pollard and W. Strauss, "Transport in Molten Salts Under Pressure. I. Glass-Forming Nitrate Melts", J. Chem. Phys., 50:2694 (1969).
12. J. H. R. Clarke and S. Miller, "The Determination of Rotational Correlation Times in Liquids from Raman Bandshapes", Chem. Phys. Lett., 13:97 (1972).
13. R. Bose, R. Weiler and P. B. Macedo, "Temperature Dependence of the Conductance of a Vitreous $KNO_3 - Ca(NO_3)_2$ Mixture", Phys. Chem. Glasses, 11:117 (1970).
14. W. T. Laughlin and D. R. Uhlmann, "Viscous Flow in Simple Organic Liquids", J. Phys. Chem., 76:2317 (1972).
15. L. V. Woodcock, C. A. Angell and P. A. Cheeseman, "Molecular Dynamics Studies of the Vitreous State: Simple Ionic Systems and Silica", J. Chem. Phys., 65:1565 (1976).
16. J. H. R. Clarke, J. Chem. Soc. Faraday Trans. II, 75, 1371 (1979)
17. C. P. Smyth, "Dielectric Behavior and Structure", McGraw Hill, New York, NY (1955).
18. C. A. Angell, J. H. R. Clarke and L. V. Woodcock, "Metastable Liquids and the Glass Transition by Computer Simulation: The Role of the Pair Potential in Real and Model Systems", (to be published).
19. P. B. Macedo, C. T. Moynihan and R. Bose, "The Role of Ionic Diffusion in Polarisation in Vitreous Ionic Conductors", Phys. Chem. Glasses, 13:171 (1972).
20. F. S. Howell, R. Bose, C. T. Moynihan and P. B. Macedo, "Electrical Relaxation in a Glass-Forming Molten Salt", J. Phys. Chem. 78:639 (1974).
21. R. E. Hester and K. Krishnan, "Vibrational Spectra of Molten Salts. II. Infrared Spectra of Some Divalent Metal Nitrates in

Alkali-Metal Nitrate Solutions", J. Chem. Phys., 47:1747 (1967).

22. J. Wong and C. A. Angell, "Spectroscopic Probing of Anion Environment in Inorganic Nitrate Glasses", J. Non-Cryst. Sol., 11:402 (1973).
23. W. F. Edgell and S. Hegde, (to be published).
24. V. Mazzacurati, M. Nardone and G. Signorelli, "Depolarized Rayleigh Scattering in 10M Amorphous and Liquid KOH Aqueous Solutions", J. Chem. Phys., 66:5380 (1977).
25. M. Eigen and L. DeMaeyer, "Technique of Organic Chemistry", Ed. A. V. Weissberger, Vol. VIII, Part II, p. 895 (1963).
26. A. Barkatt and C. A. Angell, "On the Use of Structural Probe Ions for Relaxation Studies in Glasses. I. Spectroscopic Properties of Cobalt (II) in Chloride-Doped Potassium Nitrate-Calcium Nitrate Glasses", J. Phys. Chem., 79:2192 (1975); "On the Use of Structural Probe Ions for Relaxation Studies in Glasses. II. T-Jump and T-Ramp Studies of Co(II) in Nitrate Glasses", J. Phys. Chem., 82:1972 (1978).
27. A. Barkatt and C. A. Angell, "Optical Probe Studies of Relaxation Processes in Viscous Liquids", J. Chem. Phys., 70:901 (1979).
28. C. A. Angell, G. H. Wegdam and J. van der Elsken, "FIR Spectra of Liquid, Glass, and Crystalline States of $ZnCl_2$: Order and Temperature Effects on Band Shape", Spectrochim. Acta, 30A:665 (1974).
29. J. H. R. Clarke, S. Miller and C. A. Angell (to be published).
30. V. Mazzacurati, M. Nardone and G. Signorelli, "Light Scattering From Disordered Media", J. Mol. Phys., (in press).
31. C. Reid and M. Evans, "Zero-THz Absorption Profiles in Glassy Solutions", J. Chem. Soc., Faraday Trans. II, 1218 (1979)

THE STUDY OF VIBRATIONAL RELAXATION BY ULTRASONIC AND LIGHT SCATTERING TECHNIQUES

D. Sette

Istituto di Fisica, Facoltà di Ingegneria
Università di Roma, and
Gruppo Nazionale Struttura della Materia
Rome, Italy

1 INTRODUCTION

The flow of energy between translational and internal, i.e. rotational and vibrational, degrees of freedom of polyatomic molecules occurs in gases during collisions: the average number of collisions required for the de-excitation of excited molecules ranges between a few unities and hundreds in the case of rotational degrees and between 10^3 and 10^{10} for vibrational degrees. The existence of a finite time for re-establishing equilibrium when disturbed leads to a relaxation process affecting the propagation of sound waves, and it is precisely the study of sound propagation which furnishes the method for investigating the relaxation processes.

When substances with polyatomic molecules pass from the gaseous to the liquid state, the drastic reduction in the average distance between molecules leads to the appearance of strong permanent interactions and to the creation of structures which, in most cases, hinder some molecular motion and introduce new efficient ways, other than collisions, for the energy flow to and from internal degrees of freedom. In general, the rotational motion of molecules is strongly coupled to the translational motion in liquids, and therefore the rotational-translational relaxation times become much too short to be observed at the present state of techniques. The relaxation time for the energy partition between translational motion of molecules and internal vibration are very short (10^{-12} s) for many liquids (associated) while other equilibria (structural) appear in the liquid: they may be perturbed by sound waves and cause other kinds of relaxation processes.

There are however a number of liquids, the so-called unassociated liquids, which behave at an extended degree as very dense gases; the prevalent observable relaxation process is bound to the energy partition between translational and vibrational degrees of freedom. The energy transfer occurs then, as in the case of gases, during binary collisions: because the times between collisions, however, in liquids are about 10^{-3} times those in gases, the same should roughly be expected for the relaxation times. We are interested here precisely in the study of these vibrational relaxations in the unassociated liquids, and it is clear the necessity of pushing as much as possible the high limit of the frequency range available for experiments with sound waves.

We shall consider what can be done both with the traditional ultrasonic methods where a sound wave of a fixed frequency is created by a source and studied in its propagation, and by making use of a thermally created sound wave having a selected wavelength, observed by means of light scattering.

The fundamental notions concerning the sound propagation in a liquid where a relaxation process occurs, with special regard to the case of vibrational relaxation, can be found in the literature (1,2,3).

2 ULTRASONIC MEASUREMENTS

Sound propagation measurements in the frequency (ν) range from a MHz to hundreds of MHz can be performed with various techniques. There are continuous or pulsed interferometers, optical methods making use of the light diffraction produced by sound waves. They allow accuracy better than 0.1% in the velocity (c) measurements and of 8 to 4% in absorption. Some methods have also been developed to allow measurements at higher frequencies; we quote here the method of E.S. Stewart and J.L. Stewart (4), which allows the determination of the specific acoustic impedance of the liquid placed on a quartz crystal in a microwave cavity up to 3 GHz, and the method of K.G. Plass (5) which measures the losses per cycle ($\alpha \cdot \lambda$, α being the amplitude absorption coefficient and λ the sound wave length) up to 2.8 GHz in the liquid placed between two lithium niobate crystals, each connected with an electric cavity. Recently an optical method which allows the determination of both c and α up to 1.5 GHz has been developed; it will be discussed in Section 6.

We shall present a very few selected results on vibrational relaxation in liquids obtained in the ultrasonic range. Let us first consider the case of diatomic molecules. Figure 1 gives the results for $\mu = \alpha \cdot \lambda$ obtained by H.G. Danielmeyer and

H.O. Kneser (6) with Br_2 in gaseous, liquid and solid forms; on the abscissa the parameter ν/ρ is used to eliminate the influence of the density (ρ). In the gaseous and liquid forms, a relaxation process with one relaxation time is observed: the full line is the theoretical curve obtained by using the relaxing specific heat of the only vibration present. The line for the solid form has two maxima: the lower frequency maximum is due to grain boundary absorption, while the second one is to be assigned to vibrational relaxation. The values of the product $\rho\tau$ ($g/s \cdot cm^{-3}$) (τ, relaxation time) are $5.5 \cdot 10^{-9}$ for gaseous Br_2, $3.2 \cdot 10^{-9}$ for liquid Br_2, and $2.1 (\pm 0.5) \cdot 10^{-9}$ for the solid form.

Apart from the case of diatomic molecules, vibrational relaxation involves more than one mode. The theoretical treatment of sound propagation in a system where various relaxation processes operate is not simple. The theory of a discrete number of processes occurring with different relaxation times yields different results (various shapes of dispersion and absorption curves) according to the ratio of the relaxation time, the magnitude of the relaxing terms in the specific heat and the way in which the energy flows: it may happen that the participants to each relaxation process exchange directly energy with the wave (parallel excitation) or that one degree of freedom gets energy through the disturbance of another one (series excitation), or that a mixed situation occurs.

Up to now the more common case which has been found has been

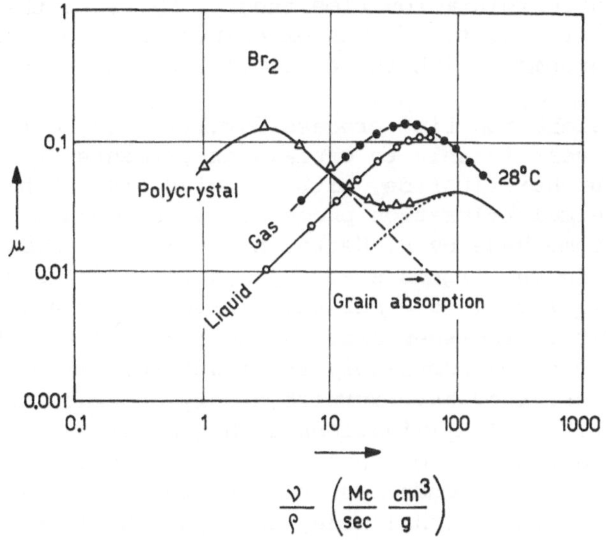

Figure 1 Sound absorption in gaseous, liquid and solid bromine (Danielmeyer) (6)

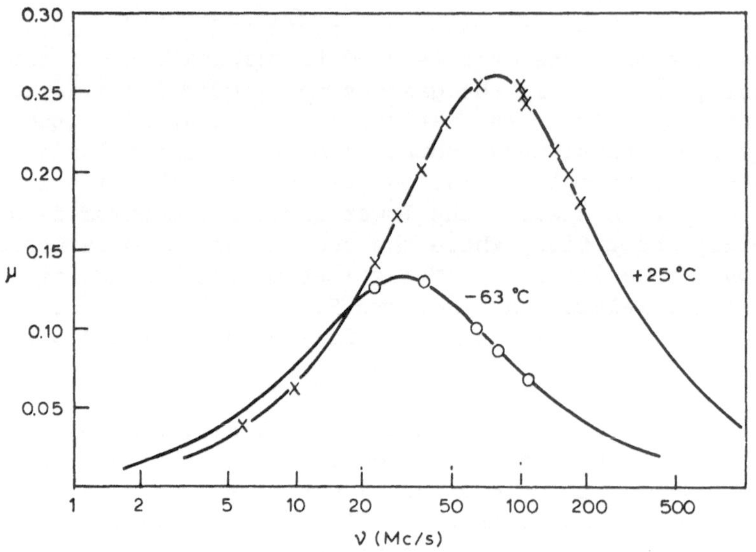

Figure 2 Sound absorption per wavelength in CS_2 (Andreae et al) (7)

the one in which all the modes relax with the same characteristic time; a few cases of double relaxation have, however, been detected, when a large frequency separation of the processes has allowed the analysis.

Figure 2 gives the results for μ of J.H. Andreae, E.L. Heasell and J. Lamb (7) in CS_2 at -63° C and 25° C. The solid curve is given by the single relaxation time theory, by using the total vibrational specific heat. The experimental values of μ_{max} are in remarkable agreement with those calculated from the theory.

A single relaxation time process involving the entire vibrational specific heat is able to explain experiments in many other liquids as sulfur hexafluoride, nitrous oxide, methylchloride. Two distinct thermal relaxation processes were found in the gaseous state of dichloromethane by D. Sette, A. Busala and J.C. Hubbard (8). The experiments in the liquid state by Andreae (9) and J.L. Hunter and H.D. Dardy (10) up to 510 MHz have allowed to confirm the presence of the two processes also in the liquid state, but while they were able to study accurately the lower relaxation frequency (∿170 MHz) process, they have furnished only general information on the other one. Determinations at higher frequencies (Section 6) are necessary for this. The two processes at lower and higher relaxation frequencies turn out to be associated with all but the lowest vibrational mode, and just the lowest one. Similar results were obtained by R. Bass and J. Lamb (11) in sulfur dioxide.

It was possible to observe that in the few cases in which it is found that all the vibrational degrees do not relax together, there is always a large difference in the wave-numbers of the lowest mode and of those of the others. One could therefore conclude that, in liquid as in gases, normally the entire vibrational specific heat relaxes with the same characteristic time, with the exception of liquids having molecules showing large differences in vibrational quanta, where multiple relaxation may be observed.

The results obtained with vibrational relaxation in liquids have allowed a deeper comparison between the behavior of the same substances in the gaseous and liquid states. Litovitz (12) in fact has used the experimental determination of relaxation times at various pressures and temperatures in various liquids, as well as measurements in the gaseous state to confirm the same nature of the process (binary collisions) of energy transfer between translational and vibrational degrees in the gaseous and liquid states, and to examine the theoretical calculation of relaxation times in liquids of this kind.

The body of experimental results obtained in the normal ultrasonic range indicates also the interest to gather information at higher frequencies, i.e. in the hypersonic range.

3 LIGHT SCATTERING AND THERMAL RELAXATION

When a monochromatic beam of light crosses a transparent medium, a small fraction of it is scattered away from the beam as a consequence of the existence of local variations of the dielectric constant. These variations are due to fluctuations in local temperature, density and particle orientation. The effect of temperature fluctuations can however usually be neglected as a consequence of the weak explicit dependence of molecular polarizability on temperature. The scattered light carries information on the motion of the molecules producing the scattering and therefore can be used, associated with a model for the fluctuations, to study bulk properties of the fluid: some aspects of the molecular dynamics, indeed, can be resolved in terms of the acoustical modes of the system allowing the study of the acoustical properties of the medium. In particular the spectrum of the diffracted light has, generally speaking, a central line and two symmetrically shifted lines (Brillouin doublet), whose origin can be traced to the diffraction by sound waves in which the adiabatic component of density fluctuations can be resolved: the frequency shift and the shape of the Brillouin lines are connected with the sound propagation characteristics. The situation however needs to be examined with a certain care in order to be able to deduce from the light scattering measurements the acoustic data. Many treatments have been proposed for which we refer to literature (13 to 17); we shall follow the general lines of Mountain's treatment (17).

The general scheme of the experiment we wish to consider is given in Figure 3.

A linearly polarized beam of light of intensity I_o and wave vector $\bar{k}_o$ impinges on the volume V; we observe the light scattered at an angle θ, at a position $\bar{R}$ from the scatterers. The orientation fluctuations give origin to a depolarized component of the scattered light; we assume that this component is negligible as it is the case when the anisotropy of molecules is small; the component is missing in the case of isotropic molecules. In such cases the scattered light has the same polarization of the impinging beam and it is entirely due to density fluctuations. The intensity of the scattered light is:

$$I(\bar{R},\omega) = I_o \left[\frac{Nk_0^4}{16\pi^2R^2}\right]\sin^2\phi < \varepsilon(k,\omega)^2 > \qquad (3.1)$$

where N is the number of molecules in the scattering volume, ϕ is the angle between the electric vector of the incident wave and $\bar{R}$, $\varepsilon(k,\omega)$ is the Fourier component of the fluctuation in the dielectric constant, ω is the angular frequency shift of the scattered light and $\bar{k}$ is the change in the light wave vector:

$$\bar{k} = \bar{k}_s - \bar{k}_o \qquad (3.2)$$

Because in Rayleigh scattering $|\bar{k}_o| \simeq |\bar{k}_s|$, the relation between k

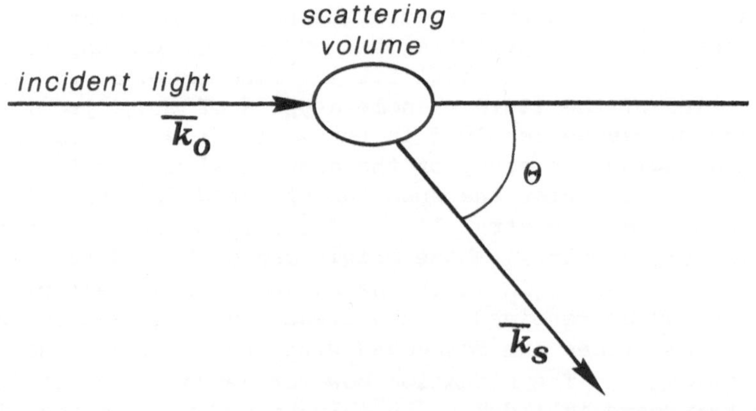

Figure 3 Scheme of light scattering experiments

and θ is:

$$k = 2 k_o \sin \frac{\theta}{2} \tag{3.3}$$

The angular brackets < . . . > indicate an overall average over the initial states. By neglecting the temperature fluctuations of ε, in relation to the density fluctuation, the intensity of the scattered light can also be put in the form:

$$I(\overline{R},\omega) = I_o \left[\frac{N\alpha^2 k^4}{16\pi^2 R^2}\right] \sin^2\phi \; S(k,\omega) \tag{3.4}$$

which makes use of the effective polarizability of the fluid α. S, the generalized structure factor, is the space and time Fourier transform of the density-density correlation function.

The calculation of S(k,ω) requires an appropriate kinetics model of the fluid. R.D. Mountain has carried out a calculation by using linearized hydrodynamic and the thermodynamic theory of fluctuations, and by assuming the existence of a single relaxation time process. The first two hypotheses are justified because many molecules are found in the length scale of the disturbance (sound wave length) and the corresponding periods are much longer than the collision times. The relaxation process is included in the hydrodynamic equation by considering a frequency-dependent part of the bulk (volume) viscosity (η'):

$$\frac{\eta'}{\rho_o} = \frac{b_1}{1 + i\omega\tau} \tag{3.5}$$

$$b_1 = (c_\infty^2 - c_0^2)\tau \tag{3.6}$$

where c_∞ and c_0 are the low frequency ($\omega << 1/\tau$) and high frequency ($\omega >> 1/\tau$) sound velocities. Mountain's treatment is valid for any kind of single relaxation time process leading to a frequency-dependent bulk viscosity. In the case of thermal relaxation equation (3.6) becomes:

$$b_1 = \frac{(C_p - C_v)C_i}{(C_v - C_i)C_p} c_0^2 \tau \tag{3.7}$$

C_v, C_p and C_i being the constant volume, constant pressure and vibrational specific heats.

The Mountain calculation for:

$$S(k,\omega) \quad = \quad < |\rho(k)|^2 > \sigma(k,\omega) \tag{3.8}$$

leads to an exact expression for $\sigma(k,\omega)$ in terms of third and fourth order polynomials. Mountain has also constructed a good approximation of $\sigma(k,\omega)$ valid for the case $(ak^2/\gamma)\tau \ll 1$, which allows the decomposition of σ in the sum of spectral lines:

$$\sigma'(k,\omega) \simeq I_c + I_M + 2I_B = \left(1 - \frac{1}{\lambda}\right)\left[\frac{2\lambda_T k^2/\rho_0 C_p}{(\lambda_T k^2/\rho_0 C_p)^2 + \omega^2}\right] +$$

$$+\left\{\frac{(c_\infty^2 - c_0^2)k^2 - \left[\frac{(c(k))^2}{c_0^2} - 1\right]\left[\frac{c_0^4}{c(k)^4\tau^2} + c_0^2 k^2\left(1 - \frac{1}{\gamma}\right)\right]}{\frac{c_0^4}{(c(k))^4\tau^2} - (c(k))^2 k^2}\right\}\frac{\frac{2c_0^2}{(c(k))^2\tau}}{\frac{c_0^2}{(c(k))^4\tau^2} + \omega^2} +$$

$$+\left\{\frac{\left[1 - \frac{c_0^2}{(c(k))^2\left(1 - \frac{1}{\gamma}\right)}\right]\left[(c(k))^2 k^2 + \frac{c_0^2}{(c(k))^2\tau^2}\right] - (c_\infty^2 - c_0^2)k^2}{\frac{c_0^4}{(c(k))^4\tau^2} + (c(k))^2 k^2}\right\} \times$$

$$\times\left\{\frac{\Gamma_B}{\Gamma_B^2 + \omega - c(k)k^2} + \frac{\Gamma_B}{\Gamma_B^2 + \omega + c(k)k^2}\right\} \tag{3.9}$$

where $a = \lambda_T/\rho_0 C_v$, λ_T the thermal conductivity, $(\Gamma_B)^{-1}$ the lifetime of the phonons and $c(k)$ their speed.

The first term represents the central component with the maximum at the frequency of the incident light; it arises entirely from the isobaric density fluctuations which decay via thermal diffusion; its shape is Lorentzian of half-width $\lambda_T k^2/\rho_0 C_p$. The second term is also centered at the frequency of the incident light; it is called the Mountain line, or the relaxation line, because it is due to the presence of the relaxation process in the bulk viscosity: it has a non-propagating type of decay. The last two terms represent the phonon modes, i.e. they are due to scattering by the adiabatic components of density fluctuations and constitute the Brillouin lines. Of course the second component would be missing in a non-relaxing fluid. The general shape of the spectrum is therefore rather complex, as it is determined by the combined effects of thermal diffusion, relaxation and sound wave propagation. For what relaxation is concerned, as already noted, it enters not only in the parameters describing the sound propagation as a function of frequency (and therefore of θ) but also in determining the Mountain line. It is useful at this point to stress the importance of this line by saying that the theory which we are describing, i.e. when relaxation occurs with only one characteristic time, has been extended to the case of longitudinal stress relaxation occurring with a distribution of relaxation times: the Mountain line has in itself all the information concerning the process.

The shape of the spectrum will depend on the relative importance of the quoted effects and shortly we shall discuss it in terms of parameters describing these effects. We wish however to make here some considerations on the connection between the characteristics of the Brillouin line and the sound parameters.

4 TEMPORAL ABSORPTION VERSUS SPATIAL ABSORPTION

In a typical ultrasonic experiment, progressive waves are used; the sound frequency imposed by the source is real and the propagating quantities, e.g. the pressure:

$$p = P \exp i\,(\omega t - kx) \tag{4.1}$$

are studied in space by assuming k as a complex parameter:

$$k = k_r + ik_i = k_r - i\alpha \tag{4.2}$$

Therefore:

$$p = P \exp k_i x \exp i(\omega t - k_r x) \tag{4.3}$$

with:

$$c = \frac{\omega}{k_r} \tag{4.4}$$

and α the spatial absorption coefficient.

In the case of Brillouin scattering, the situation is different because the wave vector is fixed by the geometry of the experiment: the situation is similar to the one encountered in the study of standing waves in solid rods. The decay to be studied is then a temporal decay. This corresponds to assuming in (4.1) k real and ω complex:

$$\omega = \omega_r + i\omega_i \tag{4.5}$$

with:

$$c = \frac{\omega_r}{k} \tag{4.6}$$

and ω_i the temporal absorption coefficient. It has been noted by J.M. Markham, R.T. Beyer and R.B. Lindsay (18) that the dispersion characteristics may be different for the two cases in the same absorbing medium.

It is important to call the attention on the need of carefully treating the results of experiments. Mountain's theory gives for $c(k)$ and Γ_B two coupled equations valid under some assumptions. They have to be used to analyze the results of Brillouin scattering experiments. It can be shown that only if one of the conditions $\omega_B >> 1$ or $\omega_B << 1$ holds, one gets:

$$c(k) = \frac{\omega_B}{k} \tag{4.7}$$

and the relation:

$$\alpha = \frac{\Gamma_B}{c} \tag{4.8}$$

between the spatial (α) and temporal absorption (Γ) coefficients.

5 THE SHAPE OF THE SPECTRUM OF SCATTERED LIGHT

Returning now to the shape of the spectrum as a combined expression of thermal diffusion, relaxation and sound propagation, various extreme regimes can be distinguished by means of the values of $\lambda_T/\rho_0 C_p$, τ, ω_B considered as parameters which characterize in order the three effects. Let us limit ourselves to the consideration of fluids where both $1/\tau$ and ω_B are much larger than $\lambda k/\rho_0 C_p$, i.e. when the adiabatic and isobaric fluctuations are uncoupled. We can distinguish various cases:

(1) $1/\tau >> \omega_B$: in such a case, the Mountain line is negligible, being in fact the relaxation τ short compared with $1/\omega_B$, the portion of the adiabatic density fluctuations associated with the relaxation process remaining in equilibrium with the phonon modes. The relaxation effects appear only in the parameters affecting the Brillouin lines. Moreover the central line is due entirely to the isobaric part of the density fluctuations: it is a Lorentzian of half-width $\lambda k/\rho_0 C$. The shape of the spectrum is given in Figure 4 (a): it is formed by three

well distinguished lines. In such a case, equations (4.7) and (4.8) are valid. The spectrum furnishes in a straightforward way the same information given by ultrasonic experiments, and can be used to extend the frequency range accessible to experiment for the study of relaxation.

From (3.3) one gets, for the wavelength and frequency of the selected elastic wave in a scattering experiment at angle θ, the expressions:

$$2n\lambda \sin\frac{\theta}{2} = \Lambda \tag{5.1}$$

$$\nu = \frac{c}{\lambda} = \frac{2nc\sin\frac{\theta}{2}}{\Lambda} \tag{5.2}$$

Λ/n being the light wavelength in the medium, and n the refraction index. Elastic waves of frequencies ranging from 0 to 10^{10} Hz would correspond to the possible values of θ (assuming $c = 1.5 \cdot 10^3 \cdot ms^{-1}$, and Λ around 4400 Å). The range of angles accessible to experiments is, however, much more limited: experiments at 0° and 180° cannot be performed because of the large parasitic light; at very small angles, ω_B is very small and this makes very difficult, or even impossible, the detection of the lines. In practice the majority of the scattering studies has been carried out in liquids at angles θ ranging between 20° and 175°, to which correspond frequencies of elastic waves of the order

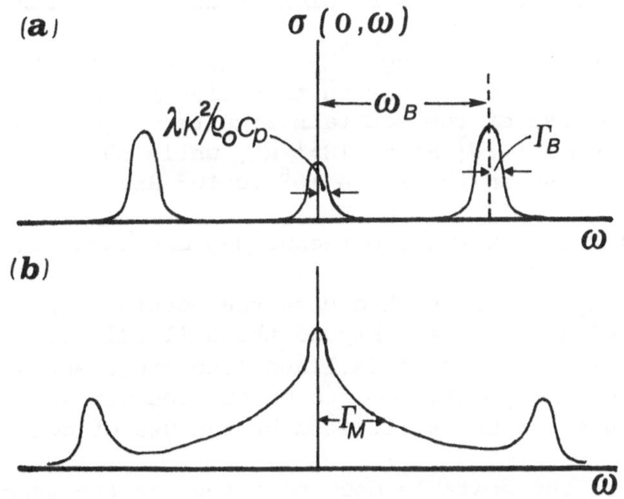

Figure 4 Typical scattering spectra: (a) case 1; (b) case 2

between $0.25 \cdot 10^{10}$ and $1 \cdot 10^{10}$ Hz. It is a field in which the frequency may change by a factor of 4, but whose position is very favorable for gathering information on relaxation processes.

(2) $\omega >> 1/\tau$: the shape of the spectrum is given in Figure 4 (b). The central line is formed by two contributions: the narrow Lorentzian term due to isobaric fluctuations just as in the previous case, and the Mountain line which has a larger width Γ_M. In such a case in fact the relaxing degrees of freedom of the density fluctuations equilibrate slowly and cannot contribute to the phonon modes. The spectrum has also the Brillouin resonance lines. The case is of relevant interest, as it has been noted by C.J. Montrose, V.A. Solovyev and T.A. Litovitz (15,19), because a single scattering experiment allows one to obtain the relaxation time, at differences of what can be done by ultrasonic techniques, where many experiments at different frequencies are required for the same purpose. It is in fact possible to show that in the case of single relaxation processes (of any kind) the Mountain half-width is:

$$\Gamma_M = \frac{c_0^2}{c_0^2} \tau \qquad (5.3)$$

while the position of the Brillouin peaks is given by:

$$c_\infty = \frac{\omega_B}{k} \qquad (5.4)$$

These considerations have been extended to the case of two or more relation times processes by W.H. Nichols and E.F. Carome (20).

The case we are examining, $\omega_B >> 1/\tau >> \lambda k^2/\rho_0 C_p$, is valid for fluids in which the relaxation time lies in the range 10^{-8} to 10^{-10} s. The use of the Mountain line technique allows studies in the range from $5 \cdot 10^7$ Hz to 10^{11} Hz, while the usual ultrasonic research can be done in the range 10^6 to 10^9 Hz.

It is evident that the two techniques are complementary.

(3) $1/\tau \simeq \omega_B$: in such a case the Mountain line extends from the central lines to and beyond the Brillouin lines. The decomposition of the spectrum in lines loses interest and the analysis of the experiments has to be carried out with the general equations: this can be facilitated by the use of computers.

Other cases and suitable decompositions of the spectra can be considered and they may assume interest in various circumstances

(19). We may however limit ourselves to the cases already examined in the light of our purpose.

6 SOME EXPERIMENTAL RESULTS ON BRILLOUIN SCATTERING AND ULTRASONIC MEASUREMENTS

In order to give an example of the study of vibrational relaxation which can be performed with Brillouin scattering, we shall report some results of Stegeman et al in CCl_4 (21); their analysis shows clearly the importance of the theoretical treatment sketched in the preceding paragraphs and the care that must be exercised in analyzing the spectrum in order to extract from the experiment accurate indications. Figure 5 gives an example of the spectra obtained at various scattering angles. The scattered light is polarized as the incident one. The authors have examined the scattering at 20 values of θ between 44° and 145°. The general shape of the spectrum clearly indicated the importance of the various components examined in Section 5 and, in particular, the influence of the Mountain line. The analysis has to be performed with the theory previously discussed. The best method would be to try to find the best least-square fit of the exact equation given by the theory to the experimental curve, with due

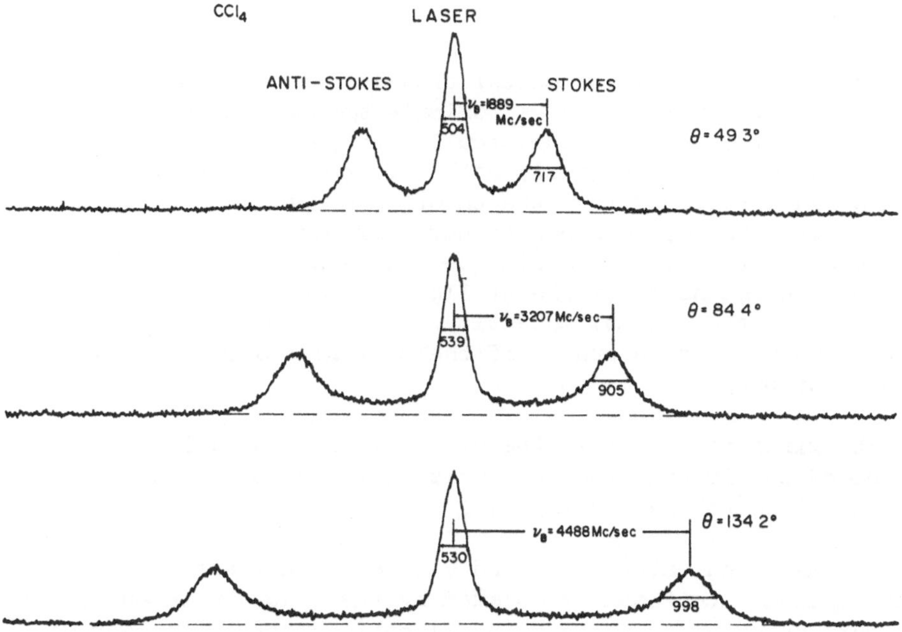

Figure 5 Scattered light spectra obtained in CCl_4 by Stegeman et al (21)

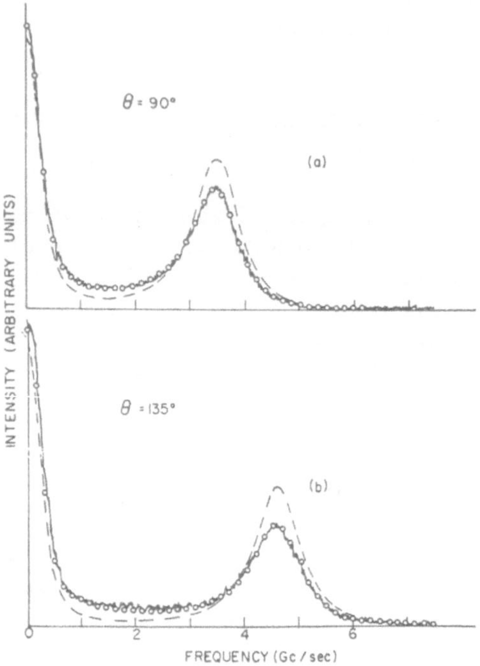

Figure 6 Observed spectra (full line) and computed intensities (circles) in CCl_4 (Stegeman et al) (21)

consideration of the instrumental profile. The authors have used a faster, sufficiently approximate method considering two parameters connected with the peculiar properties of the Brillouin spectra, which could be easily both obtained from the experimental spectra and calculated from the theory and changing the values assigned to τ and C_i in order to have agreement in the limit of experimental errors. To fix the two parameters, they have determined the intensity of the Brillouin component (I_B) measured relatively to the minimum intensity, which occurs at the midpoint between two successive orders of the central component in the Fabry-Perot output.

The first parameter is the width of the line (w) at $I_B/2$. The second one is the frequency shift (V) corresponding in the Brillouin line to w/2 and $I_B/2$.

The measured values of w and V at different scattering angles were compared with those calculated by the theory with an iterative procedure by changing τ and C_i until satisfactory agreement was found. The best results were found with $\tau = (6.5 \pm 0.5) \cdot 10^{11}$ s, $C_i = 11.6 \pm 0.3$ cal/mol°C. Figure 6 gives, for two scattering angles, the observed spectra (full line), the computed intensities (circles) and the spectra calculated by assuming no relaxation and the same value of Γ and $(\gamma - 1)$ as in observed spectra (dashed lines). The agreement between calculated and observed spectra is good. It is also evident the influence of the Mountain line, when relaxation is present. The values of C_i agree within the experimental error with the total vibrational specific heat (11.9 cal/mol°C) confirming the suggestion that in liquid CCl_4, the relaxation of all the vibrational degrees occurs with a single time as in the vapor phase. It should be noted, to stress the interest of these measurements, that they are performed in the relaxation region. The frequency-dependence of velocity and of Γ/k^2 can be calculated by using the values of τ and C_i obtained from the analysis. The agreement with the low frequency ultrasonic data is very good.

We wish to point out once again that traditional ultrasonic techniques, while able to furnish good accuracy in velocity and absorption up to a few hundred MHz, become increasingly less accurate, especially where absorption is concerned, as the frequency increases and a few isolated measurements are available in the GHz region. From the other side, the use of Brillouin scattering in the hundreds of MHz region requires an extremely accurate measurement of θ, because the values of θ are low: for instance, for measurements at 300 MHz, errors of 10" in the angle (around 3°) would lead to velocity errors of the order of 1 m. There has therefore been great interest in the development of a special optical ultrasonic method, which uses Bragg reflection (as Brillouin scattering) on injected waves of fixed frequency, and which allows high accuracy in the range from 100 MHz to 1000 MHz. The general principle was put forward by E.I. Gordon and M.G. Cohen (22): and based on this, K. Tagaki, and K. Negishi (23) have developed an apparatus for measurements up to 700 MHz. Progressive waves are sent into the liquid through a buffer rod while a laser beam is oriented to have Bragg reflection on them: changing the angle between the sound waves direction and the light beam in the proximity of the Bragg angle θ_r, one can detect a Lorentzian whose width is proportional to the space absorption coefficient, while θ_r, accurately determined, gives c.

With this method, K. Tagaki, P.K. Choi and K. Negishi (24) have accurately studied the double relaxation in dichloromethane and benzene. In benzene, their results agree with the assumption of Hunter et al (25) according to which the lowest relaxation process involves all but the slowest degenerate mode: the relaxation frequencies at 25° C would be 567 MHz and 3 GHz. M. Caloin and S. Candau (26) have also performed experiments with Brillouin scattering in both dichloromethane and benzene. Their results assign to the fastest relaxation process in dichloromethane the frequency of 4 GHz, while in the case of benzene, although supporting the existence of a double relaxation process, cast some doubt on the possibility of assigning each one of them exclusively to a particular set of vibrations. The paths of energy distribution between translational degrees of freedom and vibrational modes may be various (parallel, series, mixed).

7 CONCLUSIONS

This short review of a few experimental results has shown clearly the interest of integrating traditional ultrasonic techniques, Brillouin scattering techniques and Bragg reflection spectroscopy with imposed sound waves, in order to gather a coherent set of information on sound propagation in unassociated liquids by means of which the study of vibrational relaxation can be usefully performed.

REFERENCES

(1) K.F. Herzfeld, T.A. Litovitz, Absorption and dispersion of ultrasonic waves, Acad.Press, NY (1959)
(2) D. Sette, "Acoustic I", Handbuch der Physik, S. Flugge Ed., Springer, Berlin (1961), Vol.11
(3) J. Lamb, Thermal relaxation in liquids, Physical Acoustics, Vol.IIA, W.P. Mason Ed., Acad.Press (1965)
(4) E.S. Stewart, J.L. Stewart, JASA, 35, 975 (1963)
(5) K.G. Plass, Acustica, 19, 236 (1967/1968)
(6) H.G. Danielmeyer, Acustica, 17, 102 (1966)
(7) J.H. Andreae, E.L. Heasell, J. Lamb, Proc.Phys.Soc. (London), B29, 625 (1956)
(8) D. Sette, A. Busala, J.C. Hubbard, J.Chem.Phys., 23, 787 (1955)
(9) J.H. Andreae, Proc.Phys.Soc. (London), B70, 71 (1957)
(10) J.L. Hunter, H.D. Dardy, J.Chem.Phys., 42, 2961 (1965)
(11) R. Bass, J. Lamb, Proc.Roy.Soc. (London), A243, 94 (1957)
(12) T.A. Litovitz, J.Chem., 26, 469 (1957); also in Dispersion and absorption of sound by molecular processes, Proc. School of Physics, E. Fermi, Course XXVII, D. Sette Ed., Acad.Press (1963)
(13) L. Fabelinskii, Molecular scattering of light, English translation, Plenum, NY (1968)
(14) G. Benedek, T. Greytak, Proc.IEE 53, 1623 (1965)
(15) C.J. Montrose, V.A. Solovyev, T.A. Litovitz, JASA, 43, 117 (1968)
(16) D. McIntyre, J.V. Sengers, Study of fluids by light scattering, Physics of Simple Liquids, H.N.V. Temperley, J.S. Rowlinson, G.S. Rushbrooke Ed., North Holland Pub.Co. (1968)
(17) R.D. Mountain, Rev.Mod.Phys., 38, 205 (1966); J.Res.Nat. Bur.Std., 70A, 207 (1966)

(18) J.L. Markam, R.T. Beyer, R.B. Lindsay, Rev.Mod.Phys., 23, 353 (1951)
(19) C.J. Montrose, Light scattering and molecular acoustics, New Direction in Physical Acoustics, D. Sette Ed., North Holland Pub.Co. (1976)
(20) W.H. Nichols, E.F. Carome, J.Chem.Phys., 49, 1000 (1968)
(21) G.I. Stegeman, W.S. Garnell, V. Volterra, B.P. Stoicheff, JASA, 49, 979 (1971)
(22) E.I. Gordon, M.G. Cohen, Phys.Rev., 153, 201 (1967)
(23) K. Takagi, K. Negishi, Jap.J.Appl.Phys., 14, 29 (1975); Jap.J.Appl.Phys., 14, 149 (1975)
(24) K. Takagi, P.K. Choi, K. Negishi, JASA, 62, 354 (1977)
(25) J.L. Hunter, E.F. Carome, H.D. Dardy, J.A. Bucaro, JASA, 40, 313 (1966)
(26) M. Caloin, S. Candau, Supplement to J. de Physique, 33, C17 (1972)

THE DYNAMICS OF MOLECULAR CRYSTALS.

I. GENERAL THEORY

SALVATORE CALIFANO

ISTITUTO DI CHIMICA FISICA,LABORATORIO DI SPETTROSCOPIA MOLECOLARE,UNIVERSITA DI FIRENZE
VIA GINO CAPPONI 9,FLORENCE,ITALY

The simplest approach to the dynamics of molecular crystals is a straightforward extension of the theory developed many years ago by Born for simple atomic lattices,using crystal-fixed cartesian displacements of the atoms.[1]

Consider a piece of crystal,large enough to fulfill the Born-Von Karman cyclic conditions,containing L unit cells,with Z molecules per unit cell and N atoms per molecule.Let $X_\alpha^{a\mu i}$ represent the instantaneous cartesian coordinate in the α-th direction (α=X,Y,Z) referred to a space-fixed system,of atom i (i=1,2,..,N) of molecule μ (μ= 1,2,...,Z) in the unit cell a (a= 1,2,...,L) and $\bar{X}_\alpha^{a\mu i}$ the corresponding equilibrium position.The cartesian displacement coordinates of atom i are then defined as[2]

$$U_\alpha^{a\mu i} = X_\alpha^{a\mu i} - \bar{X}_\alpha^{a\mu i} \qquad 1.1$$

In terms of these displacement coordinates,the kinetic energy has the simple form

$$T = \sum_{a\mu}\sum T_{a\mu} = \frac{1}{2}\sum_{a\mu i\alpha}\sum\sum\sum m_i(\dot{U}_\alpha^{a\mu i})^2 \qquad 1.2$$

where m_i is the mass of atom i.

The potential energy of the crystal is expanded in a power series of the displacements

$$V = V_1 + V_2 + \ldots\ldots + V_m \qquad 1.3$$

with the generic m-th order term given by

$$V_m = \frac{1}{2} \sum\sum..\sum\sum\sum..\sum\sum\sum..\sum\sum\sum..\sum \Phi_{\alpha_1..\alpha_m}\begin{pmatrix} a & a \\ \mu .. \mu \\ i & i \end{pmatrix} U_{\alpha_1}^{a_1\mu_1 i_1} \cdots U_{\alpha_m}^{a_m\mu_m i_m} \qquad 1.4$$

where

$$\Phi_{\alpha_1..\alpha_m}\begin{pmatrix} a_1 & a_m \\ \mu_1 \cdots & \mu_m \\ i_1 & i_m \end{pmatrix} = (\partial^m V/\partial U_{\alpha_1}^{a_1\mu_1 i_1} \cdots \partial U_{\alpha_m}^{a_m\mu_m i_m}) \qquad 1.5$$

At equilibrium the crystal is assumed to be at a minimum of the energy and thus, the $U_\alpha^{a\mu i}$ being independent

$$\Phi_\alpha\begin{pmatrix} a \\ \mu \\ i \end{pmatrix} = 0 \qquad 1.6$$

For crystals at low temperatures the atomic displacements are small and we can neglect cubic and higher terms in the expansion 1.3. In this approximation, the crystal potential reduces to the second order term V_2 and this leads to the equations of motion

$$m_i \ddot{U}_\alpha^{a\mu i} + \sum_{b\nu j\beta}\sum\sum\sum \Phi_{\alpha\beta}\begin{pmatrix} ab \\ \mu\nu \\ ij \end{pmatrix} U_\beta^{b\nu j} = 0 \qquad 1.7$$

describing harmonic motions of the atoms. The occurrence of cyclic boundary conditions imposes for the system of 3NZL equations 1.7, solutions of the type [3]

$$U_\alpha^{a\mu i} = m_i^{\frac{1}{2}} E_\alpha^{\mu i} e^{-i(k.r_a - \omega(k)t)} \qquad 1.8$$

corresponding to plane waves of frequency $\omega(k)$ travelling in the crystal with wavevector k. In 1.8 r is the position vector, in the crystal fixed frame, of unit cell a. Equation 1.8, simply describes the fact that corresponding atoms in different unit cells have harmonic displacements from equilibrium which differ only by a phase factor defined by the value of k.

By substitution of 1.8, we contract the 3NZL equations 1.7 into 3NZ independent equations of the type

$$\sum_{\nu j\beta}\sum\sum (D_{\alpha\beta}^{\mu\nu ij}(k) - \omega^2(k)\delta_{\mu\nu}\delta_{ij}\delta_{\alpha\beta}) E_\beta^{\nu j} = 0 \qquad 1.9$$

where

$$D_{\alpha\beta}^{\mu\nu ij}(k) = \sum_b (m_i m_j)^{-\frac{1}{2}} \Phi_{\alpha\beta}\begin{pmatrix} ab \\ \mu\nu \\ ij \end{pmatrix} e^{-ik.r_b} \qquad 1.10$$

In matrix notation the system of equations 1.9 takes the form

$$D(k)E(k) = E(k)\Omega(k) \qquad 1.11$$

where D(k) is a 3NZx3NZ matrix with elements given by 1.10, called the " dynamical matric", $\Omega(k)$ is a diagonal matrix with elements $\omega_1^2(k)$ and E(k) is the orthonormal matrix of the eigenvectors.

If we plot the values of $\omega_1(k)$ as a function of k, we obtain 3NZ branches, called "dispersion branches".

The dispersion curves of a molecular crystal are conveniently classified into two groups.The first collects 6Z low frequency branches (external branches) corresponding to vibrations of the lattice in which the molecules perform rotational and translational motions around their equilibrium positions.The second group collects (3N -6)Z higher frequency branches corresponding to internal vibrations of the molecules.If the external and the internal dispersion curves are well separated in frequency,this formal classification assumes the direct physical meaning of a complete separation of the two types of motion.In this case the 6Z external vibrations of the crystal, for each value of k,can be treated assuming that the molecules behave as rigid bodies.In many cases,however,the molecules possess very low frequency internal vibrations which overlap in frequency with the lattice modes.In such cases the rigid body approximation breaks down and the coupling between the lattice and the internal modes cannot be neglected.

The simple treatment of the dynamics of molecular crystals outlined above presents practical and theoretical difficulties.We consider briefly the most important ones.The cartesian force constants contributing to the dynamical matrix are formal quantities,bearing no relation to the molecular or to the crystal structure.In addition the occurrence of physical entities such as the molecules in the crystal is completely obscured by the use of atomic displacement coordinates.Furthermore it is not possible with this choice of coordinates to take advantage of the rigid body approximation in which only translations and rotations of the molecules as a whole are considered.Finally some forms of intermolecular potential are not easily reduced to the cartesian force constants and on the other side, intramolecular potentials are normally expressed in terms of molecular coordinates.

For these reasons we introduce a set of "molecular coordinates" following a scheme proposed first by Wilson.For this we attach to each molecule in the crystal a local reference system with the origin at the center of mass and oriented along the principal axes of the molecular inertia tensor in the equilibrium position.We denote with small letters the cartesian coordinates of the atoms in the molecule-fixed system and define cartesian displacements [2]

$$u_{\rho}^{a\mu i} = x_{\rho}^{a\mu i} - \bar{x}_{\rho}^{a\mu i} \qquad 1.12$$

by analogy to 1.1.

The crystal-fixed and the molecule-fixed coordinates are related by the expressions

$$X_{\alpha}^{a\mu i} = X_{\alpha}^{a\mu} + \sum_{\rho} \Gamma_{\alpha\rho}\, x_{\rho}^{a\mu i} \qquad a)$$

$$\bar{X}_{\alpha}^{a\mu i} = \bar{X}_{\alpha}^{a\mu} + \sum_{\rho} \Lambda_{\alpha\rho}\, \bar{x}_{\rho}^{a\mu i} \qquad b)$$

$$U_{\alpha}^{a\mu i} = U_{\alpha}^{a\mu} + \sum_{\rho}(\Gamma_{\alpha\rho}^{\mu} x_{\rho}^{a\mu i} - \Lambda_{\alpha\rho}^{\mu} \bar{x}_{\rho}^{a\mu i}) \qquad c) \qquad 1.13$$

where $X_{\alpha}^{a\mu}, \bar{X}_{\alpha}^{a\mu}$ and $U_{\alpha}^{a\mu}$ are the instantaneous, equilibrium and displacement coordinates of the center of mass in the crystal-fixed system and $\Gamma_{\alpha\rho}$, $\Lambda_{\alpha\rho}$ are the instantaneous and the equilibrium value respectively of the direction cosine between the crystal-fixed axis α and the molecule fixed axis ρ.

We can use now the 3N molecule-fixed displacement coordinates $u_{\rho}^{a\mu i}$ to describe the 3N-6 internal degrees of freedom of each molecule and the three displacements of the center of mass plus three rotations around the molecule-fixed axes to describe the 6 external degrees of freedom. Using the displacements $u_{\rho}^{a\mu i}$ we define then 3N-6 "internal" normal coordinates for each molecule through the linear relations [4]

$$q_{t}^{a\mu} = \sum_{t}\sum_{\rho} \tilde{\mathcal{L}}_{\rho t} m_{i}^{\frac{1}{2}} u_{\rho}^{a\mu i} \qquad 1.14$$

plus six redundancy conditions.[4]

For the three degrees of translational freedom we define three translational normal coordinates as mass-weighted displacements of the center of mass along the three directions of the molecular axes by the relation [2]

$$t_{\rho}^{a\mu} = M^{\frac{1}{2}} U_{\rho}^{a\mu} \qquad 1.15$$

where M is the molecular mass. The displacements $U_{\rho}^{a\mu}$ are connected to the displacements $U_{\alpha}^{a\mu}$ in the crystal-fixed system 1.13c by the relation

$$U_{\alpha}^{a\mu} = \sum_{\rho} \Lambda_{\alpha\rho}^{\mu} U_{\rho}^{a\mu} \qquad 1.16$$

For the three degrees of rotational freedom we use in the same way three rotational normal coordinates as inertia moment-weighted rotations around the three molecular axes

$$r_{\rho}^{a\mu} = I_{\rho}^{\frac{1}{2}} \theta_{\rho} \qquad 1.17$$

The relation between the crystal-fixed cartesians and the rotations θ_{ρ} can be obtained from 1.13 once the dependence of the direction cosines $\Gamma_{\alpha\rho}^{\mu}$ on the rotations θ_{ρ} is specified. For this, consider the bases e and i formed by unit vectors along the three molecular and crystal axes respectively, connected at equilibrium by the relation

$$e = i\Lambda \qquad 1.18$$

and perform three rotations of the basis e in the order θ_z, θ_y, θ_x. We have

$$e' = eD_z$$

$$e'' = e'D_y = eD_zD_y \qquad 1.19$$

$$e''' = e''D_x = eD_zD_yD_x = i\Lambda D_zD_yD_x = i\Gamma$$

and thus

$$\Gamma = \Lambda D_zD_yD_x \qquad 1.20$$

where D_z, D_y, D_x are normal rotation matrices

$$D_z = \begin{pmatrix} \cos\theta_z & -\sin\theta_z & 0 \\ \sin\theta_z & \cos\theta_z & 0 \\ 0 & 0 & 1 \end{pmatrix} \quad D_y = \begin{pmatrix} \cos\theta_y & 0 & -\sin\theta_y \\ 0 & 1 & 0 \\ \sin\theta_y & 0 & \cos\theta_y \end{pmatrix} \quad D_x = \begin{pmatrix} 1 & 0 & 0 \\ 0 & \cos\theta_x & -\sin\theta_x \\ 0 & \sin\theta_x & \cos\theta_x \end{pmatrix}$$

We can now expand each rotation matrix in a power series of the rotation angles. We have

$$D_\rho = D_\rho + (\frac{\partial D_\rho}{\partial\theta_\rho})\theta_\rho \quad \frac{1}{2}(\frac{\partial^2 D_\rho}{\partial\theta_\rho^2})\theta_\rho^2 \qquad 1.21$$

For instance in the case of D_z we have

$$(\frac{\partial D_z}{\partial\theta_z}) = \begin{pmatrix} 0 & -1 & 0 \\ 1 & 0 & 0 \\ 0 & 0 & 0 \end{pmatrix} = M^z$$

$$1.22$$

$$(\frac{\partial^2 D_z}{\partial\theta_z^2}) = \begin{pmatrix} -1 & 0 & 0 \\ 0 & -1 & 0 \\ 0 & 0 & 0 \end{pmatrix} = M^zM^z$$

and in general we obtain

$$D_\rho = D_\rho + M^\rho\theta_\rho + \frac{1}{2}M^\rho M^\rho\theta_\rho^2 \qquad 1.23$$

with

$$M^z = \begin{pmatrix} 0 & -1 & 0 \\ 1 & 0 & 0 \\ 0 & 0 & 0 \end{pmatrix} \quad M^y = \begin{pmatrix} 0 & 0 & 1 \\ 0 & 0 & 0 \\ -1 & 0 & 0 \end{pmatrix} \quad M^x = \begin{pmatrix} 0 & 0 & 0 \\ 0 & 0 & -1 \\ 0 & 1 & 0 \end{pmatrix} \qquad 1.24$$

By substitution of 1.23 in 1.20 we obtain then

$$\Gamma = \Lambda\Big[E + \sum_\rho M^\rho\theta_\rho + \frac{1}{2}\sum_\rho M^\rho M^\rho\theta_\rho^2 + M^zM^x\theta_z\theta_x + M^zM^y\theta_z\theta_y + M^yM^x\theta_y\theta_x\Big] \qquad 1.25$$

We notice that the last part of 1.25 is not symmetric in the rotational coordinates θ_ρ but depends on the order in which the rotations are performed. To have a relation independent of the sequence of rotations we can either cut the series to the first order, in the assumption of infinitesimal rotations, or average the expression over all possible sequencies of rotation. This corresponds to a definition of Γ of the type

$$\Gamma = \frac{1}{2}\Lambda(D_zD_yD_x + D_xD_yD_z) \qquad 1.26$$

which leads to the symmetric expression

$$\Gamma = \Lambda\Big[E + \sum_{\rho} M^{\rho}\theta_{\rho} + \frac{1}{2}\sum_{\rho\sigma}\sum M^{\rho}M^{\sigma}\theta_{\rho}\theta_{\sigma}\Big] \qquad 1.27$$

From 1.31, by performing the matrix products we have

$$\Gamma_{\alpha\rho} = \Lambda_{\alpha\rho} - \sum_{\tau\chi}\sum\Lambda_{\alpha\tau}\delta_{\tau\rho\chi}\theta_{\chi} - \frac{1}{2}\sum_{\tau\omega\epsilon\chi}\sum\sum\sum\Lambda_{\alpha\tau}\delta_{\tau\omega\epsilon}\delta_{\rho\omega\chi}\theta_{\epsilon}\theta_{\chi} \qquad 1.28$$

where $\delta_{\tau\rho\chi}$ is the Levi-Civita symbol

$$\delta_{\tau\rho\chi} = \begin{cases} 0 \text{ unless } \tau \neq \rho \neq \chi \\ 1 \text{ if } \tau\rho\chi \text{ are in the cyclic order xyz} \\ -1 \text{ if } \tau\rho\chi \text{ are in the cyclic order yxz} \end{cases} \qquad 1.29$$

Once the set of molecular coordinates $q_t^{a\mu}$ $t_e^{a\mu}$ and $r_e^{a\mu}$ is defined, we can use it in place of the cartesian displacements $U_{\alpha}^{a\mu i}$ to set up the dynamical problem.

In what follows we shall use, in order to obtain more compact expressions, the generic symbol $S_L^{a\mu}$ for all molecular coordinates and, if needed, the symbol $P_m^{a\mu}$ (m= 1,2,..., 6) for the six external coordinates $t_e^{a\mu}$ and $r_e^{a\mu}$. In terms of these coordinates the kinetic and the potential energy take the form

$$T = \frac{1}{2}\sum_{a}\sum_{\mu}\Big[\sum_{m}\left(\dot{P}_m^{a\mu}\right)^2 + \sum_{t}\left(\dot{q}_t^{a\mu}\right)^2\Big] \qquad 1.30$$

$$V = \frac{1}{2}\sum_{a}\sum_{b}\sum_{\mu}\sum_{\nu}\Big[\sum_{m}\sum_{n}\Phi_{mn}^{a\mu\,b\nu}P_m^{a\mu}P_n^{b\nu} + \sum_{m}\sum_{t}f_{mt}^{a\mu\,b\nu}P_m^{a\mu}q_t^{b\nu} + \sum_{t}\sum_{s}F_{ts}^{a\mu\,b\nu}q_t^{a\mu}q_s^{b\nu}\Big] \qquad 1.31$$

where

$$\Phi_{mn}^{a\mu\,b\nu} = \left(\frac{\partial^2 V}{\partial P_m^{a\mu}\partial P_n^{b\nu}}\right)_0 \qquad f_{mt}^{a\mu\,b\nu} = \left(\frac{\partial^2 V}{\partial P_m^{a\mu}\partial q_n^{b\nu}}\right)_0 \qquad F_{ts}^{a\mu\,b\nu} = \left(\frac{\partial^2 V}{\partial q_t^{a\mu}\partial q_s^{b\nu}}\right)_0$$

We notice that the first term of V in 1.31 controls the lattice vibrations, the last term the internal vibrations wheras the second term is responsible for the coupling between them.

From the molecular coordinates we define then crystal normal coordinates through the linear transformation

$$Q_j(k) = L^{-\frac{1}{2}}\sum_{a}\sum_{\mu}\sum_{l}E_{\mu l}^{j\,*}(k)\; e^{-ik.r_a}\; S_l^{a\mu} \qquad 1.32$$

the inverse transformation being

$$S_l^{a\mu} = L^{-\frac{1}{2}}\sum_{k}\sum_{j}E_{\mu l}^{j}(k)\; e^{ik.r_a}\; Q_j(k) \qquad 1.33$$

By substitution of 1.33 in the expression of the kinetic energy we obtain

$$2T = \sum_{a}\sum_{\mu}\sum_{l} \left(\dot{S}_{l}^{a\mu}\right)^2$$

$$2T = L^{-1} \sum_{a}\sum_{\mu}\sum_{l}\sum_{k}\sum_{k'}\sum_{j}\sum_{j'} E_{\mu l}^{j}(k) E_{\mu l}^{j'}(k') e^{i(k + k').r_a} \dot{Q}_j(k)\dot{Q}_{j'}(k') \qquad 1.34$$

Using the crystal sum

$$e^{i(k + k').r_a} = \begin{cases} L \text{ if } k + k' = 0 \text{ or a reciprocal lattice vector} \\ 0 \text{ if } k + k' \neq 0 \end{cases} \qquad 1.35$$

and the orthonormality of the eigenvectors

$$\sum_{\mu}\sum_{l} E_{\mu l}^{j}(k) E_{\mu l}^{j'}(-k) = \delta_{jj'} \qquad 1.36$$

we obtain

$$2T = \sum_{k}\sum_{j} \dot{Q}_j(k)\dot{Q}_j(-k) = \sum_{k}\sum_{j} \dot{Q}_j(k)\dot{Q}_j^{*}(k) \qquad 1.37$$

By substitution of 1.33 in the expression of the potential energy we have also

$$2V = \sum_{a}\sum_{b}\sum_{\mu}\sum_{\nu}\sum_{l}\sum_{m} F_{lm}^{ab\,\mu\nu} S_{l}^{a\mu} S_{m}^{b\nu} = \qquad 1.38$$

$$= L^{-1} \sum_{a}\sum_{b}\sum_{\mu}\sum_{\nu}\sum_{l}\sum_{m}\sum_{k}\sum_{k'}\sum_{j}\sum_{j'} F_{lm}^{ab\,\mu\nu} E_{\mu l}^{j}(k) E_{\nu m}^{j'}(k')\; e^{i(k.r_a + k'.r_b)} Q_j(k) Q_{j'}(k')$$

where

$$F_{lm}^{ab\,\mu\nu} = (\partial^2 V / \partial S_{l}^{a\mu} \partial S_{m}^{b\nu})_0$$

By using the translational symmetry relation

$$F_{lm}^{ab\,\mu\nu} = F_{lm}^{(b-a)\,\mu\nu}_{\;0} \qquad 1.39$$

together with 1.35, we obtain

$$2V = \sum_{\mu}\sum_{\nu}\sum_{l}\sum_{m}\sum_{j}\sum_{j'}\sum_{k} D_{lm}^{\mu\nu}(k) E_{\mu l}^{j}(k) E_{\nu m}^{j'}(-k) Q_j(k) Q_{j'}(-k) \qquad 1.40$$

where the elements of the dynamical matrix are given by

$$D_{lm}^{\mu\nu}(k) = \sum_{b} F_{lm}^{0b\,\mu\nu} e^{ik.r_b} \qquad 1.41$$

Finally, using the relation, analogous to 1.11

$$\sum_{\mu}\sum_{\nu}\sum_{l}\sum_{m} D_{lm}^{\mu\nu}(k) E_{\mu l}^{j}(k) E_{\nu m}^{j'}(-k) = \omega_j^2(k)\delta_{jj'} \qquad 1.42$$

we obtain

$$2V = \sum_j \sum_k \omega_j^2(k) Q_j(k) Q_j(k) \qquad 1.43$$

Consider now the dynamical matrix elements 1.41. For their construction we need to specify the form of the crystal potential. Due to the specific nature of the molecular crystals we can write V in the form [2]

$$V = V_M + V_I \qquad 1.44$$

where V represents the intramolecular potential of the molecules in the crystal

$$V_M = \sum_c \sum_\lambda V^{c\lambda} \qquad 1.45$$

and V_I the interaction potential between different molecules. Neglecting multibody interactions, the intermolecular potential is expressed as the sum of all pairwise interactions between molecules

$$V_I = \sum_c \sum_d \sum_\lambda \sum_\pi{}' V_{d\pi}^{c\lambda} \qquad 1.46$$

where the symbol $\sum'$ means that $\lambda \neq \pi$ if c=d.

The intramolecular potential is normally expanded in a Taylor [4] series of the internal coordinates and the internal force constants are determined with well established techniques of molecular spectroscopy.Analytical forms of the molecule-molecule interaction potential are instead used in lattice dynamical calculations.We shall discuss the nature of these interactions in the next lecture. For the moment we concentrate our attention on the form of the elements of the dynamical matrix,when a crystal potential of the type 1.44 is used.

By substitution of 1.44 in 1.41 and by remembering that the coordinates $S_l^{a\mu}$ are normal coordinates of the isolated molecules,we obtain [5]

$$D_{lm}^{\mu\nu}(k) = \sum_b \sum_\lambda{}' \left(\frac{\partial^2 V^{o\mu}_{b\lambda}}{\partial S_l^{o\mu} \partial S_m^{o\mu}} \right)_o \delta_{\mu\nu} + \omega_l^2 \delta_{\mu\nu} \delta_{lm} + \sum_b{}' \left(\frac{\partial^2 V^{o\mu}_{b\nu}}{\partial S_l^{o\mu} \partial S_m^{b\nu}} \right)_o e^{-ik.r_b} \qquad 1.47$$

where $\sum_\lambda{}'$ means $\lambda \neq \mu$ if b = 0 and $\sum_b{}'$ means that b $\neq$ 0 if $\mu=\nu$.

References.

1. M.Born and K.Huang,"Dynamical theory of crystal lattices", Clarendon Press,Oxford (1954)
2. N.Neto,R.Righini,S.Califano and S.H.Walmsley,Chem.Phys.29:167(1978)
3. B.Donovan and J.F.Angress,"Lattice Vibrations",Chapman,London(1971)
4. S.Califano,"Vibrational States",J.Wiley,London(1976)
5. G.Taddei,H.Bonadeo,M.Marzocchi and S.Califano,J.Chem.Phys.58:167 (1973)

THE DYNAMICS OF MOLECULAR CRYSTALS

II. INTERMOLECULAR POTENTIALS

Salvatore Califano

ISTITUTO DI CHIMICA FISICA, LABORATORIO DI SPETTROSCOPIA
MOLECOLARE, UNIVERSITA DI FIRENZE
VIA GINO CAPPONI 9, FLORENCE, ITALY

From the theory of intermolecular forces we know that analytical forms are available for the intermolecular potential $V_{b\nu}^{a\mu}$ discussed in the previous lecture. This offers the advantage of making possible the calculation of a wide range of crystal properties from the same intermolecular potential, such as for instance the crystal structure, the thermodynamic and several dynamical properties.

Many different physical effects contribute to the total interaction between two molecules. A detailed discussion of the nature of the intermolecular forces and of the functional forms used, has been given by Prof. Buckingham in his lectures. We shall limit ourselves here to those aspects of the problem that are relevant for the lattice dynamics calculations.

The intermolecular forces are normally classified in terms of their interaction radius as short, medium and long range forces or, on the basis of their nature, as repulsive, dispersion, induction, electrostatic and polarization forces. For simplicity we shall collect under the generic name of Van der Waals forces, the short range repulsive, dispersion and induction forces and discuss them altogether.

A direct calculation of the Van der Waals energy from the knowledge of the electronic structure of the molecules is a formidable task even for small molecules and has been actually performed only for very simple systems. Phenomenological potentials are therefore invariably used in lattice dynamics calculations to account for these interactions. The main approximation made is to assume a complete additivity of atomic contributions in the molecule-molecule interaction. The intermolecular potential $V_{b\nu}^{a\mu}$ is thus expressed in the form [1]

$$V_{VW}{}_{b\nu}^{a\mu} = \sum_{ij} V_{b\nu j}^{a\mu i} \qquad 2.1$$

where the index i counts the atoms of molecule $a\mu$ and the index j the atoms of molecule $b\nu$. $V_{vw}^{a\mu i}_{b\nu j}$ represents then the Van der Waals interaction between two atoms of different molecules.

A number of analytical functions have been proposed in the literature to represent the atom-atom potential. The most known are the 6 - exp (modified Buckingham)

$$V^{a\mu i}_{b\nu j} = A\exp(-Br_{ij}) - Cr_{ij}^{-6} \qquad 2.2$$

and the Lennard-Jones potential

$$V^{a\mu i}_{b\nu j} = \alpha r_{ij}^{-n} - Cr_{ij}^{-6} \qquad 2.3$$

In these expressions A,B,C, and α are empirical constants characteristic of each kind of atom pair and r_{ij} is the atom-atom distance. The exponent n in 2.3 is normally taken equal to 12 but values ranging from 9 to 13 have been used in some cases. The exponential dependence on r_{ij} of the repulsive term, as well as the r_{ij}^{-6} dependence of the attractive term, taken as representative of the London dispersion forces, are derived from the theoretical quantum-mechanical treatment of the interaction between neutral atoms.[2]

The atom-atom potential is undoubtedly a simple and crude approximation. Despite this, it has been used often to represent the total intermolecular interaction and in many cases, specially for molecules containing only hydrogen and carbon atoms, it has been found that it works extremely well in reproducing crystal structures, energies and vibrational frequencies.[3]

The central nature of the atom-atom potential 2.2 or 2.3 has as consequence an isotropy of the interaction around the atom, that is clearly unrealistic for atoms linked by chemical bonds and even more so, for atoms with lone pairs or involved in delocalized π electron systems. Attempts to obtain anisotropic atom-atom potentials have been made very recently, but their possibilities are not yet fully exploited. The simplest solution seems the use of atom-atom coefficients including an angular dependence, but this introduces some complications in the construction of the dynamical matrix that have been ignored until now.

Other variations from the simple atom-atom potentials of equations 2.2 and 2.3 concern the attractive part. The r^{-6} term accounts actually only for part of the London dispersion forces. Additional terms, which depend on r^{-8} and r^{-10} and which include also the contribution of the induction forces, should in principle be considered. Since these terms would overestimate the attraction at very short distances, a damping function must be introduced in the potential. A more sophisticated atom-atom potential is then of the form [4]

$$V_{vw}^{a\mu i b\nu j} = A\exp(-Br_{ij}) - f(r_{ij})\left[C_6 r_{ij}^{-6} + C_8 r_{ij}^{-8} + C_{10} r_{ij}^{-10}\right] \quad 2.4$$

where C_6, C_8 and C_{10} are empirical constants and $f(r_{ij})$ is a damping function. A convenient form for $f(r_{ij})$ is

$$f(r_{ij}) = 1 \quad \text{for } r_{ij} > r_{ij}^{o}$$

$$f(r_{ij}) = \exp\left[-(r_{ij}^{o} r_{ij}^{-1} - 1)^2\right] \text{for } r_{ij} < r_{ij}^{o}$$

where r_{ij}^{o} is a limiting distance which depends on the type of atom pair.

In order to use the atom-atom potentials in the construction of the dynamical matrix elements, one needs the derivatives of the potential with respect to the molecular coordinates defined in the first lecture. For this it is only necessary to establish the relationship between the atom-atom distances and the molecular coordinates. The atom-atom distance is simply given by [1]

$$r_{ij} = \left[\sum_{\alpha} (X_{\alpha}^{b\nu j} - X_{\alpha}^{a\mu i})^2\right]^{\frac{1}{2}} \quad 2.5$$

and the crystal fixed cartesian coordinates are connected to the molecular coordinates by simple relations that have been discussed in the first lecture. It is therefore easy to express the derivatives of the atom-atom potential with respect to the molecular coordinates in terms of partial derivatives with respect to the cartesian coordinates. The details of the calculations are illustrated in reference 1 .

Another important contribution to the intermolecular potential is the electrostatic interaction between the charge distributions of the molecules. Two main approaches have been used to represent this type of interaction. The first is to localize charges,[4] fractions of charges or even dipoles[18] on the atoms or in the bonds. The second is to represent the charge distribution by an expansion in terms of point multipoles[1] at the center of charges. Both approaches have their advantages and disadvantages. The use of discrete charges is for instance simple, but there is no unique way of partitioning the charge distribution and this introduces a considerable amount of uncertainty in the problem. Furthermore serious problems of convergence arise in the lattice sums. Similar considerations can be made for the point dipoles localized on the atoms or in the bonds. In this case the convergence problem is less serious, but the uncertainty in the values of these formal dipoles remains a major problem.

the multipole expansion[1] has the advantage that a precise functional form is furnished by the theory and that the molecular multipoles are often available as experimental quantities or, if not, can be calculated with sufficient accuracy. It has however the drawback of being not valid for close-packed neighbouring molecules whose charge distributions overlap.

If discrete charges on the atoms or in the bonds are used, the interaction is simply given by the Coulomb potential $e^i e^j / r_{ij}$ which can be added directly to the atom-atom potential.

For the multipole expansion, the electrostatic interaction potential can be expressed in the compact form [1]

$$V_{el}^{a\mu b\nu} = \sum_{m=0}^{\infty} \sum_{n=0}^{\infty} (-1)^m \left[(2m-1)!!(2n-1)!!\right]^{-1} V_{mn}^{a\mu b\nu} \qquad 2.6$$

where

$$(2m-1)!! = (2m-1)(2m-3)\ldots 1 \qquad 2.7$$

Each term of this expansion represents the interaction potential between the multipole of order m on molecule $A=a\mu$ and the multipole of order n on molecule $B=b\nu$.

The multipole-multipole potential V_{mn}^{AB} is given by [1]

$$V_{mn}^{AB} = M_{(m)}^{A} \cdot T_{AB}^{(m+n)} \cdot M_{(n)}^{B} \qquad 2.8$$

where M and T are tensors of rank given by the indices in brackets and where the dots indicate the inner product of the tensors.

The T tensor is defined as [1]

$$T^{(m+n)} = \nabla^{(m+n)} R_{AB}^{-1} \qquad 2.9$$

where R_{AB} is the vector distance between the centers of mass of the molecules. The nomenclature used for the first few terms of the expansion for neutral molecules and their dependence on R are shown below

m	n	M^A	M^B		
1	1	μ	μ	dipole-dipole	R^{-3}
1	2	μ	Θ	dipole-quadrupole	R^{-4}
1	3	μ	Ω	dipole-octopole	R^{-5}
2	2	Θ	Θ	quadrupole-quadrupole	R^{-5}

Higher terms, depending on higher inverse powers of R, should be included in principle in the expansion of the electrostatic potential . In practice, however, since they overlap in the R dependence with the phenomenological atom-atom potential, that accounts for the short range dispersion forces, their inclusion simply amounts to a different choice of the empirical parameters A, B, and C of eq. 2.2. For this reason, terms higher than those depending on R^{-5}, can be neglected when the electrostatic potential 2.6 is used in combination with the atom-atom potential.

For the sake of clarity we write in full the electrostatic potential 2.6 for a pair of neutral molecules A and B

$$V_{el}^{AB} = -\mu^A.T_{AB}^{(2)}.\mu^B + (1/3)(\Theta^A.T_{AB}^{(3)}.\mu^B - \mu^A.T_{AB}^{(3)}.\Theta^B) + \frac{1}{9}(\Theta^A.T_{AB}^{(4)}.\Theta^B) - \frac{1}{15}(\mu^A.T_{AB}^{(4)}.\Omega^B + \Omega^A.T_{AB}^{(4)}.\mu^B) + \cdots \qquad 2.10$$

We consider now each multipole-multipole term V_{mn}^{AB}. By performing all inner tensor products, we obtain from 2.8, in crystal-fixed cartesian coordinates

$$V_{mn}^{AB} = \sum_{\alpha}\sum_{\beta}\cdots\sum_{\alpha'}\sum_{\beta'} M_{\alpha\beta\ldots}^{A} M_{\alpha'\beta'\ldots}^{B} T_{\alpha\beta\ldots\alpha'\beta'\ldots}^{AB} \qquad 2.11$$

where the dots mean m different sums over the cartesian components of molecule A and n different sums over the cartesian components of molecule B and where [1]

$$T_{\alpha\beta\gamma\ldots}^{AB} = \frac{\partial}{\partial R_\alpha}\frac{\partial}{\partial R_\beta}\frac{\partial}{\partial R_\gamma}\cdots\frac{1}{R_{AB}} \qquad 2.12$$

with

$$R_{AB} = \left[\sum_\alpha R_\alpha^2\right]^{\frac{1}{2}} \quad ; \; R_\alpha = X_\alpha^B - X_\alpha^A \qquad 2.13$$

The molecular multipoles are, however, known in the molecule-fixed reference system. By projecting them on the crystal axes, we have

$$V_{mn}^{AB} = \sum_{\alpha}\sum_{\beta}\cdots\sum_{\alpha'}\sum_{\beta'}\cdots\sum_{\rho}\sum_{\sigma}\cdots\sum_{\rho'}\sum_{\sigma'} T_{\alpha\beta\ldots\alpha'\beta'\ldots}^{AB}\,\Gamma_{\alpha\rho}^{A}\Gamma_{\beta\sigma}^{A}\cdots\Gamma_{\alpha'\rho'}^{B}\Gamma_{\beta'\sigma'}^{B}\cdots M_{\rho\sigma\ldots}^{A} M_{\rho'\sigma'\ldots}^{B} \qquad 2.14$$

In this expression the indices $\alpha,\beta,\ldots$ refer to the crystal-fixed and the indices $\rho,\sigma,\ldots$ to the molecule-fixed axes. The product involves m direction cosines of molecule A and n direction cosines of molecule B. As an example we specify eq. 2.14 in the case of the quadrupole-quadrupole interaction

$$V^{AB} = \frac{1}{9}\sum_{\alpha\beta\gamma\delta\rho\sigma\tau\pi} T_{\alpha\beta\gamma\delta}^{AB}\,\Gamma_{\alpha\rho}^{A}\Gamma_{\beta\sigma}^{A}\Gamma_{\gamma\tau}^{B}\Gamma_{\delta\pi}^{B}\,\Theta_{\rho\sigma}\Theta_{\tau\pi} \qquad 2.15$$

The number of independent components of a molecular multipole of order m is equal to 2m+1. The molecular symmetry reduces, however this number and this is of great help in simplifying the calculations. A table of the true independent components of the dipole, quadrupole and octopole moments, for several molecular symmetries, can be found in reference 1.

We consider now the use of 2.14 in lattice dynamical calculations. From the definition 2.9, the T tensor components depend clearly only on the distance between the centers of mass of the molecules and are thus sensitive only to translational molecular coordinates which change the position of the centers of mass.

In the same way,the direction cosines depend only on the molecular orientation and thus change only with the rotation coordinates. Finally the molecular multipoles depend on the molecular structure and change only with the internal coordinates.The variation of a molecular multipole with an internal coordinates is given by the expansion[1]

$$M_{\rho\sigma\ldots} = M^{o}_{\rho\sigma\ldots} + \sum_{t}(\partial M_{\rho\sigma\ldots}/\partial q_t)q_t + \ldots \qquad 2.16$$

The multipole-multipole potential is therefore well adapted to the coordinate definition given in the first lecture.For reasons of space we do not discuss here the construction of the dynamical matrix elements.For this one needs first and second derivatives of the potential with respect to the molecular coordinates.The general method of construction of these derivatives can be found in reference 1 .

A further contribution to the intermolecular potential arises from the polarization forces.These are typical three body forces in the sense that a multipole on one molecule interacts with a multipole on a second molecule,induced in turn by the multipole on a third molecule.Owing to the fact that these forces are weaker than the Van der Waals and the electrostatic forces,we can limit here the discussion to the interaction with induced dipoles. Using the same notation as before,we express the polarization potential as[5]

$$V_{pol} = \sum\sum\sum_{ABC}{}' \; \mu^{A}.T^{(2)}_{AB}.\alpha^{B}.T^{(2)}_{BC}.\mu^{C} - \frac{1}{3}(\mu^{A}.T^{(2)}_{AB}.\alpha^{B}.T^{(3)}_{BC}.\Theta^{C} + \Theta^{A}.T^{(3)}_{AB}.\alpha^{B}.T^{(2)}_{BC}.\mu^{C}) + \ldots \qquad 2.17$$

By performing the inner tensor products and by using,as before, molecular multipole components,we obtain

$$V_{pol} = \sum\sum\sum_{ABC}{}' \sum\sum\sum\sum\sum\sum\sum\sum_{\alpha\beta\gamma\delta\rho\sigma\tau\lambda} \Big[\mu^{A}_{\rho}\alpha^{B}_{\tau\lambda}\mu^{C}_{\sigma}T^{AB}_{\alpha\beta}T^{BC}_{\gamma\delta}\Gamma^{A}_{\alpha\rho}\Gamma^{B}_{\beta\tau}\Gamma^{B}_{\gamma\lambda}\Gamma^{C}_{\delta\sigma} - \frac{1}{3}\sum\sum_{\varepsilon\nu}(\mu^{A}_{\rho}\alpha^{B}_{\tau\lambda}\Theta^{C}_{\sigma\nu}T^{AB}_{\alpha\beta}T^{BC}_{\gamma\delta\varepsilon}\Gamma^{A}_{\alpha\rho}\Gamma^{B}_{\beta\tau}\Gamma^{B}_{\gamma\lambda}\Gamma^{C}_{\delta\sigma}\Gamma^{C}_{\varepsilon\nu} + \Theta^{A}_{\sigma\nu}\alpha^{B}_{\tau\lambda}\mu^{C}_{\rho}T^{AB}_{\alpha\beta\gamma}T^{BC}_{\delta\varepsilon}\Gamma^{A}_{\alpha\sigma}\Gamma^{A}_{\beta\nu}\Gamma^{B}_{\gamma\tau}\Gamma^{B}_{\delta\lambda}\Gamma^{C}_{\varepsilon\rho})\Big] \qquad 2.18$$

The polarization potential 2.18 has not yet been used for the calculation of the lattice vibrations,whereas has been often added to the electrostatic dipole-dipole potential for the calculation of dispersion curves of internal modes of molecular crystals.In most of the cases it has been found that its influence on the calculated frequencies is small but not negligible.[6 7]

The construction of the derivatives of the polarization potential with respect to the molecular coordinates follows the same pattern outlined before for the electrostatic potential.We shall not discuss this technical problem here.

A more general form of the intermolecular potential can be then obtained by combining together the contributions of the Van der Waals, electrostatic and polarization interactions

$$V^{AB} = V^{AB}_{vw} + V^{AB}_{el} + V^{AB}_{pol} \qquad 2.19$$

Before analyzing the performances of the potential 2.19, we discuss briefly some complications that arise in summing over the crystal the force constants derived from 2.19 for the construction of the dynamical matrix elements 1.47.

The Van der Waals potential depends on r^{-6} or on higher inverse powers of the atom-atom distance. Crystal sums converge thus rapidly to the limiting value and it has been actually found that an interaction radius of 6 - 8 A is sufficient to ensure complete convergence. The same is not true for the first terms of the electrostatic potential. For instance the charge-charge interaction depends on 1/R and the crystal sum is very slowly convergent. The dipole-dipole interaction, which depends on $1/R^3$, presents the same difficulty. Higher terms of the electrostatic potential depend on higher inverse powers of R and the corresponding crystal sums are again well convergent. A brute force approach can be used, extending the crystal sums over a large radius of interaction, but this is very expensive with large computers and never ensures a complete convergence. A number of powerful techniques exist, which permit to overcome this difficulty and to increase greatly the rapidity of convergence. The most known method is the Ewald-Kornfeld technique that is normally used in conjunction with the electrostatic potential.[8]

A large number of calculations of crystal vibrations has been made in recent years. References to the original papers can be found in the review by Schnepp and Jacobi[9] and in the book by Hexter[10] and Decius. A list of more recent works is given in reference 19. For reasons of space we shall limit ourselves here to a short discussion of a few selected examples, taken from the work done in our laboratory. More detailed discussions can be found in the original papers.

Internal vibrations of molecular crystals are shifted from the corresponding gas frequency and split in a number of components, depending on the symmetry of the unit cell and the number of molecules per cell. They have normally a very small dispersion, except in cases of vibrations giving rise to very strong infrared bands. In such cases the transition dipole-transition dipole interaction is almost completely responsible for the large dispersion of the mode. In addition, since the dipole-dipole is a long range interaction, it produces significant LO - TO (longitudinal - transverse) splittings at k = 0, that in some crystals can be directly observed in the vibrational spectrum.

An interesting example is that of the ω_3 mode of crystalline

CO_2,which gives rise to a very strong infrared band at 2345 cm^{-1}, with a $(\partial\mu/\partial q)$ of 206 e.s.u. .Crystalline CO_2 belongs to the face centered cubic system (space group T_h^6),with four molecules per unit cell,oriented along the diagonals of the cube.The dispersion curves of this mode,calculated using an intermolecular potential of the type [11]

$$V = -\frac{1}{2}\mu'.T^{(2)}.(E - \alpha T^{(2)})^{-1}.\mu' \qquad 2.20$$

which collects,in a more compact form,the electrostatic and the polarization interactions discussed before,are shown in fig.1.The symbol μ' in 2.20 represents the transition dipole contribution to the electrostatic potential,i.e. the second term of the expansion 2.16. On the right hand side of the figure we have added the histogram of the calculated density of phonon states and the shape of the observed absorption band for two different sample thicknesses.The frequency scale is given in positive and negative shifts from the gas phase frequency.Since there are four molecules per unit cell,there are four different crystal modes for ω_3.These are seen in the points of the Brillouin region without symmetry (direction Γ - X in fig.1). In other directions and in specific points of high symmetry,some of the modes are degenerate.For instance at k = 0 (Γ point),there are only three modes,one of which is doubly degenerate.Only one of these vibrations is infrared active and thus we obtain only one infrared band.The asymmetric broadening of the band,when the sample thickness increases, is a characteristic feature of modes with large LO - TO splittings and is due to the increase in reflectivity in the LO - TO frequency gap.

Another interesting example[12] is that of the ω_3 triply degenerate internal vibration of SiF_4.In this case the longitudinal mode at k = 0 is directly observable in the Raman spectrum and thus the comparison between experiments and calculations is even more significant.Silicon tetrafluoride crystallizes in the cubic system,space group T_d^3,with one molecule per unit cell.The only infrared active crystal mode at k = 0,gives again rise to a very strong absorption band and thus the potential 2.20 can be taken again as the dominant

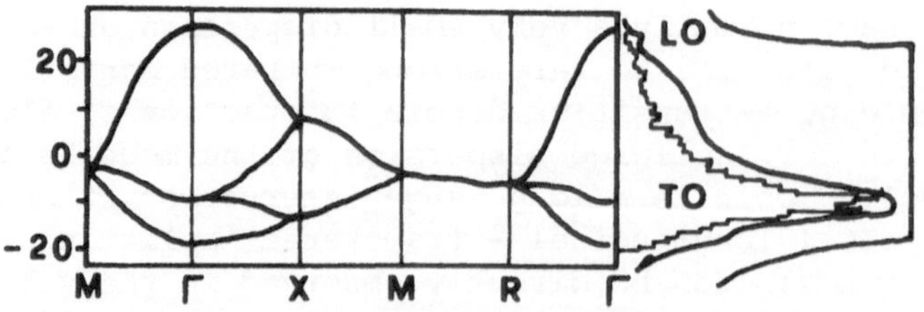

Fig. 1.Dispersion curves,density of phonon states and infrared absorption band shape for ω_3 of crystalline CO_2.

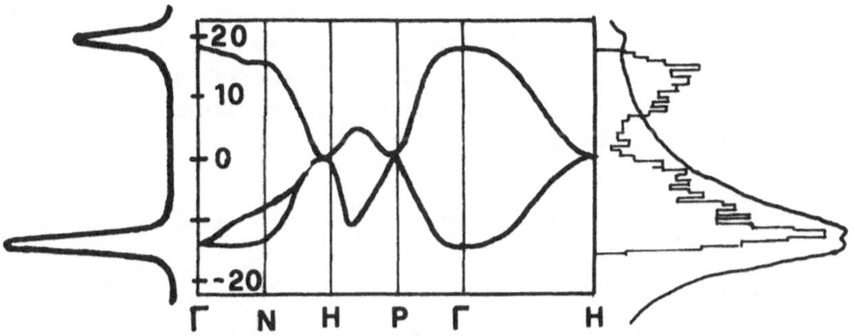

Fig. 2.Dispersion curves,infrared spectrum,Raman spectrum and density of phonon states for the ω_3 internal mode of crystalline SiF_4.

interaction of long range type.The dispersion curves,the infrared (right) and the Raman spectrum (left),together with the histogram of the density of phonon states,are shown in fig.2.In this case we have one infrared and two Raman active modes.

In the two examples discussed above there is an excellent agreement between the observed and calculated spectra and this proves the importance of the electrostatic terms of the potential,in particular of the dipole-dipole interaction,in the interpretation of the internal spectra of the molecular crystals.

The static multipole-multipole interactions,that are not involved in the internal vibrations,play an important role in the lattice vibrations of molecular solids,specially for polar molecules.A detailed analysis of the relative importance of the different terms of the intermolecular potential 2.19,has been carried out recently in our laboratory for crystalline ammonia.We shall briefly review here the main results obtained.[13]

Ammonia has a simple cubic structure,space group T^4,with four molecules per unit cell,located at sites of symmetry C_3.Each molecule has six neighbouring molecules to which is weakly hydrogen-bonded. The crystal possesses 21 optically active lattice modes that,because of the high site symmetry,are groupped into two non-degenerate modes of A species, two doubly degenerate modes of species E and five triply degenerate modes of species F.The optically active lattice vibrations have been calculated using an intermolecular potential including an atom-atom part of the type 2.2 and an electrostatic potential of the type 2.10.Owing to the weakness of the hydrogen bonds in this crystal,the only effort made to simulate this kind of interaction,was to use two different N...H atom-atom potentials,one for the hydrogen-bonded contacts and one for all others.

Because of the C_{3v} molecular symmetry,the electrostatic potential includes only one dipole, one quadrupole and two octopole independent components. By choosing the z axis along the symmetry axis of the molecule, these components can be designed as $\mu_z, \Theta_{zz}, \Omega_{zzz}$ and Ω_{xxx}. The relationships with other non-zero components of the quadrupole and of the octopole moments are

$$\Theta_{xx} = \Theta_{yy} = -\frac{1}{2}\Theta_{zz}$$

$$\Omega_{zxx} = \Omega_{zyy} = -\frac{1}{2}\Omega_{zzz} \qquad \Omega_{xyy} = -\Omega_{xxx}$$

The observed and calculated lattice frequencies of crystalline NH_3 are collected in table 1. Four sets of calculations are included in the table, in order to show the relative influence on the lattice modes, of the different terms of the potential. The first set was obtained using only the atom-atom part of the potential. The other sets were obtained by adding, one at a time, the dipole, the quadrupole and the octopole contributions. The table includes also the contributions of the different potential terms to the energy and to the equilibrium conditions of the crystal.

Table 1. Observed and calculated lattice modes of NH_3 crystal.

Symmetry	I	II	III	IV	exp.	mode
A	95	84	117	143	...	T_z
A	196	194	203	306	310	R_z
E	122	122	119	122	107	T_{xy}
E	192	195	278	299	298	R_{xy}
F	132	135	125	130	140	T
F	186	183	187	181	183	T
F	128	129	130	273	260	R_z
F	203	206	327	360	358	R_{xy}
F	265	316	483	524	533	R_{xy}

I. Atom-atom. II = I + dipole. III = II + quadrupole. IV = III + octopole.

Crystal energy (Kcal./mole)

at-at	μ-μ	μ-Θ	Θ-Θ	μ-Ω	tot	exp.
-3.19	-1.21	-1.09	-1.14	-1.50	-8.13	-8.5 $\mp$ 1

Equilibrium conditions (Kcal/mole)/A or /rad.

-0.40	0.34	0.77	-0.48	-0.05	0.18	0.0	$\partial V/\partial t_z$
0.06	0.00	0.00	0.00	0.20	0.26	0.0	$\partial V/\partial r_z$

Table 1 shows very clearly that the electrostatic potential affects only specific crystal frequencies.In addition the different multipole moments have different effects on the lattice frequencies. For instance ω_9 is sensitive to all multipole,ω_3 and ω_8 only to the quadrupole and octopole and ω_2 and ω_7 only to the octopole.

Very extensive calculations of the lattice modes of crystalline NH_3 and ND_3,including the full set of dispersion curves,the corresponding one- and two-phonon density of states,the width of the lattice bands and the elastic constants of the crystal,have been performed to test the validity of the mixed atom-atom plus multipole-multipole potential.The results obtained are in excellent agreement with the experiments and confirm the validity of this model potential.[4,13,14]

As a last example we consider the crystal of naphthalene for which a very complete set of experimental data exists.The crystal structure is monoclinic,space group C^5_{2h},with two molecules per unit cell,located at sites of C_i symmetry.The crystal structure is known at 300° and 120° K and the cell constants have been measured recently also at 6° K.The k = 0 lattice vibrations have been measured at several temperatures in infrared and Raman,with detailed polarization measurements,while the phonon dispersion curves have been recently obtained at 6° K by an international group at the nuclear reactor in Grenoble.[15]

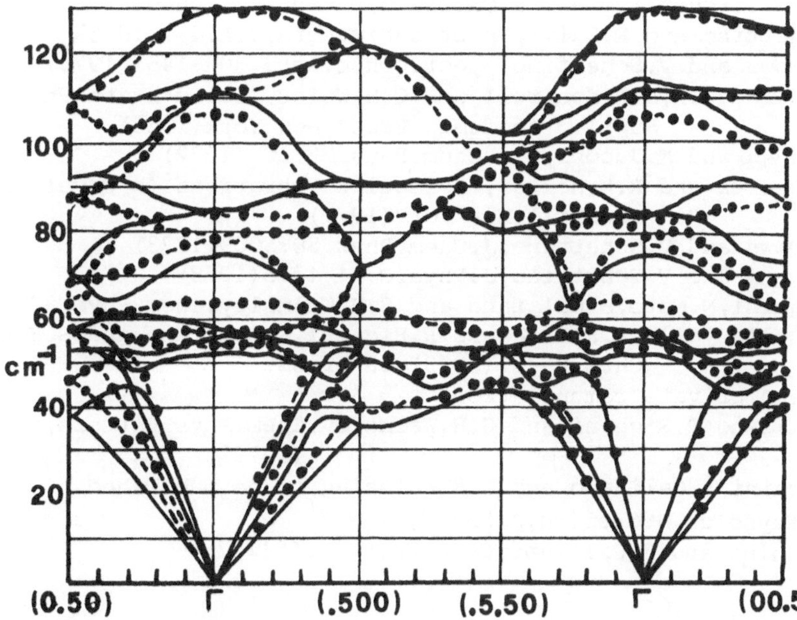

Fig.3.Observed (dotted line) and calculated (full line) dispersion curves of naphthalene at 6°K.Experimental points shown by dots.

A number of calculations of the lattice frequencies of naphthalene have been made in the past using only the atom-atom potential. The agreement between observed and calculated frequencies is normally good, except for the highest A_g and B_g lattice modes. Experimentally it is found that the B_g mode is about 20 cm^{-1} higher than the A_g mode, whereas the calculations predict the A_g mode at higher frequency.

An intermolecular potential which includes, in addition to the atom-atom, the electrostatic interaction, furnishes the correct frequency sequence for these two modes. Since the molecule has a center of symmetry, the dipole and octopole moments are zero and the only electrostatic interaction that one needs to consider is the quadrupole-quadrupole term. Calculations have been made in our laboratory for the 300°K optical frequencies and for the complete phonon dispersion curves at 6°K. The results of the latter calculations are compared with the experimental data in fig.3.[16,17]

References

1. N.Neto, R.Righini, S.Califano and S.H.Walmsley, Chem.Phys.29:167(1978)
2. H.Margenau and N.R.Kestner, "Theory of intermolecular forces", Pergamon Press, Oxford (1969)
3. D.E.Williams, J.Chem.Phys.45:3770(1966)
H.Bonadeo and E.D'Alessio, in "Lattice Dynamics and intermolecular Forces", ed.by S.Califano, Academic Press (1975)
4. R.Righini and M.Klein, J.Chem.Phys.68:5553 (1978)
5. R.Giua, V.Schettino and S.Califano, to be published.
6. V.Schettino and R.Salvi, Spectrochim.Acta, 31A:411(1975)
7. D.A.Dows and V.Schettino, Spectrochim.Acta, 30A:1451(1974)
8. A.A.Maradudin, E.W.Montroll and G.H.Weiss, "Solid state Physics, Suppl.3, Academic Press, New York(1963)
9. O.Schnepp and N.Jacobi, Adv.Chem.Phys.22:205(1972)
10. J.C.Decius and R.M.Hexter, "Molecular Vibrations in Crystals", Mc Graw-Hill, New York(1977)
11. D.A.Dows and V.Schettino, J.Chem.Phys.58:5009(1973)
12. F.Bogani and V.Schettino, J.Phys.C 11:1275(1978)
13. R.Righini, N.Neto, S.Califano and S.H.Walmsley, Chem.Phys.33:345(1978)
14. R.G.Della Valle, P.F.Fracassi, R.Righini, S.Califano and S.H.Walmsley, Chem.Phys.(1979) in press.
15. E.Sheka, private comunication.
16. S.Califano, R.Righini and S.H.Walmsley, Chem.Phys.Letters, 64:491 (1979)
17. R.Righini, S.Califano and S.H.Walmsley, to be published.
18. P.A.Reynolds, J.Chem.Phys.59:2777(1973)
19. R.Righini and S.Califano, Chem.Phys.17:45(1976)

THE DYNAMICS OF MOLECULAR CRYSTALS:

III. INFRARED AND RAMAN INTENSITY OF LATTICE BANDS

Salvatore Califano

ISTITUTO DI CHIMICA FISICA, LABORATORIO DI SPETTROSCOPIA MOLECOLARE, UNIVERSITA DI FIRENZE
VIA GINO CAPPONI 9, FLORENCE, ITALY

The lattice vibrational frequencies, observed in the infrared and Raman spectra or in neutron scattering experiments, represent the most important body of experimental data for the study of the dynamical behaviour of the molecular crystals. From them one obtains precise and detailed information about the intermolecular potentials and about the cooperative motions of the molecules in the solid phase.

The intensity and the shape of the infrared and Raman lattice bands constitute another valid source of information on the dynamics of the molecules in the crystal and on their mutual interactions. The potentialities of intensity data have been well recognized in the past by many authors, but have not yet been sufficiently investigated in comparison to the extensive work accumulated on the intensity of the internal bands of gases and liquids. The main reason for this is that in gases and liquids the lattice frequencies are completely missing and thus one is forced to make virtue out of necessity, using the only valid information at disposal on the intermolecular interactions and on the dynamics of the interacting molecules.

The type of information that one can extract from intensity measurements in the crystalline phase is complementary to that obtained from the lattice frequencies. The vibrational intensities are in fact governed by the molecular moments and polarizabilities and these quantities change from the gas or liquid to the solid because of the intermolecular interactions. In addition, band intensities are much more sensitive than the frequencies themselves, to the form of the crystal eigenvectors and therefore furnish a very powerful test for the validity of the intermolecular potentials used in lattice dynamics calculations. Finally they are connected to the optical and dielectric properties of the materials. Intensity theories furnish then a bridge between microscopic theories of the dielectric and of the dynamical behaviour of the molecular crystals.

The approach normally utilized in the interpretation of lattice band intensities[1] is based on the so called "oriented gas model". The infrared intensity of a lattice mode is assumed to originate from the change in orientation of the permanent molecular moments, essentially the molecular dipoles, during the lattice vibrations. In the same way, the Raman intensity of a lattice band is attributed to the change in orientation of the anisotropic molecular tensor.

A direct consequence of these assumptions is that translational motions of the molecules in the crystal give rise to infrared or Raman bands with vanishing intensities, since in a molecular translation the orientation of the molecular dipole or polarizability tensor does not change. The intensity in infrared and Raman of a lattice band is thus entirely attributed to the effect of the rotational types of motions.

The calculations of the intensities of lattice vibrations by means of the oriented gas model, have been reviewed recently[2]. The failures of the oriented gas model have been clearly pointed out by many authors, since it has been often observed experimentally that modes with prevailing translational character give rise to very strong Raman bands, sometimes even stronger than those of bands due to molecular rotations.

We have recently developed[3] a theory of the infrared and Raman intensity of lattice bands, that takes into account the effect of the crystal field produced on each molecule by all other molecules in the crystal. In this theory, the translational motions of the molecules in the crystal field gradient at each site, give rise to strong variations of the molecular dipole and polarizability and thus originate infrared and Raman bands of noticeable intensity. We shall briefly describe here the main features of the theory and illustrate some applications to molecular crystals. Since the formalism used is similar to that utilized in the previous lectures we shall not repeat here definitions already given before unless not strictly necessary for the understanding of the text.

The crystal electric moment can be defined as the sum over all molecules of effective molecular moments

$$P = \sum_{a}\sum_{\mu} P^{a\mu} \qquad 3.1$$

In the same way the crystal polarizability can be defined as the sum of effective molecular polarizabilities

$$A = \sum_{a}\sum_{\mu} A^{a\mu} \qquad 3.2$$

where P is a 3NZ dimensional vector and A is a 3NZ x 3NZ tensor.

The crystal moment P can be considered as the sum of the static crystal moment μ and of the moment induced by the crystal field. We can thus write P as

$$P = \mu + \alpha F \tag{3.3}$$

where F is the local field acting on each site. The field F is produced by the charge distribution of the molecules in the crystal. By expanding this charge distribution in terms of molecular multipoles and by neglecting terms higher than the quadrupole, we have

$$F = T^{(2)}.P + \frac{1}{3} T^{(3)}.\Theta + \varepsilon \tag{3.4}$$

In eq. 3.4 we have added to the crystal field an external field ε, for generality. We recall from the previous lecture that $T^{(2)}$ and $T^{(3)}$ are second and third rank tensors, whose elements, in crystal fixed coordinates, are given by

$$T^{AB}_{\alpha\beta} = \frac{\partial}{\partial R_\alpha}\frac{\partial}{\partial R_\beta}\frac{1}{R_{AB}} \tag{3.5}$$

$$T^{AB}_{\alpha\beta\gamma} = \frac{\partial}{\partial R_\alpha}\frac{\partial}{\partial R_\beta}\frac{\partial}{\partial R_\gamma}\frac{1}{R_{AB}} \tag{3.6}$$

where R_{AB} is the vector distance between the centers of mass of the molecules A and B and $R_\alpha, R_\beta, R_\gamma$ are the cartesian components. The T tensor elements have the properties

$$T^{AB}_{\alpha\beta} = T^{BA}_{\alpha\beta} \qquad T^{AB}_{\alpha\beta\gamma} = -T^{BA}_{\alpha\beta\gamma} \tag{3.7}$$

By introduction of 3.4 in 3.3 we obtain

$$P = (E - \alpha T^{(2)})^{-1}.(\mu + \frac{1}{3}\alpha T^{(3)}\Theta + \alpha\varepsilon) \tag{3.8}$$

and by using the expansion

$$(E - \alpha T^{(2)})^{-1} = E + \alpha T^{(2)} - \alpha T^{(2)}\alpha T^{(2)} + \ldots.$$

up to the second term, we have

$$P = (\mu + \alpha T^{(2)}\mu + \frac{1}{3}\alpha T^{(3)}\Theta) + (\mu + \alpha T^{(2)}\alpha)\varepsilon \tag{3.9}$$

The first term of 3.9 is the dipole operator that controls the infrared intensity and the second term is the field induced dipole operator that produces the Raman scattering.

Consider first the infrared intensity. By performing all the inner tensor products we obtain from 3.9

$$P_{\alpha} = \sum_{a\mu}\sum (\mu_{\alpha}^{a\mu} + \sum_{b\nu}\sum{}' \sum_{\beta\gamma}\sum \alpha_{\alpha\beta}^{a\mu} T_{\beta\gamma}^{a\mu\ b\nu} \mu_{\gamma}^{b\nu} \quad \frac{1}{3} \sum_{b\nu}\sum{}' \sum_{\beta\gamma\delta}\sum\sum \alpha_{\alpha\beta}^{a\mu} T_{\alpha\beta\gamma}^{a\mu\ b\nu} \Theta_{\gamma\delta}^{b\nu}) \qquad 3.10$$

In the same way we obtain from 3.9, for the Raman operator

$$A_{\alpha\beta} = \sum_{a\mu}\sum (\alpha_{\alpha\beta}^{a\mu} + \sum_{b\nu}\sum{}' \sum_{\gamma\delta}\sum \alpha_{\alpha\gamma}^{a\mu} T_{\gamma\delta}^{a\mu\ b\nu} \alpha_{\delta\beta}^{b\nu}) \qquad 3.11$$

For the calculation of the infrared and Raman intensities we need the derivatives of 3.10 and 3.11 with respect to the crystal normal coordinates at k = 0, since only these modes are optically active. For this we define symmetrized local coordinates

$$S_{l}^{\mu}(0) = L^{-\frac{1}{2}} \sum_{a} S_{l}^{a\mu} \qquad 3.12$$

and from these, crystal normal coordinates

$$Q_{j}(0) = \sum_{\mu l}\sum E_{\mu l}^{j}(0) S_{l}^{\mu}(0) \qquad 3.13$$

The inverse transformations of 3.12 and 3.13 are easily obtained (see eq. 1.33 of the first lecture). From these we obtain, for the derivative of an operator

$$\left(\frac{\partial \zeta}{\partial Q_{j}(0)}\right) = L^{-\frac{1}{2}} \sum_{a\mu l}\sum\sum E_{\mu l}^{j} \left(\frac{\partial \zeta}{\partial S_{l}^{a\mu}}\right) \qquad 3.14$$

Introducing then the lattice sums

$$S_{\beta\gamma}^{\mu\nu} = \sum_{b}{}' T_{\beta\gamma}^{a\mu\ b\nu}$$
$$S_{\beta\gamma\delta}^{\mu\nu} = \sum_{b}{}' T_{\beta\gamma\delta}^{a\mu\ b\nu} \qquad 3.15$$

we obtain for the dipole and for the polarizability derivatives

$$\left(\frac{\partial P_{\alpha}}{\partial S_{l}^{\mu}(0)}\right) = \left(\frac{\partial \mu_{\alpha}^{a\mu}}{\partial S_{l}^{a\mu}}\right) + \sum_{\nu\beta\gamma}\sum\sum \frac{\partial}{\partial S_{l}^{a\mu}} (\alpha_{\alpha\beta}^{\mu} S_{\beta\gamma}^{\mu\nu} \mu_{\gamma}^{\nu} + \alpha_{\alpha\beta}^{\nu} S_{\beta\gamma}^{\nu\mu} \mu_{\gamma}^{\mu})$$
$$+ \frac{1}{3} \sum_{\nu\beta\gamma\delta}\sum\sum\sum \frac{\partial}{\partial S_{l}^{a\mu}} (\alpha_{\alpha\beta}^{\mu} S_{\beta\gamma\delta}^{\mu\nu} \Theta_{\gamma\delta}^{\nu} + \alpha_{\alpha\beta}^{\nu} S_{\beta\gamma\delta}^{\nu\mu} \Theta_{\gamma\delta}^{\mu}) \qquad 3.16$$

$$\left(\frac{\partial A_{\alpha\beta}}{\partial S_{l}^{a\mu}(0)}\right) = \left(\frac{\partial \alpha_{\alpha\beta}^{a\mu}}{\partial S_{l}^{a\mu}}\right) + \sum_{\nu\gamma\delta}\sum\sum \frac{\partial}{\partial S_{l}^{a\mu}} (\alpha_{\alpha\gamma}^{\mu} S_{\gamma\delta}^{\mu\nu} \alpha_{\delta\beta}^{\nu} + \alpha_{\alpha\gamma}^{\nu} S_{\gamma\delta}^{\nu\mu} \alpha_{\delta\beta}^{\mu}) \qquad 3.17$$

In these expressions we have not included the normalizing factors, since they will cancel out when the intensities normalized per unit cell are computed. As discussed in the previous lecture, the molecular moments and polarizability are known in terms of molecule-fixed coordinates and not of crystal-fixed coordinates as required by the equations 3.16 and 3.17. The transformation between the two bases can be easily made by means of the matrix Γ of the direction cosines. Without loss of generality the molecular axes can be oriented so that the dipole moment lies along the z axis and the molecular polarizability and quadrupole tensors are diagonal. In this way we have

$$\mu^{\mu} = \Gamma^{\mu}\mu \quad \text{a)}$$
$$\alpha^{\mu} = \Gamma^{\mu}\alpha\tilde{\Gamma}^{\mu} \quad \text{b)} \qquad 3.18$$
$$\Theta^{\mu} = \Gamma^{\mu}\Theta\tilde{\Gamma}^{\mu} \quad \text{c)}$$

The molecular coordinates $S_{1}^{a\mu}$ can be either translational coordinates $t_{\rho}^{a\mu}$ or rotational coordinates $r_{\rho}^{a\mu}$ or internal coordinates $q_{t}^{a\mu}$. Using their definition, given in the first lecture, and their relationships with the crystal-fixed cartesian coordinates, condensed in the relations

$$X_{\delta}^{a\mu} = \bar{X}_{\delta}^{a\mu} + M^{-\frac{1}{2}}\sum_{\rho}\Gamma_{\delta\rho}^{\mu}t_{\rho}^{a\mu} \qquad 3.19$$

$$\Gamma_{\delta\rho}^{\mu} = \Lambda_{\delta\rho}^{\mu} + I_{\rho}^{-\frac{1}{2}}\sum_{\sigma\tau}\sum\Lambda_{\delta\tau}^{\mu}\delta_{\tau\rho\sigma}t_{\sigma}^{a\mu} \qquad 3.20$$

we obtain for the dipole operator

$$(\partial P_{\alpha}/\partial t_{\lambda}^{\mu}(0)) = -M^{-\frac{1}{2}}\sum_{\nu\beta\gamma\delta\tau}\sum\sum\sum\sum\Big[\alpha_{\tau\tau}\mu_{z}S_{\beta\gamma\delta}^{\mu\nu}\Lambda_{\delta\lambda}^{\mu}(\Lambda_{\alpha\tau}^{\mu}\Lambda_{\beta\tau}^{\mu}\Lambda_{\gamma z}^{\nu} + \Lambda_{\alpha\tau}^{\nu}\Lambda_{\beta\tau}^{\nu}\Lambda_{\gamma z}^{\mu}) + \frac{1}{3}\sum_{\epsilon\rho}\sum\alpha_{\tau\tau}\Theta_{\rho\rho}S_{\beta\gamma\delta\epsilon}^{\mu\nu}\Lambda_{\epsilon\lambda}^{\mu}(\Lambda_{\alpha\tau}^{\mu}\Lambda_{\beta\tau}^{\mu}\Lambda_{\gamma\rho}^{\nu}\Lambda_{\delta\rho}^{\nu} + - \Lambda_{\alpha\tau}^{\nu}\Lambda_{\beta\tau}^{\nu}\Lambda_{\gamma\rho}^{\mu}\Lambda_{\delta\rho}^{\mu})\Big] \qquad 3.21$$

$$(\partial P_{\alpha}/\partial r_{\lambda}^{\mu}(0)) = I^{-\frac{1}{2}}\sum_{\sigma}\Big\{\Lambda_{\alpha\sigma}^{\mu}\mu_{z}\delta_{\sigma z\lambda} + \sum_{\nu\beta\gamma\tau}\sum\sum\sum\alpha_{\tau\tau}\mu_{z}S_{\beta\gamma}^{\mu\nu}\Big[\Lambda_{\alpha\tau}^{\nu}\Lambda_{\beta\tau}^{\nu}\Lambda_{\gamma\rho}^{\mu}\delta_{\sigma z\lambda} +-\Lambda_{\gamma z}^{\nu}(\Lambda_{\alpha\sigma}^{\mu}\Lambda_{\beta\tau}^{\mu} + \Lambda_{\alpha\tau}^{\mu}\Lambda_{\beta\sigma}^{\mu})\delta_{\sigma\tau\lambda}\Big] + \frac{1}{3}\sum_{\nu\beta\gamma\delta\tau\rho}\sum\sum\sum\sum\sum\alpha_{\tau\tau}\Theta_{\rho\rho}S_{\beta\gamma\delta}^{\mu\nu}\Big[\Lambda_{\gamma\rho}^{\nu}\Lambda_{\delta\rho}^{\nu}(\Lambda_{\alpha\sigma}^{\mu}\Lambda_{\beta\tau}^{\mu} + \Lambda_{\alpha\tau}^{\mu}\Lambda_{\beta\sigma}^{\mu})\delta_{\sigma\tau\lambda} + - \Lambda_{\alpha\tau}^{\nu}\Lambda_{\beta\tau}^{\nu}(\Lambda_{\gamma\sigma}^{\mu}\Lambda_{\delta\rho}^{\mu} + \Lambda_{\gamma\rho}^{\mu}\Lambda_{\delta\sigma}^{\mu})\delta_{\sigma\rho\lambda}\Big]\Big\} \qquad 3.22$$

In the same way we obtain for the polarizability operator

$$(\partial A_{\alpha\beta}/\partial t_{\lambda}^{\mu}(0)) = -M^{-\frac{1}{2}}\sum_{\nu\gamma\delta\epsilon\tau\sigma}\sum\sum\sum\sum\sum\alpha_{\tau\tau}\alpha_{\sigma\sigma}(\Lambda_{\alpha\tau}^{\mu}\Lambda_{\gamma\tau}^{\mu}\Lambda_{\delta\sigma}^{\nu}\Lambda_{\beta\sigma}^{\nu} + \Lambda_{\alpha\tau}^{\nu}\Lambda_{\gamma\tau}^{\nu}\Lambda_{\delta\sigma}^{\mu}\Lambda_{\beta\sigma}^{\mu})\Lambda_{\epsilon\lambda}^{\mu}S_{\gamma\delta\epsilon}^{\mu\nu} \qquad 3.23$$

$$(\partial A_{\alpha\beta}/\partial r_{\lambda}^{\mu}(0)) = I_{\lambda}^{-\frac{1}{2}}\sum_{\tau\rho}\sum\Big\{\alpha_{\tau\tau}(\Lambda_{\alpha\rho}^{\mu}\Lambda_{\beta\tau}^{\mu} + \Lambda_{\alpha\tau}^{\mu}\Lambda_{\beta\rho}^{\mu})\delta_{\rho\tau\lambda} +$$

$$+ \sum_{\nu\gamma\delta\sigma}\sum\sum\sum \alpha_{\tau\tau}\alpha_{\sigma\sigma} S^{\mu\nu}_{\gamma\delta} \Big[\Lambda^{\nu}_{\delta\sigma}\Lambda^{\nu}_{\beta\sigma}(\Lambda^{\mu}_{\alpha\rho}\Lambda^{\mu}_{\gamma\tau} + \Lambda^{\mu}_{\alpha\tau}\Lambda^{\mu}_{\gamma\rho})\,\delta_{\rho\tau\lambda} +$$

$$+ \Lambda^{\nu}_{\alpha\tau}\Lambda^{\nu}_{\gamma\tau}(\Lambda^{\mu}_{\delta\rho}\Lambda^{\mu}_{\beta\sigma} + \Lambda^{\mu}_{\delta\sigma}\Lambda^{\mu}_{\beta\rho})\,\delta_{\rho\sigma\lambda}\Big]\Big\} \qquad 3.24$$

If the molecules have low symmetry, the principal axes of the polarizability and quadrupole tensors may not coincide.In such cases different Λ matrices must be used.The changes in the above given formulas are straightforward.We have not given here the final expressions for the derivatives of the moment and of the polarizability with respect to internal coordinates.These are however easily obtained from the equations 3.16 and 3.17 by expanding the moment and the polarizability in terms of the internal coordinates and by making then the axis transformation described above.

We consider now some applications of the theory to specific cases.For this we notice that,as far as the infrared intensities are concerned,the squares of the dipole derivatives 3.21 and 3.22 can be used directly,since the intensities are proportional to them. For the Raman intensities instead,one needs to consider the scattering geometry used in each case.

For measurements on single crystals in polarized light,the intensity of a Stokes line at frequency ν_m is given by

$$I_{\alpha\beta} = \frac{C(\nu_0 - \nu_m)^4}{(1 - \exp[h\nu_m/KT])}\,\frac{1}{\nu_m}\,A'^{\,2}_{\alpha\beta} \qquad 3.25$$

where C is a constant,ν is the frequency of the exciting line and A' is the polarizability derivative given by 3.23 or 3.24.For vapor deposited samples,i.e.for most of the measurements done on crystals of simple molecules at low temperature,the scattering geometry is not defined.Assuming that they consist of randomly oriented crystallites,we can use the intensity expression for liquids averaged on the parallel and perpendicular observation,in order to take into account the depolarization of the laser beam,caused by the multiple reflections within the sample.We have therefore used for this case the expression[4]

$$I = \frac{C(\nu_0 - \nu_m)^4}{(1 - \exp[h\nu_m/KT])}\,\frac{1}{\nu_m}(4.5A'^2 + \Gamma'^2) \qquad 3.26$$

where

$$A' = \frac{1}{3}(A'_{xx} + A'_{yy} + A'_{zz}) \qquad 3.27$$

and

$$\Gamma'^2 = \frac{1}{2}\,(A'_{xx} - A'_{yy})^2 + (A'_{yy} - A'_{zz})^2 + (A'_{zz} - A'_{xx})^2 \quad +$$

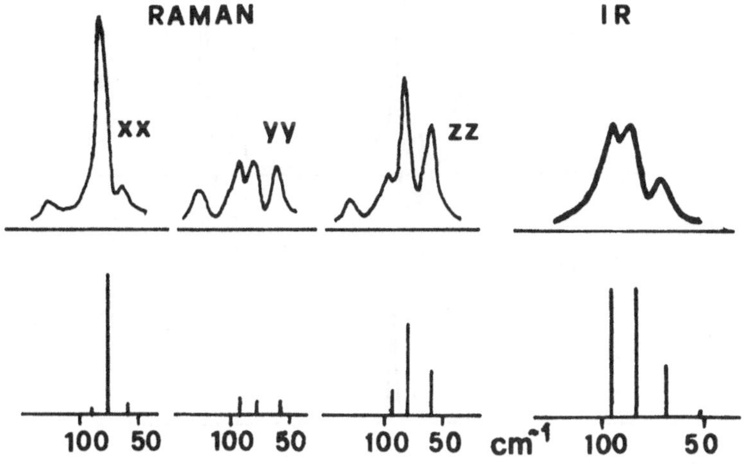

Fig.1.Observed (top) and calculated (bottom)Raman and infrared spectra of crystalline benzene in the region of the lattice vibrations.Single crystal spectrum in polarized light for Raman.Pölycrystalline sample for infrared.

$$+ 6(A'^2_{xy} + A'^2_{xz} + A'^2_{yz}) \qquad 3.28$$

In fig.1 we have reported the Raman spectrum in polarized lightt of crystalline benzene (single crystal),measured some years ago in our laboratory[5]and the far infrared spectrum obtained by Harada et all[6] on a polycrystalline sample.At the bottom of the figure we have schematically represented the band intensities obtained with our model.The infrared intensity arises only from moments induced by the molecular quadrupoles,since the benzene molecule has no dipole moment.The excellent agreement found between experiment and theory, proves clearly the importance of the quadrupole induced intensity in the case of centrosymmetrical molecules.The agreement between the calculated and the observed Raman spectrum of the single crystal is also very satisfactory,if one takes into account the incomplete experimental polarizations of the sample,due to the cracks formed in the crystal at low temperature.

As a second example we consider the Raman spectrum of crystalline ammonia,a molecular crystal that we have investigated in details in the last two years.The crystal eigenvectors,necessary for the construction of the crystal normal coordinates,were obtained using the mixed atom-atom plus multipole-multipole potential,described in the previous lecture.The relative intensities were obtained from the spectrum of a polycrystalline sample obtained from vapor deposition at 130° K and cooled successively to 10° K.The calculated Raman spectra ,using the oriented gas model and the more general model described here,are compared to the experimental spectrum in fig.2.

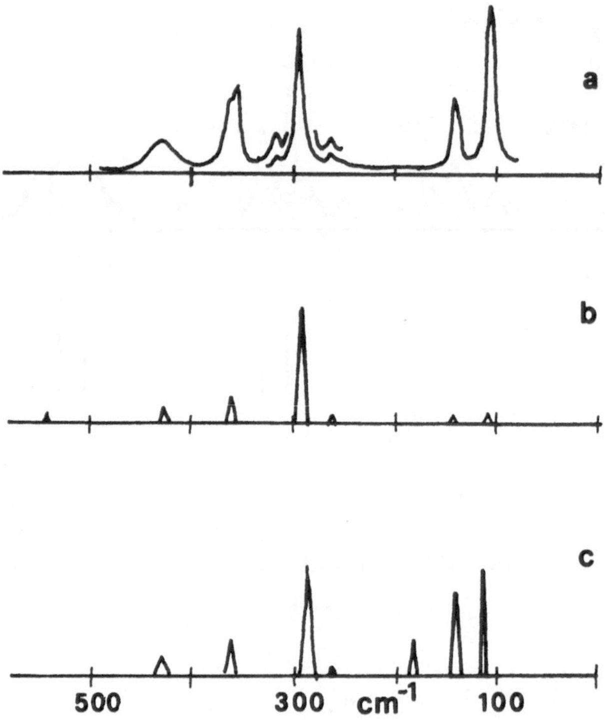

Fig.2.Calculated and observed Raman spectrum of crystalline NH_3. a)observed spectrum.b)oriented gas model.c)spectrum calculated with our theory.

The following values of the molecular parameters were used:

$$\mu = 1.53 \times 10^{-18} \text{ e.s.u.}$$
$$\Theta_{xx} = \Theta_{yy} = + 1.18 \times 10^{-26} \text{ e.s.u.} \qquad \Theta_{zz} = - 2\Theta_{xx}$$
$$\alpha_{xx} = \alpha_{yy} = 2.18 \times 10^{-24} \text{cm}^3 \qquad \alpha_{zz} = 2.42 \times 10^{-24} \text{cm}^3$$

From fig.2 it is easily seen that the oriented gas model is unable to reproduce correctly the observed pattern of the Raman spectrum of crystalline NH_3.In particular the two low frequency translational modes have a very strong intensity in the observed spectrum,while have no appreciable intensity in the spectrum calculated with the oriented gas model.Introduction of the local field effects,has a drastic effect on the calculated intensities in the low frequency region.The two translational modes acquire a very strong intensity,comparable to that of the band at about 300 cm^{-1}, which is the only strong band in the spectrum calculated with the oriented gas model.We have performed several types of calculations with our model,with different eigenvectors obtained in different calculations of the crystal frequencies.It has been found,as expected,that the intensities are sensitive to the form of the normal coordinates.It is therefore possible to foresee that a refinement of the intermolecular potential,in which attention is payed to the form of the crystal normal coordinates,could give rise to calculated intensities,in the molecular crystal spectra,in perfect agreement with the experimental results.

References

1. A.Kastler and A.Rousset ,J.Phys.Radium 2:49 (1941)
2. H.Bonadeo and E.D'Alessio,J.Chem.Phys.63:38 (1975)
3. V.Schettino and S.Califano,J.Chim.Phys.76:197 (1979)
4. J.E.Cahill and G.E.Leroi,J.Chem.Phys.51:97 (1969)
5. H.Bonadeo,M.P.Marzocchi,E.Castellucci and S.Califano,J.Chem.Phys. 57:4299 (1972)
6. I.Harada and T.Shimanouchi,J.Chem.Phys.46:2708 (1967)

ANHARMONIC EFFECTS IN MOLECULAR CRYSTALS:

HAMILTONIAN PERTURBATION TREATMENT

Salvatore Califano

Istituto Di Chimica-Fisica, Laboratorio Di Spettroscopia
Molecolare, Universita Di Firenze
Via Gino Capponi 9, Italy

In the previous lectures we have developed the general theory of the harmonic vibrations of molecular crystals and discussed the intermolecular interactions which control their structure and their dynamical properties.We extend now the treatment to anharmonic vibrations,in order to cover features of the lattice dynamics that were not considered before.

For solid state spectroscopy two anharmonic effects are of special interest: a)the shift in frequency of overtones and combination bands from the predicted harmonic value;b) the width of infrared and Raman bands.

The first effect represents the major source of information on the anharmonicity of the internal vibrations,since overtones or combinations of internal modes are easily observed in the vibrational spectra.This type of information is however missing for lattice modes for which such bands are normally not observed.For external vibrations therefore only the second effect is experimentally detectable.

The band width of a lattice mode is directly connected to the finite lifetime of the phonon state.A complete treatment should include all possible multiphonon processes which contribute to the phonon decay and this leads to calculations of enormous complexity. In the low frequency region however,and at very low temperatures, the phonon decay will be essentially controlled by three phonon processes and thus one can reasonably expect that a simplified treatment,in which higher order processes are neglected,should furnish an adequate approximation to the bandwidth of molecular crystals.

In the lectures of Professor Michel the anharmonic problem has been treated using the Green' function formalism. In this lecture I shall instead utilize an hamiltonian perturbation method developed by Wallace[1] which leads to the same formal results but has the advantage of a greater simplicity.

Using the formalism of the previous lectures we expand the Hamiltonian to the m-th order in series of molecular coordinates $S_l^{a\mu}$

$$H = H_o + H_1 + H_2 + \ldots + H_m \qquad 4\text{-}1$$

where H_o is the harmonic hamiltonian

$$H_o = T + V_2$$

and

$$H_1 = V_3, \quad H_2 = V_4, \quad \ldots . \; H_m = V_{m+2} = V_n$$

with

$$T = \frac{1}{2}\sum_a \sum_\mu \sum_l (\dot{S}_l^{a\mu})^2$$

$$V_2 = \frac{1}{2}\sum_{a_1}\sum_{a_2}\sum_{\mu_1}\sum_{\mu_2}\sum_{l_1}\sum_{l_2} F_{l_1 l_2}\begin{pmatrix} a_1 & a_2 \\ \mu_1 & \mu_2 \end{pmatrix} S_{l_1}^{a_1\mu_1} S_{l_2}^{a_2\mu_2} \qquad 4\text{-}2$$

$$V_n = \frac{1}{n!}\sum_{a_1}\sum_{a_2}\cdots\sum_{a_n}\sum_{\mu_1}\sum_{\mu_2}\cdots\sum_{\mu_n}\sum_{l_1}\sum_{l_2}\cdots\sum_{l_n} F_{l_1 l_2 \cdots l_n}\begin{pmatrix} a_1 & a_2 \cdots a_n \\ \mu_1 & \mu_2 \cdots \mu_n \end{pmatrix} S_{l_1}^{a_1\mu_1} S_{l_2}^{a_2\mu_2} \ldots S_{l_n}^{a_n\mu_n}$$

where

$$F_{l_1 l_2 \ldots l_n} = (\partial^n V / \partial S_{l_1}^{a_1\mu_1} \partial S_{l_2}^{a_2\mu_2} \ldots \partial S_{l_n}^{a_n\mu_n})$$

Upon transformation to crystal normal coordinates

$$S_l^{a\mu} = L^{-\frac{1}{2}}\sum_k\sum_j E_{\mu l}^j(k) e^{ik.r_a} Q_j(k) \qquad 4\text{-}3$$

The kinetic and potential energy operators assume the form

$$T = \sum_k\sum_j \dot{Q}_j(k)\dot{Q}_j^*(k)$$

$$V_2 = \frac{1}{2}\sum_k\sum_j \omega_j^2(k) Q_j(k) Q_j^*(k) \qquad 4\text{-}4$$

and

$$V_n = \sum_{k_1}\sum_{k_2}\cdots\sum_{k_n}\sum_{j_1}\sum_{j_2}\cdots\sum_{j_n} C_n\begin{pmatrix} j_1 & j_2 \cdots j_n \\ k_1 & k_2 \cdots k_n \end{pmatrix} Q_{j_1}(k_1) Q_{j_2}(k_2) \ldots Q_{j_n}(k_n) \qquad 4\text{-}5$$

where

$$C_n\begin{pmatrix} j_1 j_2 \cdots j_n \\ k_1 k_2 \cdots k_n \end{pmatrix} = L^{-\frac{n}{2}}\sum_{a_1}\cdots\sum_{a_n}\sum_{\mu_1}\cdots\sum_{\mu_n}\sum_{l_1}\cdots\sum_{l_n} F_{l_1 \cdots l_n}\begin{pmatrix} a_1 \cdots a_n \\ \mu_1 \cdots \mu_n \end{pmatrix} E_{\mu_1 l_1}^{j_1}(k_1) \ldots E_{\mu_n l_n}^{j_n}(k_n) \times$$

$$\times\; e^{i(k_1 \cdot r_{a_1} + \cdots + k_n \cdot r_{a_n})}$$

Going over to second quantization we express now the normal coordinates $Q_j(k)$ in terms of phonon-creation and annihilation operators $a_j^{o\dagger}(k)$ and $a_j^o(k)$

$$Q_j(k) = g_j(k) A_j^o(k) \qquad 4\text{-}6$$
$$\dot{Q}_j(k) = \frac{i\hbar}{2} g_j^{-1}(k) B_j^o(k)$$

where

$$g_j(k) = \left(\frac{\hbar}{2\,\omega_j(k)}\right)^{1/2}$$
$$A_j^o(k) = a_j^{o\dagger}(-k) + a_j^o(k)$$
$$B_j^o(k) = a_j^{o\dagger}(-k) - a_j^o(k)$$

The phonon-creation and annihilation operators have well-known transformation properties

$$a_j^{o\dagger}(k)\left|\ldots n_j(k)\ldots\right\rangle = (n_j(k)+1)^{\frac{1}{2}}\left|\ldots n_j(k)+1\ldots\right\rangle$$
$$a_{j'}^o(k)\left|\ldots n_j(k)\ldots\right\rangle = n_j(k)^{\frac{1}{2}}\left|\ldots n_j(k)-1\ldots\right\rangle \qquad 4\text{-}7$$
$$a_j^{o\dagger}(k)a_j^o(k)\left|\ldots n_j(k)\ldots\right\rangle = n_j(k)\left|\ldots n_j(k)\ldots\right\rangle$$

and obey the basic commutation relations

$$\left[H_o, a_j^{o\dagger}(k)\right] = \hbar\omega_j^o(k)\, a_j^{o\dagger}(k) \qquad a)$$
$$\left[a_{j_1}^{o\dagger}(k_1), a_{j_2}^o(k_2)\right] = \delta_{j_1 j_2}\delta_{k_1 k_2} \qquad b) \quad 4\text{-}8$$
$$\left[a_{j_1}^{o\dagger}(k_1), a_{j_2}^{o\dagger}(k_2)\right] = \left[a_{j_1}^o(k_1), a_{j_2}^o(k_2)\right] = 0 \qquad c)$$

From these, additional commutators can be obtained. We list below some of them.

$$\left[A_j^o(k), a_j^{o\dagger}(k)\right] = 1 \qquad a)$$
$$\sum_{K_1}\cdots\sum_{K_n}\sum_{J_1}\cdots\sum_{J_n}\left[A_{j_1}^o(k_1)\,A_{j_2}^o(k_2)\ldots A_{j_n}^o(k_n), a_j^{o\dagger}(k)\right] = \qquad 4\text{-}9$$
$$= n\sum_{K_1}\cdots\sum_{K_{n-1}}\sum_{j_1}\cdots\sum_{j_{n-1}} A_{j_1}^o(k_1)\,A_{j_2}^o(k_2)\ldots A_{j_{n-1}}^o(k_{n-1}) \qquad b)$$

Using these operators, the terms of the Hamiltonian become

$$H = G_o + \hbar\sum_K\sum_j \omega_j(k)\, a_j^{o\dagger}(k)\, a_j^o(k) \qquad a)$$
$$H_m = V_n = \sum_{K_1}\cdots\sum_{K_n}\sum_{j_1}\cdots\sum_{j_n} B_n\begin{pmatrix} j_1 \ldots j_n \\ k_1 \ldots k_n\end{pmatrix} A_{j_1}^o(k_1)\,A_{j_2}^o(k_2)\ldots A_{j_n}^o(k_n) \qquad b) \quad 4\text{-}10$$

where

$$G_o = \frac{\hbar}{2} \sum_k \sum_j \omega_j^o (k)$$

is the zero point energy and

$$B_n \begin{pmatrix} j_1 \cdots j_n \\ k_1 \cdots k_n \end{pmatrix} = g_{j_1} (k_1) \cdots g_{j_n} (k_n) \; C_n \begin{pmatrix} j_1 \cdots j_n \\ k_1 \cdots k_n \end{pmatrix}$$

The harmonic Hamiltonian H_o describes a system of non-interacting phonons with energies $\omega_j^o(k)$ and occupation numbers $n_j(k)$ and obeys the commutation relation 4-8.

We treat now the crystal as a system of interacting phonons using the Hamiltonian perturbation method of Wallace.[1] The method amounts to a step-wise diagonalization of the perturbed Hamiltonian to the m-th order, through a renormalization procedure of the creation and annihilation operators for the anharmonic phonons.

Consider again the crystal Hamiltonian 4-1. As customary in perturbation treatments we assume that if λ is a positive number $\lambda \ll 1$ taken as expansion parameter, then

$$H_1 \sim \lambda H_o \qquad H_2 \sim \lambda^2 H_o \; \cdots \; H_m \sim \lambda^m H_o$$

where $\sim$ means "of the order of". We assume also that all crystal eigenstates are analytical functions of λ. In order to apply the renormalization procedure we define m-th order correct creation and annihilation operators for the anharmonic phonons

$$\begin{aligned} a_j^\dagger(k) &= a_j^{o\dagger}(k) + a_j^{1\dagger}(k) + \dots + a_j^{m\dagger}(k) \\ a_j(k) &= a_j^o(k) + a_j^1(k) + \dots + a_j^m(k) \end{aligned} \qquad \text{4-11}$$

and we require that the commutation relations 4-8 are satisfied to the m-th order by the operators 4-11

$$\left[H, a_j^\dagger(k)\right] = \hbar\, \omega_j(k)\, a_j^\dagger(k) + O(\lambda^{m+1}) \qquad \text{a)}$$

$$\left[a_{j_1}^\dagger(k_1), a_{j_2}(k_2)\right] = \delta_{j_1 j_2} \delta_{k_1 k_2} + O(\lambda^{m+1}) \qquad \text{b)} \quad \text{4-12}$$

$$\left[a_{j_1}^\dagger(k_1), a_{j_2}^\dagger(k_2)\right] = \left[a_j(k_1), a_j(k_2)\right] = 0 + O(\lambda^{m+1}) \qquad \text{c)}$$

where $O(\lambda^{m+1})$ means a quantity of order m+1 that can be neglected at the order m. If the conditions 4-12 are satisfied, the Hamiltonian is diagonal to the m-th order

$$H = G(m) + \sum_k \sum_j \hbar\, \omega_j(k)\, a_j^\dagger(k)\, a_j(k) + O(\lambda^{m+1}) \qquad \text{4-13}$$

with m-th order correct anharmonic phonon frequencies

$$\omega_j(k) = \omega_j^o(k) + \omega_j^1(k) + \dots + \omega_j^m(k)$$

and m-th order corrected zero-point energy

$$G(m) = G_o + G_1 + \ldots + G_m$$

where

$$G_m = \frac{\hbar}{2} \sum_k \sum_j \omega_j^m (k)$$

is the m-th order correction to G(m).

An important feature of the method is that one can obtain the (m + 1)th correction to the phonon energy from the m-th order corrected operators, through the commutation relation 4-12 a). To the zero order 4-12 a) yields 4-8 a). To the first order, using the zero order result an neglecting second order terms

$$\left[H_o + H_1,\ a_j^{o\dagger}(k) + a_j^{1\dagger}(k)\right] = \hbar\,(\omega_j^o(k) + \omega_j^1(k))(a_j^{o\dagger}(k)+a_j^{1\dagger}(k))$$

$$\left[H_o,\ a_j^{1\dagger}(k)\right] + \left[H_1,\ a_j^{o\dagger}(k)\right] = \hbar\,\omega_j^o(k)\ a_j^{1\dagger}(k) + \hbar\,\omega_j^{(1)}(k)\ a_j^{o\dagger}(k)$$

In the same way, to the second order

$$\left[H_o,\ a_j^{2\dagger}(k)\right] + \left[H_1,\ a_j^{1\dagger}(k)\right] + \left[H_2,\ a_j^{o\dagger}(k)\right] =$$
$$= \hbar\omega_j^o(k)\ a_j^{2\dagger}(k) + \hbar\omega_j^{(1)}(k)\ a_j^{(1)\dagger}(k) + \hbar\omega_j^{(2)}(k)\ a_j^{o\dagger}(k) \qquad 4\text{-}15$$

To find the corrections to the single phonon energy we evaluate upper-diagonal matrix elements of the commutator 4-12. In doing this we make use of the relation

$$\left\langle \ldots n_j(k) + 1 \ldots \right| \left[H_o,\ a_j^{m\dagger}(k)\right] \left| \ldots n_j(k) \ldots \right\rangle =$$
$$= \hbar\omega_j^o(k) \left\langle \ldots n_j(k) + 1 \ldots \right| a_j^{m\dagger}(k) \left| \ldots n_j(k) \ldots \right\rangle \qquad 4\text{-}16$$

which is easily proved.

1) zero order

$$\left\langle \ldots n_j(k) + 1 \ldots \right| \left[H_o,\ a_j^{o\dagger}(k)\right] \left| \ldots n_j(k) \ldots \right\rangle = \hbar\omega_j^o(k)\,(n_j(k)+1)^{\frac{1}{2}}$$

2) first order

$$\left\langle \ldots n_j(k)+1 \ldots \right| \left[H_1,\ a_j^{o\dagger}(k)\right] \left| \ldots n_j(k) \ldots \right\rangle = \hbar\omega_j^1(k)\,(n_j(k)+1)^{\frac{1}{2}}$$

It is easily shown, by direct evaluation of the matrix element above, that $\omega_j^1(k) = 0$. Therefore 4-14 becomes

$$\left[H_o,\ a_j^{1\dagger}(k)\right] + \left[H_1,\ a_j^{o\dagger}(k)\right] = \hbar\omega_j^o(k)\ a_j^{1\dagger}(k) \qquad 4\text{-}17$$

The analytical expression of $\left[H_1,\ a_j^{o\dagger}(k)\right]$ is found from 4-9 b) and 4-10 b)

$$\left[H_1, a_j^{o\dagger}(k)\right] = 3 \sum_{kk_1 j j_1}\sum\sum\sum B_3 \begin{pmatrix} j & j_1 & j_2 \\ k & k_1 & k_2 \end{pmatrix} A_{j_1}^{o}(k_1)\, A_{j_2}^{o}(k_2) \qquad 4\text{-}18$$

and from this expression and from 4-17 we expect $a_j^{1\dagger}(k)$ to be also of second order in $a_j^{o\dagger}(k)$ and $a_j^{o}(k)$. We assume then for $a_j^{1\dagger}(k)$ the general form

$$a_j^{1\dagger}(k) = \sum_{kk_1 j_1 j_2}\sum\sum\sum B_3 \begin{pmatrix} j & j_1 & j_2 \\ k & k_1 & k_2 \end{pmatrix} \Big\{ \alpha\, a_{j_1}^{o}(k_1)\, a_{j_2}^{o}(k_2) + \beta a_{j_1}^{o}(k_1) a_{j_2}^{o\dagger}(-k_2) + \gamma\, a_{j_1}^{o\dagger}(-k_1) a_{j_2}^{o}(k_2) + \delta\, a_{j_1}^{o\dagger}(-k_1) a_{j_2}^{o\dagger}(-k_2) \Big\} \qquad 4\text{-}19$$

and we determine the coefficients α, β , γ and δ from 4-17. This is achieved by substitution of 4-18 and 4-19 in 4-17 and by equating the coefficients of like operators. To avoid vanishing denominators, we introduce, as customary in anharmonic treatments, a damping factor in the phonon frequencies and thus we require that the commutators satisfy, instead of 4-17 the relation

$$\left[H_o, a_j^{1\dagger}(k)\right] + \left[H_1, a_j^{o\dagger}(k)\right] = \hbar\, (\omega_j^{o}(k) - i\varepsilon_j(k))\, a_j^{1\dagger}(k) \qquad 4\text{-}20$$

we obtain then

$$\begin{aligned}
\alpha &= 3\left[\hbar(\omega_j^{o}(k) + \omega_{j_1}^{o}(k_1) + \omega_{j_2}^{o}(k_2)) - i\varepsilon_j(k)\right]^{-1} \\
\beta &= 3\left[\hbar(\omega_j^{o}(k) + \omega_{j_1}^{o}(k_1) - \omega_{j_2}^{o}(k_2)) - i\varepsilon_j(k)\right]^{-1} \\
\gamma &= 3\left[\hbar(\omega_j^{o}(k) - \omega_{j_1}^{o}(k_1) + \omega_{j_2}^{o}(k_2)) - i\varepsilon_j(k)\right]^{-1} \\
\delta &= 3\left[\hbar(\omega_j^{o}(k) - \omega_{j_1}^{o}(k_1) - \omega_{j_2}^{o}(k_2)) - i\,\varepsilon_j(k)\right]^{-1}
\end{aligned} \qquad 4\text{-}21$$

3) second order.

We compute now the second order correction to the single phonon energy, using the results obtained for the first order case. The upper-diagonal matrix element of the commutator 4-12 a), using 4-15 4-16 and the condition $\omega_j^{1}(k) = 0$, gives

$$\left\langle \ldots n_j(k)+1 \ldots \right| \left[H_2, a_j^{o\dagger}(k)\right] + \left[H_1, a_j^{1\dagger}(k)\right] \left| \ldots n_j(k) \ldots \right\rangle =$$

$$= \hbar\, \omega_j^{2}(k)\,(n_j(k) + 1)^{1/2} \qquad 4\text{-}22$$

The calculation of the matrix elements on the left hand side of 4-22 is straightforward. We summarize here the essential steps.

$$\left[H_2, a_j^{o\dagger}(k)\right] = 4 \sum_{k_1 k_2 k_3 j_1 j_2 j_3}\sum\sum\sum\sum\sum B_4 \begin{pmatrix} j & j_1 & j_2 & j_3 \\ k & k_1 & k_2 & k_3 \end{pmatrix} A_{j_1}^{o}(k_1) A_{j_2}^{o}(k_2) A_{j_3}^{o}(k_3) \qquad 4\text{-}23$$

$$\left\langle \ldots n_j(k)+1 \ldots \right| \left[H_2, a_j^{o\dagger}(k)\right] \left| \ldots n_j(k) \ldots \right\rangle =$$

$$= 12 \sum_{k_1 j_1}\sum B_4 \begin{pmatrix} j & j & j_1 & j_1 \\ k & -k & k_1 & -k_1 \end{pmatrix} (n_j(k)+1)^{\frac{1}{2}} (n_{j_1}(k_1) + n_{j_1}(-k_1) + 1) \qquad 4\text{-}24$$

$$\langle \ldots n_j(k)+1 \ldots | [H_1, a_j^{1\dagger}(k)] | \ldots n_j(k) \ldots \rangle = \qquad 4.25$$

$$-6(n_j(k)+1)^{\frac{1}{2}} \sum_{k_1 k_2 j_1 j_2} \Big\{ B_3\begin{pmatrix} j & j_1 & j_2 \\ k & k_1 & k_2 \end{pmatrix} B_3\begin{pmatrix} j & j_1 & j_2 \\ -k & -k_1 & -k_2 \end{pmatrix} \Big[\alpha(n_{j_1}(k_1)+n_{j_2}(k_2)+1) +$$

$$\beta(n_{j_2}(-k_2) - n_{j_1}(k_1)) + \gamma(n_{j_1}(-k_1) - n_{j_2}(k_2)) - \delta(n_{j_1}(-k_1)+n_{j_2}(-k_2)+1) \Big]$$

$$+ B_3\begin{pmatrix} j & j & j_1 \\ k & k & k_1 \end{pmatrix} B_3\begin{pmatrix} j_2 & j_2 & j_1 \\ k_2 & -k_2 & -k_1 \end{pmatrix} (\gamma - \delta)(n_{j_2}(k_2) + n_{j_2}(-k_2) + 1) \Big\}$$

By summing 4.24 and 4.25 and by taking statistical averages, we obtain the second order correction to the phonon energy.We recall that the statistical average of the occupation number $n_j(k)$ for bosons is given by

$$\overline{n_j}(k) = \langle a_j^{o\dagger}(k)\; a_j^{o}(k) \rangle = (e^{\hbar\omega_j^o(k)/KT} - 1)^{-1} \qquad 4.26$$

and that

$$\overline{n_j}(k) = \overline{n_j}(-k) \qquad 4.27$$

Taking then the limit for $\varepsilon_j \to 0$ of the coefficients 4.21,through the symbolic identity

$$\lim_{\varepsilon \to 0^+} \frac{1}{x \pm i\varepsilon} = \left(\frac{1}{x}\right)_p \mp i\pi\delta(x) \qquad 4.28$$

where the subscript p means principal part,we obtain the final result

$$\hbar\omega_j^2(k) = \hbar\Delta_{jk} + i\hbar\Gamma_{jk} \qquad 4.29$$

where the real part $\hbar\Delta_{jk}$ corresponds to the frequency shift and the imaginary part $\hbar\Gamma_{jk}$ is related to the phonon lifetime.The explicit expressions for them are[2]

$$\hbar\Delta_{jk} = 12\sum_{j_1 k_1} B_4\begin{pmatrix} j & j & j_1 & j_1 \\ k & -k & k_1 & -k_1 \end{pmatrix}(2\overline{n}_{j_1}(k_1)+1) +$$

$$- 18\hbar^{-1}\sum_{j_1}\sum_{j_2}\sum_{K_1}\sum_{K_2} B_3\begin{pmatrix} j & j_1 & j_2 \\ k & k_1 & k_2 \end{pmatrix} B_3\begin{pmatrix} j & j_1 & j_2 \\ -k & -k_1 & -k_2 \end{pmatrix} \times$$

$$\Big\{\Big[\frac{\overline{n}_{j_1}(k_1)+\overline{n}_{j_2}(k_2)+1}{(\omega_j(k)+\omega_{j_1}(k_1)+\omega_{j_2}(k_2))p} + \frac{\overline{n}_{j_1}(k_2) - \overline{n}_{j_1}(k_1)}{(\omega_j(k)+\omega_{j_1}(k_1)-\omega_{j_2}(k_2))p}$$

$$+ \frac{\overline{n}_{j_1}(k_1) - \overline{n}_{j_2}(k_2)}{(\omega_j(k)-\omega_{j_1}(k_1)+\omega_{j_2}(k_2))p} - \frac{\overline{n}_{j_1}(k_1)+\overline{n}_{j_2}(k_2)+1}{(\omega_j(k)-\omega_{j_1}(k_1)-\omega_{j_2}(k_2))p}\Big]$$

$$+ 2B_3\begin{pmatrix} j & j & j_1 \\ k & -k & k_1 \end{pmatrix} B_3\begin{pmatrix} j_2 & j_2 & j_1 \\ k_2 & -k_2 & -k_1 \end{pmatrix} \frac{2\overline{n}_{j_2}(k_2)+1}{(\omega_{j_1}(k_1))p} \Big\} \qquad 4.30$$

$$\hbar\Gamma_{jk} = 18\pi\hbar^{-1}\sum_{j_1}\sum_{j_2}\sum_{K_1}\sum_{K_2} B_3\begin{pmatrix} j & j_1 & j_2 \\ k & k_1 & k_2 \end{pmatrix} B_3\begin{pmatrix} j & j_1 & j_2 \\ -k & -k_1 & -k_2 \end{pmatrix} \times$$

$$\Big\{(\overline{n}_{j_1}(k_1)+\overline{n}_{j_2}(k_2)+1)\Big[\delta(\omega_j(k)-\omega_{j_1}(k_1)-\omega_{j_2}(k_2)) +$$

$$\delta(\omega_j(k) + \omega_{j_1}(k_1) + \omega_{j_2}(k_2))\Big] + (\bar{n}_j(k_1) + \bar{n}_j(k_2)\,x \qquad 4.31$$
$$\Big[\delta(\omega_j(k) + \omega_{j_1}(k_1) - \omega_{j_2}(k_2)) - \delta(\omega_j(k) - \omega_{j_1}(k_1) - \omega_{j_2}(k_2))\Big]\Big\}$$

The calculation of the frequency shifts and of the phonon lifetimes by means of equations 4.30 and 4.31 is by no means a simple matter. The most complex part of these calculations concerns the coefficients B_3 and B_4, which involve long and complicated crystal sums. The translational symmetry introduces a $\Delta(k_1 + k_2 + k_3)$ factor in each B_3 coefficient and a $\Delta(k_1 + k_2 + k_3 + k_4)$ factor in each B_4 coefficient. This eliminates a crystal sum from them and thus simplifies the calculations. Another crystal sum is eliminated by the use of the pairwise potentials. Without entering in the details of the calculations, we specify here, as an example the final expression for B_3

$$B_3\begin{pmatrix} k & k_1 & k_2 \\ j & j_1 & j_2 \end{pmatrix} = L^{-\frac{1}{2}}\, g_j(k)\, g_{j_1}(k_1)\, g_{j_2}(k_2)\, \Delta(k + k_1 + k_2)\, x$$
$$\sum_{\mu\nu lmn} E^{j}_{\mu l}(k)\, E^{j_1}_{\mu m}(k_1)\, E^{j_2}_{\nu n}(k_2) \Big[F_{lmn}\begin{pmatrix} o & \mu \\ o & \mu \\ o & \nu \end{pmatrix} + \sum_{a}^{a\neq o} \Big\{ F_{lmn}\begin{pmatrix} a & \nu \\ o & \mu \\ o & \mu \end{pmatrix} e^{ik.r_a} +$$
$$F_{lmn}\begin{pmatrix} o & \mu \\ a & \nu \\ o & \mu \end{pmatrix} e^{ik_1.r_a} + F_{lmn}\begin{pmatrix} o & \mu \\ o & \mu \\ a & \nu \end{pmatrix} e^{ik_2.r_a} \Big\}\Big] \qquad 4.32$$

Further simplifications can be obtained by use of the full symmetry of the unit cell. Despite this, the calculation of the coefficients B_4, involving two crystal sums, remains a very complex problem and for the moment has not been solved in full details.

In our laboratory, attention has been confined to the calculation of the phonon lifetimes, involving only the B_3 coefficients. This offers the additional advantage that the phonon lifetimes can be compared with experimental bandwidth, in order to check the model used for the phonon decay.

For the calculations we have adopted the following representation of the Dirac delta function[3]

$$\delta(x) \simeq \frac{1}{\pi} \frac{\varepsilon}{x^2 + \varepsilon^2} \qquad 4.33$$

and we have used the known approximation[3]

$$\left(\frac{1}{x}\right)_p \simeq \frac{x}{x^2 + \varepsilon^2} \qquad 4.34$$

where ε is a positive parameter that must be large with respect to the spacing between adjacent eigenfrequencies. A value of $\varepsilon = 3\ cm^{1}$ was chosen as a compromise between keeping the number of points in the Brillouin region as low as practical and keeping the value of ε small, for crystalline ammonia.

In order to calculate the phonon lifetimes for a molecular crystal like ammonia, we need the coefficients B_3 and for them we need the crystal eigenvectors over the complete Brillouin zone.

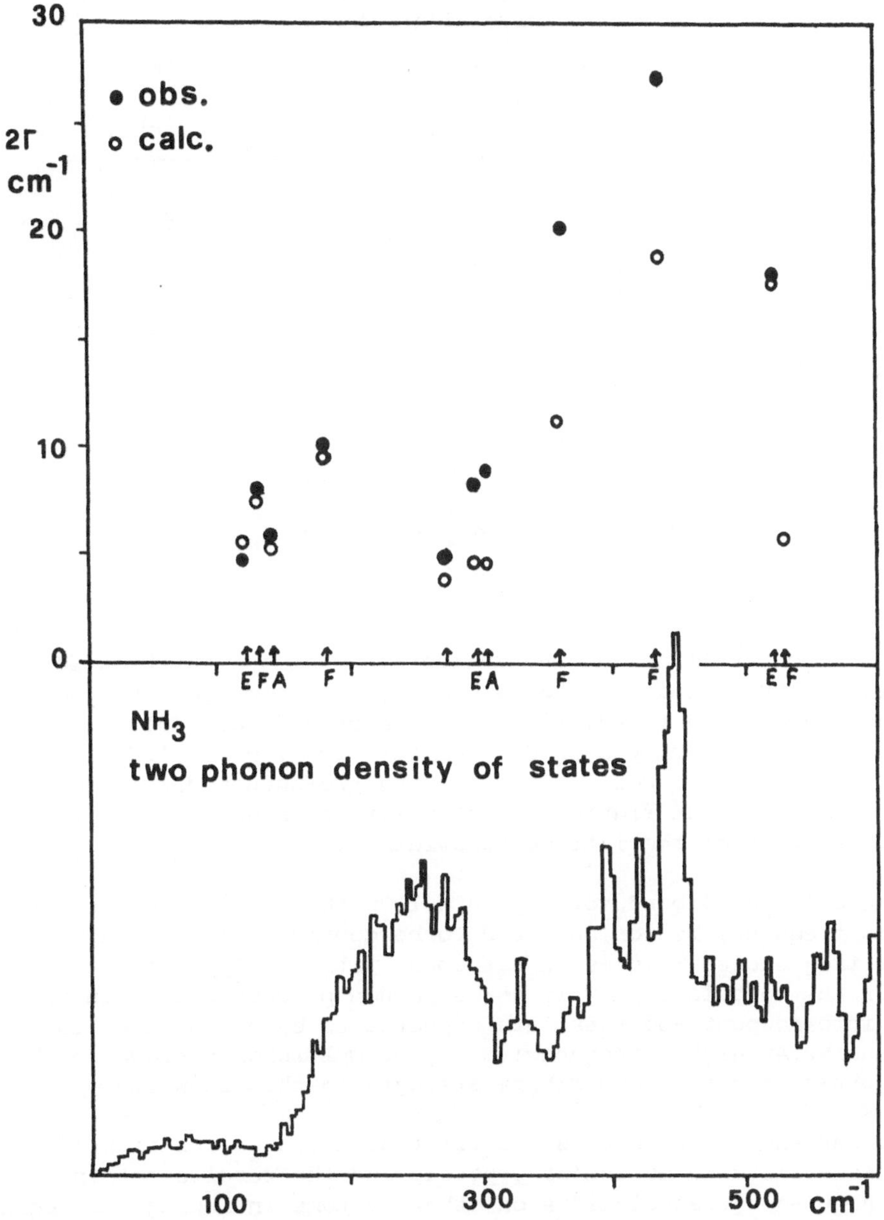

Fig.1.Two phonon density of states of crystalline ammonia(bottom) and width of the lattice bands in the frequency range 100-500 cm^{-1} (top).

For this,using the atom-atom plus multipole-multipole potential[4] discussed in the previous lecture,we have calculated the complete set of phonon dispersion curves,the density of one- and two phonon

Table 1. Calculated and observed bandwidth of the lattice modes at k = 0 of crystalline NH_3.

Symm.	$\nu_{calc.}$	$2\Gamma_{calc.}$	$2\Gamma_{obs.}$	$\nu_{obs.}$
F	530 LO	6.6	18	532
F	523 TO	17.6		
F	435 LO	18.8	27	427
F	360 TO	11.2	21	358
A	305	4.1	9	313
E	299	4.1	8	298
F	274 LO	2.0	5	260
F	273 TO	3.4		
F	181 LO + TO	9.6	10	184
A	143	5.0	6	138
F	131 LO	5.1	8	141
F	129 TO	7.2		
E	122	5.2	5	107

states for crystalline NH_3 and the corresponding quantities for crystalline ND_3. Using these data we have then computed[4] the phonon lifetimes according to equation 4.31. The results are collected in table 1 and are displayed, in diagrammatic form, in fig. 1. For comparison we have also reported in fig. 1 the two-phonon density of states. It is clear from the figure that this latter is qualitatively similar to the plot of the lattice bandwidth.

The calculated bandwidth agree well with the observed ones in the low frequency region, but tend to be considerably smaller at higher frequencies. This is in agreement with the type of assumption made in the calculations. Only three phonon processes were in fact taken into account and these are expected to be dominant at low frequencies. At higher frequencies higher multiphonon processes become important and they contribute strongly to the bandwidth.

To our knowledge this is the first detailed calculation of anharmonic effects in molecular crystals and the results obtained are very encouraging. Calculations on other systems, including four-phonon processes, are in progress in our laboratory.

References.

1. D.C.Wallace, Phys.Rev.152:247(1966)
2. R.G.Della Valle, P.F. Fracassi, R.Righini, S.Califano and S.H.Walmsley, Chem.Phys.(1979) in press.
3. A.A.Maradudin and A.E.Fein, Phys.Rev.128:2589(1962)
4. R.Righini, N.Neto, S.Califano and S.H.Walmsley, Chem.Phys.33:345(1978)

LARGE AMPLITUDE MOTIONS IN MOLECULAR CRYSTALS

K.H. Michel*

Theoretische Physik, Universität

des Saarlandes, Saarbrücken (W-Germany)

1. INTRODUCTION (1) (2) (3)

1.1 The Crystalline State

The main characteristic of the crystalline state is the arrangement of the atoms or atomic ions in a periodic array. In order to describe such an array we need a lattice plus a basis.

The lattice consists of $N \approx 10^{23}$ unit cells labelled by integer vectors $\vec{n} = (n_1, n_2, n_3)$, $n_i = 0, \pm 1, \ldots$. The position of the $\vec{n}$-th lattice point is then defined by the vector

$$\vec{X}(\vec{n}) = \vec{a}_1 n_1 + \vec{a}_2 n_2 + \vec{a}_3 n_3 \quad . \tag{1.1}$$

The $\vec{a}_i$, i=1,2,3 are non coplanar primitive lattice vectors. They characterize the shape of the unit cells. The volume of the unit cell is: $V_z = \vec{a}_1.(\vec{a}_2 x \vec{a}_3)$. Inside a unit cell, we may have one of several atoms, they form a basis. The position of the κ-th atom in the $\vec{n}$ unit cell is then given by:

$$\vec{X}^{\kappa}(\vec{n}) = \vec{X}(\vec{n}) + \vec{r}^{\kappa}, \quad \kappa = 1, \ldots s. \tag{1.2}$$

$\vec{r}^{\kappa}$ denotes the position within the unit cell. The crystal is called a Bravais lattice if s=1. Then we can take $\vec{r}$=o. So far we have considered a rigid crystal. In reality, due to thermal

* Present address: Natuurkunde, Universiteit Antwerpen,
2610, Wilrijk, Belgium.

agitation and to zero point motion the particles carry out oscillations around their equilibrium positions. The actual position of a particle then reads

$$\vec{R}(\vec{n}) = \vec{X}^{\kappa}(\vec{n}) + \vec{u}^{\kappa}(n)/\sqrt{m_\kappa} \tag{1.3}$$

Here $\vec{u}^{\kappa}(\vec{n})/\sqrt{m_\kappa}$ denotes the momentaneous displacement of particle κ in cell $\vec{n}$ away from its equilibrium position. The conjugate momentum variable reads $\vec{p}^{\kappa}(\vec{n})\sqrt{m_\kappa}$.

1.2 The Hamiltonian

We assume that there exists an Hamiltonian

$$H = T + V \,. \tag{1.4}$$

Here the kinetic energy T is given by

$$T = \sum_\lambda \frac{p^2(\lambda)}{2} \quad , \tag{1.5}$$

with $\lambda=\vec{n},\kappa,i$ (obviously $m(\lambda) = m_\kappa$). The potential energy depends on the momentaneous positions of the particles:

$$V = V(R(\lambda_1)\ldots R(\lambda_N)). \tag{1.6}$$

In the case of two body interactions, we have: $V = \frac{1}{2} \Sigma' \, v(|\vec{R}^{\kappa}(\vec{n})-\vec{R}^{\kappa'}(\vec{n}')|)$. Since the particles oscillate around their equilibrium position $\vec{X}(\vec{\lambda})$, it is convenient to perform a series expansion:

$$V = \sum_{\nu=0}^{\infty} \frac{1}{\nu!} \Phi_{(\nu)}(\lambda_1\ldots\lambda_\nu)u(\lambda_1)\ldots u(\lambda_\nu) \,. \tag{1.7a}$$

where

$$\Phi_{(\nu)}(\lambda_1\ldots\lambda_\nu)= m(\lambda_1)\ldots m(\lambda_\nu)^{-1/2} \frac{\partial}{\partial R(\lambda_1)} \cdots\cdot \frac{\partial}{\partial R(\lambda_\nu)} \times$$

$$\times V(\vec{R}(\lambda_1)\ldots))_{\vec{R}(\lambda)=\vec{X}(\lambda)} \tag{1.7b}$$

The quantities Φ are called coupling matrices. Here $\Phi_{(o)} = V(X(\lambda_1)\ldots X(\lambda_N))$ is the rigid lattice energy. This term is irrelevant for the dynamics. The first order term $\Phi_{(1)}(\lambda_1)$ represents the force acting on particle (λ_1) when all particles of the lattice are in their equilibrium positions. Therefore, this term vanishes. The term of second order is called the harmonic term, while the remaining coefficients are anharmonic terms of order 3,4, etc... .

The coupling coefficients have some important properties, resulting from the symmetry of the crystal. The potential energy does not change if we change all particle positions by a same vector $\vec{a}$:

$$V(R(\lambda_1)\ldots R(\lambda_N)) = V(R(\lambda_1)+\vec{a},\ldots,\vec{R}(\lambda_N)+\vec{a}) \quad . \tag{1.8}$$

Differentiating this equation ν-times with respect to $R(\lambda_i)$, $i=1-\nu$, and expanding the result up to first order in a, we find

$$\sum_{\vec{n},\kappa} \Phi_{\nu+1},(\lambda_1\ldots\lambda_\nu,\vec{n},\kappa,i)a_i\sqrt{m_\kappa} = 0 \tag{1.9a}$$

Since $\vec{a}$ is arbitrary, the translational invariance leads to

$$\sum_{\vec{n},\kappa} \Phi_{\nu+1},(\lambda_1\ldots\lambda_\nu,\vec{n},\kappa,i)\sqrt{m_\kappa} = 0 \quad . \tag{1.9b}$$

In particular for the harmonic coefficient one has:

$$\sum_{\vec{n},\kappa} \Phi\begin{matrix} n & m \\ \kappa & \nu \\ i & j \end{matrix} \sqrt{m_\kappa} = 0 \tag{1.10}$$

Another symmetry properties reads:

$$\Phi\begin{matrix} n & m \\ \kappa & \nu \\ i & j \end{matrix} = \Phi\begin{matrix} n-m & o \\ \kappa & \nu \\ i & j \end{matrix} \tag{1.11}$$

The displacements and their conjugate momenta are considered as quantum mechanical operators that satisfy the communication relations:

$$[u(\lambda),\, p(\lambda')] = i\,\delta_{\lambda,\lambda'}; \qquad (\hbar = 1) \tag{1.12}$$

$[u,u] = 0$, $[p,p] = 0$. u and p are considered as time dependent operators in the Heisenberg representation with Hamiltonian H:

$$u(\lambda,t) = e^{iHt}\,u(\lambda)e^{-iHt},\; p(\lambda,t) = e^{iHt}p(\lambda)e^{-iHt} \quad .$$

These operators are Hermitian: $u(\lambda) = u^{\dagger}(\lambda)$.

1.3 Reciprocal Space and Fourier transforms

Given the Bravais lattice of a crystal with a set of primitive vectors $\vec{a}_1,\vec{a}_2,\vec{a}_3$, the corresponding reciprocal lattice is generated by three primitive vectors

$$\vec{b}_1 = 2\pi \frac{\vec{a}_2 x \vec{a}_3}{\vec{a}_1 . (\vec{a}_2 x \vec{a}_3)} \tag{1.13a}$$

$$\vec{b}_2 = 2\pi \frac{\vec{a}_3 x \vec{a}_1}{\vec{a}_1 . (\vec{a}_2 x \vec{a}_3)} \tag{1.13b}$$

$$\vec{b}_3 = 2\pi \frac{\vec{a}_1 x \vec{a}_2}{\vec{a}_1 . (\vec{a}_2 x \vec{a}_3)} \tag{1.13c}$$

Obviously the $\vec{b}_i$ satisfy

$$\vec{b}_i . \vec{a}_j = 2\pi\delta_{ij} \tag{1.14}$$

where δ_{ij} is the Kronecker delta symbol. We now take the $\vec{b}_i$ as primitive lattice vectors of a Bravais lattice in reciprocal space: A reciprocal lattice vector is then given by

$$\vec{K} = h_1\vec{b}_1 + h_2\vec{b}_2 + h_3\vec{b}_3 \tag{1.15}$$

where the h_i are integers. Consider now a plane wave $e^{i\vec{K}.\vec{r}}$. Since $\vec{K}$ is a vector of the reciprocal lattice, we have

$$\vec{K}.\vec{R} = 2\pi\ h_i n_i \tag{1.16a}$$

and therefore

$$e^{i\vec{K}.(\vec{r}+\vec{R})} = e^{i\vec{K}.\vec{r}} \ . \tag{1.16b}$$

The reciprocal lattice is directly accessible to experiment by X-ray and neutron scattering.

It is convenient for the mathematical formulation of lattice dynamics to introduce <u>periodic boundary conditions</u>. One divides an infinite crystal in equivalent bloc crystals of N particles of dimensions $N_1a, N_2a_2N_3a_3$ such that $N_1N_2N_3 = N$ and assumes that after N_i steps in direction a_i, the displacements repeat themselves. This implies

$$e^{i\vec{q}.\vec{R}} = e^{i\vec{q}.(\vec{R}+\vec{N}_i\vec{a}_i)} \quad \text{for each } i = 1,2,3. \tag{1.17}$$

This imposes for the vector $\vec{q}$ the restriction to values:

$$q_j = \frac{n_i b_{ij}}{N_i} \quad , \text{ with } n_i = 0, 1, 2, \ldots N_i - 1 \; . \tag{1.18}$$

Equivalently, we could also take the interval:

$$n_i = -\left(\frac{N_i - 1}{2}\right) , \quad \ldots \quad -1, 0, 1 \ldots \quad \frac{N_i}{2} \; .$$

There are $N = N_1 N_2 N_3$ different $\vec{q}$-vectors, they form the (first) Brillouin zone. Any $\vec{q}$ vector beyond the first Brillouin zone can be written as

$$Q = \vec{q}(1) + \vec{K}$$

where $\vec{q}(1)$ belongs to the first Brillouin zone and where $\vec{K}$ is a vector of the reciprocal lattice. In chosing periodic boundary conditions one neglects surface effects.

Now we are ready to define Fourier transforms (F.T) according to

$$F^{\kappa_1 \ldots \kappa_\nu}_{i_1 \ldots i_\nu}(q_1 \ldots q_\nu) = \frac{1}{N^{\nu/2}} \sum_{\vec{n}_1 \ldots \vec{n}_\nu} e^{-i(\vec{q}_1 . \vec{X}_1(n_1) + \ldots + \vec{q}_\nu . \vec{X}(n_\nu))} F(\lambda_1, \ldots, \lambda_\nu) \tag{1.19a}$$

and inversely

$$F(\lambda_1 \ldots \lambda_\nu) = \frac{1}{N^{\nu/2}} \sum_{q_1 \ldots q_\nu} e^{i(\vec{q}_1 . \vec{X}(n_1) + \ldots \vec{q}_\nu . \vec{X}(n_\nu))} F^{\mu_1 \ldots \mu_\nu}_{i_1 \ldots i_\nu}(\vec{q}_1 \ldots \vec{q}_\nu) . \tag{1.19b}$$

In particular,

$$\Phi^{\kappa_1 \kappa_2}_{i_1 i_2}(q_1, q_2) = \frac{1}{N} \sum_{n_1 n_2} e^{-i(q_1 . X(n_1) + q_2 . X(n_2))} \Phi^{\kappa_1 \kappa_2}_{i_1 i_2}{}^{\vec{n}_1 \vec{n}_2} \; . \tag{1.20}$$

Since the coupling coefficients depend only on the difference $\vec{n}_1 - \vec{n}_2$, we make a change of variables $\vec{n}_1 - \vec{n}_2 = \vec{h}$, and use

$$\sum_{\vec{n}_2} e^{-i(\vec{q}_1+\vec{q}_2).\vec{X}(\vec{n}_2)} = N\Delta(\vec{q}_1+\vec{q}_2) \quad , \tag{1.21}$$

with

$$\Delta(\vec{q}) = \sum_{\vec{K}} \delta_{\vec{q},\vec{K}} \quad , \tag{1.22}$$

where $\vec{K}$ runs over the reciprocal lattice. We finally obtain

$$\Phi^{\kappa_1\kappa_2}_{i_1i_2}(\vec{q}_1\vec{q}_2) = N\Delta\,(\vec{q}_1+\vec{q}_2)\Omega^{\kappa_1\kappa_2}_{i_1i_2}(\vec{q}_1) \quad , \tag{1.23}$$

where

$$\Omega^{\kappa_1\kappa_2}_{i_1i_2}(\vec{q}) = \frac{1}{N}\sum_{\vec{h}} e^{-i\vec{q}.\vec{X}(\vec{h})}\Phi\begin{pmatrix}\vec{h} & o\\ \kappa_1 & \kappa_2\\ i_1 & i_2\end{pmatrix} \tag{1.24}$$

The quantity $\Omega^{\kappa_1\kappa_2}_{i_1i_2}(\vec{q})$ plays a central role in lattice dynamics.

It is called <u>dynamical matrix</u> (in harmonic approximation). It is convenient to introduce the F.T. of the field operators:

$$u^{\kappa}_{i}(\vec{q}) = \frac{1}{\sqrt{N}}\sum_{n} e^{-i\vec{q}.\vec{X}(\vec{n})}u(\lambda) \quad ; \qquad \lambda = (\bar{n},\kappa,i) \quad ; \tag{1.25a}$$

$$p^{\kappa}_{i}\;(\vec{q}) = \frac{1}{\sqrt{N}}\sum_{m'} e^{-i\vec{q}.\vec{X}(\vec{n})}p(\lambda\;). \tag{1.25b}$$

In the following we write one index ρ for $\rho = (\kappa,i)$. The hermiticity condition now reads

$$(u_\rho(\vec{q}))^\dagger = u_\rho(-\vec{q}) \tag{1.26a}$$

and

$$(p_\rho(\vec{q}))^\dagger = p_\rho(-\vec{q}).$$

From the commutator (1.12) we readily obtain:

$$[\,u^+_\rho(\vec{q}),p_{\rho'}(\vec{k})] = i\delta_{\rho'\rho}\Delta(\vec{q}-\vec{k}); \tag{1.27}$$

$$[\,u,u\,] = 0,\ [\,p,p] = 0\ .$$

The Hamiltonian reads:

$$H=\sum_{\vec{\rho},\vec{q}}\frac{p^{\dagger}_{\rho}(q)p_{\rho}(\vec{q})}{2}+\sum_{\substack{\rho_1\ldots\rho_\nu\\ q_1\ldots q_\nu}}\frac{1}{\nu!}\Phi_{\rho_1\ldots\rho_\nu}(-\vec{q}_1,\ldots,-\vec{q}_\nu)u_{\rho'}(\vec{q}_1)\ldots u_{\rho\nu}(\vec{q}_\nu). \tag{1.28}$$

The symmetry properties (1.10) and (1.11) now read

$$\sum_{\vec{\kappa}}\Phi_{\kappa i,\rho_2\ldots\rho\nu}(0,\ldots\vec{q}_\nu)\ m_\nu=0 \tag{1.29}$$

and

$$\Phi_{\rho_1\ldots\rho_\nu}(\vec{q}_1\ldots\vec{q}_\nu)=Nv_{\rho_1\ldots\rho\nu}(\vec{q}_1\ldots\vec{q}_\nu)\Delta(\vec{q}_1+\ldots\vec{q}_\nu) \tag{1.30}$$

where

$$v_{\rho_1\ldots\rho\nu}(\vec{q}_1\ldots\vec{q}_\nu)=\frac{1}{N^{\nu/2}}\sum_{h_1\ldots h_\nu}e^{-i(\vec{q}_1.\vec{X}(\vec{h}_1)\ldots+\vec{q}_{\nu-1}.\vec{X}(\vec{h}_{\nu-i}))}\ \Phi\binom{\vec{h}_1\ldots\vec{h}_{1-1}o}{\rho_1\ldots\rho_{\nu-1}\rho_\nu} \tag{1.31}$$

Note that Eq.(1.24) is a special case of this relation.

2. HARMONIC AND ANHARMONIC PHONONS

2.1 Green's Functions (4),(5).

We define the retarded Green's functions for two operators A and B by

$$\langle\langle A(t);B(o)\rangle\rangle=-i\,\Theta(t)\langle[A(t),B(o)]\rangle\quad, \tag{2.1}$$

$$\Theta(t)=0,\quad t<o$$
$$\qquad 1,\quad t\geq o\ .$$

Here < > on the right hand side denote the thermal averages $\langle A\rangle = (\mathrm{Tr}\,e^{-\beta H}A)/\mathrm{Tr}\,e^{-\beta H}$, and $A(t)=e^{iHt}A\,e^{-iHt}$, where H is the Hamiltonian of the many body system without external perturbation and $\beta=1/(k_BT)$ is the inverse temperature.

The equation of motion reads:

$$i\frac{d}{dt}\langle\langle A(t);B(o)\rangle\rangle=\delta(t)\langle[A,B]\rangle+i\langle\langle[H,A(t)];B(o)\rangle\rangle\ . \tag{2.2}$$

Defining the Fourier transform:

$$\chi_{AB}(z) = - \langle\langle A;B \rangle\rangle_z = - \int dt\ e^{izt} \langle\langle A(t);B(o) \rangle\rangle , \qquad (2.3)$$

for Im z>o, one obtains the equation of motion

$$z \langle\langle A;B \rangle\rangle_z = [A,B] + \langle\langle [A,H];B \rangle\rangle z . \qquad (2.4)$$

The function $\chi(z)$ can be written as a spectral integral

$$\chi(z) = \int \frac{d\omega}{\pi} \frac{\chi''(\omega)}{\omega - z} \qquad (2.5)$$

where $\chi(\omega)$ is the discontinuity of $\chi(z)$ across the real axis:

$$\chi(\omega \pm i\ o) = \chi'(\omega) \pm i\ \chi''(\omega) . \qquad (2.6)$$

From Eq. (2.3) we find the spectral function

$$\chi''_{AB}(\omega) = \frac{1}{2} \int dt\ e^{i\omega t} \langle [A(t),B(o)] \rangle . \qquad (2.7)$$

This function is of great interest because it is directly related to the time dependent correlation function

$$\langle A(t)B(o) \rangle = \int \frac{d\omega}{\pi} e^{-i\omega t} [1+n(\omega)] \chi''_{AB}(\omega) \qquad (2.8a)$$

where $n(\omega) = [\exp(\beta\omega)-1]^{-1}$, $\beta = 1/k_B T$.

Eq. (2.8) expresses the fluctuation-dissipation theorem. The quantity

$$S_{AB}(\omega) = (1+n(\omega))\chi''_{AB}(\omega) \qquad (2.8b)$$

enters the neutron and Brillouin scattering law. We finally remind that $\chi(z)$ and thus $\chi''(\omega)$ is directly related to the <u>linear response</u> (6) of the system to an external perturbation:

$$\delta F = - B\ b(t) , \qquad (2.9)$$

where b(t) is an external field that couples to the operator B referring to the many body system. One then finds that the corresponding change of the quantity A is given by

$$\delta \langle A \rangle(\omega) = \chi_{AB}(\omega + i\ o)\ b\ (\omega) \qquad (2.10)$$

Eqs. (2.10) and (2.8) demonstrate the usefulness of Green's functions, for the calculation of experimentally accessible quantities. We will use these results in calculating the displacement-displacement response functions.

2.2 Harmonic Phonons.

We consider the Green's function

$$D_{\rho\rho'}(\vec{q},z) = \langle\langle u_\rho(\vec{q});\ u^\dagger_{\rho'}(\vec{q})\rangle\rangle_z \quad . \tag{2.11}$$

Using $[u,u] = 0$; and

$$[u_\rho(\vec{q}),H] = i\ p_\rho(\vec{q}) \ ,$$

we find from Eq. (2.4)

$$zD_{\rho\rho'}(\vec{q},z) = i\langle\langle p_\rho(\vec{q});u^\dagger_{\rho'}(\vec{q})\rangle\rangle_z \quad .$$

Applying again Eq. (2.4), we obtain:

$$z^2D_{\rho\rho'}(\vec{q},z) = \delta_{\rho\rho'} + \sum_{\substack{\nu \\ \vec{q}_2\ldots-\vec{q}_\nu \\ \rho_2\ldots\rho_\nu}} \frac{1}{(\nu-1)!}\,\Phi_{\rho\rho_2\ldots\rho_\nu}(\vec{q},-\vec{q}_2\ldots-\vec{q}_\nu) +$$

$$+ \langle\langle u_{\rho_2}(\vec{q}_2)\ldots u_{\rho_\nu}(\vec{q}_\nu)\ ;\ u^\dagger_{\rho'}(\vec{q})\rangle\rangle_z \tag{2.12a}$$

In particular, the contribution $\nu=2$ on the r.h.s. of this equation reads:

$$\sum_{\substack{\vec{q}_2 \\ \rho_2}} \Phi_{\rho\rho_2}(\vec{q},-\vec{q}_2)\langle\langle u_{\rho_2}(\vec{q}_2);\ u_{\rho'}(\vec{q})\rangle\rangle_z \ .$$

The first factor is further simplified by using Eq. (1.23). Since $\vec{q},\vec{q}_2$ are restricted to the first Brillouin zone, $\Delta(\vec{q}-\vec{q}_2) = \delta_{\vec{q},\vec{q}_2}$. Then we get

$$\Phi_{\rho\rho_2}(\vec{q},-\vec{q}_2) = \delta_{\vec{q},\vec{q}_2}\Omega_{\rho\rho_2}(\vec{q}) \tag{2.13}$$

and consequently Eq. (2.12) can be rewritten as:

$$(z^2\delta_{\rho,\rho_2}-\Omega_{\rho\rho_2}(\vec{q}))D_{\rho_2\rho'}(\vec{q},z) = \delta_{\rho\rho'} \quad +$$

$$+\sum_{\nu=3}\frac{1}{(\nu-1)!}\Phi_{\rho\rho_2\ldots\rho_\nu}(\vec{q},-\vec{q}_2,\ldots-\vec{q}_\nu)\langle\langle u_{\rho_2}(\vec{q}_2)\ldots u_{\rho_\nu}(\vec{q}_\nu);u^{+}_{\rho'}(\vec{q})\rangle\rangle_z . \tag{2.12b}$$

This equation plays a central role in lattice dynamics. If we take $\Phi_{(\nu)} = 0$ for all $\nu \geq 3$, we have the harmonic approximation and thus a closed expression for D(q,z):

$$(z^2\delta_{\rho\rho_2} - \Omega_{\rho\rho_2}(\vec{q}))D_{\rho_2\rho'}(\vec{q},z) = \delta_{\rho\rho'} \quad . \tag{2.14}$$

The resonances of D(q,z) are called phonons. From the properties of $\Phi(\lambda,\lambda')$ it follows that the dynamical matrix is hermitian:

$$\Omega^{\star}_{\rho_1\rho_2}(\vec{q}) = \Omega_{\rho_2\rho_1}(\vec{q}) \tag{2.15a}$$

and that

$$\Omega_{\rho_1\rho_2}(\vec{q}) = \Omega^{\star}_{\rho_1\rho_2}(-\vec{q}). \tag{2.15b}$$

Consequently the dynamical matrix has 3s real eigenvalues $(\omega(\alpha,q))^2$, α=1-3s and corresponding eigenvectors $e_\rho(\alpha,\vec{q})$, where ρ runs over the 3s components. The orthonormality and closure relations read:

$$\sum_\rho e_\rho(\alpha,\vec{q})e^{\star}_\rho(\beta,\vec{q}) = \delta_{\alpha,\beta} \tag{2.16a}$$

$$\sum_\alpha e_\rho(\alpha,\vec{q})e^{\star}_{\rho'}(\alpha,\vec{q}) = \delta_{\rho,\rho'} \quad . \tag{2.16b}$$

The eigenvalues are solutions of:

$$|\Omega_{\rho\rho'}(\vec{q}) - \omega^2\delta_{\rho\rho'}| = 0. \tag{2.17}$$

We quote

$$\Omega_{\rho\rho'}(\vec{q}) = \sum_\alpha e^{\star}_\rho(\alpha,\vec{q})\omega^2(\alpha,\vec{q})e_{\rho'}(\alpha,\vec{q}) \quad . \tag{2.18}$$

and

$$D_{\rho\rho'}(\vec{q},z) = \sum_{\alpha} e_{\rho}(\alpha,\vec{q}) \frac{1}{z^2-\omega^2(\alpha,\vec{q})} e^{\star}_{\rho'}(\alpha,\vec{q}) \tag{2.19}$$

Stability requires that all eigenvalues of Ω are positive, or equivalently, all $\omega(\alpha,\vec{q})$ are real. If this is not the case, the harmonic approximation is insufficient. Such a situation arises in quantum crystals. Anharmonic corrections then can stabilize the situation (7), (8).

We call the resonances of Eq. (2.19) phonons. Generally we have 3s phonon branches (3s diff. values of α). Translational invariance, Eq. (1.10) implies that the dynamical matrix, for $\vec{q}$=o, has the rank (3s-3). Consequently there exist three branches $\omega_i(\vec{q})$, i = 1,2,3 where $\omega_i(\vec{q}\to o)\to o$. These are called the three acoustical branches, while the remaining 3s-3 are called optical branches. Applying Eq. (2.6), we use Eq. (2.19) to calculate the spectral function (z = ω ± i o):

$$\chi''(\omega) = \frac{1}{2i}(D(\omega + i\,o) - D(\omega - i\,o)) \quad , \tag{2.20a}$$

$$\chi''_{\rho\rho'}(\vec{q},\omega) = \sum_{\alpha} e_{\rho}(\alpha,\vec{q}) \frac{\pi}{\omega(\alpha,\vec{q})} [\,\delta(\omega-\omega(\alpha,\vec{q})) - \delta(\omega+\omega(\alpha,\vec{q}))]\, e^{\star}_{\rho'}(\alpha,\vec{q}) \,. \tag{2.20b}$$

The inelastic neutron and Brillouin scattering law is given by

$$S_{\rho\rho'}(\vec{q},\omega) = (1 + n(\omega))\chi''_{\rho\rho'}(\vec{q},\omega) \quad ,$$

and consequently the harmonic phonons enter as δ-peak resonances the inelastic neutron scattering law.

2.3 Anharmonic Phonons

Experimentally one finds that the phonon peaks have a certain width. This width and also the position of the peaks depends on temperature.

These phenomena can be explained by phonon damping. In pure crystals, the phonon damping is due to lattice anharmonicities. In order to calculate anharmonic effects, it is convenient to use phonon creation and annihilation operators. These are related to the displacements and the conjugate momenta by transformations that are familiar from the harmonic oscillator problem in quantum mechanics.

We define

$$b_{\vec{k}} = \frac{1}{2} \sum_{\rho} e^{\star}_{\rho}(\vec{k}) \, [\sqrt{2\omega(k)} \, u_{\rho}(\vec{k}) + i \sqrt{\frac{2}{\omega(k)}} \, p_{\rho}(\vec{k})] \, , \tag{2.21a}$$

$$b^{\dagger}_{\vec{k}} = \frac{1}{2} \sum_{\alpha} e_{\rho}(\vec{k}) \, [\sqrt{2\omega(k)} \, u^{\dagger}_{\rho}(\vec{k}) - i \sqrt{\frac{2}{\omega(k)}} \, p^{\dagger}_{\rho}(\vec{k})] \, . \tag{2.21b}$$

Here we write k for $(\vec{k},\alpha)$ and -k for $(-\vec{k},\alpha)$. The commutation relations (1.29) for u and p imply

$$[b_k, b^{\dagger}_p] = \delta_{kp} \tag{2.22}$$

$[b,b] = o, [b^{\dagger},b^{\dagger}] = o$. Equations (2.21a,b) are equivalent to

$$u_{\rho}(\vec{k}) = \sum_{\alpha} \frac{1}{2\omega(\vec{k},\alpha)} e_{\rho}(\alpha,\vec{k})[b(\alpha,\vec{k}) + b^{\dagger}(\alpha,-\vec{k})] \tag{2.23}$$

$$p_{\rho}(\vec{k}) = - i\sum_{\alpha} \frac{\omega(k,\alpha)}{2} e^{\star}_{\rho}(\alpha,\vec{k})[b(\alpha,\vec{k}) - b^{\dagger}(\alpha,-\vec{k})] \, . \tag{2.4}$$

The Hamiltonian (1.28) is now rewritten as

$$H = H^h + H^a \tag{2.25}$$

where the harmonic part H^h reads:

$$H^h = \sum_{\rho,q} \frac{p^{\dagger}_{\rho}(\vec{k}) p_e(\vec{k})}{2} + \sum_{\substack{\rho_1\rho_2 \\ k_1k_2}} \frac{1}{2} \Phi_{\rho_1\rho_2} (-\vec{k}_1,-\vec{k}_2) u_{\rho_1}(\vec{k}_1) u_{\rho_2}(\vec{k}_2) \tag{2.26a}$$

or equivalently

$$H^h = \sum_{k} \omega(k) \, b^{\dagger}_{k} k_b \quad . \tag{2.26b}$$

The anharmonic part is given by

$$H^a = \sum_{\nu=3} \frac{1}{\nu!} \sum_{k_1 \dots k_{\nu}} \Phi(-k_1,\dots -k_{\nu}) \, B_{k_1} \dots B_{k_{\nu}} \tag{2.27}$$

with

$$B_k = b_k + b^{\dagger}_{-k} \tag{2.28a}$$

and

$$\Phi(-k_1\ldots-k_\nu)=\sum_{\rho_1\ldots\rho_\nu}\Phi_{\rho_1\ldots\rho_\nu}(-\vec{k}_1\ldots-\vec{k}_\nu)\ \times$$

$$\times\ [\,2^\nu\omega(k_1)\ldots\omega(k_\nu)]^{-1/2}e_{\rho_1}(k_1)\ldots e_{\rho_\nu}(k_\nu)\ . \tag{2.28b}$$

We have to calculate the anharmonic term on the right hand side of Eq. (2.12b). In particular the third order anharmonic contribution reads:

$$D^{(3)}_{\rho\rho'}(\vec{q},z)=\frac{1}{2}\,\Phi_{\rho\rho_2\rho_3}(\vec{q},-\vec{q}_2,-\vec{q}_3)\ \langle\langle u_{\rho_2}(\vec{q}_2)u^\dagger_{\rho'}(\vec{q}_3);u^\dagger_{\rho'}(\vec{q})\rangle\rangle_z =$$

$$=\frac{1}{2}\,\Phi_{\rho\rho_2\rho_3}(\vec{q},-\vec{q}_2,-\vec{q}_3)\ \frac{1}{\sqrt{2\omega(q_2)}}\ \frac{1}{\sqrt{2\omega(q_3)}}\ e_{\rho_2}(q_2)e_{\rho_3}(q_3)\ \times$$

$$\times\ \langle\langle B_{q_2}B_{q_3};\ u^\dagger_{\rho'}(\vec{q})\rangle\rangle_z\ . \tag{2.29}$$

The problem consists in the evaluation of $\langle\langle B_{q_2}B_{q_3};u^\dagger_{\rho'}(\vec{q})\rangle\rangle_z$. The calculation is quite lengthy. It can be found in Ref. 11. The final result reads:

$$D^{(3)}_{\rho\rho'}(\vec{q},z)\ =\ \sum_{\rho''}\ M^{(3)}_{\rho\rho''}(\vec{q},z)D_{\rho''\rho'}(\vec{q},z) \tag{2.30}$$

where

$$M^{(3)}_{\rho\rho''}(q,z)\ =\ \frac{1}{2}\sum_{\substack{\vec{p},\vec{k}\\ \alpha,\beta}}\Phi_{\rho,\alpha\beta}(\vec{q},-\vec{k},-\vec{p})\Phi^{\star}_{\rho'',\alpha\beta}(\vec{q},-\vec{k}-\vec{p})\quad\times$$

$$\times\ \{[\,1+n(\vec{k},\alpha)+\ n(\vec{p},\beta)][\ \frac{1}{z-\omega(\vec{k},\alpha)-\omega(\vec{p},\beta)}\ -\ \frac{1}{z+\omega(k,\alpha)+\omega(\vec{p},\beta)}\,]\ +$$

$$+\ [\,n(\vec{k},\alpha)\ -\ n(\vec{p},\beta)][\frac{1}{z+\omega(\vec{k},\alpha)-\omega(\vec{p},\beta)}\ -\ \frac{1}{z-\omega(\vec{k},\alpha)+\omega(\vec{p},\beta)}\,]\} \tag{2.31}$$

Here $n(\vec{k},\alpha)\ =\ [\,\exp(\omega(\vec{k},\alpha)/T)-1]^{-1}$ is the Bose-Einstein distribution

and

$$\Phi_{\rho,\alpha\beta}(\vec{q},-\vec{p},-\vec{k})=\Phi_{\rho\rho_1\rho_2}(\vec{q},-\vec{k},-\vec{p})e_{\rho_1}(\vec{k},\alpha)e_{\rho_2}(\vec{p},\beta)[4\omega(\vec{k},\alpha)\omega(\vec{p},\beta)]^{-1/2}. \tag{2.32}$$

For $z=\omega \pm i\,o$, we separate real and imaginary part in $M(\vec{q},z)$ and write:

$$M^{(3)}_{\rho\rho''}(\vec{q},\omega\pm io) = \Delta^{(3)}_{\rho\rho''}(\vec{q},\omega) \mp \Gamma^{(3)}_{\rho\rho''}(\vec{q},\omega) \quad , \tag{2.33}$$

where the real part is given by

$$\Delta^{(3)}_{\rho\rho''}(\vec{q},\omega) = \frac{1}{2}\sum_{\substack{\vec{p},\vec{k}\\ \alpha,\beta}}\Phi_{\rho,\alpha\beta}(\vec{q},-\vec{k},-\vec{p})\Phi^{\star}_{\rho'',\alpha\beta}(\vec{q},-\vec{k},-\vec{p}) \times$$

$$\times \{[1+n(\vec{k},\alpha)+n(\vec{p},\beta)][P(\frac{1}{\omega-\omega(\vec{k},\alpha)-\omega(\vec{p},\beta)}) - P(\frac{1}{\omega+\omega(\vec{k},\alpha)+\omega(\vec{p},\beta)}) +$$

$$+ [n(\vec{k},\alpha)-n(\vec{p},\beta)][P(\frac{1}{\omega+\omega(\vec{k},\alpha)-\omega(\vec{p},\beta)}) - P(\frac{1}{\omega-\omega(\vec{k},\alpha)+\omega(\vec{p},\beta)})]\} . \tag{2.34}$$

Here P denotes the principal part of the integrand. Similarly the imaginary part reads:

$$\Gamma^{(3)}_{\rho\rho''}(\vec{q},\omega) = \frac{1}{2}\sum_{\substack{\vec{p},\vec{k}\\ \alpha,\beta}}\Phi^{(3)}_{\rho,\alpha\beta}(\vec{q},-\vec{k},-\vec{p})\Phi^{(3)\star}_{\rho'',\alpha\beta}(\vec{q},-\vec{k},-\vec{p}) \times$$

$$\times \{[1+n(\vec{k},\alpha)+n(\vec{p},\beta)]\pi[\delta(\omega-\omega(k\alpha)-\omega(\vec{p},\beta))-\delta(\omega+\omega(\vec{k},\alpha)+\omega(\vec{p},\beta))] +$$

$$+ [n(\vec{k},\alpha)-n(\vec{p},\beta)]\pi[\delta(\omega+\omega(\vec{k},\alpha)-\omega(\vec{p},\beta))-\delta(\omega-\omega(\vec{k},\alpha)+\omega(\vec{p},\beta))]\} . \tag{2.35}$$

Finally we quote the result for the fourth order anharmonic contribution to Eq. (2.13). In lowest order decoupling approximation this contribution can be written as

$$D^{(4)}_{\rho\rho'}(\vec{q},z) = \Omega^{(4)}_{\rho\rho''}(\vec{q})D_{\rho''\rho'}(\vec{q},z) \quad , \tag{2.36}$$

where

$$\Omega^{(4)}_{\rho\rho''}(\vec{q}) = \frac{1}{2}\sum_{\vec{k},\alpha,\rho''} \Phi^{(4)}_{\rho\alpha\alpha\rho''}(\vec{q},-\vec{k},+\vec{k},-\vec{q})[1+2n(\vec{k},\alpha)] \quad , \tag{2.37}$$

with

$$\Phi^{(4)}_{\rho\alpha\alpha\rho''}(\vec{q},-\vec{k},\vec{k},-\vec{q}) = \sum_{\rho_2\rho_3} \Phi_{\rho\rho_2\rho_3''}(\vec{q},-\vec{k},+\vec{k},-\vec{q})e_{\rho_2}(\vec{k},\alpha)e_{\rho_3}(\vec{k},\alpha) \times$$

$$\times [2\omega(\vec{k},\alpha)]^{-1} \quad . \tag{2.38}$$

2.4 <u>Renormalized Harmonic Approximation (RHA)</u> (7) (8)

We first discuss the physical implications of the fourth order anharmonicities. Approximating the second term on the r.h.s. of Eq. (2.13) by Eq. (2.36), we obtain:

$$(z^2\delta_{\rho\rho_2} - \tilde{\Omega}_{\rho\rho_2}(\vec{q}))D_{\rho_2\rho'}(\vec{q},z) = \delta_{\rho\rho'} \quad . \tag{2.39}$$

Here we see that the fourth order anharmonicities merely shift the phonon frequencies in Eq. (2.14) from the harmonic values to the renormalized values

$$\tilde{\omega}^2(\vec{q},\alpha) = \sum_{\rho,\rho'} e_\rho(\vec{q},\alpha)\, \tilde{\Omega}_{\rho\rho'}(\vec{q})e^*_{\rho'}(\vec{q},\alpha) \quad , \tag{2.40}$$

with

$$\tilde{\Omega}_{\rho\rho'}(\vec{q}) = \Omega_{\rho\rho'}(\vec{q}) + \Omega^{(4)}_{\rho\rho'}(\vec{q}) \quad . \tag{2.41}$$

Note that $\Omega^{(4)}$ depends on temperature. It is easy to show that all even order term contribute to a renormalization of the phonon frequencies but do not change essentially the structure of Eq. (2.14).

A theory where one takes into account in the most simple approximate way all even order anharmonicities but where one neglects completely the third order and all other uneven order anharmonicities, is called <u>renormalized harmonic approximation</u> (RHA). It is necessary to use the RHA instead of the simple harmonic theory whenever the amplitudes of the lattice displacements are large. These amplitudes generally increase with temperature and they are large in crystals where the inter-atomic forces are weak. This is the case for instance in the heavier rare gas crystals.
A more extreme situation is even

reached in solid He (quantum crystals) (9). There the harmonic approximation fails to given real phonon frequencies, $\Omega_{\rho\rho'}(q)<0$. Then the RHA is essential to obtain a stable crystal, i.e. $\Omega>0$.

We also mention at this place already that in molecular (10) crystals the interatomic forces are often weak. Then even in an orientationally ordered phase the amplitudes of the orientational motions (librations) are large. A simple harmonic theory is not adequate and one has to use some form of RHA, applied to the angular displacements. In the disordered phase of molecular crystals or near an orientational phase transition, the RHA becomes completely inadequate.

2.5 Phonon Damping (11)

By taking into account fourth order anharmonicities and in addition the third order contribution Eq. (2.30), we obtain as an approximation to Eq. (2.13):

$$(z^2\delta_{\rho\rho_2} - \tilde{\Omega}_{\rho\rho_2}(\vec{q}) - M^{(3)}_{\rho\rho_2}(\vec{q},z))D_{\rho_2\rho'}(\vec{q},z) = \delta_{\rho_2\rho'} \quad . \tag{2.42}$$

According to Eq. (2.33), $M^{(3)}$ plits into a real and an imaginary part. The real part, $\Delta^{(3)}$ produces an additional temperature-dependent frequency shift of $\tilde{\Omega}$. The imaginary part produces as a quantitative new feature phonon damping. Taking into account only diagonal elements, we rewrite Eq. (2.42) for fixed $\vec{q}$

$$(z^2-\omega^2-\Delta(z) + i\Gamma(z))D(z) = 1 \quad .$$

In order to examine the resonances of D(z), we treat $\Delta(z)$ and $\Gamma(z)$ as small quantities in comparison to $\tilde{\omega}$. We determine the roots of $z^2-\tilde{\omega}^2-\Delta(z)+i\Gamma(z) = 0$ by perturbation theory and find for $z = \omega + io$:

$$D(\omega+io) = \frac{1}{2\varepsilon}\left[\frac{1}{\omega-\varepsilon-\frac{i\gamma}{2}} - \frac{1}{\omega+\varepsilon+\frac{i\gamma}{2}}\right] \tag{2.43}$$

where we have defined

$$\varepsilon = \tilde{\omega} + \frac{\Delta(\tilde{\omega})}{2\,\tilde{\omega}} \quad , \tag{2.44a}$$

$$\gamma = \frac{\Gamma(\tilde{\omega})}{2\tilde{\omega}} \quad . \tag{2.44b}$$

A similar expression is found for $D(\omega-io)$ where only the sign of γ is changed in comparison with Eq. (2.43). Calculating then the spectral function according to Eq. (2.20a), we find

$$\chi''(\omega) = \frac{1}{2\varepsilon}\left\{\frac{\gamma}{(\omega-\varepsilon)^2 + (\frac{\gamma}{2})^2} - \frac{\gamma}{(\omega+\varepsilon)^2 + (\frac{\gamma}{2})^2}\right\} . \tag{2.45}$$

The resonances are now given by Lorentizians with maxima located at the shifted frequencies $\pm\ \varepsilon$ instead of $\tilde{\omega}$ and with a finite linewidth γ instead of δ-peaks (compare Eq. (2.20b)). By Fourier transforming back to time space we find

$$\int \chi''(\omega)e^{-i\omega t}d\omega \propto \exp\left[\pm i\ \varepsilon\ t - \frac{\gamma}{2}\ |t|\right] . \tag{2.46}$$

We conclude that a finite linewidth leads to phonon damping.

Before leaving the subject of anharmonic phonons, we would like to mention that the results of section 3 of the present chapter are valid in the high frequency regime. In the limit of low frequencies ($\omega\to o$), the last term on the r.h.s. of Eq. (2.31) exhibits the so called hydrodynamic singularities. They can be avoided by going beyond the simple decoupling approximation.

The physical origin of the breakdown of simple perturbation theory is due to the fact that lattice anharmonicities couple the longwave length phonons or sound waves to head conduction. (11), (12). The coupling to heat conduction leads to an additional resonance in the phonon spectrum located at ω=o. This central resonance is known as Landau-Placzeck peak or Payleigh peak.

3. MOLECULAR ORIENTATIONS IN SOLIDS

3.1 Introduction (13)

In principle we are able to treat the displacements of atoms in a crystal by the methods of lattice dynamics.(14) In practice the complexity of the approach increases tremendously as soon as the number of particles per unit cell is of the order of 5 - 10. It becomes prohibitive if the amplitude of the displacements are large. In order to make any progress, one has to reexamine the situation.

As an example, we consider the crystal of NH_4Cl. Above 242K, this substance exhibits a CsCl-type structure.(15)There are two sterically different orientations for a NH_4^+ ion in the cubic unit cell where the hydrogene atoms point toward the nearest neighbor Cl^- ions. In the cubic phase, the ammonium ion reorients among these two sterically different positions I and II. In addition there are in each of the states I and II, twelve sterically equivalent positions, obtained by reorientations around the threefold tetrahedral axes. Obviously with respect to these reorientations the ammonium group behaves as a quasi rigid body. This is due to the fact that the chemical bounds leading to the NH_4^+-"molecule" are stronger than the bounds of the H-protons with the surrounding Cl-ions. Strictly speaking the NH_4^+ ion is not completely rigid. But as long as we are not interested in the fast internal modes of the NH_4^+ ion (fast in comparison with the reorientations or external modes), it is a meaningful approximation to consider the NH_4^+ as a rigid tetrahedron. This point of view greatly simplifies the theoretical description. While we would need three separate coordinates for each of the four protons in a lattice dynamical approach, we need only three coordinates, namely the three Euler angles (α,β,γ) to describe the orientational position of the tetrahedron. Nevertheless the situation is not simple. Indeed the two different tetrahedral positions are not close together. It is not possible to reach $(\alpha,\beta,\gamma)_{II}$ by small deviations from $(\alpha,\beta,\gamma)_I$. In fact the tetrahedral reorientations I - II correspond to an extreme case of large amplitude motion ($\Delta\alpha \approx \pi/2$).

On the other hand, when the tetrahedron is for a certain time τ in one of its rest positions, say I, it is still able to perform small angular oscillations $\delta\varphi = \varphi_I - \varphi_I^\circ, \varphi = (\alpha,\beta,\gamma)$ around its equilibrium position φ_I°. These angular oscillations are called librations. After a time τ, this oscillatory motion is interrupted by a reorientation. The description in terms of small angular displacements breaks down. The problem now consists in finding a description where we are able to describe small and large amplitude motion within a same framework. This can be achieved by describing the orientational positions of the molecules in terms of symmetry adapted functions.

3.2 Examples of symmetry adapted functions

In elementary quantum mechanics of the hydrogen atom, the electronic wave function is given by

$$\psi_{\ell,m}(\vec{r}) = F_\ell(r)Y_\ell^m(\Omega) \quad , \tag{3.1}$$

with $\Omega = (\Phi,\varphi)$.

While the radial part F of the wave function measures the probability amplitude of finding the electron at a certain distance r from the center, the spherical harmonics Y_ℓ^m specify the angular distribution of the electron amplitude. For a given ℓ, there are $2\ell + 1$ function Y_ℓ^m corresponding to different values of m. These values of m specify the orientation in space. Obviously $Y^\circ = 1/\sqrt{4\pi}$ represents the spherical symmetric distribution. If we define the linear combinations (16)

$$Y_{\ell}^{m,c} = \frac{(Y_{\ell}^{m} + Y_{\ell}^{-m})}{\sqrt{2}} \tag{3.2a}$$

and

$$Y_{\ell}^{m,s} = - i \frac{(Y_{\ell}^{m} - Y_{\ell}^{-m})}{\sqrt{2}} \quad , \tag{3.2b}$$

then with $z = \cos\Theta$, $y = \sin\Theta \;\; \varphi$, $x = \sin\Theta \cos\varphi$,

$$Y_{1}^{1,c}(\Omega) = \frac{3}{4\pi} \;\; x \tag{3.3a}$$

$$Y_{s}^{1,s}(\Omega) = \frac{3}{4\pi} \;\; y \tag{3.3b}$$

$$y_{1}^{\circ}(\Omega) \;\; \frac{3}{4\pi} \; z \tag{3.3c}$$

represent the well know p-orbitals:

The corresponding 5 functions for $\ell=2$ are known as d-orbitals:

$$d_{z^2},\ d_{x^2-y^2},\ d_{xz}, d_{yz} \quad .$$

Next we consider an example taken from ligand field theory. Take an ion with electrons in d orbitals, assume this ion to be located in the center of an octahedron formed by six negative charges. Since the electronic orbitals of the central ion will experience repulsive forces from the 6 negative ions on the axes, the energy of the two d_{z^2}, $d_{x^2-y^2}$ orbitals which have their maxima along the coordinate axes is larger than the energy of the three d_{xy}, d_{yz}, d_{xz} orbitals which have their maxima along the (xy), (yz), (xz)-plane diagonals. These five orbitals form a two and a three dimensional representation respectively of the cubic group. The first two belong two the E_g-representation, the last three of the T_{2g} representation. They are examples of symmetry adapted functions. (17)

3.3 Single particle orientational distribution (18), (19)

To be specific, we will consider the case of a molecule in a field of cubic symmetry (O_h). We assume that the center of mass of the molecule is fixed in the center of the field and that the molecule itself is a rigid body. Then only the orientational degrees of freedom of the molecule are relevant. We consider the mass distribution $a(\vec{r})$ on a spherical shell at distance d from the center. Taking the center as origin, we have

$$a(\vec{r}) = \frac{\delta(r-d)}{r^2} f(\Omega) \quad . \tag{3.4}$$

Here $f(\Omega)$, $\Omega = (\Theta,\varphi)$, stands for the orientational distribution function. Since the surrounding potential has the full O_h-symmetry, the density distribution has this symmetry.
Expanding $f(\Omega)$ in terms of spherical harmonics:

$$f(\Omega) = \Sigma a_{\ell m} Y_\ell^m(\Omega) \quad , \tag{3.5}$$

we find that only those combinations of Y_ℓ^m contribute which transform itself as the unit representation of the cubic group. These functions belong to the A_{1g} representation. Therefore Eq. (3.4) becomes

$$f(\Omega) = \Sigma a_{\ell n} K_{\ell n}(\Omega) \quad , \tag{3.6a}$$

in particular,

$$f(\Omega) = a_{01}K_{01}(\Omega) + a_{41}K_{41} + a_{61}K_{61} + \dots \quad . \tag{3.6b}$$

These functions $K_{\ell n}$ are cubic harmonics (17), (16) belonging to the A_{1g} representation. In particular

$$K_{01}(\Omega) = Y_\circ^\circ(\Omega) = 1/\sqrt{4\pi} \tag{3.7a}$$

$$K_{41}(\Omega) = \frac{7}{12} Y_\circ^\circ(\Omega) + \frac{5}{12} Y_4^{4,c}(\Omega) \tag{3.7b}$$

where (see def. (3.2a)):

$$Y_4^{4,c} = (Y_4^4 + Y_4^{-4})/\sqrt{2} \quad . \tag{3.8}$$

It is most instructive to examine K_{41} in Cartesian coordinates:

$$K_{41}(x,y,z) = \frac{5}{4}\sqrt{\frac{21}{4}}\left[x^4 + y^4 + z^4 - \frac{3}{5}\right] \quad . \tag{3.9}$$

This function has eight equivalent minima in ⟨111⟩ directions, six maxima in ⟨100⟩ directions and twelve saddle points in ⟨110⟩ directions. The orientational distribution function depends on temperature. This follows intuitively from

$$f(\Omega) \propto \exp\,(-\beta V(\Omega)) \;, \tag{3.10}$$

$\beta=(k_BT)^{-1}$, where $V(\Omega)$ is the octahedral potential. Since the cubic harmonics form a coupled set of orthonormalized functions, one has

$$a_{\ell n} = \int d\,\Omega\; K_{\ell n}(\Omega) f(\Omega) = \langle K_{\ell n}\rangle \quad . \tag{3.11}$$

Equivalently we can define

$$f(\Omega) = \langle\delta(\Omega-\Omega')\rangle \tag{3.12}$$

where the thermal average is defined according to

$$\langle A(\Omega')\rangle = \frac{\int d\,\Omega'\, \exp(-V(\Omega')\beta)\; A(\Omega')}{\int d\,\Omega'\, \exp(-\beta V(\Omega'))} \quad , \tag{3.13}$$

with $d\Omega' = \sin\Theta'\, d\Theta'\, d\varphi'$.
We use the identity

$$\delta(\Omega-\Omega') \because \frac{\delta(\Theta-\Theta')\delta(\varphi-\varphi')}{\sin\,\Theta} = \sum_{\ell,m} Y_\ell^m\,(\Omega)\; Y_\ell^m(\Omega') \tag{3.14}$$

and arrive at

$$f(\Omega) = \sum_{\ell,m} \langle Y_\ell^m(\Omega')\rangle Y_\ell^m(\Omega) \tag{3.15}$$

Since the potential has O_h symmetry, only those combinations of $Y_\ell^m(\Omega')$ subsist under the thermal average that transform according to the unit representation of the group O_h. Therefore the Y_ℓ^m-functions combine to $K_{\ell n}$ and Eq. (3.15) reduces to Eq. (3.6a).

Generally speaking, symmetry adapted functions are appropriate linear combinations of spherical harmonics with the same ℓ. They are all real and normalized and each of them belongs to a certain column (or row) of a given irreducible representation. For a given ℓ, the $2\ell+1$ spherical harmonics are related to the $(2\ell+1)$ symmetry adapted functions $U_\ell^\nu(\Omega)$ by the unitary transformation

$$Y_\ell^m(\Omega) = \sum_\nu a_\ell^{m,\nu} U_\ell^\nu(\Omega) \quad . \tag{3.16}$$

Here the index ν labels the column and the representation. Unitarity implies

$$\sum_m (a_\ell^{m,\nu})^\star a_\Omega^{m,\nu} = \delta_{\rho\rho'} \quad . \tag{3.17}$$

Eq. (1.7c) is a particular example of

$$U_\ell^\nu = \sum_m (a_\ell^{m,\nu})^\star \, Y_\ell^m \quad . \tag{3.18}$$

Having defined the orientational distribution function, it is straightforward to define the orientational form factor by

$$F(\vec{k}) = \int d^3r \, e^{i\vec{k}.\vec{r}} \, a(\vec{r}) \quad . \tag{3.19}$$

Here the integration runs over the space occupied by the given molecule. Following Refs. (18), (19), we rewrite the density distribution as

$$a(\vec{r}) = \sum_{\ell,n} \bar{a}_{\ell n}(r) \, K_{\ell n}(\Omega) \tag{3.20}$$

where

$$\bar{a}_{\ell n} = \frac{\delta(r-d)}{r^2} \, a_{\ell n} \quad . \tag{3.21}$$

Next we use the expansion

$$e^{i\vec{k}.\vec{r}} = 4 \sum_{\ell,m} i^\ell j_\ell(kr) K_{\ell m}(\Omega_k) K_{\ell' m'}(\Omega) \quad , \tag{3.22}$$

where Ω_k refers to the polar coordinates of the wave veetor $\vec{k}$ and Ω to those of $\vec{r}$, while j_ℓ denote the spherical Bessel functions. Inserting Eq. (1.22) into Eq. (1.19), integrating over d^3r and using the orthonormality of the cubic harmonics, we find for the orientational form factor

$$F(\vec{k}) = 4\pi \sum_{\ell,n} i^\ell j_\ell(kd) a_{\ell n} K_{\ell n}(\Omega_k) \quad . \tag{3.23}$$

As said previously, the coefficients $a_{\ell n}$ depend on temperature. In addition, the weight of the various terms in Eq. (3.23) is largely determined by the value of the Bessel functions $j_\ell(kd)$ and by the direction of the $\vec{k}$-vector Ω_k. It is known that for small wave vectors or small molecular dimensions d such that $kd<1$, the asymptotic form of the Bessel function $j_\ell(kd)$ is $(kd)^\ell/(2\ell+1)!!$ and therefore the lowest order terms in ℓ are the most important in Eq. (3.23).

In complete analogy to the treatment of the orientational distribution function, all quantities $A(\Omega)$ which refer to time independent orientational properties of a single molecule in a certain field admit an expansion of the type

$$A(\Omega) = \sum_{\ell,\nu} a_\ell^{\nu,1} U_\ell^{\nu,1}(\Omega) \tag{3.24}$$

where the functions $U_\ell^{\nu,1}$ transform according to the unit representation of the given symmetry group.

4. ORIENTATIONAL CORRELATION FUNCTIONS

4.1 Orientational correlations of a single molecule (20)

So far we have investigated time independent orientational properties. We now consider time dependent correlation functions of a single molecule (selfcorrelation). We are interested in correlations

$$I^s(\vec{k},t) = <\rho^s(\vec{k},t)\rho^s(-\vec{k},o) \tag{4.1}$$

of the Fourier transformed densities

$$\rho^s(\vec{k},t) = \int e^{i\vec{k}.\vec{r}} \sum_i \delta(\vec{r}-\vec{r}_i(t))d^3r = \sum_i e^{i\vec{k}.\vec{r}_i(t)} \quad , \tag{4.2}$$

where i runs over the scattering centers in the molecule. Note that $<\rho(\vec{r})> = <\sum_i\delta(\vec{r}-\vec{r}_i)> = a(\vec{r})$, where $a(\vec{r})$ has been studied in the previous chapter. As a concrete example we consider a dumbbell-shaped molecule in an octahedral field. Then the index i runs over the two ends ± of the dumbbell:

$$\vec{r}_\pm = \vec{X} \pm \vec{d}(\Omega) \quad . \tag{4.3}$$

Here $\vec{X}$ is the center of mass position of the dumbbell, 2d its length and $\Omega=(\Theta,\varphi)$ and its polar angles. We rewrite Eq. (4.1) explicitely as:

$$I^s(\vec{k},t) = \sum_{i,j} <e^{i\vec{k}.\vec{r}_i(t)} e^{-i\vec{k}.\vec{r}_j(o)}> =$$

$$= <(e^{i\vec{k}.\vec{d}(t)}+e^{-i\vec{k}.\vec{d}(t)})(e^{-i\vec{k}.\vec{d}(o)}+e^{i\vec{k}.\vec{d}(o)})> \quad . \tag{4.4}$$

The thermal average is taken with a Hamiltonian having a full octahedral symmetry. Although the products of exponentials entering the thermal averages in Eq. (4.4) should then have full cubic symmetry, this needs not to be the case for the factors like exp. $(i\vec{k}.\vec{d})$. They only need to transform according to certain irreducible representations of the cubic group O_h, (21), (22). Therefore we expand the exponentials in terms of symmetry adapted functions U_ℓ^ν:

$$e^{i\vec{k}.\vec{d}} = 4\pi \sum_{\ell,\nu} i^\ell j_\ell(kd) U_\ell^\nu(\hat{k}) U_\ell^\nu(\Omega(t)) \tag{4.5}$$

where $\hat{k}$ stands for the polar coordinates Ω_k. We note this expansion follows directly from the well known result

$$e^{i\vec{k}.\vec{d}} = 4\pi \sum_{\ell,m} i^\ell j_\ell(kd) Y_\ell^{m*}(\hat{k}) Y_\ell^m(\Omega(t)) \tag{4.6}$$

when we make use of Eqs. (3.16) and (3.17). The inversion symmetry

$$Y_\ell^m(\Theta,\varphi) = (-1)^\ell Y_\ell^m(\pi-\Theta,\varphi+\pi) \tag{4.7}$$

of the spherical harmonics implies a similar symmetry property for the symmetry adapted functions U_ℓ^ν. Inserting expansion (4.5) into the last member of Eq. (4.4) we find:

$$I^s(k,t) = 64\pi^2 \sum_{\ell,\ell'} \sum_{\nu\nu'} j_\ell(kd) j_{\ell'}(kd) i \, i^{\ell'\ell} U_\ell^\nu(\hat{k}) U_{\ell'}^{\nu'}(\hat{k}) \times$$

$$\times \langle U_\ell^\nu(\Omega(t)) U_{\ell'}^{\nu'}(\Omega(t'))\rangle \quad . \tag{4.8}$$

As a consequence of inversion symmetry, only even order terms in ℓ,ℓ' contribute: $\ell,\ell' = 0,2,4...$. We note that expression (4.8) is in fact still quite general and valid for any symmetry. At this stage the number of expectation values is greatly reduced by using symmetry arguments or equivalently speaking, by using a fundamental theorem of group theory (23). Indeed the expectation value, at the same lattice site, of two symmetry adapted functions belonging to different irreducible representations or different columns (rows) of the same irreducible representation vanishes. In our case this means that only expectation values with $\nu=\nu'$ are different from zero: $\langle U_\ell^\nu(\Omega(t)) U_{\ell'}^\nu(\Omega(o))\rangle$. We now write down explicitly the relevant functions up to order $\ell=4$ corresponding to irreducible representations of the cubic group.

For $\ell=0$ we have one single function with A_{1g} symmetry $(\nu=1, A_{1g})$:

$$U_o^{1,a} = Y_o^o(\Omega) = 1/\sqrt{4\pi} \qquad (4.9)$$

where a stands for A_{1g}.
For ℓ=2, there are two functions belonging to the $E_g \equiv e$ representation:

$$U^{1,e} = Y_2^o(\Omega) = \frac{5}{16\pi}(3z^2-1) \qquad (4.10a)$$

$$U_2^{2,e} = Y_2^{2,c}(\Omega) = \frac{15}{16\pi}(x^2-y^2) \qquad (4.10b)$$

and three functions belonging to the $T_{2g} \equiv f$ representation:

$$U_2^{1,f} \equiv Y_2^{2,s}(\Omega) = \frac{15}{4} xy \qquad (4.11a)$$

$$U^{2,f} \equiv Y_2^{1,c}(\Omega) = \frac{15}{4} xz \qquad (4.11b)$$

$$U_2^{3,f} \equiv Y_2^{1,s}(\Omega) = \frac{15}{4} yz \qquad (4.11c)$$

The functions (4.10a) - (4.11c) represent the 5d-orbitals mentioned previously.
Since we restrict ourselves to terms up to order $\ell+\ell'$=4, we need only one function belonging to the A_{1g} representation for ℓ=4:

$$U_4^{1,a} \equiv K_4(\Omega) \quad . \qquad (4.12)$$

After using the fundamental theorem, we rewrite the selfcorrelation function as:

$$I^s(\vec{k},t) = I_s^{oo} + I_s^{22} + I_5^{04} + \ldots \qquad (4.13)$$

where

$$I_s^{oo}(\vec{k},t) = (8\pi j_o(kd))^2 (U_o^{1,a}(\hat{k}))^2 A_{1g}^{(o)}(t) \qquad (4.14a)$$

with

$$A_{1g}^{(o)}(t) = \langle U_o^{1,a}(t)\rangle = 1/4\pi \quad , \qquad (4.14b)$$

$$I_s^{22}(k,t) = (8\pi j_2(kd))^2 \{ \sum_{i=1}^{2} (U_2^{i,e}(\hat{k}))^2 E_g{}^{(2)}(t) \; +$$

$$+ \sum_{i=1}^{3} (U_2^{i,f}(\hat{k}))^2 T_{2g}(t)\} \tag{4.15}$$

where

$$E_g^{(2)}(t) = \langle U_2^{1,e}(t)U_2^{1,e}(o)\rangle = \langle U_2^{2,e}(t)U_2^{2,e}(o)\rangle \quad , \tag{4.16a}$$

and

$$T_{2g}^{(2)}(t) = \langle U_2^{1,f}(t)U_2^{1,f}(o)\rangle = \langle U_2^{2,f}(t)U_2^{2,f}(o)\rangle =$$

$$= \langle U_2^{3,f}(t)U_2^{3,f}(o)\rangle \quad . \tag{4.16b}$$

Finally we quote

$$I_s^{o,4} = 128\pi^2 j_o(kd) j_4(kd) U_4^{1,a}(\hat{k}) U_o^{1,a}(\hat{k}) A_{1g}^{(o,4)} \quad , \tag{4.17}$$

with

$$A_{1g}^{(o,4)} = \langle U_4^{1,a}(t)U_o^{1,a}(o)\rangle = \frac{1}{\sqrt{4\pi}} \langle K_4\rangle \quad . \tag{4.18}$$

Since we know the explicite expression for the function $\underset{\sim}{U}(\hat{k})$, we are able to calculate explicitly the $\hat{k}$-dependence of $I^s(\vec{k},t)$.

For the interpretation of neutron and Raman scattering experiments, one needs the Fourier transformed expression

$$I^s(\vec{k},\omega) = \frac{1}{2\pi} \int_{-\infty}^{+\infty} dt\, e^{i\omega t} I^s(\vec{k},t) \quad . \tag{4.19}$$

Here $\hbar\omega$ is the energy transfer between the experimental probe and the molecule. Collecting all terms on the r.h.s. of Eq. (4.13), we finally obtain:

$$I^s(k,\omega) = 4[\,(j_o(kd))^2 + 8\pi j_o(kd) j_4(kd) K_4(\Omega_k)\langle K_4\rangle]\,\delta(\omega) \; +$$

$$+120\pi(j_2(kd))^2[\,(\hat{k}_x^4\hat{k}_y^4+\hat{k}_z^4-\tfrac{1}{3})E_g^{(2)}(\omega) + (1-\hat{k}_x^4-\hat{k}_y^4-\hat{k}_z^4)T_{2g}^{(2)}(\omega)] \quad , \tag{4.20}$$

where $\hat{k}_x, \hat{k}_y, \hat{k}_z$ stands for the vector components k_x/k etc., while $E_g^{(2)}(\omega)$ and $T_{2g}^{(2)}(\omega)$ are the Fourier transforms of the correlation functions $E_g^{(2)}(t)$ and $T_{2g}^{(2)}$. These functions are directly accessible to Raman scattering (22). The coefficient of $\delta(\omega)$ on the r.h.s. of Eq. (4.20) corresponds to the elastic part of the scattering law. It has the A_{1g} symmetry. The correlation functions $E_g^{(2)}(\omega)$ and $T_{2g}^{(2)}(\omega)$ represent inelastic contributions.

4.2 Orientational correlations on a lattice

We consider a correlation function

$$C_{\ell\ell'}^{\nu\nu'}(\vec{k},\vec{q};t) = \frac{1}{N}\sum_{n,n'} \langle U_\ell^\nu(\vec{n},t)U_{\ell'}^{\nu'}(\vec{n}',o)\rangle\, e^{-i\vec{k}.\vec{X}(\vec{n})} e^{-i\vec{q}.\vec{X}(\vec{n}')} =$$

$$= \langle U_\ell^\nu(\vec{k},t)U_\ell^{\nu'}(\vec{q},o)\rangle \;. \qquad (4.21)$$

Here $\vec{n}$ and $\vec{n}'$ label the rigid lattice points $\vec{X}(\vec{n})$ and $\vec{X}(\vec{n}')$ corresponding to the center of mass positions of the molecules. We have abbreviated the notation by writing $(\vec{n},t)$ for $\Omega(\vec{n},t)$. Making use of translation symmetry, we write

$$C_{\rho\rho'}^{\nu\nu'}(\vec{k},\vec{q};t) = \frac{1}{N}\sum_{\vec{n},\vec{n}'} \langle U_\rho^\nu(\vec{n}-\vec{n}',t)U_{\rho'}^{\nu'}(\vec{0},o)\rangle \quad \times$$

$$\times\; e^{i(\vec{k}+\vec{q}).\vec{X}(\vec{n}')}\; e^{-i\vec{k}.(\vec{X}(\vec{n})-\vec{X}(\vec{n}'))}$$

$$= N\Delta(\vec{k}+\vec{q})C_{\rho\rho'}^{\nu\nu'}(\vec{k},t) \qquad (4.22)$$

where

$$C_{\ell\ell'}^{\nu\nu'}(\vec{k},t) = \frac{1}{N}\sum_{\vec{n}} \langle U_\ell^\nu(\vec{n},t)U_{\ell'}^{\nu'}(\vec{0},o)\rangle e^{-i\vec{k}.\vec{X}(\vec{n})} \;, \qquad (4.23)$$

In general, the fundamental theorem used in the previous section cannot be applied to the function $C_{\ell\ell'}^{\nu\nu'}(\vec{k},t)$ because the correlation runs over different lattice sites. An exception however is provided by symmetry points in $\vec{k}$ space where

$$R\vec{k} = \vec{k} + \vec{K} \;, \qquad (4.24)$$

R being an operation of the given point group. All the symmetry related lattice vectors are identical through reciprocal lattice vector displacements. Then the $\vec{k}$-vectors must be considered invariant under all the operations of the group, just as it had been $\vec{k}$=o.

As an example of a correlation function on a lattice we consider a simple cubic lattice of dumbbells. Since $\vec{X}(\vec{n})$ denotes the center of mass positions, the two end of the dumbbell are located at

$$\vec{r}(\vec{n},\pm) = \vec{X}(\vec{n}) + \vec{d}(\vec{n}) \quad . \tag{4.25}$$

We consider the density

$$\rho(\vec{r}(t)) = \sum_i \delta(\vec{r}-\vec{r}_i(t)) \quad , \tag{4.26}$$

where i runs over $\vec{n}$ and over the two ends of each dumbbell. The corresponding Fourier transform is given by

$$\rho(\vec{k},t) = \frac{1}{\sqrt{N}} \int d^3r \; e^{-i\vec{k}.\vec{r}} \rho(\vec{r},t)) =$$

$$= \frac{1}{\sqrt{N}} \sum_n e^{-i\vec{k}.\vec{X}(\vec{n})} (e^{i\vec{k}.\vec{d}(\vec{n},t)} + e^{-i\vec{k}.\vec{d}(\vec{n},t)}) \tag{4.27}$$

We are interested in the density-density correlation function

$$S(\vec{k},t) = \frac{1}{N} <\rho(\vec{k},t)\rho(-\vec{k},o)> = \frac{1}{N} \sum_n e^{-i\vec{k}.\vec{x}(\vec{n})} \quad \times$$

$$\times <(e^{i\vec{k}.\vec{d}(\vec{n},t)} + e^{-i\vec{k}.\vec{d}(\vec{n},t)}) \; (e^{i\vec{k}.\vec{d}(\vec{o},o)} + e^{-i\vec{k}.\vec{d}(\vec{0},o)})> \; . \tag{4.28}$$

We now proceed as in the last section where we derived Eq. (4.8). In complete analogy we now find

$$S(\vec{k},t) = 64\pi^2 \sum_{\ell,\ell'} \sum_{\nu,\nu'} j_\ell(kd) j_{\ell'}(kd) i^{\ell} i^{\ell'} U_\ell^\nu(\hat{k}) U_\ell^{\nu'}(\hat{k}) \; C_{\ell\ell}^{\nu\nu'}(\vec{k},t) \tag{4.29}$$

where $C_{\ell\ell'}^{\nu\nu'}(k,t)$ is defind by Eq. (4.23). Symmetry of the dumbbell implies that only even order terms in ℓ,ℓ' contribute to the sum (4.29). Octahedral symmetry of the lattice sites implies that only function U_ℓ^ν occur which transform according to the irreducible representations of the octahedral group. Further simplifications are only possible at certain symmetry points in $\vec{k}$-space (see Eq. (4.24).

The concept of symmetry adapted functions is also applicable to three dimensional molecules as for instance CH_4, Ref. (24). Then the molecular orientation Ω has to be specified by the Euler angles α,β,γ. Although the details of the theoretical treatment and the interpretation of experiments is more involved than in the case of linear molecules, general expressions as Eq. (4.8) or Eqs. (4.21), (4.23) and (4.29) remain valid.

5. COLLECTIVE ORIENTATIONAL DYNAMICS (25)

5.1 Introduction

The previous considerations have shown that the formalism of symmetry adapted functions is well suited to describe orientationally disordered states as well as orientational order. In addition we have formulated the static and dynamic correlation functions in terms of symmetry adapted functions. The remaining goal is the calculation of dynamic correlations functions in terms of these variables. Such a description is of essential importance in order to treat the dynamics of orientational phase transitions. We would like to mention that all treatments based on an expansion in terms of angular coordinates (26) are not adequate to handle the problems just mentioned.

Here we want to present a method which is based on the systematic use of symmetry adapted functions and the angular momenta of the molecules as dynamic variables (25). We will demonstrate that this approach allows to treat the dynamics of orientationally ordered, disordered and partially ordered phases in a unified way.

5.2 Basic relations

We start with recalling some known concepts of dynamic response theory. This section is an extension of the content of Sect. 2.1.

Generally we are only interested in the time evolution of some slow or secular variables. Obviously in hydrodynamics the conserved densities are slow variables. From a more general point of view the concept of secular variable is determined by the time scale of the experiment one wants to explain.

Following Mori (27) and Zwanzig (28), we separate the space of dynamic variables $F = \{A_\alpha\}$, $\alpha=1,\ldots,$ in a relevant (secular) part $E = \{A_\nu\}$, $E \in F$ and a non relevant (non secular) part. The static susceptibility (compare Eq. (2.3)

$$\chi_{\alpha\beta}(o) = - \langle\langle A^\dagger;A_\beta\rangle\rangle_o \equiv (A_\alpha,A_\beta) \qquad (5.1)$$

has the properties of a scalar product. Following refs. (27), (28), we introduce an operator P which projects onto the space of secular variables:

$$P\,A_\alpha = \sum_{\nu,\mu} A_\nu \chi^{-1}_{\nu\mu}(A_\mu, A_\alpha) \quad . \qquad (5.2)$$

The operator Q projects onto the space of non secular variables:

$$Q\,A_\alpha = A_\alpha - P\,A_\alpha \quad . \qquad (5.3)$$

The time evolution of an operator A_α is governed by the Liouville operator $\mathcal{L}$:

$$A_\alpha(t) = e^{i\mathcal{L}t}A_\alpha \quad , \quad \mathcal{L}A_\alpha = [H, A_\alpha] , \qquad (5.4)$$

where H is the Hamiltonian. It is convenient to introduce the relaxation function

$$\Phi_{\alpha\beta}(t) = \int_t^\infty dt'\; e^{-\varepsilon t'} < [A^\dagger(t'), A_\beta(o)] > \quad , \qquad (5.5)$$

the Laplace transform of which is directly related to the dynamic response function by

$$\Phi_{\alpha\beta}(z) = \int_0^\infty dt\; e^{izt}\Phi_{\alpha\beta}(t) = \frac{1}{z}[\chi_{\alpha\beta}(o) - \chi_{\alpha\beta}(z)] \qquad (5.6)$$

Having chosen the secular variables $\{A_\nu\}$, one can show (see e.g. Ref. (11)) that the relaxation function $\Phi_{\mu\nu}(z)$ obeys the following matrix equation

$$[z\,\tilde{1} - \tilde{\Omega} + \tilde{\Sigma}(z)]\,\tilde{\Phi}\,(z) = \tilde{\chi}(o). \qquad (5.7)$$

Here the tilde denotes the matrix character.
In Eq. (5.7) $\tilde{\Omega}$ represents the matrix of restoring forces:

$$\Omega_{\mu\lambda} = (A_\mu, \mathcal{L}A_\nu)\chi^{-1}_{\nu\mu}(o) \quad , \qquad (5.8)$$

while $\Sigma_{\mu\lambda}$ is called the dissipation or memory matrix:

$$\Sigma_{\mu\lambda}\,(z) = -\,(Q\mathcal{L}A_\mu, (z - Q\mathcal{L}Q)^{-1}\;Q\mathcal{L}A_\nu)\chi^{-1}_{\nu\lambda}(o) \qquad (5.9)$$

Equation (2.42) for the displacement-displacement response function is a particular example of Eq. (5.7). In that particular case Ω corresponds to the dynamical matrix and Σ is given by $M^{(3)}(z)$.

5.3 The Model

As a model system, we will treat the case of solid methane (CD_4). In first approximation we are able to neglect the coupling between translational and rotational degrees of freedom.

The orientational coupling between CD_4 tetrahedra is provided by the octopole-octopole interaction (24).

The Hamiltonian is then written as

$$H = \sum_{n,i} \frac{L_i^\dagger(\vec{n})L_i(\vec{n})}{2I} + \sum_{\substack{n,n' \\ \mu,\nu'}} U_{\mu 1}^{(3)}(\varphi(\vec{n}))C_{\mu\nu}^{\vec{n}\vec{n}'} U_{\nu 1}^{(3)}(\varphi(\vec{n}')) \ . \tag{5.10}$$

Here L_α^i denotes the components of the angular momentum operators, I is the moment of inertia of the methane molecule and $C_{\nu\nu}^{nn'}$ represents a geometrial coupling tensor. The indices $\vec{n}$ and $\vec{n}'$ label the rigid lattice sites. The orientation of the molecule at site $\vec{n}$ is given by the Euler angles $\varphi(n) = (\alpha,\beta,\gamma)$. The functions $U_{\mu 1}^{(3)}$; $\mu = 1,\ldots 7$ are linear combinations of the Wigner D-functions for $\ell=3$. They are called tetrahedral rotator functions. They take into account the symmetry of the methane molecule.

From theory (24), (29) and experiment (30) we know the different orientational phases of CD_4. In phase I, above 27K, the molecules are orientationally disordered (plastic phase). The orientational motion should correspond to hindered rotations. A partially ordered phase II exists in the temperature range 21.5 - 27K. There are three out of four molecules are orientationally ordered while the fourth one behaves as weakly hindered rotator (See fig. 3 in Ref. (31)). This phase is of particular interest because we expect two types of orientational motion: orientational oscillations for the ordered molecules and rotations for the disordered molecules.

Considering the ordered molecules in phase II, we see that all transformations of the group D_{2d} are symmetry operations. Group theoretical considerations then yield

$$\langle s\rangle \equiv U_{11}^{(3)} \neq 0 \quad , \quad \langle U_{\mu 1}^{(3)}\rangle = 0 \quad , \quad \mu = 2,\ldots 7. \tag{5.11}$$

where we write s for $U_{11}^{(3)}$.

On the other hand, the disordered molecules have an octahedral surrounding and therefore

$$U_{\mu 1}^{(3)} = 0, \qquad \mu = 1,\ldots .7 \ . \tag{5.12}$$

Consequently $\langle U_{11}^{(3)}\rangle$ is an order parameter (31). We will restrict ourselves to wave vectors which satisfy condition (4.24). It then follows that

$$(U_{\mu 1}^{(3)}(\vec{k}),\ U_{\nu 1}^{(3)}(\vec{k})) \equiv (U_{\mu 1}^{(3)},\ U_{\nu 1}^{(3)})_{\vec{k}} = \delta_{\mu\nu}(U_{\mu 1}^{(3)},\ U_{\nu 1}^{(3)})_{\vec{k}} \ . \tag{5.13}$$

and

$$(U_{\mu 1}^{(3)},\ (z-\mathcal{L})^{-1}U_{\nu 1}^{(3)})_{\vec{k}} = \delta_{\mu\nu}(U_{\mu 1}^{(3)},\ (z-\mathcal{L})^{-1}\ U_{\mu 1}^{(3)})_{\vec{k}} \ . \tag{5.14}$$

Therefore it follows that an appropriate choice of the wave vectors considerably simplifies the matrix equations (32).

Since the components of the angular momentum are the generators of infinitesimal rotations, they break the symmetry of a given orientational state. We calculate

$$L^{\alpha}U^{(3)}_{\mu 1} = \sum_{\nu} b^{\alpha}_{\mu\nu} U^{(3)}_{\nu 1} \quad ; \qquad \alpha = \pm , 0 \quad , \tag{5.15}$$

and using Eq. (5.11), we arrive at

$$<[L^{\alpha}, U^{(3)}_{\mu 1}]> = b^{\alpha}_{\mu 1} <U^{(3)}_{11}> = b^{\alpha}_{\mu 1} <s> \quad . \tag{5.16}$$

The coefficients $b^{\alpha}_{\mu 1}$ are complex numbers, they are calculate explicitly by using the linear relations between U and the Wigner D-functions. It is found that L^{+}, L^{-}, L^{0} break the symmetry of the functions $U^{(3)}_{\mu 1}$, $\mu = 5,6,7$ while $U^{(3)}_{11} = s$ is the order parameter.

By using the transformation properties of $U^{(3)}_{\mu 1}$, $\mu = 5,6,7$ and of the components of the angular momentum one can show that it is sufficient to take

$$E \equiv \{L^{o}, s\} \tag{5.17}$$

as set of relevant variables because all other combinations are equivalent.

5.4 Dynamic equations

Having chosen the set of relevant variables, we consider explicitely the general equation (5.7). The static susceptibility matrix is given by

$$\tilde{\chi}(o) = \begin{pmatrix} (L^{o},L^{o})_{\vec{k}} & 0 \\ 0 & (s,s)_{\vec{k}} \end{pmatrix} . \tag{5.18}$$

Here we have used time reversal symmetry to show that

$$(L^{z}, U^{(3)}_{71})_{k} = 0 \quad .$$

The frequency Ω is calculated by using the identity

$$(A, \mathcal{L}B) = <[A^{\dagger}, B]> \quad . \tag{5.19}$$

Taking into account Eq. (3.16) and determining explicitely $b^{o}_{71} = - 2i$, we arrive at

$$\Omega = -2\ i\ \langle s\rangle \begin{pmatrix} 0 & 1 \\ -1 & 0 \end{pmatrix} \chi^{-1} . \tag{5.20}$$

The memory matrix Σ depends on the frequency z. Under the assumption that the time behaviour of all non secular modes is much faster than the one of the secular modes, we take $\Sigma(z)$ at $z = io$. Then time reveral symmetry also implies that $\Sigma_{UL} = \Sigma_{LU} = 0$ and we obtain:

$$\tilde{\Sigma}(0) = i \begin{pmatrix} \Lambda_{LL} & 0 \\ 0 & \Lambda_{UU} \end{pmatrix} \chi^{-1} \equiv i \begin{pmatrix} \eta & o \\ 0 & \lambda \end{pmatrix}. \tag{5.21}$$

where

$$\Lambda_{LL} = (Q\mathcal{L}L^{o},(io-Q\mathcal{L}Q)^{-1}Q\mathcal{L}L^{o}) = \eta\ \chi_{LL}(o) \tag{5.22}$$

$$\Lambda_{UU} = (Q\mathcal{L}U,(io-Q\mathcal{L}Q)^{-1}Q\mathcal{L}U) = \lambda\ \chi_{UU}(o) \quad . \tag{5.23}$$

Equation (3.7) now becomes:

$$\begin{pmatrix} z + i\ \eta & 2\ i\langle s\rangle/(s,s) \\ -2\ i\langle s\rangle/(L^{o},L^{o}) & z + i\ \lambda \end{pmatrix} \Phi(z) = \chi(o). \tag{5.24}$$

Inverting this expression, we obtain

$$[\, z+i\ \lambda - \frac{4\langle s\rangle^{2}}{(L,L)(s,s)}\frac{1}{(z+i\ \ )}]\ \ \Phi_{ss}(z) = (s,s)\ . \tag{5.25}$$

The poles of $\Phi_{ss}(z)$ are found to be

$$z_{\pm} \doteq \pm\ \omega_{o} + i\ \Delta \quad , \tag{5.26}$$

with

$$\omega_{o} = 2\sqrt{\frac{\langle s\rangle^{2}}{(L,L)(s,s)} + \frac{(\lambda-\eta)^{2}}{16}} \tag{5.27}$$

and

$$\Delta = -\ (\lambda+\eta)/2\ . \tag{5.28}$$

As it is to be expected, the transport coefficients λ and η determine the damping. The quantity $\lambda=\Lambda_{UU}/\chi_{UU}(o)$ describes the orientational relaxation while $\eta = \Lambda_{LL}/\chi_{LL}(o)$ describes the relaxation of angular momentum.

In the absence of damping, the modes $z_{\pm}$ are purely oscilla-

tory with frequency $\omega_o \propto \sqrt{<s>}$. These collective excitations are called librations. Here we see explicitely that the temperature dependence of the libration frequency is determined by the order parameter $<s>$ and by the order parameter susceptibility. By approaching the orientational phase transition, $<s>\to o$ and $(s,s)\to\infty$. Consequently the librational frequencies vanish. This is completely analogous to the case of spin waves in magnetic systems. Note that no description based on an expansion in terms of small angular displacements would yield such a behaviour.
The basic equation (5.25) applies also to a description of the disordered phase. Putting therein $<s>=0$, we are left with

$$(z+i\ \lambda)\Phi_{ss}(z) = (s,s) \quad . \qquad (5.29)$$

This equation is of relaxational type. The orientational molecular interaction causes as a collective motion rotational "diffusion", or better rotational relaxation. By approaching the phase transition from the disordered phase, the temperature variation of $\lambda=\Lambda_{UU}/(s,s)$ is determined by the diverging order parameter susceptibility. Therefore λ becomes increasingly small and this shows up as or critical slowing down of orientational correlations (33) (34).

Consequently the present theory is able to describe the ordered and the disordered phases of plastic crystals in an unified way (25).

6. SINGLE PARTICLE DYNAMICS (20)

6.1 Introduction

Experimental methods such as nuclear magnetic resonance, incoherent neutron scattering and Raman scattering measure the Fourier-transformed orientational autocorrelation function of molecular motion in solids and liquids. These experiments give direct information about the single particle reorientational and librational motion in the effective potential built up by the surrounding medium. Generally the molecule can have several equivalent equilibrium orientations which are determined by the minima of the potential and the shape of the molecule. At low temperature (low in comparison with the height of the potential barrier), the molecule stays in one of the minima and performs small high-frequency torsional oscillations (librations). Reorientations by means of thermal activation are rather improbable. In addition, quantum mechanical tunneling between equivalent sites is possible. The tunneling regime is rather well understood, both in the case of molecular crystals (35) and of defect states in solids (36).

On the other hand, the regime of very low potentials and relatively high temperature is also very accessible to a theoretical treatment. In fact one is then in the situation of almost isotropic rotational diffusion (37). Equation of motion for the orientational distribution function have been derived in Refs. (38)

In contradistinction to these two extreme cases, the intermediate regime, where the potential barriers and the temperature are of the same order of magnitude, presents great theoretical difficulties. In the case of relatively strong potentials, such that the molecular position is well defined in one of the potential minima, reorientations among equivalent positions occur as a consequence of thermal activation. For a theoretical description one uses generally the so called jump models (39),(40) which are based on stochastic rate equations for a discrete orientational distribution function. Note that in the framework of the jump models, the time of reorientations is assumed to be infinitely short. In addition, librational motion around a given equilibrium position is neglected. Attempts to combine reorientations among potential minima with rotational diffusion (41) and torsional oscillations (42) are based on numerous phenomenological assumptions. Note that these different aspects of molecular motion are also exhibited by numerical calculations (43).

Recently Dianoux and Volino have extended the rotational diffusion model (44) for a uniaxial rotator by adding a N-fold cosine potential. Their theory constitutes a considerable progress in comparison with previous approaches.

Here we will present a theory (20) which is an extension of the concepts we used in Sect. 5 for the description of collective orientational dynamics.

In the absence of long range orientational order, the static restoring matrix Ω vanishes, all dynamic structure has to be extracted from the memory function $\Sigma(z)$.

6.2 The model

Again we consider a molecule of dumbbell shape in a field of octahedral symmetry. This case is realized in practice by molecular impurities (e.g. N_2, O_2, CN^-) that occupy a halide-site in the alkali-halides (45). The concentration of impurities is kept sufficiently low such that their mutual interaction is negligible. We want to investigate the motion of one single molecule and therefore we pick out one particular unit cell which contains in its center such an impurity. The surrounding lattice creates at the molecular site an average effective field of octahedral symmetry. The corresponding potential is assumed to be static and temperature independent. All other degrees of freedom of the surrounding lattice are assumed to play the role of a heat bath, in thermal contact with the chosen unit cell. Furthermore, we assume that translational motion of the molecule as well as deformations of the unit cell can be neglected. The model then consists of a rigid rotator of dumbbell shape in a lattice potential V (of octahedral symmetry). The corresponding Hamiltonian reads

$$H = \frac{L^2}{2I} + V(\Omega) \quad , \quad \Omega \equiv (\Theta,\varphi). \tag{6.1}$$

Here $\vec{L}$ is the angular momentum of the molecule and I its moment of inertia. The orientation of the linear molecule is again specified by the polar angles (Θ,φ). The effect of the heat bath will cause a permanent redistribution of the molecule over its different energy levels. We assume that the temperature is sufficiently high such that quantum mechanical effects as tunneling are unimportant. For CN^- impurities in KCl we expect that classical considerations apply for $k_BT \geq 20K$. In the following considerations we will restrict ourselves to the classical regime.

We expand the octahedral potential in terms of cubic surface harmonics belonging to the identity representation A_{1g}. (see Eq. 3.6b).

Since the coefficients a_{41} will allways occur in the form a_{41}/k_BT, we will use a single parameter $\beta = a_{41}/k_BT$ to scale the relative magnitude of the potential with respect to temperature. As an additional parameter we will consider the sign of the potential:

$$V \propto \sigma K_{41} \quad ; \quad \sigma = \pm\ 1\ . \tag{6.2}$$

A change of sign reverses the role of maxima and minima in the potential. In Fig. 1 we give the temperature behavior of $\langle K_{41}\rangle$ as a function of the potential:

$$\langle K_{41}\rangle = \mathrm{Tr}\ K_{41}e^{-\beta\sigma K_{41}}/\mathrm{Tr}\ e^{-\beta\sigma K_{41}} \quad ; \tag{6.3}$$

where the symbol Tr (trace) stands for the integral over the orientations: $d\Omega = d\Theta si\Theta d\varphi$.

In Fig. 1 we have plotted $\langle x^4\rangle$ or equivalently $\langle K_{41}\rangle$ as a function of inverse temperature.

6.3 Dynamic equations

In chapter 4 we have seen that the correlation functions

$$E_g^{(2)}(\omega) = \frac{1}{2\pi}\int_{-\infty}^{+\infty} dt\ e^{i\omega t}\langle U_2^{1,e}(t)U_2^{1,e}(0)\rangle \tag{6.4}$$

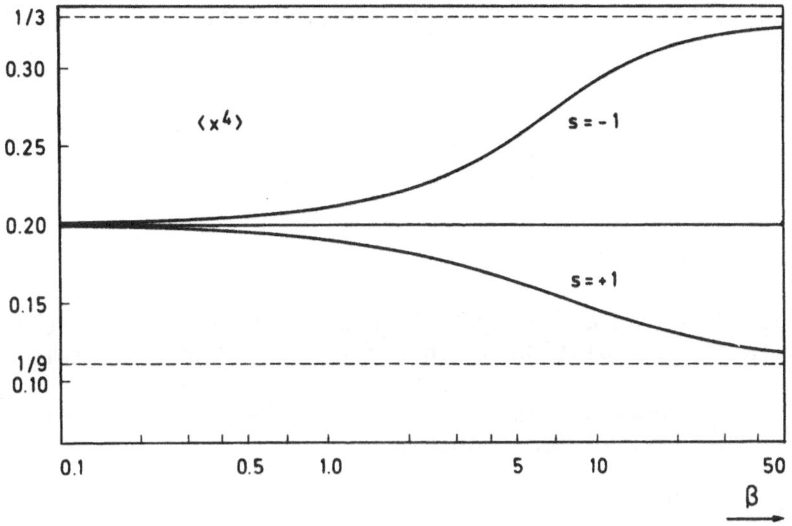

Fig. 1 Thermal expectation value of x^4 as a function of inverse temperature β for opposite signs of the potential.

and

$$T_{2g}^{(2)}(\omega) = \frac{1}{2\pi}\int_{-\infty}^{+\infty} dt\ e^{i\omega t} <U_2^{1,f}(t)U_2^{1,f}(o)> \qquad (6.5)$$

are the relevant dynamic quantities that enter the scattering law (4.20). Here we will calculate explicitly these correlation functions for the Hamiltonian (6.1). We will make use of a new self-consistent method for the calculation of frequency dependent transport coefficients (46). We rewrite the functions (6.4) and (6.5) as

$$C^{\alpha\alpha}(\omega) = \frac{1}{2\pi}\int dt\ e^{i\omega t} <U^{\alpha}(t),U^{\alpha}(o)> \qquad (6.6)$$

where α stands for (1,e;2) or (1,f;2). In the classical limit the fluctuation-dissipation theorem yields:

$$C_{\alpha\alpha}(\omega) = -T\Phi''_{\alpha\alpha}(\omega) \tag{6.7}$$

where $\Phi''(\omega)$ has been defined as in Eq. (2.20a).

We apply the concepts of the general theory and choose $E = \{U_\alpha, \dot{U}_\alpha\}$ as a set of dynamic variables with $\dot{U}=(d/dt)U$. Group theory again implies that only diagonal functions in α occur. The method is based on a continued fraction expansion (46). We find (20)

$$\Phi(\omega) = \frac{\beta\langle U^\alpha U^\alpha\rangle}{\omega - \dfrac{\langle\omega_\alpha^2\rangle}{\omega + \Sigma_{\alpha\alpha}(\omega)}} \quad , \tag{6.8}$$

where

$$\langle\omega_\alpha^2\rangle = \langle\dot{U}^\alpha\dot{U}^\alpha\rangle/\langle U^\alpha U^\alpha\rangle \tag{6.9}$$

is the second moment and $\Sigma_{\alpha\alpha}(\omega)$ the memory kernel. The problem consists in the calculation of this frequency dependent memory kernel.

In lowest approximation, assuming fast relaxation such that $\omega \ll \Sigma(\omega = io) \equiv i\gamma$, we would obtain

$$\Phi(\omega) = \frac{\beta\langle U^\alpha U^\alpha\rangle}{\omega + i\lambda} \quad , \tag{6.10}$$

with $\lambda = \langle\omega_\alpha^2\rangle/\gamma$. Expression (6.10) leads to a Lorentzian for $C_{\alpha\alpha}(\omega)$.

In the next approximation we find

$$\Sigma_{\alpha\alpha}(\omega) = -\frac{1}{\langle\omega_\alpha^2\rangle} \frac{(\langle\omega_\alpha^4\rangle - \langle\omega_\alpha^2\rangle^2)}{\omega + i\left(\dfrac{\langle\omega_\alpha^4\rangle}{\langle\omega_\alpha^2\rangle}\right)^{2}} \quad . \tag{6.11}$$

Putting ω=o, this expression determines γ. Generally $\omega \neq$o in $\Sigma(\omega)$ means that additional resonances occur in $\Phi_{\alpha\alpha}(\omega)$.

We have calculated the moments $\langle\omega_\alpha^2\rangle$ and $\langle\omega_\alpha^4\rangle$ = $\langle\ddot{U}^\alpha\ddot{U}^\alpha\rangle/\langle U^\alpha U^\alpha\rangle$ by means of the Hamiltonian (6.1). Then we have evaluated numerically all thermal averages as a function of temperature. Finally we have discussed $\Phi''(\omega)$ as a function of frequency for different temperatures and for the two signs of the potential. The results are as follows.

For σ=-1 and at low temperatures, the T_{2g} mode shows well defined off center resonances (Fig. 2). These resonances are symmetric with respect to the origin and correspond to oscillatory motion. Consequently the T_{2g} mode has librational character for σ=-1.

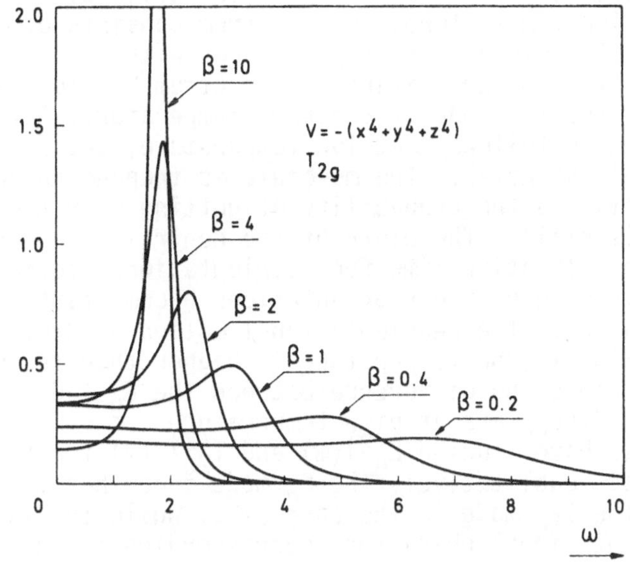

Fig. 2 Normalized dynamic correlation function $T_{2gN}(\omega)$ as a function of frequency for different temperatures, case s = -1.

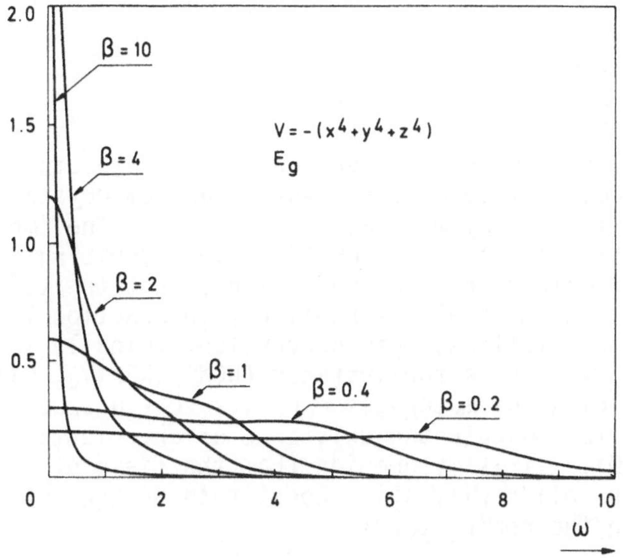

Fig. 3 Normalized dynamic correlation $E_{g,N}(\omega)$ as a function of frequency for different temperatures, case s = -1.

For the same potential, the E_g mode behaves quite differently (Fig. 3). At low temperature, the spectrum consists of a sharp central peak. The width of this peak grows with increasing temperature. At high temperature the spectrum becomes very similar to that of the T_{2g} mode at the same temperature. These results are interpreted as follows. At low temperature, the E_g mode has reorientational character. The molecule is trapped in one of the potential minima and the probability of getting over the potential barrier is small. The width of the central peak corresponds to the inverse relaxation time for reorientational motion. With increasing temperature, the reorientations become faster and the linewidth broadens. The reorientational motion goes over in a hindered rotation if the temperature is higher than the potential barrier. Therefore the difference between the E_g and the T_{2g} representation disappears at high temperature.

Finally we have studied $E_g^{(2)}(\omega)$ and $T_{2g}^{(2)}(\omega)$ for the potential with $\sigma=+1$. Qualitatively the E_g mode now has the same character as the T_{2g} mode in the case $\sigma=-1$, while The T_{2g} mode now has reorientational character, corresponding to the E_g mode in the case $\sigma=-1$. This inversion of the character of the modes corresponds to the exchange of the potential maxima and minima with changing sign of σ.

The present theory is in qualitative agreement with Raman scattering experiments on CN impurities in KCl, KBr and NaCl (47) (48).

CONCLUDING REMARKS

I have given a subjective version of some aspects of the rotational dynamics of molecular crystals and of crystals with molecular impurities. My main aim was to stress the importance of the use of symmetry adapted functions in describing the statics and dynamics of molecular rotations. I have not treated here such important topics as molecular tunnelling, interaction between translations and rotations, spin conversion. Finally I have omitted mixed crystals as for instance $(KCN)_x(KBr)_{1-x}$ with $x \geqslant 0.3$. This class of materials exhibits very unusual properties that seem to indicate the existence of an orientational glass state (49). It is my opinion that the field of orientational disorder in solids will still constitute an area of fruitful research during the coming years.

REFERENCES

(1) M. Born and K. Huang, Dynamical theory of crystal lattices, Oxford University Press, Oxford (1954).
(2) G. Leibfried and W. Ludwig, in Solid State Physics, 12, 275, ed. by F. Seitz and D. Turnbull, Academic Press, New York (1961).
(3) A.A. Maradudin, in Dynamical properties of solids, vol. I,1, ed. by G.K. Horton and A.A. Maradudin, North-Holland (1974).
(4) D.N. Zubarev, Soviet Physics Usp. 3, 320 (1960).
(5) L.P. Kadanoff and P.C. Martin, Ann. Phys. (N.Y.), 24, 419 (1963).
(6) R. Kubo, J. Phys. Soc. Japan 12, 570 (1957).
(7) M. Born, in Festschrift d. Akad. d. Wiss. Göttingen (1951).
(8) D.J. Hooton, Phil. Mag. 46, 422 (1955), idem Z. Phys. 142, 42 (1955), idem Phil. Mag. 3, 49 (1958).
(9) H.Horner, in Dynamical properties of solids, vol. I,8, same as Ref. (3); T.R. Koehler, in Dynamical properties of solids, vol. II, 1, same as Ref. (3).
(10) O. Schnepp, in Dynamical properties of solids, vol. II, 3, same as Ref. (3); J.C. Raich, N.S. Gillis, and A.B. Anderson, J. Chem. Phys. 61, 1399 (1974).
(11) W. Götze and K.H. Michel, in Dynamical properties of solids, vol. I, 9.
(12) H. Beck, in Dynamical properties of solids, vol. II, 4.
(13) G. Dolling, in Neutron Scattering, vol. I, p. 263, Proceedings of the Conference on Neutron Scattering, Gatlinburg 1976, R. Moon editor.
(14) G. Venkataraman and V.C. Sahni, Revs. Mod. Phys. 42, 409 (1970).
(15) H.A. Levy and S.W. Peterson, Phys. Rev. 83, 1270 (1951); 86, 766 (1952); E.L. Wagner and D.F. Hornig, J. Chem. Phys. 18, 296 (1950).
(16) C.J. Bradley and A.P. Cracknell, The Mathematical Theory of Symmetry in Solids, (Clarendon, Oxford 1972).
(17) F.C. von der Lage and H.A. Bethe, Phys. Rev. 71, 612 (1947).
(18) K. Kurki-Suonio, Ann. Acad. Sci. Fenn. A VI 263 (1967); R.S. Seymour and A.W. Preyor, Acta Cryst. B26, 1487 (1970).
(19) W. Press and A. Hüller, Acta Cryst. A29, 252 (1973); W. Press, Acta Cryst. A29, 257 (1973).
(20) B. De Raedt and K.H. Michel, Phys. Rev. B19, 767 (1979).
(21) For a complete group theoretical treatment of symmetry adapted functions, see lectures by R.M. Pick, present course.
(22) D. Fontaine, R. Pick, and M. Yvinec, Solid State Commun. 21, 1095 (1977); M. Yvinec and R. Pick, to be published.
(23) E. Wigner, Gruppentheorie und ihre Anwendung auf die Quantummechanik der Atomspektren (Vieweg, Braunschweig, 1931), p. 123.

(24) H.M. James and T.A. Keenan, J. Chem. Phys. 31, 12 (1959); T. Yamamoto, J. Chem. Phys. 48, 3139 (1968).
(25) K.H. Michel and H. De Raedt, J. Chem. Phys. 65, 977 (1976).
(26) K.H. Michel and D.M. Kroll, J. Chem. Phys. 64, 1300 (1976).
(27) H. Mori, Progr. Theor. Phys. 23, 423 (1965).
(28) R.W. Zwanzig, J. Chem. Phys. 33, 1388 (1960).
(29) T. Yamamoto, Y. Kataoka, and K. Okada, J. Chem. Phys. 66, 270 (1977).
(30) W. Press, J. Chem. Phys. 56, 2597 (1972).
(31) W. Press and A. Hüller, in Anharmonic lattices, structural transitions and melting, p. 185, ed. by T. Riste, Noordhoff, Leiden (1974); W. Press and A. Hüller, Phys. Rev. Lett. 30, 1207 (1973).
(32) H. De Raedt, private communication.
(33) W. Press. A. Hüller, H. Stiller, W. Stirling, and P. Currat, Phys. Rev. Lett. 32, 1354 (1974).
(34) D.M. Kroll and K.H. Michel, Phys. Rev. B15, 1136 (1977).
(35) A. Hüller and D.M. Kroll, J. Chem. Phys. 63, 4495 (1975).
(36) V. Narayanamurti and R.O. Pohl, Revs. Mod. Phys.
(37) A. Einstein, Ann. Phys. (Leipzig) 19, 372 (1906); P. Debije, Polar Molecules (Dover, New York, 1929).
(38) W.H. Furry, Phys. Rev. 107, 7 (1957); P.S. Hubbard, ibid. 109, 1153 (1958); M. Shimizu, J. Chem. Phys. 37, 765 (1962).
(39) R.C. Livingston, J.M. Rowe, and J.J. Rush, J. Chem. Phys. 60, 4541 (1974); K.H. Michel, ibid. 58, 1143 (1973).
(40) R. Stockmeyer and H. Stiller, Phys. Status. Solidi 27, 269 (1968); K. Sköld, J. Chem. Phys. 49, 2443 (1968).
(41) E.N. Ivanov, Sov. Phys. JETP 18, 104 (1964).
(42) K.E. Larsson, J. Chem. Phys. 59, 4612 (1973).
(43) C. Brot and I. Darmon, Mol. Phys. 21, 785 (1971); C. Brot and B. Lassier - Gevers, Ber. Bunsenges. Phys. Chem. 80, 31 (1976).
(44) A.J. Dianoux and F. Volino, Mol. Phys. 34, 1263 (1977); A.J. Dianoux, H. Hervet, and F. Volino, J. Phys. (Paris) 38, 809 (1977).
(45) F. Lüty, Phys. Rev. B10, 3677 (1974); H.U. Beyeler, ibid. 11, 3078 (1975).
(46) H. De Raedt and B. De Raedt, Phys. Rev. B15, 5379 (1977).
(47) R. Callender and P.S. Pershan, Phys. Rev. A2, 672 (1970).
(48) D. Durand and F. Lüty, Phys. Status Solidi B81, 443 (1977).
(49) J.M. Rowe, J.J. Rush, D.G. Hinks and S. Susman, Phys. Rev. Letters, in press (1979).

RAMAN AND INFRARED LINESHAPES OF

INTERNAL MODES IN ODIC PHASES

Robert M. Pick

Département de Recherches Physiques*,
Université P. et M. Curie
4 Place Jussieu, 75005 Paris (France)

1 INTRODUCTION

1.1 Presentation of plastic crystals

ODIC (Orientational Disorder in Crystal), ideally refers to a state of matter (formerly and still frequently called plastic crystals) which can be considered as an assembly of molecules, the centers of mass of which form a regular lattice while the orientation of those molecules displays some degree of disorder. This state, as the liquid crystal one (which is its counter part: center of mass disorder combined with some orientational order) appears as an intermediate state between the completely ordered, low temperature state, and the high temperature liquid one. Many crystals display such an effect, and the molecules which enter into such crystals may have either a two dimensional character or a three dimensional one.

Examples of linear molecules entering into plastic crystals are N_2 (which can form the βN_2 phase)(1) or CN^- (with the two plastic phases I and II of CNNa and CNK)(2). Three dimensional molecules with very odd shapes usually do not form plastic crystals so that the molecules of interest can conveniently be classified following some steric character.

- The pseudo planar molecules (such as NO_3^- in NO_3Na (3), cyclohexane (4), or thiophene (5), have a type of disorder which affects essentially the orientation of the molecule within its

* Laboratoire associé au C.N.R.S.

own plane. The same is true, presumably, for some spherical top molecules, such as $C(CH_3)_3CN$ (6) for instance.

- Pseudo spherical molecules also form rather easily plastic crystals, and one can put into this category, molecules such as CF_4, CBr_4, $C(CH_3)_4$, adamantane, $C(CH_3)_3Cl$, CF_3Cl...
- Finally, there exists also cases of plastic crystals formed with molecules which can exhibit different steric forms (e.g. succinonitrile (7)) or are very deformable (e.g. paraterphenyl (8)).

Properties of crystals exhibiting ODIC characters vary much from one system to another

- The number of successive plastic phases can be larger than one (three in thiophene e.g.)
- The stability range of plastic phases varies from a few degrees in some van der Waals solids (e.g. 5° in $C(CH_3)_2(CN)_2$ to hundreds of degrees for ionic systems (740° in phase I of CNK)
- The transition between the ordered and the ODIC phase is generally first order but cases are known ($ClNH_4$, CNK) where the transition is a second order one.
- The plastic phase, which always has a higher symmetry than the ordered one is frequently cubic (or hexagonal) for pseudo spherical or linear molecules, but has a much lower symmetry for pseudo planar molecules.

In these lectures, we shall specialize ourselves in rigid molecules, (excluding such cases as succinonitrile or paraterphenyl) and focus our attention on the orientational dynamics of the molecules. It is thus necessary to specify the orientation of a molecule, which is done in the following way. If {X, Y, Z} is an orthonormal coordinate system attached to the crystal, the orientation of a molecule with respect to the crystal will be described by Ω which has one of the two following meanings.

- For a linear molecule, Ω will represent the set of the two spherical coordinates which define the direction of the axis of the molecule;
- for a three dimensional molecule, if {x, y, z} is an orthornormal coordinate system attached to the molecule, Ω will represent the set of three Euler angles associated with the rotation which brings {X, Y, Z} in coincidence with {x, y, z}.

For reason of simplicity we shall always consider one molecule per cell crystals, so that Ω_L represents the orientation of the molecule in the Lth cell, the center of which is at $\underset{\sim}{R}_L$. Quantities of interest for our study are then:

- The probability distribution function (p.d.f.)

$$P_0(\Omega) = \frac{1}{N} \sum_L \delta(\Omega - \Omega_L) \qquad (1\text{-}1)$$

- The two particle orientational correlation function

$$P_0(\Omega_1, \Omega_2, i) = \frac{1}{N} \sum_{L_1, L_2} \delta(\Omega_1 - \Omega_{L_1}) \, \delta(\Omega_2 - \Omega_{L_2}) \, \delta(\underset{\sim}{R}_{L_1} - \underset{\sim}{R}_{L_2} - R_i) \quad (1\text{-}2)$$

etc... as well as time correlation functions such as the self correlation:

$$P(\Omega_1, \Omega_2, t) = \frac{1}{N} \sum_{L} \delta(\Omega_1 - \Omega_L(0)) \, \delta(\Omega_2 - \Omega_L(t)) \quad (1\text{-}3a)$$

which is the joint probability of finding the same molecule with orientation Ω_1 at time 0, and Ω_2 at time t, or the generalized two molecule correlation function

$$P(\Omega_1, \Omega_2, i, t) = \frac{1}{N} \sum_{L_1, L_2} \delta(\Omega_1 - \Omega_{L_1}(0)) \delta(\Omega_2 - \Omega_{L_2}(t)) \delta(\underset{\sim}{R}_{L_1} - \underset{\sim}{R}_{L_2} - \underset{\sim}{R}_i) \quad (1\text{-}3b)$$

Such quantities clearly enter into the study of the optical spectra of plastic crystals. We shall show this point in discussing the particular case of the Raman spectrum of an internal molecular mode.

1.2 Raman Lineshape of an Internal Mode

In the study of such a problem, two different aspects must be considered: the dynamical problem and the detection process.

1.2.1. Dynamical Problem.

Let us consider an isolated molecule (which will eventually enter into a plastic crystal). This molecule has a given symmetry, so that its internal modes can be classified by an index λ which represents at the same time

- the irreducible representation Γ of this mode,
- an index μ which labels the various modes which are in the same representation Γ
- an index n which labels (when Γ is a multidimensional representation) the various basis vectors for given Γ and μ.

One can thus write

$$\lambda \equiv \Gamma, \mu, n \quad \text{or } \lambda \equiv j, n \text{ with } j \equiv \Gamma, \mu . \quad (1\text{-}4)$$

In a condensed phase, it is usually reasonable to consider that the dynamics of modes with different j do not couple. In a plastic crystal with one molecule per cell the corresponding potential energy may thus be written as

$$E_j = \sum_{L,n} [U_j^n(\Omega_L)]^2 \omega_j^2 + \frac{1}{2} \sum_{\substack{L_1, L_2 \\ n_1, n_2}} U_j^{n_1}(\Omega_{L_1}) V_{n_1 n_2}^{L_1 L_2}(\Omega_{L_1}, \Omega_{L_2}) U_j^{n_2}(\Omega_{L_2}) \quad (1\text{-}5)$$

where

- $U^n_j(\Omega_L)$ represents the amplitude of the vibration of the nth component (of the jth mode) of molecule L, when it has the orientation Ω ;
- ω_j is the frequency of this mode in the isolated molecule;
- $V^{L_1L_2}_{n_1n_2}$ is the coupling interaction between these components, and has been written in a way which emphasizes its dependence both on Ω_{L_1} and Ω_{L_2} but can be, in fact, a function of the orientation of the other molecules of the crystal.

The existence of the coupling interaction implies that the solutions of (1-5) are still harmonic phonons but, that the corresponding eigen vectors are no longer plane waves $\exp(-i\underset{\sim}{k}.\underset{\sim}{R}_L)$, as would be the case in a regular lattice. Furthermore, the dispersion of the frequencies will depend on the importance of this interaction and of its dependence on the various molecular orientations.

1.2.2. Detection Aspect.

In molecular crystals, it is usually assumed that the individual Raman tensors of the internal modes are not affected by the existence of the other molecules. In other words, if the Raman tensor of nth component of the jth mode is, in the molecular axis, $R^{\alpha\beta n}_j$, where α,β are cartesian coordinates in the {x, y, z} axis, the same tensor will be written, for the Lth molecule, with orientation Ω_L

$$R^{ABn}_j(\Omega_L) = \sum_{\alpha,\beta} M^{AB\alpha\beta}(\Omega_L)\, R^{\alpha\beta n}_j \tag{1-6}$$

where $M^{AB\alpha\beta}(\Omega_L)$ is the rotation matrix which projects the $\alpha\beta$ tensor of {x, y, z} into the AB tensor of {X, Y, Z} .

1.2.3. Raman lineshape and Infrared lineshape.

With the preceding notations, the various Raman spectra which are related to this mode j are proportional to

$$I^{AB,CD}(\omega) = \int_{-\infty}^{+\infty} e^{i\omega t} \sum_{\substack{L_1L_2\\ n_1n_2}} \langle R^{ABn_1}_j(\Omega_{L_1}(0))U^{n_1}_j(\Omega_{L_1}(0))\, U^{n_2}_j(\Omega_{L_2}(t))^* \times R^{CDn_2}_j(\Omega_{L_2}(t))^* \rangle\, dt \tag{1-7}$$

where $\langle \quad \rangle$ means that an expectation value has been taken on the orientation of the molecules at time 0.

The number of independent tensors $I^{AB,CD}(\omega)$ depends only on the crystal symmetry, and ranges from 3 (for cubic crystals) to 21 for triclinic ones. In the same way, the infrared lineshape of this mode is given by a formula similar to (1-7) in which the Raman tensors R^{ABn}_j (Ω) are simply replaced by the Infra induced dipole tensor I^{An}_j (Ω) which are defined as

$$I^{An}_j (\Omega) = \sum_\alpha M^{A\alpha} (\Omega) I^{\alpha n}_j \qquad (1\text{-}8)$$

where $M^{A\alpha}$ (Ω) is the rotation matrix which projects the α coordinate of {x, y, z} on the A coordinate of {X, Y, Z} , and $I^{\alpha n}_j$ is the dipole induced in the α direction by the nth component of the jth mode of the isolated molecule. Depending on the symmetry the number of independent Infrared spectra will range from 1 to 6.

The evaluations of (1-7) or (1-8) are very complicated problems and many approximations will have to be done in order to use this expression. It clearly implies the use of probability distribution functions introduced in 1.1, which explains why we shall study them first in the two following sections. Nevertheless, a qualitative discussion of these formula can already be done, and will help understanding some aspects of this paper.

1.3 Qualitative discussion of an internal mode band shape and Remarks

1.3.1. Band shape origins.

(1-7), or (1-8), explicitly contains two time scales, the internal mode period, $T_j = 2\pi/\omega_j$, and an orientational correlation time, $T_r \gg T_j$ which, very roughly, describes the time after which a molecular orientation has been largely modified. It also contains, implicitly, a third time through the second term of the r-h-s of (1-5) which gives rise , for a given value of the molecular orientations Ω_L, to a distribution of frequencies with a band width $\Delta\omega$.

When $\Delta\omega T_r \gg 1$, one is in the static limit, where the reorientational dynamics does not play any role. Nevertheless, even in this case, and though the molecular centers of mass form a lattice, the corresponding spectrum does not have a δ line shape because the disorder enters both in the detection mechanism (R^{ABn}_j (Ω))and in the dynamics of the mode. (In fact, this is true, even if only one of the two mechanisms (detection or potential) is sensitive to disorder effects.)

In the general case, the measured spectrum thus contains informations on the total density of states of the mode, modulated, on the one hand, by the detection mechanism, and, on the other, by the reorientational dynamics of the system*. The disentangling of these effects is, in general, a difficult task, which we shall partly cover in this paper, through the help of various approxima-

* see footnote on next page

tions, that we shall now briefly described.

1.3.2. Plan of this paper.

One way of looking at the problem of band shapes consists in considering the role of the second term of the r-h-s of (1-5).

- When both the $L_1 \neq L_2$ terms are negligible, and the $L_1 = L_2$ ones do not depend much, either on Ω_L, or on the orientation of the neighbouring molecules, the spectrum directly reflects the one particle reorientational dynamics. This will be shown at the end of section 3, and models for this dynamics will be discussed in section 4.
- When only the $L_1 \neq L_2$ terms are negligible, both local field effects (i.e. fluctuations of $V^{L_1 L_1}_{n_1 n_2}(\Omega_{L_1})$ with Ω_{L_1} and Ω_{L_2}) and reorientational effects may be simultaneously present. This case will be briefly discussed in 5.2.
- When the $L_1 \neq L_2$ terms are not negligible (strong correlation case) the situation becomes extremely complex, and we shall only discuss the very simple case when $\Delta\omega T_r >> 1$. This will be the subject of 5.3 and 5.4.

Before going into these discussions, we must present a systematic way of dealing with the p.d.f. $P_0(\Omega)$ and the self correlation orientational distribution function (s.o.d.f.) $P(\Omega_1, \Omega_2, t)$. This will be the subject of sections 2 and 3, in which we shall also show under which circumstances some aspects of these functions can be directly measured by an internal mode band shape study.

2 PROBABILITY DISTRIBUTION FUNCTION $P_o(\Omega)$

2.1 Introduction

In this section we shall focus our interest on $P_0(\Omega)$, its formal development and the optical measurement of some of its coefficients. In principle, $P_0(\Omega)$ is a quantity which can vary between two extremes

- Uniform distribution function: $P_0(\Omega)$ = cte. There is an equal probability of finding a molecule in any direction (which means that the molecule does not feel any orientational potential). This is e.g. the case for CD_4(10) between 89.7K and 27.0K.
- Pseudo spin case: $P_0(\Omega) = \sum_i \delta(\Omega - \Omega_i)$ where Ω_i is a discrete set of orientations (which, of course, reflects the symmetry of the crystal). Though such extreme cases never really exist,

* We have not, here, taken into account the fact that the jth mode has a frequency ω_j only in an inertial framework, while the existence of $P(\Omega_1, \Omega_2, t)$ implies that the molecule is rotating. Though this effect may be not negligible (9), we shall not consider it here.

adamantane, $ClNH_4$, phase II of CNK are good approximations of them. The existence of such cases implies that for such crystals, the characteristic time T_r (introduced in 1.3) is much longer than the period of most of their external modes; the name "pseudo spin" comes from the fact that, in the treatment of the thermodynamics and dynamics of such systems (11), the various discrete orientations are usually described as the various sublevels of a fictitious spin.

The existence of a non uniform p.d.f. implies the existence of a non zero orientational potential for the molecules, which will notably influence the orientational dynamics of the molecule. The systematic study of $P_0(\Omega)$ in the pseudo spin case and in the intermediate cases is thus a prerequisite for the dynamical study. As a consequence, the dependence of $P_0(\Omega)$ on the symmetry of both the crystal and the molecule will be the subject of 2.2 for the case of linear molecules, and 2.3 for the 3 d case; finally 2.4 will briefly deal with the optical measurement of this quantity.

2.2 The Linear Molecule case

2.2.1. Symmetry considerations.

Let us consider a linear molecule which sits on a site which has a point group symmetry $\mathcal{S}$; as we deal, here, with Bravais lattices, $\mathcal{S}$ will also be the symmetry group $\mathcal{C}$ of the crystal, though in general $\mathcal{S}$ is only a subgroup of $\mathcal{C}$. $\mathcal{S}$ always contains a subgroup, $\mathcal{S}_r$, of proper rotations, but, if $\mathcal{S}$ also contains improper rotations, it is always possible to write $\mathcal{S}$ under the form

$$\mathcal{S} = \mathcal{S}_r \otimes (1 + ir_s) \qquad (2\text{-}1)$$

where i is the inversion operation,

$r_s \in \mathcal{S}_r$ if the group contains the inversion operation,

$r_s \notin \mathcal{S}_r$ otherwise.

It is clear that $P_0(\Omega)$ must be invariant under all the operations of $\mathcal{S}$. Nevertheless, the only physical operations which can be performed on a molecule are pure rotations. But, as an inversion on a 2 d object is equivalent to a 180° rotation with respect to any axis perpendicular to the axis of this object, we do not have here to worry about this distinction which will become important in 3 d cases.

$P_0(\Omega)$ being only a function of the spherical coordinates θ, and ϕ, it is natural to expand it on the basis of the spherical harmonics $Y_\ell^m(\theta, \phi)$ and to write, for a given temperature

$$P_0(\Omega) = \sum_{\ell,m} a_\ell^m \, Y_\ell^m \, (\theta, \phi) \qquad (2\text{-}2)$$

where the a_ℓ^m are temperature dependent coefficients which specify, for a given temperature, the p.d.f.. Nevertheless, these a_ℓ^m coefficients are redundant, because, as already mentioned, $P_0(\Omega)$ must be invariant with respect to all the operations of $\mathcal{S}$. To avoid this redundancy it is convenient to use the symmetry adapted surface harmonics which have been introduced e.g. by Bradley and Cracknell(12).

2.2.2. Symmetry adapted surface harmonics.

If Ω is any 3 d rotation (defined e.g. by its three Euler angles) $Y_\ell^m \, (\Omega^{-1}(\theta, \phi)$ and $Y_\ell^m \, (\theta, \phi)$ are related through the Wigner $D_\ell^{m'm} \, (\Omega)$ matrices through the equation

$$Y_\ell^m \, (\Omega^{-1}(\theta, \phi)) = \sum_{m'} D_\ell^{m'm} \, (\Omega) \, Y_\ell^{m'} \, (\theta, \phi) \qquad (2\text{-}3)$$

When one is dealing only with rotations which belong to a point group, $\mathcal{S}$, it is possible to avoid using (2-3) by defining symmetry adapted surface harmonics $Y_\ell^\lambda \, (\theta, \phi)$ which have the following properties

$$Y_\ell^\lambda \, (\theta, \phi) = \sum_m \alpha_\ell^{\lambda m} \; Y_\ell^m \, (\theta, \phi) \qquad (2\text{-}4)$$

where λ is the same type of composite index $\lambda \equiv \Gamma, \mu, n$ as defined in (1-4);

$$Y_\ell^{\Gamma,\mu,n} \, (s^{-1}(\theta, \phi)) = \sum_{n'} M_\Gamma^{n'n}(s) Y_\ell^{\Gamma,\mu,n'} \, (\theta, \phi) \qquad (2\text{-}5)$$

where $s \in \mathcal{S}$ and $M_\Gamma^{n'n}(s)$ is the representation matrix related to the operation s of the irreducible representation Γ of $\mathcal{S}$. (Note that s does not need to be a proper rotation).

Before proceeding further, we can simply note that the definition of the matrix $\alpha_\ell^{\lambda m}$, and the comparison between (2-3) and (2-5) yield the following relation (which we shall use in section 2.3)

$$\forall s \in \mathcal{S}_r$$

$$\sum_{m,m'} \alpha_\ell^{\lambda' m' *} \; D_\ell^{m'm}(s) \, \alpha_\ell^{\lambda m} = \delta_{\Gamma\Gamma'} \, \delta_{\mu\mu'} \, M_\Gamma^{n'n}(s) \qquad (2\text{-}6)$$

2.2.3. Development of $P_0(\Omega)$.

We can now use (2-4) to write the p.d.f. in the alternative form

$$P_0(\Omega) = \sum_{\ell,\lambda} a_\ell^\lambda \; Y_\ell^\lambda \, (\theta, \phi) \qquad (2\text{-}7)$$

By definition, $P_0(\Omega)$ is subjected to the condition

$$\forall s \in \mathcal{S} \qquad P_0(s\Omega) = P_0(\Omega) \qquad (2\text{-}8)$$

Use of (2-5) shows that the only non zero coefficients of (2-7) are those for which the index Γ represents the unity representation of $\mathcal{S}$, and table 1 gives for a certain number of point groups of practical interest the number of a priori non zero coefficients of (2-7) for $\ell \leqslant 4$.

In the case of O_h symmetry (CNK, CNNa) the surface harmonics are the usual Kubic harmonics and up to $\ell = 8$ it exists only three non trivial coefficients, one for $\ell = 4$, 6 and 8. The two first ones have been measured for T = 293K by neutron diffraction by Rush and Rowe (13) and the $\ell = 4$ has been measured at the same temperature (14) by Raman spectroscopy.

2.3 The three dimensional case

With 3d molecules, the orientations of which are now specified by the three Euler angles contained in Ω, the situation is somewhat more complex, and has been treated by Huller and Press (15) and in (16). The difficulty arises from the fact that on the one hand the symmetry of the molecule has now to be taken into account, and on the other hand, that the molecules can perform only proper rotations.

In the same way as in 2.2, if $\mathcal{M}$ is the symmetry point group of the molecule we shall write

$$\mathcal{M} = \mathcal{M}_r \otimes (1 + i\, m_s) \qquad (2\text{-}9)$$

Table 1. Number of independent coefficients entering into the development of $P_0(\Omega)$ for linear molecules forming crystals of various symmetries.

$\mathcal{S}$ \ ℓ	0	1	2	3	4
O_h	1				1
T_d	1			1	1
D_{4h}	1		1		2
C_{3v}	1	1	1	2	2

where the second term exists only if $\mathcal{M}$ contains improper rotations. (Let us simply point out (cf. (16)) that, if $\mathcal{M}$ does not contain improper rotations, $\mathcal{S}$ cannot contain such rotations either).

2.3.1. Symmetry properties of $P_0(\Omega)$.

A first obvious property of $P_0(\Omega)$ is

$$\forall s \in \mathcal{S}_r \qquad P_0(s\Omega) = P_0(\Omega) \tag{2-10}$$

Let us now consider the effect of the proper rotations of $\mathcal{M}$. If m represents, in the molecular axis, the three Euler angles of a rotation which changes the molecule into itself, as this molecule has an orientation Ω in the crystal, the same rotation is represented in the crystal axis by the product of the three rotations $\Omega\, m\, \Omega^{-1}$. This leads to the relation

$$\forall m \in \mathcal{M}_r \quad P_0(\Omega) = P_0(\Omega m\, \Omega^{-1}\Omega) = P_0(\Omega m) \tag{2-11}$$

Finally, in the practically important case when both $\mathcal{S}$ and $\mathcal{M}$ contain improper rotations, a third relation comes from the fact that the product of two improper rotations is a proper rotation. $P_0(\Omega)$ must then be invariant under any rotation which is the product of any improper rotation of $\mathcal{S}$ by any improper rotation of $\mathcal{M}$. When one takes (2-10 and 11) into account this yields

$$P_0(\Omega) = P_0(\, r_s\, \Omega r_m\,) \tag{2-12}$$

2.3.2. Symmetry adapted functions.

An obvious basis for developing $P_0(\Omega)$ are the Wigner $D_\ell^{mm'}(\Omega)$ which form a complete set of orthonormal functions of the Euler angles Ω . Nevertheless, in order to fully use the relations (2-10 to 12) it is useful to note that

$$D_\ell^{m\, m_1}(\Omega_0^{-1}\Omega) = \sum_{m_2} D_\ell^{m\, m_2}(\Omega_0^{-1}) \quad D_\ell^{m_2 m_1}(\Omega)$$

$$= \sum_{m_2} D_\ell^{m_2 m *}(\Omega_0)\; D_\ell^{m_2 m_1}(\Omega) \tag{2-13}$$

a relation which shows that the Wigner matrices transform nearly as the spherical harmonics. In view of the form of (2-10 to 12), it is then natural to define a new set of basis functions by

$$\Delta_\ell^{\lambda\lambda'}\ (\Omega) \equiv \sum_{m,m'} \alpha_\ell^{\lambda m *}\ D_\ell^{m\, m'}(\Omega)\ \ \beta_\ell^{\lambda'\, m'} \tag{2-14}$$

where the $\alpha_\ell^{\lambda m}$ are the coefficients of the symmetry adapted surface harmonics related to the $\mathcal{S}$ symmetry group and $\beta^{\lambda'}_{\ell}{}^{m}$ are those coefficients related to the $\mathcal{M}$ symmetry group. One is just led to write, in accordance with (2-7)

$$P_0(\Omega) = \sum_{\substack{\lambda\lambda' \\ \ell}} A_\ell^{\lambda\lambda'} \Delta_\ell^{\lambda\lambda'}(\Omega) \tag{2-15}$$

where the $A_\ell^{\lambda\lambda'}$ are, in fact, temperature dependent, and, as in section 2.2, the purpose of this transformation is to use the symmetry properties of the $\Delta_\ell^{\lambda\lambda'}(\Omega)$ in order to reduce the number of independent coefficients $A_\ell^{\lambda\lambda'}$ to its minimum value.

In the simple case when both $\mathcal{S}$ and $\mathcal{M}$ do not contain improper rotations, i.e. when $\mathcal{S} \equiv \mathcal{S}_r$, $\mathcal{M} \equiv \mathcal{M}_r$, $P_0(\Omega)$ is only subjected to (2-10 and 11); direct applications of (2-6) and (2-13, 14) show that the $A_\ell^{\lambda\lambda'}$ must be zero unless the irreducible representation Γ contained in λ is the identity representation Γ_0 of $\mathcal{S}$ and, the irreducible representation Γ' contained in λ' is the identity representation Γ_0' of $\mathcal{M}$, whence

$$P_0(\Omega) = \sum_{\substack{\lambda\lambda' \\ \ell}} A_\ell^{\lambda\lambda'} \Delta_\ell^{\lambda\lambda'}(\Omega) \qquad \begin{array}{l} \lambda = \Gamma_0, \mu, 1 \\ \lambda' = \Gamma_0', \mu', 1 \end{array} \tag{2-16}$$

The case when both $\mathcal{S}$ and $\mathcal{M}$ contain improper rotations is more complicated because, on the one hand, (2-10) applies only to the proper rotations of $\mathcal{S}$ (which form the subgroup $\mathcal{S}_r$) and the same is true for (2-11), on the other hand, (2-12) has also to be fulfilled. It is clear that, in order to satisfy (2-10) the irreducible representation Γ contained in λ must induce the identity representation of $\mathcal{S}_r$ and it is easily verified that, for any point group $\mathcal{S}$, there exists only two such representations. One is the identity representation Γ_0, and the other is another one dimensional representation Γ_0^-, which is listed for every point group in (16). The same is true for (2-11) which leaves for Γ' only the two representations, Γ_0' and $\Gamma_0'^-$. When (2-12) is also taken into account it is found that the development (2-15) is limited to the two sets

$$\begin{Bmatrix} \lambda = \Gamma_0, \mu, 1 \\ \lambda' = \Gamma_0', \mu', 1 \end{Bmatrix} \quad \text{and} \quad \begin{Bmatrix} \lambda = \Gamma_0^-, \mu, 1 \\ \lambda' = \Gamma_0'^-, \mu', 1 \end{Bmatrix} \tag{2-17}$$

The first type of coefficients are the same as in (2-16) and have been identified by Huller and Press (15). The second type of coefficients have been shown in (16) also to exist and table 2 gives, for $\ell \leqslant 9$, for three cases related respectively to the plastic phase of CBr_4 and the high and low temperature plastic phases of $C(CH_3)_3Cl$, the number of a priori non zero coefficients of the two kinds for each value of ℓ.

Finally, the rules for the coefficients of the second kind for the case when $\mathcal{M}$ contains improper rotations but $\mathcal{S}$ does not are also given in (16).

Table 2. Number of independent coefficients entering into the development of $P_0(\Omega)$ for 3 d molecules; this number is given, up to $\ell = 9$, for the coefficients of the first kind (1) and of the second kind (2) for the following cases:
A: $\mathcal{S} = O_h$ $\mathcal{M} = T_d$; B: $\mathcal{S} = O_h$ $\mathcal{M} = C_{3v}$;
C: $\mathcal{S} = D_{4h}$ $\mathcal{M} = C_{3v}$. In these three cases,
$\Gamma_0 \equiv A_{1g}$, $\Gamma_0' \equiv A_1$, $\Gamma_0^- \equiv A_{1u}$, $\Gamma_0'^- \equiv A_2$.

ℓ	A		B		C	
	1	2	1	2	1	2
0	1		1		1	
1						
2					1	
3						
4	1		2		4	
5						1
6	1		3		6	
7						2
8	1		3		9	
9		1		3		6

For 3 d molecules, coefficients of the first kind have been measured, by neutron scattering, up to $\ell = 3$ for CD_4 in its plastic phase (10), $\ell = 6$ for CBr_4 (17) and $\ell = 10$ for Adamantane (18), but it exists no optical measurement of them up to now. Furthermore, coefficients of the second kind, which can, in principle, be measured by elastic incoherent neutron scattering (19), have not yet been measured.

2.4 Optical measurement of $P_0(\Omega)$

Let us show how some coefficients, a_ℓ^λ for 2 d molecules or $A_\ell^{\lambda\lambda'}$ for 3 d molecules, can, at least in principle, be obtained by measuring the integrated intensities of internal modes. For reasons which will become clear in 2.4.3., we shall restrict ourselves to the treatment of Raman spectra, that of Infrared spectra being an obvious extension of this calculation.

2.4.1. The decoupling approximation, and integrated intensities.

Let us start with the general formula (1-7) which gives the expression of the individual Raman tensor spectra for a mode j of symmetry Γ .

$$I^{AB,CD}(\omega) = \int_{-\infty}^{+\infty} e^{i\omega t}\, dt \sum_{\substack{L_1 L_2 \\ n_1 n_2}}$$

$$< R_j^{ABn_1}(\Omega_{L_1}(0))\, U_j^{n_1}(\Omega_{L_1}(0))\, U_j^{n_2}(\Omega_{L_2}(t))^* \, R_j^{CDn_2}(\Omega_{L_2}(t))^* >$$

This formula simplifies if the individual phonon lifetime is shorter than the characteristic time T_r of the phonon dynamics; in this case the mean value appearing in the r-h-s of (1-7) may be transformed into

$$P(\Omega_{L_1}, \Omega_{L_2}, t)\, R_j^{ABn_1}(\Omega_{L_1}(0))$$

$$< U_j^{n_1}(\Omega_{L_1}(0)\, U_j^{n_2}(\Omega_{L_2}(t))^* > \; R_j^{CDn_2}(\Omega_{L_2}(t))^* \tag{2-18}$$

where $P(\Omega_{L_1}, \Omega_{L_2}, t)$ is the quantity introduced in (1-3b).

On the other hand, quite generally, we have

$$\int I^{AB,CD}(\omega)\, d\omega = \sum_{\substack{L_1 L_2 \\ n_1 n_2}}$$

$$< R_j^{ABn_1}(\Omega_{L_1}(0))\, U_j^{n_1}(\Omega_{L_1}(0))\, U_j^{n_2}(\Omega_{L_2}(0))^* \, R_j^{CDn_2}(\Omega_{L_2}(0))^* > \tag{2-19}$$

When the decoupling approximation is valid, because of the incoherence of the instantaneous internal displacements relative to the same mode, one gets

$$< U_j^{n_1}(\Omega_{L_1}(0))\, U_j^{n_2}(\Omega_{L_2}(0) > = \delta_{L_1 L_2}\, \delta_{n_1 n_2} < U_j^{n_1}(\Omega_{L_1})^2 > \tag{2-20}$$

Finally, because those modes have, in fact, nearly the same frequency, $< U_j^{n_1}(\Omega_{L_1})^2 >$ is also independent of Ω_{L_1} and n_1. Thus, when the decoupling approximation (2-18) is valid, one finally obtains

$$\int I^{AB,CD}(\omega)\, d\omega \simeq \int P_0(\Omega) \sum_n R_j^{ABn}(\Omega)\, R_j^{CDn}(\Omega)^* \, d\Omega \tag{2-21}$$

Let us stress that this formula does not entirely rely on (2-18). On the one hand, its derivation shows that it is <u>not</u> restricted to the case when one can neglect the dynamical coupling between different molecules. On the other hand, when this coupling is negligible, (2-19) directly reduces to a sum of monomolecular terms and it is then very likely that, for the self term at time 0, the decoupling approximation is valid, even if correlation effects cannot be neglected in the dynamical problem.

2.4.2. Evaluation of the Raman tensor.

The Raman tensor $R^{ABn}_{j}(\Omega)$ can easily be expressed with the help of the $\Delta^{\lambda\lambda'}_{\ell}$. Indeed, one has

$$M^{AB,\alpha\beta}(\Omega) = \sum_{\ell,mm'} d^{ABm}_{\ell}\, D^{mm'}_{\ell}(\Omega)\, d^{\alpha\beta m'*}_{\ell} \quad (2\text{-}22)$$

where d^{ABm}_{ℓ} is the projection operator of the AB tensor on the spherical harmonics Y^m_{ℓ}. (2-22) can also be written as

$$M^{AB,\alpha\beta}(\Omega) = \sum_{\ell,\lambda\lambda'} e^{AB\lambda}_{\ell}\, \Delta^{\lambda\lambda'}_{\ell}(\Omega)\, g^{\alpha\beta\lambda'*}_{\ell} \quad (2\text{-}23)$$

with $$e^{AB\lambda}_{\ell} = \sum_{m} d^{ABm}_{\ell}\, \alpha^{\lambda m}_{\ell} \quad (2\text{-}24a)$$

$$g^{\alpha\beta\lambda'}_{\ell} = \sum_{m'} d^{\alpha\beta m'}_{\ell}\, \beta^{\lambda'm'}_{\ell} \quad (2\text{-}24b)$$

Thus, one obtains

$$R^{ABn}_{j}(\Omega) = \sum_{\ell,\lambda\lambda'} e^{AB\lambda}_{\ell}\, \Delta^{\lambda\lambda'}_{\ell}(\Omega)\, r^{\lambda'n}_{\ell\, j} \quad (2\text{-}25a)$$

with

$$r^{\lambda'n}_{\ell\, j} = \sum_{\alpha\beta} R^{\alpha\beta n}_{j}\, g^{\alpha\beta\lambda'}_{\ell} \quad (2\text{-}25b)$$

Now, because a Raman tensor is a second order symmetrical tensor, $r^{\lambda'n}_{\ell\, j}$ is restricted to $\ell = 0$ and 2, and to $\ell = 2$ only if the internal mode is not in the unity representation Γ_0' of $\mathcal{M}$. Furthermore, if $\lambda' \equiv \Gamma', \mu', n'$ and $j \equiv \Gamma, \mu$, one has

$$r^{\lambda'n}_{\ell\, j} = \delta_{\Gamma\Gamma'}\, \delta_{nn'}\, \bar{r}^{\lambda',\mu}_{\ell} \quad (2\text{-}26)$$

and the number of different r coefficients which enter into (2-25a) for a given j mode is equal to the number of independent coefficients of the Raman tensor which exist for this mode.

2.4.3. Integrated tensorial intensity.

Inserting (2-25a) into (2-21) and integrating over Ω we see that the integrated intensity tensor $\int I^{AB,CD}(\omega)\,d\omega$ will be a linear function of the coefficients $A^{\lambda\lambda'}_{\ell}$, where the sum will be restricted to those terms for which the integral of the product of three $\Delta^{\lambda\lambda'}_{\ell}(\Omega)$ functions (one coming from $P_0(\Omega)$ and two from (2-25a)), are not zero.

As the product of two Δ functions with ℓ_1 and ℓ_2 can be expanded in a sum of Δ functions the ℓ of which vary from $|\ell_1 - \ell_2|$ to $(\ell_1 + \ell_2)$, we see, from the $\ell \leqslant 2$ restriction on (2-25a), that one can determine only $A^{\lambda\lambda'}_{\ell}$ for $\ell \leqslant 4$. The same argument would give $\ell \leqslant 2$ for I.R. spectroscopy, which, in view of tables 1 and 2, shows the interest, here, of Raman studies.

It is, in fact, advantageous to define the independent tensorial spectra

$$I^{\lambda_1\lambda_2}_{\ell_1\ell_2}(\omega) = \sum_{AB,CD} I^{AB,CD}(\omega)\, e^{AB\lambda_1}_{\ell_1}\, e^{CD\lambda_2 *}_{\ell_2} \tag{2-27}$$

As a Raman tensor is an observable of the crystal, it is in the unity representation of this crystal; thus, if $\lambda_1 \equiv \Gamma_1, \mu_1, n_1$; $\lambda_2 \equiv \Gamma_2, \mu_2, n_2$, this spectrum exists only if $\Gamma_1 = \Gamma_2$ and $n_1 = n_2$.

With the help of (2-27) one obtains the final formula

$$\int I^{\lambda_1\lambda_2}_{\ell_1\ell_2}(\omega)\, d\,\omega = \sum_{\ell,\lambda\lambda'} A^{\lambda\lambda'}_{\ell}\; D^{\lambda\lambda'\;\lambda_1\;\lambda_2}_{\ell\;\;\;\ell_1\;\ell_2} \tag{2-28}$$

with

$$D^{\lambda\lambda'\;\lambda_1\;\lambda_2}_{\ell\;\;\;\ell_1\;\ell_2} = \sum_{\substack{\lambda'_1,\lambda'_2\\ n'}} C(\ell\ell_1\ell_2|\lambda\lambda',\lambda_1\lambda'_1,\lambda_2\lambda'_2)\; r^{\lambda'_1 n'}_{\ell_1 j}\; r^{\lambda'_2 n' *}_{\ell_2 j} \tag{2-29}$$

$$C(\ell\ell_1\ell_2|\lambda\lambda',\lambda_1\lambda'_1,\lambda_2\lambda'_2) = \int d\,\Omega\; \Delta^{\lambda\lambda'}_{\ell}(\Omega)\; \Delta^{\lambda_1\lambda'_1}_{\ell_1}(\Omega)\; \Delta^{\lambda_2\lambda'_2 *}_{\ell_2}(\Omega) \tag{2-30}$$

2.4.4. Discussion and examples.

In general, (2-28) is of no practical use, because the Raman tensors, or the $r^{\lambda n}_{\ell j}$ tensors, or even their relative values for one mode are not known. Nevertheless it may happen that, for modes of a given symmetry, two independent tensorial spectra depend only on one and the same $r^{\lambda n}_{\ell j}$.

For instance, in an O_h Bravais lattice with molecules of C_{3v} symmetry (plastic phase of $C(CH_3)_3Cl$) the internal modes of A_1 symmetry gives an E_g and a F_{2g} spectrum which depend only on the traceless part of the molecular Raman tensor. From (2-29) it follows that the D coefficients of (2-28) are completely known up to a coefficient which is the same for the two spectra. A measure of the ratio of their integrated intensity thus gives a quantity which depends only on the two unknown coefficients with $\ell = 4$ (see table 2).

The same is true if the molecule has a T_d symmetry, in which case, either E or F_2 internal modes can be used. In this case, as the ratio depends only on the single $\ell = 4$ coefficient (see table 1), this coefficient can uniquely determined, as one obtains (20)

$$R = \frac{2A_0 + A_4}{6A_0 - 2A_4} \qquad \text{for E modes}$$

$$R = \frac{3A_0 - A_4}{9A_0 + 2A_4} \qquad \text{for } F_2 \text{ modes} \qquad (2\text{-}31)$$

where $$R = \frac{\int I^{E_g}(\omega)\, d\omega}{\int I^{F_{2g}}(\omega)\, d\omega}$$

(here A_0, and A_4 stand for the unique coefficients with $\ell = 0$ and $\ell = 4$.)

The same technique can be applied to linear molecules for which case the calculation is simpler, but of the same type as presented here. In the special case of a crystal with O_h symmetry, one obtains for the A_1 modes of $C_{\infty v}$ molecules (20)

$$R = 3 \times \frac{a_0 - {}^4/_7 a_4}{a_0 + {}^6/_7 a_4} \qquad (2\text{-}32)$$

with notations similar to that of (2-31). This technique has been used to measure a_4 for KCN (21) and NaCN (14) and it has been verified that it gives values in agreement with the corresponding neutron measurements.

3 SELF CORRELATION ORIENTATIONAL DISTRIBUTION FUNCTION $P(\Omega_1, \Omega_2, t)$

This section is devoted to the formal development and to the optical measurement of the self correlation orientational distribution function (in short s.o.d.f.) P (Ω_1, Ω_2, t) which, as defined in (1-3a), is the joint probability of finding the same molecule with orientation Ω_1 at time 0, and Ω_2 at time t. We shall follow the same plan as in section 2, discussing first the 2d molecule case, then the 3d molecule one, and finally discussing their Raman measurements when both the decoupling approximation and the absence of local field effect approximation are simultaneously valid.

3.1 The linear molecule case

As the s.o.d.f. depends on two orientations of the same molecule, one can clearly write

$$P(\Omega_1, \Omega_2, t) = \sum_{\substack{\ell_1,\ell_2 \\ \lambda_1,\lambda_2}} b^{\lambda_1\ \lambda_2}_{\ell_1\ \ell_2}(t)\ Y^{\lambda_1}_{\ell_1}(\Omega_1)\ Y^{\lambda_2}_{\ell_2}(\Omega_2)^* \qquad (3\text{-}1)$$

Furthermore, the only symmetry requirement on this function is that it belongs to the unity representation of $\mathcal{S}$, the symmetry point group of the site where the molecule stands:

$$\forall s \in \mathcal{S} \qquad P(s\Omega_1, s\Omega_2, t) = P(\Omega_1, \Omega_2, t). \qquad (3\text{-}2)$$

This immediately sets up the selection rules on λ_1, and λ_2 :

if $\lambda_1 \equiv \Gamma_1, \mu_1, n_1$ and $\lambda_2 = \Gamma_2, \mu_2, n_2$ then $\Gamma_1 = \Gamma_2$; $n_1 = n_2$.

Furthermore, because

$$\int P(\Omega_1, \Omega_2, t)\ d\Omega_2 = P_0(\Omega_1) \qquad (3\text{-}3a)$$

$$\int P(\Omega_1, \Omega_2, t)\ d\Omega_1 = P_0(\Omega_2)\ , \qquad (3\text{-}3b)$$

as the $Y^\lambda_\ell(\Omega)$ form an orthonormal complete set, and Y^λ_0 = cte, the $b^{\lambda_1\ \lambda_2}_{\ell_1\ \ell_2}(t)$ coefficients are constant if ℓ_1 or $\ell_2 = 0$.

In the case of a Bravais lattice, the various different time dependent $b^{\lambda_1\ \lambda_2}_{\ell_1\ \ell_2}(t)$ which can be obtained for $\ell_1, \ell_2 \leqslant 4$ are tabulated in table 3 for the important case of a site with O_h symmetry.

3.2 The three dimensional case

In a manner similar to (3-1) one may write

$$P(\Omega_1,\Omega_2,t) = \sum_{\substack{\ell_1,\lambda_1,\lambda'_1 \\ \ell_2,\lambda_2,\lambda'_2}} B^{\lambda_1\lambda'_1\lambda_2\lambda'_2}_{\ell_1\quad \ell_2}(t)\ \Delta^{\lambda_1\lambda'_1}_{\ell_1}(\Omega_1)\ \Delta^{\lambda_2\lambda'_2}_{\ell_2}(\Omega_2)^* \qquad (3\text{-}4)$$

Table 3. Names of the representations $\Gamma_1 (=\Gamma_2)$ which appear, for a site with O_h symmetry, in $b^{\lambda_1 \lambda_2}_{\ell_1 \ell_2}(t)$ for ℓ_1 and $\ell_2 \leqslant 4$; each representation appears at most once, in the present case, for given ℓ_1 and ℓ_2.

ℓ_2 \ ℓ_1	1	2	3	4
1	F_{1u}		F_{1u}	
2		E_g, F_{2g}		F_{2g}, E_g
3	F_{1u}		A_{2u}, F_{1u}, F_{2u}	
4		F_{2g}, E_g		A_{1g}, E_g F_{1g}, F_{2g}

The three symmetry properties that are met by $P(\Omega_1, \Omega_2, t)$ directly stem from section 2 and can be written

$$\forall s \in \mathcal{S}_r \quad P(s\Omega_1, s\Omega_2, t) = P(\Omega_1, \Omega_2, t) \tag{3-5a}$$

$$\forall m \in \mathcal{M}_r \quad P(\Omega_1 m, \Omega_2 m, t) = P(\Omega_1, \Omega_2, t) \tag{3-5b}$$

$$P(r_s \Omega_1 r_m, r_s \Omega_2 r_m, t) = P(\Omega_1, \Omega_2, t) \tag{3-5c}$$

this last relation existing only when both groups $\mathcal{S}$ and $\mathcal{M}$ contain improper rotations (see 2.3 for notations).

Group theory arguments, as well as the use of the symmetry properties of the Δ functions, show that the selection rules for (3-4) are (19)

$$\left\{ \begin{matrix} \lambda_1 = \Gamma_1, \mu_1, n_1 \\ \\ \lambda_2 = \Gamma_1, \mu_2, n_1 \end{matrix} \right\} \quad \text{and} \quad \left\{ \begin{matrix} \lambda'_1 = \Gamma'_1, \mu'_1, n'_1 \\ \\ \lambda'_2 = \Gamma'_1, \mu'_2, n'_1 \end{matrix} \right\} \tag{3-6}$$

which are called coefficients of the first kind, or

$$\left\{ \begin{matrix} \lambda_1 = \Gamma_1, \mu_1, n_1 \\ \\ \lambda_2 = \bar{\Gamma}_1, \mu_2, n_1 \end{matrix} \right\} \quad \text{and} \quad \left\{ \begin{matrix} \lambda'_1 = \Gamma'_1, \mu'_1, n'_1 \\ \\ \lambda'_2 = \Gamma'^{-}_1, \mu'_2, n'_1 \end{matrix} \right\} \tag{3-7a}$$

which are called coefficients of the second kind and exist only when both $\mathcal{S}$ and $\mathcal{M}$ contain improper rotations, with the definition

$$\Gamma_1^- = \Gamma_1 \times \Gamma_0^- \qquad \Gamma_1'^- = \Gamma_1' \times \Gamma_0^- \tag{3-7b}$$

(see (2-17) and comments above this equation for the definition of Γ_0^-, $\Gamma_0'^-$).

We shall here concentrate only on the coefficients of the first kind, which are the only ones that we are presently able to measure. As (3-3) does not depend on the dimensionality of the molecule, the only self correlation coefficients which depend on time are again those for which ℓ_1 and $\ell_2 \neq 0$. Table 4 gives, in the case of a Bravais lattice with O_h symmetry, for molecules with T_d and C_{3v} symmetry, the various correlation functions which can be obtained for ℓ_1 and $\ell_2 \leqslant 2$.

3.3 Raman measurement of the self correlation coefficients.

Optical spectroscopy offers, in certain cases, a direct way of measuring some of the s.o.d.f. coefficients we have just defined. Indeed, some of their Fourier transforms directly appear in the optical band shapes if individual dynamics of the internal modes is assumed, i.e. if at the same time,

- it exists no dynamical coupling between the internal modes of two different molecules,
- the local field effects are negligible, so that the decoupling approximation is valid.

We shall show this by considering Raman scattering experiments, I.R. absorption being able to be treated along the same lines.

3.3.1. Raman scattering formulation.

With the preceding approximations, one can simplify the correlation function entering into (1-7) as

$$< R_j^{ABn_1}(\Omega_{L_1}(0))\, U_j^{n_1}(\Omega_{L_1}(0))\, U_j^{n_2}(\Omega_{L_2}(t))^X\, R_j^{n_2}(\Omega_{L_2}(t))^* > \simeq$$

$$\delta_{L_1L_2} P(\Omega_1,\Omega_2,t)\; R_j^{ABn_1}(\Omega_1) < U_j^{n_1}(\Omega_1(0))\, U_j^{n_2}(\Omega_2(t))^* > R_j^{CDn}(\Omega_2) \tag{3-8}$$

Furthermore, as we have neglected the Coriolis effect and the local field effect

$$< U_j^{n_1}(\Omega_1(0))\, U_j^{n_2}(\Omega_2(t)) > = \delta n_1 n_2 e^{-i\bar{\omega}_j t} \tag{3-9}$$

where $\bar{\omega}_j$ differs from the free molecule frequency, ω_j, due to the

Table 4. Names of the representations $\Gamma_1 (= \Gamma_2)$, $\Gamma_1' (= \Gamma_2')$ which appear in $B^{\lambda_1\lambda_1'\,\lambda_2\lambda_2'}(t)$ for a site with O_h symmetry and a molecule $\ell_1\ \ell_2$ with T_d, and C_{3v}, symmetry for ℓ_1 and $\ell_2 \leqslant 2$. The number of coefficients which correspond to a given Γ_1, Γ_1' is given in parenthesis when larger than 1.

	T_d		C_{3v}	
ℓ_2 \ ℓ_1	1	2	1	2
1	F_{1u}, F_2		F_{1u},A_1;F_{1u},E	
2		F_{2g},E;F_{2g},F_2 E_g,E;E_g,F_2		F_{2g},A_1;F_{2g},E(4) E_g,A_1;E_g,E(4)

influence of the mean value of the potential created by the other molecules. Using the definitions of the tensorial spectra (2-27) we easily gets

$$I^{\lambda_1\lambda_2}_{\ell_1\ell_2}(\omega) = \int dt e^{i(\omega - \bar{\omega}_j)t} \sum_{\substack{n \\ \lambda_1'\lambda_2'}} \int d\Omega_1 d\Omega_2\ \Delta^{\lambda_1\lambda_1'}_{\ell_1}(\Omega_1)\, r^{\lambda_1' n}_{\ell_1 j}\ \times$$

$$\Delta^{\lambda_2\lambda_2'}_{\ell_2}(\Omega_2)^{x} r^{\lambda_2' n^*}_{\ell_2 j}\quad P(\Omega_1, \Omega_2, t).$$

$$I^{\lambda_1\lambda_2}_{\ell_1\ell_2}(\omega) = \int dt e^{i(\omega - \bar{\omega}_j)t} \sum_{\substack{n \\ \lambda_1'\lambda_2'}} B^{\lambda_1\lambda_1'\lambda_2\lambda_2'}_{\ell_1\ \ell_2}(t)^* r^{\lambda_1' n}_{\ell_1 j}\ r^{\lambda_2' n^*}_{\ell_2 j} \quad (3\text{-}10)$$

where the selection rules are given for the B in (3-6) and for the r in (2-26).
(Note that the selection rule for $I^{\lambda_1\lambda_2}_{\ell_1\ell_2}(\omega)$, given below (2-27) directly stem from those of the B).

(3-10) shows that the lineshape analysis can only give informations on $\ell_1 = \ell_2 = 2$ coefficients. In fact, as in (2-4), the application of (3-10) to the measurement of the individual coefficients of the s.o.d.f. is, practically, restricted to the case when only one coefficient r enters the formula. This is the case for molecules with T_d symmetry forming a cubic Bravais lattice. In this case, internal modes with E and F_2 symmetry must be used, and their spectra recorded in the "E_g" and the "F_{2g}" geometry. Examination of Table 4 shows that the four $\ell_1 = \ell_2$ independent coefficients of the s.o.d.f. can thus be individually measured. On the other hand, for a molecule with C_{3v} symmetry (plastic phase I of $C(CH_3)_3Cl$) independent spectra will be obtained from the traceless part of A_1 modes, while E modes will give spectra the shape of which depend on the ratio of the two independent Raman tensors of the mode, a quantity which is not, usually,available.

(3-10) can be applied to the stretching modes of linear molecules. Such modes belong to the identity representation of the molecular group $C_{\infty v}$ or $D_{\infty h}$ and one gets from (3-10) and (3-1)

$$I_{\ell_1\ell_2}^{\lambda_1\lambda_2}(\omega) = \int e^{i(\omega - \bar{\omega}_j)t} \, d\omega \sum_{\lambda'_0} b_{\ell_1\ell_2}^{\lambda_1\lambda_2}(t) \, r_{\ell_1 j}^{\lambda'_0} \, r_{\ell_2 j}^{\lambda'_0} \qquad (3\text{-}11)$$

This formula has beenapplied to the stretching mode of CN^- in KCN and NaCN (21), where, from table 3, it is clear that it exists only for $\ell_1 = \ell_2 = 2$, two independent coefficients which can be obtained by simply recording the E_g and F_{2g} spectra.

4 MODELS FOR $P(\Omega_1, \Omega_2, t)$

4.1 Introduction

In principle, the s.o.d.f. has to be computed from an Hamiltonian of the crystal which takes into account all the external dynamic properties. In other words, the dynamical variables of the problem are the center of mass displacements on the one hand, and the molecular orientations on the other hand; furthermore, a general potential energy written in terms of those variables has to be written down, and the equation of motion solved, at finite temperature, in the plastic phase. Very little is yet known on this problem, and the present state of art is described in the papers contained in this volume, by K. Michel (22) from the theoretical point of view and G. Jaccuci (and references herein) (23) from the molecular dynamics point of view.

In the absence of clear theoretical predictions, one is led to set up phenomenological models of one sort or another to compare with the experiments. When $P_0(\Omega)$ is anisotropic, the minimum requirements that one would expect for the s.o.d.f. are that it should

- satisfy equations (3-3) thus reflect the anisotropy of the p.d.f.

- contain some description of the librations the molecules perform around the maxima of the p.d.f. as well as of a time of residence in the vicinity of such maxima,
- describe the reorientational dynamics of the molecule.

In fact, it exists presently no model which simultaneously meet the three preceding requirements, and all experimental data have been up to now analysed using simpler models which incorporate only some of those aspects. We shall here review some of these models, express their result in terms of the formalism given in section 3, and analyse some of their features and drawbacks.

4.2 Spherical models

These models mathematically neglect the anisotropy of $P_0(\Omega)$ and thus, physically, the sterical hindrance which forces $P_0(\Omega)$ to be anisotropic in many plastic crystals. We shall briefly describe, first the Gordon J model (24) which only takes into account the reorientation of the molecules, and second, the Larsson model (25) which adds some aspects of the libration dynamics to the previous model.

4.2.1. The Gordon J model.

This model, which has been often used to analyse vibrational line shapes (26), is directly derived from the idea of a molecular gas, in which the molecules perform free rotations, interrupted by stochastic instantaneous collisions which are such that there is no correlations either between the kinetic energies of the molecule or between the directions of its axis of rotation before and after the collision. The mathematics of the problem can then be splitted in two parts: an intermediate $\bar{P}(\Omega_1, \Omega_2, t)$ for free rotation is first computed, the stochastic collisions are then convoluted with this intermediate s.o.d.f. to obtain the final $P(\Omega_1, \Omega_2, t)$. We shall just here summarize this calculation for the case of molecules with cubic symmetry.

First, if $\Omega_2 = \Omega_1 R$, where R describes the rotation in the molecular axis, $P(\Omega_1, \Omega_2, t)$ must be independent of Ω_1 in a spherical model so that

$$P(\Omega_1, \Omega_2, t) = P(0, R, t); \quad R = \Omega_1^{-1}\Omega_2 \tag{4-1}$$

Second, if R is a rotation performed at the angular velocity ω_0 around an axis with spherical coordinates α, β in the molecular framework, R is defined by the three Euler angles α, β, $\omega = \omega_0 t$, and one has

$$\bar{P}(0,R,t) = \int p(\alpha, \beta, \omega_0)\ \delta(R - \{\alpha, \beta, \omega_0 t\})\ d\alpha \sin\beta d\beta d\omega_0 \tag{4-2}$$

where $p(\alpha, \beta, \omega_0)$ is the probability of having a rotation of angular frequency ω_0 along the axis α, β. As

$$\delta(R - R') = \sum_{\ell_1 m_1 m_1'} D_{\ell_1}^{m\, m'}(R)\, D_{\ell_1}^{m\, m'}(R')^* \tag{4-3}$$

and as $p(\alpha, \beta, \omega_0)$ does not depend on α or β for a spherical molecule, one easily performs the integration on α and β and gets

$$\bar{P}(0,R,t) = \int p(\omega_0) \sum_{\ell_1, m_1} D_{\ell_1}^{m_1 m_1}(R) \sum_{s=-\ell}^{s=\ell} e^{is\omega_0 t} d\omega_0 \tag{4-4}$$

(4-4) shows that the time functions depend only on ℓ, and can be computed easily once a Maxwellian distribution for $p(\omega_0)$ is included. Using the definition of R one gets

$$\bar{P}(\Omega_1, \Omega_2, t) = \sum_{\ell_1 m_1 m_1'} \bar{C}_{\ell_1}(t)\, D_{\ell_1}^{m_1 m_1'}(\Omega_1)\, D_{\ell_1}^{m_1 m_1'}(\Omega_2)^* \tag{4-5}$$

$$\text{with } \bar{C}_\ell(t) = \sum_{s=-\ell}^{s=\ell} \int_0^\infty p(\omega_0)\, e^{-is\omega_0 t}\, d\omega_0 \quad , \tag{4-6}$$

a formula in which m_1' can be easily changed into λ_1', if one would use functions adapted to the spherical symmetry of the site, and the point symmetry $\mathcal{M}$ of the molecule. One sees that, in such a case, where no attention has been paid to the $\mathcal{M}$ symmetry, the $B_{\ell_1\ \ell_2}^{\lambda_1 \lambda_1' \lambda_2 \lambda_2'}(t)$ coefficients which, for a general liquid, do not depend on λ_1 an λ_2 and satisfy an $\ell_1 = \ell_2$ selection rule, here, have the additional property of not depending either on λ_1' or on λ_2'.

Third, $P(\Omega_1, \Omega_2, t)$ is obtained by solving the integral equation

$$P(\Omega_1, \Omega_2, t) = r(t)\bar{P}(\Omega_1, \Omega_2, t) - \int_0^t r'(t_1)\bar{P}(\Omega_1, \Omega', t_1) P(\Omega', \Omega_2, t - t_1)\, dt_1\, d\Omega' \tag{4-7}$$

where $-r'(t)dt$ is the probability that the first collision takes place between time t and t + dt. Making a Fourier transform of (4-7) and using (4-5) one finds that the s.o.d.f. has the same form as (4-5) with coefficients $C_\ell(t)$ the Fourier transform of which are given by

$$C_\ell(\omega) = f_\ell(\omega)\ \left[1 + g_\ell(\omega)\right]^{-1} \tag{4-8}$$

with

$$f_\ell(\omega) = \int_0^\infty e^{-i\omega t}\, r(t)\, \bar{C}_\ell(t)\, d\,t \qquad (4\text{-}9a)$$

$$g_\ell(\omega) = \int_0^\infty e^{-i\omega t}\, r'(t)\, \bar{C}_\ell(t)\, d\,t \qquad (4\text{-}9b)$$

The basic interest of this formula is that for small ℓ , $C_\ell(\omega)$ always has a pseudo Lorentzian shape, and that, if $r(t) = \exp(-t/\tau)$ and ω_r is the free rotation frequency, the half width of this Lorentzian has a maximum for $\omega_r\tau \sim 1$. In practical cases, the values of $\omega_r\tau$ deduced from such an analysis are of the order of a tenth or a few tenths of a radian and decrease with T. The value deduced from I.R. and Raman experiments for τ do not always agree (27). Furthermore, this form of s.o.d.f. predicts the same band shape for modes of different symmetry. This does not agree with experiment, in CF_4 for instance, but it has been argued (26b) that this could result from the neglect of Coriolis effect in the formulae of section 3.

4.2.2. The Larsson model.

This model is a generalization of the preceding one in which, between two reorientations, a libration at frequency $\bar{\omega}$ takes place around an axis which has a random orientation with respect to the molecule. Each of the two processes is interrupted in a stochastic way with a probability law, $r(t)$ for the reorientation, and $v(t)$ for the libration. Thus if $\bar{P}_{\ell ib}(\Omega_1, \Omega_2, t)$ is the s.o.d.f. for the pure libration the integral equation (4-7) is replaced by the set of equations

$$P_r(\Omega_1,\Omega_2,t) = r(t)\, \bar{P}(\Omega_1,\Omega_2,t) - \int_0^t r'(t_1)\, \bar{P}(\Omega_1,\Omega',t_1) \times$$

$$P_v(\Omega',\Omega_2,t - t_1)\, d\,t_1\, d\,\Omega' \qquad (4\text{-}10)$$

$$P_v(\Omega_1,\Omega_2,t) = v(t)\, \bar{P}_{\ell ib}(\Omega_1,\Omega_2,t) - \int_0^t v'(t_1)\, \bar{P}_{\ell ib}(\Omega_1, \Omega',t_1) \times$$

$$P_r(\Omega',\Omega_2,t - t_1)\, d\,t_1\, d\,\Omega' \qquad (4\text{-}11)$$

where $P_r(\Omega_1, \Omega_2, t)$ is the s.o.d.f. of the molecule if at time 0 the molecule is freely rotating, and $P_v(\Omega_1, \Omega_2, t)$ the same quantity for a molecule which librates at time 0. If $\bar{r}$ is the probability for the molecule to rotate, and $\bar{v} = 1 - \bar{r}$, the probability for rotating

$$P(\Omega_1, \Omega_2, t) = \bar{r}\, P_r(\Omega_1, \Omega_2, t) + \bar{v}\, P_v(\Omega_1, \Omega_2, t) \qquad (4\text{-}12)$$

The system of equations (4-10,11) is easily solved, by the same technique as (4-7), because, from the very definition of the model for librations, $\bar{P}_{\ell ib}(\Omega_1, \Omega_2, t)$ must have the form of the r-h-s of (4-5). Larsson proposed that the corresponding time dependent coefficient should be written as

$$\bar{C}_{\ell ib,\ell}(t) = \frac{1}{8\pi^2}\left[F + (1-F)\exp -\frac{\sigma^2}{2}t^2\ell(\ell+1)\cos\sqrt{\ell(\ell+1)}\,\bar{\omega}t\right] \quad (4\text{-}13)$$

$$\text{with } (\bar{\omega}^2 + \sigma^2)(1-F) = \omega_r^2 \quad (4\text{-}14)$$

ω_r being the free rotation frequency.

One finally gets

$$P(\Omega_1, \Omega_2, t) = \sum_{\ell,mm'} C_{L\ell}(t)\, D_\ell^{mm'}(\Omega_1)\, D_\ell^{mm'}(\Omega_2)^* \quad (4\text{-}15)$$

with

$$C_{L\ell}(\omega) = \frac{\bar{v}(h_\ell(\omega) - k_\ell(\omega)f_\ell(\omega)) + \bar{r}(f_\ell(\omega) - g_\ell(\omega)h_\ell(\omega))}{1 - g_\ell(\omega)k_\ell(\omega)} \quad (4\text{-}16)$$

where $f_\ell(\omega)$ and $g_\ell(\omega)$ are defined in (4-9) and

$$h_\ell(\omega) = \int_0^\infty e^{-i\omega t}\, v(t)\, \bar{C}_{\ell ib,\ell}(t)\, dt \quad (4\text{-}17a)$$

$$k_\ell(\omega) = \int_0^\infty e^{-i\omega t}\, v'(t)\, \bar{C}_{\ell ib,\ell}(t)\, dt \quad (4\text{-}17b)$$

This model, which has four parameters if r(t) and v(t) are given an exponential form, has been used up to now only to analyse neutron incoherent inelastic data (25). Though it has more flexibility in the line shape, it nevertheless predicts again the same spectral shape for lines of different symmetries.

4.3 Discrete Jump model

This model has been used in conjunction with pseudo spin type p.d.f.. It thus fully takes into account the symmetry of the crystal and of the molecule. On the other hand, this model consists in supposing that the molecule has an initial orientation Ω_i and suddenly jumps into an orientation Ω_j. It thus neglects all the reorientational dynamics as well as the librational one. In fact, this model has been used in cases where all the orientations Ω_i have an equal occupation probability, which implies, that any orientation Ω_j can be deduced from Ω_i by a proper rotation which is a product of an operation of $\mathcal{S}$ by an operation of $\mathcal{M}$, and, for simplicity we call this group $(\mathcal{S} \otimes \mathcal{M})_r$.

If $P_i(t)$ is the probability occupation of orientation Ω_i at time t, by definition of the model, $P_i(t)$ satisfies the rate equation

$$\frac{d P_i}{dt} = \sum_j [- P_i(t)\, c_{ij} + P_j(t)\, c_{ji}] \qquad (4\text{-}18)$$

$$\equiv - \sum_j A_{ij} P_j(t)$$

As the matrix A_{ij} is globally invariant under the operations of the group $(\mathcal{S} \otimes \mathcal{M})_r$, its eigenvalues can be classified according to the irreducible representation of this group, thus of the two groups $\mathcal{S}$ and $\mathcal{M}$ (28). The solutions of (4-1) appear thus like sums of exponential functions of the time t, one exponential for each irreducible representation of $(\mathcal{S} \times \mathcal{M})_r$ founded in the A matrix.

With initial condition $P_i(0) = 1 \quad P_{j \neq i}(0) = 0$, (4-18)

provides the joint probability $P_{ij}(t)$ of having orientation Ω_i at time 0 and Ω_j at time t, in terms of which the s.o.d.f. can be written

$$P(\Omega_1, \Omega_2, t) = \frac{1}{n} \sum_{ij} \delta(\Omega_1 - \Omega_i) P_{ij}(t)\, \delta(\Omega_2 - \Omega_j)$$

(where n is the number of possible orientation of the molecule) or

$$P(\Omega_1, \Omega_2, t) = \sum_{\substack{\ell_1 \ell_2 \\ \lambda_1 \lambda_2 \\ \lambda'_1 \lambda'_2}} \Delta^{\lambda_1 \lambda'_1}_{\ell_1}(\Omega_1) \left(\frac{1}{n} \sum_{ij} \Delta^{\lambda_1 \lambda'_1}_{\ell_1} (\Omega_i)^* P_{ij}(t) \Delta^{\lambda_2 \lambda'_2}_{\ell_2}(\Omega_j) \right) \times$$

$$\Delta^{\lambda_2 \lambda'_2}_{\ell_2} (\Omega_2)^* \qquad (4\text{-}19)$$

In (4-19) the terms in parenthesis define the usual coefficients $B^{\lambda_1 \lambda'_1 \lambda_2 \lambda'_2}_{\ell_1 \quad \ell_2}(t)$. As one sums over all i (and j) orientations, it is easy to show that the relationship between the representations contained in A and those contained in $\lambda_1 \lambda'_1$ (or $\lambda_2 \lambda'_2$) can be only those given in (3-6) or (3-7); this gives, of course, the relations (3-6) or (3-7) for the correnpondence between λ_1 and λ_2, and λ'_1 and λ'_2. The number of exponential terms contained in the B coefficients is thus always small. In practice, one usually takes only a small number of c_{ij} coefficients to be different from zero so that the number of eigenvalues and their symmetry is easy to determine. The symmetry of the modes which are visible, and the tensorial symmetry in which they can be seen is thus immediately known.

4.4 The Diffusion Model

4.4.1. Presentation and discussion.

We shall be very brief on this model which have only been used, up to now, only for one dimensional reorientation (29), but can be, in principle, generalized to more complicated situations. This model takes into account the symmetry of the crystal and of the molecule and allows for a finite reorientation time, as it considers the reorientation as a diffusion process inside a fixed potential well which is expanded in terms of $\Delta_{\ell}^{\lambda\lambda'}(\Omega)$ appropriate for the description of $P_0(\Omega)$. It suffers, nevertheless from some weaknesses. On the one hand, the validity of the diffusion equation is questionable, as it does not allow for librations. On the other hand, the very concept of a fixed potential may be doubtful in many cases where the steric hindrance is large and the reorientational process must be accompanied by some center of mass displacements.

4.4.2. Mathematical aspect.

The diffusion problem is written in terms of the conditional probability s.o.d.f. p (Ω_1, Ω_2, t) where

$$P(\Omega_1, \Omega_2, t) = P_0(\Omega_1)\, p(\Omega_1, \Omega_2, t) \tag{4-20}$$

This conditional probability s.o.d.f., considered as a function of Ω_2, satisfies the diffusion equation

$$\frac{1}{D}\frac{\partial p}{\partial t} = \mathrm{div}(\vec{\mathrm{grad}}\, p + \beta(\vec{\mathrm{grad}}\, V(\Omega_2)\, p) \tag{4-21}$$

where D is a diffusion constant and $V(\Omega)$ a potential such as

$$P_0(\Omega) = e^{-\beta V(\Omega)} \tag{4-22}$$

(4-21) has clearly the long time correct behaviour as its time independent solution is

$$p(\Omega_1, \Omega_2, \infty) = e^{-\beta V(\Omega_2)} = P_0(\Omega_2)$$

$p(\Omega_1, \Omega_2, t)$ can be expanded, as the s.o.d.f., on the basis of

$\Delta_{\ell}^{\lambda\lambda'}(\Omega_1)$ and $\Delta_{\ell}^{\lambda\lambda'}(\Omega_2)$. Furthermore, all the operations on the r-h-s of (4-15) transform a $\Delta_{\ell}^{\lambda\lambda'}(\Omega_2)$ into a sum of other functions with the same symmetry. Thus, if one writes, for shortness

$$p(\Omega_1, \Omega_2, t) = \sum_{a,b} C_{ab}(t)\, \Delta_a(\Omega_1)\, \Delta_b(\Omega_2) \tag{4-23}$$

with $a \equiv \ell_1, \lambda_1', \lambda_1$, $b \equiv \ell_2, \lambda_2, \lambda_2'$, one can project out (4-21) on the basis ab and obtain the first order differential equation

$$\frac{1}{D}\frac{\partial}{\partial t} C_{ab}(t) = \sum_c C_{ac}(t)\, H_{cb} \tag{4-24}$$

where H_{cb} is the time independent matrix which represents the result of the operations of the r-h-s of (4-21) on the various $\Delta_b(\Omega_2)$. (4-24) tells us that all the coefficients $C_{ab}(t)$ have an exponential character, which is the result obtained in 1 dimension by Volino and Dianoux (29), who numerically solved (4-21) in a specific case (30).

4.5 The Discrete reorientation model

We shall end up this brief review by summarizing a model which combines some aspects of free rotation, with a discrete jump model. The basic idea is to replace the instantaneous jumps between Ω_i and Ω_j by a free planar rotation which brings the molecule from its first to its second orientation. As in the Gordon J model, the distribution of angular frequency around the (unique) axis of rotation is taken to be Maxwellian, but the rotation is terminated when the molecule has reached the orientation Ω_j and the probability law to start a new rotation towards orientation Ω_k between t and t + dt after having reached the orientation Ω_j is given by $-p'_{jk}(t)$ dt where $p_{jk}(t)$ is supposed to have an exponential character. The physical idea which underlies this model is opposite to that of the diffusion model: the molecules can never reoriente in the potential in which they librate because of infinite sterical hindrances. The latter can be suppressed by some movements of the centers of mass of the neighbouring molecules, in which case the potential barrier against rotation can be neglected. Finally, the rearrangement of the centers of mass after the reorientation creates a new potential barrier which traps the molecule in the orientation Ω_j.

This model does not take into account any libration movements. It has been used to analyse the I.R. and Raman spectrum of the CN^- internal mode in NaCN, plastic phase I, assuming for the discrete orientations Ω_i the six 100 directions. In that case the librations appear only in the F_{2g} Raman spectrum (see table 3) while the reorientations contribute only to the E_g Raman and to the I.R. spectrum

Assuming a long enough time of residence τ, it turns out that the part of the s.o.d.f. which is relevant for the E_g spectrum, in the frequency range $\omega\tau >> 1$ can be written

$$P(\Omega_1,\Omega_2,t) = \sum_{ij}{}' \int_{\omega_c}^{\infty} p(\omega_0)d\omega_0 H(\tfrac{\pi}{2} - \omega_0 t)\delta(\Omega_1-\Omega_i)\delta(\Omega_2-R_{ij}(\omega_0 t)\Omega_i) \quad (4\text{-}25a)$$

which gives for the b coefficients of (3-1)

$$b_{\ell_1\ell_2}^{\lambda_1\lambda_2}(t) = \int_{\omega_c}^{\infty} p(\omega_0)d\omega_0 H(\tfrac{\pi}{2} - \omega_0 t) \sum_{ij}{}' Y_{\ell_1}^{\lambda_1 *}(\Omega_i)\Delta_{\ell_2}^{\lambda_2\lambda_2'}(R_{ij}(\omega_0 t))^* Y_{\ell_2}^{\lambda_2'}(\Omega_i)$$

(4-25b)

In (4-25), the summation Σ' involves all the possible Ω_j orientation which can be reached from the Ω_i orientation; $R_{ij}(\omega_0 t)$ is the fraction $\frac{\pi}{2\omega_0 t}$ of the rotation $R_{ij}(\pi/2)$ which brings the molecule from the orientation Ω_i to the orientation j; ω_0 is the angular velocity, and $p(\omega_0)$ the Maxwellian distribution which appeared in (4-4). Finally, H is a Heaviside function which truncates $b^{\lambda_1\lambda_2}_{\ell_1\ell_2}(t)$ once the 90° reorientation is terminated, and $\omega_c \simeq 5\ \text{cm}^{-1}$ is a cut-off frequency to avoid reorientations which would take much longer time than a typical phonon frequency.

(4-19) which contains no adjustable parameter predicts the correct line shape for both $\ell_1 = \ell_2 = 1$, $\lambda_1 = \lambda_2 = F_{1u}$, and $\ell_1 = \ell_2 = 2$, $\lambda_1 = \lambda_2 = E_g$ for a 200° range of temperature (21). Nevertheless this model[9] does not contain any libration, as most of the models presented here.

5 BEYOND THE ONE PARTICLE DYNAMICS

5.1 Introduction

This section is devoted to a brief review of our present understanding of internal mode line shapes when the hypothesis of individual dynamics, summarized in (3-3) is not valid. As mentioned in 1.3.2. , the effect of the other molecule orientations and positions on the dynamics of an internal mode of a given molecule can be split up into two terms, a local field term ($L_1 = L_2$ term of (1-5)) and a coupling term ($L_1 \neq L_2$ term of the same equation). We shall very briefly discuss here the role of these terms, emphasizing only those aspects which differ from the liquid state and are thus not covered by the paper by S. Bratos (31). We shall first discuss the role of the local field term, in 5.2, then the influence of the coupling terms in 5.3, analysing the different line shapes in 5.4.

5.2 Local field effects

We are interested here in the case where the energy of the internal modes can be reduced to

$$E_j = \sum_{L,n_1 n_2} U_j^{n_1}(\Omega_L)\, U_j^{n_2}(\Omega_L)\left[\delta_{n_1 n_2}\omega_j^2 + V_{n_1,n_2}^{L,L}\right] \qquad (5\text{-}1)$$

$V_{n_1,n_2}^{L,L}$ being a function of Ω_L and of the orientations (and positions of the center of mass) of the neighbouring molecules. If we neglect first the existence of the external dynamics of the crystal the effect of $V_{n_1,n_2}^{L,L}$ is to produce a shift of frequency (if the mean value of $V_{n_1,n_2}^{L,L}$ is not zero) as well as a diagonal width $\Delta\omega$ (31). This $\Delta\omega$ has to be compared with the mean residence time of the molecule, T_r, within (or around) a given orientation. T_r, which can be deduced from the line shapes when the local field effects are negligible, must be obtained, here, by other techniques such as neutron inelastic incoherent or coherent scattering.

If $\Delta\omega T_r \gg 1$ (which is the case in some plastic crystals), the effect of the environment and of the molecular reorientation can be decoupled in two steps. On the one hand, the spectrum is made of two terms, a static part, of width $\Delta\omega$, due to the molecules at rest, and a dynamic part due to the reorienting molecules. On the other hand, if $V_{n_1,n_2}^{L,L}$ does not depend very much on Ω_L, the latter part can be obtained by computing first the line shape due to a molecular reorientation in a fixed environment and then averaging over the environment. Nevertheless, if $V_{n_1,n_2}^{L,L}(\Omega_L)$ depends itself noticably on Ω_L, the reorientation will affect not only the Raman tensor orientation but also the internal mode frequency. The line shape is then very much model dependent and no general statements can be made.

If $\Delta\omega T_r \lesssim 1$, the situation is very reminiscent of the liquid case: a motional narrowing is expected, due to the averaging of the local field effect, which brings back to the situation described in the previous sections. But it is not clear that such a situation does exist in plastic crystals.

5.3 Dynamical coupling between the molecules

5.3.1. Symplifying hypothesis.

Many new problems arise when the coupling term

$$U_j^{n_1}(\Omega_{L_1})\, V_{n_1,n_2}^{L_1,L_2}(\Omega_{L_1},\Omega_{L_2})\, U_j^{n_2}(\Omega_{L_2}) \qquad (L_1 \neq L_2)$$

cannot be neglected in (1-5). We shall discuss them very briefly, concentrating our remarks on the case when $\Delta\omega T_r >> 1$ where $\Delta\omega$ is the inhomogeneous broadening due to both the local field effects, and the dynamical coupling.

Even in the case when the decoupling hypothesis is valid, the Raman line shape of the mode is the Fourier transform of

$$\sum_{\substack{L_1,L_2\\ n_1,n_2}} \int d\Omega_{L_1} d\Omega_{L_2}\, P(\Omega_{L_1},\Omega_{L_2},t) R^{ABn_1}_j(\Omega_{L_1}(0))\, R^{CDn_2 *}_j(\Omega_{L_2}(t))$$

$$< U^{n_1}_j(\Omega_{L_1}(0))\; U^{n_2}_j(\Omega_{L_2}(t))^* > \quad . \qquad (5\text{-}2)$$

Because, in (5-2), the time average over the internal motion does not reduce any more to a $L_1 = L_2$ term, one sees that the band shape contains, in principle, information related to the pair correlation function $P(\Omega_{L_1}, \Omega_{L_2}, t)$. Such information is, in fact, presently impossible to extract because of our lack of knowledge on $< U^{n_1}_j(\Omega_{L_1}(0))\; U^{n_2}_j(\Omega_{L_2}(t))^* >$. Conversely, one may simplify the problem by assuming that the orientational correlation function between neighbouring molecules is negligible, an assumption we shall make in the rest of this section.

As this assumption leads to

$$P(\Omega_{L_1}, \Omega_{L_2}, t) = \delta_{L_1L_2} P(\Omega_1, \Omega_2, t) + P_0(\Omega_{L_1}) P_0(\Omega_{L_2}), \qquad (5\text{-}3)$$

(5-2) splits into two terms: a $\delta_{L_1L_2}$ term and a term which involves integrations of the form $\int P_0(\Omega) R^{ABn}_j(\Omega)\, d\Omega$. Due to the properties of $P_0(\Omega)$ and $R^{ABn}_j(\Omega)$ given in 2, this last term gives a non zero contribution only for tensorial spectra which are in the identity representation of the crystal, and are related to internal modes in the identity representation of the molecule.

In summary, using the tensorial notations and the results of 2, we have

$$I^{\lambda_1\lambda_2}_{\ell_1\ell_2}(\omega) = \int dt e^{i\omega t} \sum_{\substack{n'_1 n'_2\\ \lambda'_1\lambda'_2}} B^{\lambda_1\lambda'_1,\lambda_2\lambda'_2}_{\ell_1\quad \ell_2}(t)\, r^{\lambda'_1 n'_1}_{\ell_1 j}\, r^{\lambda'_2 n'_2 *}_{\ell_2 j} \sum_L < U^{n'_1}_{jL}(0) U^{n'^*_2}_{jL}(t)>$$

$$+\delta_{\Gamma_1\Gamma_0}\delta_{\Gamma_2\Gamma_0}\delta_{\Gamma'_1\Gamma'_0}\delta_{\Gamma'_2\Gamma'_0} \int dt e^{i\omega t} \sum_{\substack{L_1L_2\\ \lambda'_1\lambda'_2}} A^{\lambda_1\lambda'_1}_{\ell_1} r^{\lambda'_1}_{\ell_1 j} (A^{\lambda_2\lambda'_2}_{\ell_2} r^{\lambda'_2}_{\ell_2 j})^* < U_{jL_1}(0) U^*_{jL_2}(t)>$$

$$(5\text{-}4)$$

where, as usual, $\lambda_1 \equiv \Gamma_1, n_1, \mu_1 \ldots$; Γ_0 being the identity representation of the crystal; $\lambda'_1 \equiv \Gamma'_1, n'_1, \mu'_1 \ldots$; Γ'_0 being the identity representation of the molecule symmetry group.

(5-4) generalizes the results of section 3 for the case where the frequency of the normal mode is no longer ω_j. In general, the band shape gives information on the density of states of the internal mode and on the s.o.d.f. but this information is formally absent from the identity spectra of internal modes which belong to the identity representation of the molecule.

5.3.2. Local field effects versus dynamical coupling.

Though, as in paragraph 5.2, we have reduced the summation in the first term of (5-4) to the $L_1 = L_2$ terms, the problem differs from the former case because, for a given orientation of all the molecules of the crystal, the internal mode of one molecule is now coupled to that of the neighbouring molecules. The eigenvalues of the dynamical matrix related to the energy (1-5) have the form $\omega_j + \Delta\omega_j^p$, and the eigen vector corresponding to an eigen value $\omega_j + \Delta\omega_j^p$ is $a_{nL}^p \; \hat{U}_j^n(\Omega_L)$ where $\hat{U}_j^n(\Omega_L)$ is the unit vector corresponding to the mode j, n, when the molecule L has an orientation Ω_L. Then, the displacement vector of this mode, has, for this molecule, a time evolution given by:

$$U_j^n(\Omega_L, t) = e^{i\omega_j t} \sum_p a_{nL}^p \exp (i\Delta\omega_j^p t + \phi^p) \qquad (5\text{-}5)$$

:the internal motion of the nth component of the mode j of a given molecule is a sum of harmonic motions.

The effect of one molecular reorientation is to partially change all the eigen vectors and eigen values, and, at least in principle, this effect depends on the range of a_{nL}^p. If the fluctuations of the local field (or diagonal) terms $V_{n_1 n_2}^{L,L}$ are larger than the coupling (or off diagonal (31)) terms, there is an Anderson localization (32). It means that, if a molecular orientation is changed at $\underset{\sim}{R}_L = 0$, only a few number of eigen values and eigen vectors are affected, and the influence of this change on a_{nL}^p decreases faster than $e^{-b\|\underset{\sim}{R}_L\|}$ for large $\|\underset{\sim}{R}_L\|$. On the contrary, if the local field terms are smaller than the coupling ones, all the eigen values and eigen vectors are affected.

In practice, nothing is presently known on the possible influence of such localization effects on $< U_{jL_1}^{n_1}(0) \; U_{jL_2}^{n_2}(t)^* >$ for $L_1 = L_2$ or $L_1 \neq L_2$. In any case, the existence of many frequencies and eigen vectors in (5-5) makes the influence of reorientations on such functions more difficult to compute than in the case of local field effects only, and in the next paragraph, we shall largely neglect them.

5.4 Line shape of the various modes

5.4.1. Identity representation modes.

As discussed in the previous paragraph, modes which belong to the identity representation of the molecule appear in the two terms of the r-h-s of (5-4). The first term depends on the s.o.d.f. as well as on the total dynamics of the mode and will be discussed in 5.4.2.. The second, on the other hand, does not contain direct orientational effects, and depends on $< U^{n_1}_{L_1}(0)\ U^{n_2}_{L_2}(t)>$. When neither the local field term nor the coupling terms of (1-5) depend very much on the orientation of the molecules, it is possible to simplify partly the problem by computing first, a mean value of (1-5):

$$\bar{E}_j = \sum_L U_L^2\ \bar{V} + \frac{1}{2} \sum_{L_1 L_2} U_{L_1}\ \bar{V}^{L_1,L_2}\ U_{L_2} \tag{5-6}$$

$$\text{with } \bar{V} = \int P_0(\Omega)\ V^{L,L}\ (\Omega)\ d\,\Omega$$

$$\bar{V}^{L_1,L_2} = \int P_0(\Omega_{L_1})\ P_0(\Omega_{L_2})\ V^{L_1,L_2}(\Omega_{L_1}, \Omega_{L_2})\ d\,\Omega_{L_1}\ d\,\Omega_{L_2}\ . \tag{5-7}$$

The total spectrum, for a fixed orientation of the molecules, as well as the role of the change of orientations of the molecules, may then be obtained by perturbation methods. One can show that, for a mode in the identity representation, both $\bar{V}$ and $\bar{V}\ L_1 L_2$ are, a priori, different from zero (while $\bar{V}^{L_1,L_2}_{n_1,n_2}$ is zero by symmetry for modes belonging to other representations (33)). (5-7) leads thus to a dispersion of the mode frequencies, but to eigen vectors which are plane waves. The summation on L_1 and L_2 in the second term of (5-4) yields the usual k = 0 selection rule so that the spectrum, even in the presence of interactions between the molecules, consists, in the lowest order, of a $\delta(\omega - \omega_j(k = 0))$ line. This line is actually broadened by the (time dependent) fluctuations of both $V(\Omega_{L_1})$ and $V^{L_1,L_2}(\Omega_{L_1}, \Omega_{L_2})$, but this broadening may be less important than the total band width due to (5-6). Moreover, the $\omega_j(k = 0)$ frequency is usually an extremum of the spectrum, so that, even in the case of strong dynamical coupling, the identity Raman spectrum of such modes will, in many cases consist, of a sharp line located at one edge of the larger spectra obtained in the other representations of the same mode (34).

5.4.2. Other representation modes. Strong coupling.

No general statements can be made on the band shape of modes of other symmetries, except if the frequency dependence of one of the two factors which enter into the first term of (5-4) can be neglected with respect to the other. As we are studying here the influence of the dynamical coupling, we need to consider only the case when the reorientational dynamics can be neglected. It means that $B_{\ell_1\ \ell_2}^{\lambda_1\lambda'_1\lambda_2\lambda'_2}(t)$ can be considered as a constant which can be taken at its t = 0 value.

Furthermore, if one uses (5-5) and the absence of correlation between the phases ϕ^p of the various eigen vectors, one gets:

$$\sum_L < U_{jL}^{n_1}(0)\, U_{jL}^{n_2}(t) > = \sum_p e^{i(\omega_j + \Delta\omega_j p)t} \sum_L a_{n_1L}^{p}\, a_{n_2L}^{p*} \qquad (5\text{-}8)$$

This shows that, in the limit where the t dependence of $B_{\ell_1\ \ell_2}^{\lambda_1\lambda'_1\lambda_2\lambda'_2}(t)$ can be neglected, the line shape represents simply a projected density of states of the frequencies $\Delta\omega_j{}^p$. In the case where the representation of the mode is one dimensional, $n_1 = n_2$. As

$$\sum_{n,L} a_{nL}^{p}\, (a_{nL}^{p})^* = 1,$$

in this case one exactly gets the density of states of the mode in every spectrum.

The same is true in the case of a cubic crystal with molecules with T_d symmetry. Indeed, it stems from table 2 and (19) that, for $\ell_1 = \ell_2 = 2$,

$$B_{\ell_1\ \ell_2}^{\lambda_1\lambda'_1\lambda_2\lambda'_2}(0) = b_2^{\Gamma_1\Gamma'_1}\, \delta_{\lambda_1\lambda_2}\, \delta_{\lambda'_1\lambda'_2}\, \delta_{n_1n_2} \qquad (5\text{-}9)$$

with $\lambda \equiv \Gamma, n, \mu$. The two E_g and F_{2g} spectra of any E or F_2 mode are thus exactly proportional to the density of states of this mode, and as the formula for I.R. absorption have the same form as (5-4), this result is also true for the F_{1u} spectrum of the F_2 modes.

The preceding result has been partly checked in the case of the strongly I.R. active mode ν_3 of the plastic phase of CF_4 (35). The coupling between the mode has been assumed to be due to the interaction between the induced dipoles. As the ν_3 mode is triply degenerate, the density of states is actually the same as if all the molecules would have the same orientation, with , e.g. their 3 S_4 axes parallel to the 100 axes of the cubic crystal. For a given value of ω_j and of the induced dipole, the computed density of states has, indeed, the same shape with a total width of ≃ 120 cm^{-1} and a two peak structure, as the one measured by Raman scattering on a powder sample by Gilbert and Drifford (36).

Finally, when the role of the reorientational dynamics can be confined to the $B_{\ell_1\ \ell_2}^{\lambda_1\lambda'_1\lambda_2\lambda'_2}(t)$ term (cf. paragraph 5.3.2.) and when this dynamics is slow enough for the starting hypothesis $\Delta\omega T_r >> 1$ to be correct, the total band shape can be computed by a convolution of the two processes.

When $\Delta\omega T_r \lesssim 1$, the motional narrowing processes must be taken into account, and the band shape has to be computed by techniques similar to those used in the liquid case.

ACKNOWLEDGMENTS

These lecture notes have largely benefited from long discussions with M. Yvinec. It is also a pleasure to thank the hospitality of Ecole Polytechnique, Lausanne, where a part of the material presented here was elaborated, and of Brookhaven National Laboratory where this paper was written.

REFERENCES

(1) W.E. Streib, T.H. Jordan and W.N. Lipscomb, J. Chem. Phys. 37, 2962 (1962).
(2) P.A. Cooper, Nature, 107, 745 (1921); H.J. Verweel and J.M. Bijvoet, Z. Krist. 100, 201 (1939).
(3) J.M. Bijvoet and J.A.A. Ketlaar, J. Am. Chem. Soc. 54, 625 (1932).
(4) G.S.R. Krishna Murti, Ind. J. Phys. 32, 460 (1958).
(5) F. Tranchant and R. Guerin, C.R.A.S. Paris, 274B,795 (1972).
(6) A.J. Leadbetter, private communication.
(7) H. Fontaine, W. Longueville and F. Wallart, C.R.A.S. Paris, 274B, 641 (1972).
(8) J.L. Baudour et al., Mol. Crys. and Liq. Crys. 32, 5 (1976).
(9) M. Gilbert and M. Drifford, J. Chem. Phys. 65, 923 (1976).
(10) W. Press, Acta Cryst. A29, 257 (1973).
(11) Y. Yamada, M. Mori, Y. Noda, J. Phys. Soc. Japan 32, 1565 (1972); R.M. Pick, in Physics of Impurity Centres in Crystals, ed. Eston, Acad. Scien., Tallinn. p. 293 (1972).
(12) C.J. Bradley and A.P. Cracknell, The Mathematical Theory of Symmetry in Solids, Clarendon, Oxford, (1972).
(13) J.M. Rowe, D.G. Hinks, D.L. Price, S.Susman and J.J. Rush, J. Chem. Phys. 58, 2039 (1973).
(14) D. Fontaine, R.M. Pick and M. Yvinec, Sol. Sta. Com. 21, 1095 (1977).
(15) W. Press and A. Huller, Acta Cryst. A29, 252 (1973).
(16) M. Yvinec and R.M. Pick, to be published.
(17) M. More, J. Lefebvre and R. Fouret, Acta Cryst. B33, 3862 (1977).
(18) J.P. Amoureux and M. Bee, to appear in Act. Cryst. B (1979).
(19) R.M. Pick and M. Yvinec, to be published.
(20) M. Yvinec, Thèse de 3ème Cycle, Univ. P. et M. Curie (1977), unpublished.
(21) D. Fontaine and R.M. Pick, to appear in Jour. de Phys. (1979).

(22) K. Michel, this volume.
(23) G. Jacucci, this volume.
(24) R.G. Gordon, J. Chem. Phys. 44, 1830 (1966); R.E.D. Mc Clung, J. Chem. Phys. 51, 3842 (1969); 57, 5478 (1972).
(25) K.E.J. Larsson, J. Chem. Phys. 59, 4612 (1973); T. Mansonn, L.G. Olson and K.E.J. Larsson, J. Chem. Phys. 66, 5817 (1977).
(26) see e.g.
a - M. Gilbert and M. Drifford, in Mol. Mot. in Liquids, ed. J. Lascombe, Reidel, Dordrecht, p. 279 (1974).
b - ibid J. Chem. Phys. 65, 923 (1976).
c - M.L. Bansal and A.P. Roy, to appear in Mol. Phys. (1976).
(27) R.C. Livingston, W.G. Rothschild and J.J. Rush, J. Chem. Phys. 59, 2498 (1973).
(28) P. Rigny, Physica 59, 707 (1972); C. Thibaudier and F. Volino, Mol. Phys. 26, 1281 (1973).
(29) A.J. Dianoux and F. Volino, Mol. Phys. 34, 1263 (1977).
(30) A.J. Dianoux, H. Hervet and F. Volino, Jour. de Phys. 38, 809 (1977).
(31) E.N. Economou and P.P. Antoniou, Sol. Stat. Com. 21, 285 (1977)
(32) P.W. Anderson, Phys. Rev. 109, 1492 (1958); ibid, Comments Sol. Stat. Phys. 2, 193 (1970).
(33) M. Yvinec and R.M. Pick, to be published.
(34) The same effect has been already shown to exist and analysed in the case of tetrahedrally bonded glasses by R.M. Martin and F.L. Galeener, Bul. Am. Phys. Soc. 24, 495 (1979); ibid, to be published.
(35) M. Yvinec and R.M. Pick, to appear in J. Chem. Phys. (1979).
(36) M. Gilbert and M. Drifford, J. Chem. Phys. 66, 3205 (1977).

VIBRATIONAL BAND SHAPES OF AMORPHOUS SOLIDS

M. F. Thorpe

Department of Physics
Michigan State University
East Lansing, Michigan 48824 USA

1. INTRODUCTION

The vibrational modes of molecules and the lattice vibrations, or phonons, in crystalline solids have been studied since the beginning of the century.[1,2] These two classes of materials are now understood rather completely in terms of the force constants between atoms; although it is still all but impossible to calculate these force constants from first principles in any but the very simplest systems. However using phenomenological force constants it is possible to solve for the vibrational modes in the two classes of materials above because:

1. In molecules, only a small number of atoms and degrees of freedom are involved and the dynamical matrix can be diagonalised directly.
2. In crystalline solids, the periodicity of the structure and Bloch's theorem means that only the degrees of freedom within a single unit cell need be considered and again the resulting dynamical matrix can be diagonalised.

Neither of these situations holds in an amorphous solid and so the problem presented to us is much more difficult. The number of degrees of freedom in an amorphous solid is $O(N)$ where $N \sim 10^{23}$, whereas in molecules and crystalline solids it is effectively a number $O(1)$, typically less than 100. It is our purpose in these lectures to establish a sound theoretical framework within which to discuss these systems. Some of the more important experiments will also be mentioned. The techniques for observing the vibrations in amorphous solids are the same as those used in molecules and crystalline solids; namely infrared absorption, Raman scattering and inelastic neutron scattering. As a theoretician, I can say that these experiments are not particularly difficult (with the possible exception of inelastic neutron scattering) and so we have an added impetus to get the theory in good shape.

Since the general acceptance of the continuous random network model of amorphous semiconductors,[3] it has been realised that amorphous materials are not "messed-up crystals" but intrinsic structures that are very appealing intellectually. This has the important consequence that it is not possible to derive the properties of amorphous solids by perturbing the corresponding crystal. It is necessary to have a good understanding of the structure before the phonons, which are small deviations from the equilibrium structure, can be understood. We discuss this in the next section, where we also discuss the general principles involved in constructing a potential energy for use in the calculation of phonon frequencies. Those principles are the same as for molecules and crystalline solids. In Section 3, we review some numerical techniques for diagonalising the dynamical matrix and in Section 4 we discuss the use of the Bethe lattice and its relevance to amorphous solids. In Section 5 some general spectral theorems are developed that lead to the positions of the vibrational bands but do not give the shapes of the bands. Finally in the conclusions we suggest some directions in which this work might go in the future.

2. STRUCTURE AND VIBRATIONAL POTENTIALS

It is essential to have a good understanding of the equilibrium structure of an amorphous solid before proceding to calculate the dynamic properties. In these lectures we will concentrate on covalently bonded materials like Si and SiO_2. From studies of the x-ray diffraction pattern, these are believed to be well described by a continuous random network[3] in which the valence requirements of each atom are satisfied locally. A two-dimensional piece of such a structure is shown in Fig. 1. Each atom forms three bonds of roughly

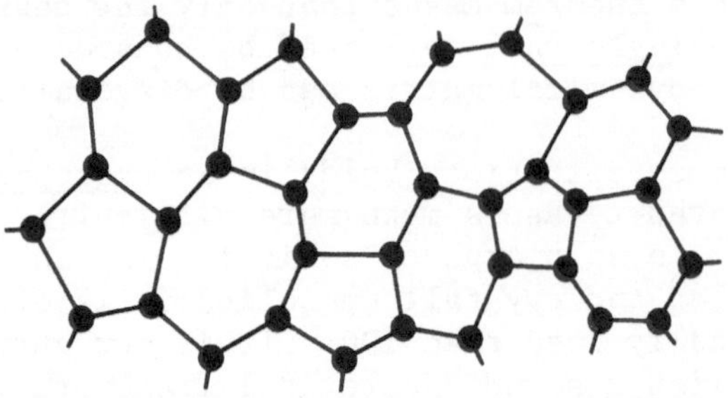

Fig. 1. A piece of a two-dimensional covalent amorphous solid. Notice that the local valency requirements are satisfied everywhere but there is no long range order.

equal length separated by roughly 120°. However there is no long range order and we have an amorphous material. There is considerable short range order. Amorphous Si is similar to this except that it is three-dimensional and each Si has four bonds associated with it rather than three. Because of the nature of the forces, the bond lengths are more or less fixed in length but there are variations of a few degrees in the bond angles around the tetrahedral angle $\cos^{-1}(1/3) = 109°$. This has been established by building large models and calculating the radial distribution function of the model and comparing with the radial distribution function obtained by Fourier transforming the x-ray data.[4] Fused silica (SiO_2) may be thought of as being derived from amorphous Si by placing an additional O atom between each pair of nearest neighbor Si atoms--so that each Si has four O neighbors and each O has two Si neighbors. The structure is then deformed, without changing the topology, so that the Si-O-Si bond angles are around 135°.[5]

These amorphous solids can usefully be regarded as giant covalently bonded molecules. They are idealisations in much the same way that perfect crystalline solids are: a few imperfectly bonded atoms (e.g. only three bonds in Si) will always exist. Nevertheless under the right preparation conditions, one can get very close to these ideal networks and we will use them in our discussion of the vibrational modes.

Other types of amorphous solids exists--for example amorphous metals of low temperatures are well described by the Bernal model of a dense random packing of hard spheres.[6] In some sense these solids are like frozen liquids. There is much more variation in the nearest neighbor bond distances than in the covalently bonded solids. This means that variations of force constants with position, as well as the topological aspect of the problem, are important. As a consequence much less progress has been made on understanding the vibrational properties of these materials and we will not discuss them further in these lectures.

The dynamics of a solid, crystalline or amorphous, is a complicated problem and it must be simplified if progress is to be made. The first of these simplifications is the adiabatic or Born-Oppenheimer approximation which allows us to write down a potential energy containing only the nuclear coordinates, the second is the harmonic approximation which truncates this potential after the second order terms.

The Hamiltonian, or energy, of any solid is a function of the nuclear coordinates $\underline{u}_i$ where i labels the ion, and the electron coordinates $\underline{r}_e$ where e labels the electron. A curly bracket denotes a set

$$H = H(\{\underline{u}_i\} , \{\underline{r}_e\}) \qquad (1)$$

So for example in GaAs, there are 2 nuclear and 64 electronic labels per molecular unit and each one of these is a vector that contributes three degrees of freedom.

The adiabatic approximation is possible because the electron mass m is much smaller than the nuclear mass M,

$$m/M \simeq 10^{-3}$$

The electronic motion is therefore very rapid compared with the nuclear motion and so when a nucleus has completed only a small fraction of a cycle, the electrons have been through many cycles. Because the nuclear motion is so slow, the electrons come into equilibrium at each nuclear configuration. Therefore the electron coordinates may effectively be eliminated from (1)

$$H = H(\{\underline{u}_i\}) \tag{1a}$$

Another way to state the approximation is that the electronic excitations (~1 eV) are much higher in energy than the phonon energies (~10^{-3} eV) it is desired to calculate. One must, however, be careful with this approximation in metals where there is no energy gap between occupied and unoccupied electron states so that electronic excitations with arbitrarily small energy are possible. This is also true in amorphous semiconductors where there is finite density of states at the Fermi level. However, in this case the number of states in the "gap region" is so small (<1% of total states) that one can make the adiabatic approximation with some confidence.

In solids, the equilibrium positions of the atoms are usually well defined, and so it is reasonable to expand (1a) as a Taylor series in the $\{\underline{u}_i\}$ about the equilibrium positions. The linear terms vanish as the net force on each atom must be zero in equilibrium. In the harmonic approximation the series is truncated at the second order terms

$$H = \sum_i \frac{p_i^2}{2m_i} + V \tag{2}$$

where

$$V = \frac{1}{2} \sum_{ij} \underline{u}_i \underline{\underline{V}}_{ij} \underline{u}_j \tag{3}$$

This can be shown, a posteriori, to be a good approximation if the mean square displacement is small compared to the nearest neighbor atomic separation a,

$$\langle u^2 \rangle \ll a^2$$

This is usually so, but breaks down as the melting temperature of the solid is approached, and also in the quantum crystals H and He where there is a large zero point motion due to the light nuclear masses. It also breaks down if the potential energy V has two neighboring minima between which tunnelling can take place. The best example of this is a ferroelectric material like KH_2PO_4 where the proton can tunnel between the two minima associated with the hydrogen bond. A similar phenomena may occur in glasses, that manifests itself at low temperatures.

The word phonon as used in these lectures means simply a quantum of energy in a vibrational mode. The fact that the phonon is an eigenstate presupposes that we are working with a harmonic Hamiltonian. It is of course possible to write down Hamiltonians in both crystalline and amorphous solids if the structure is locally stable. Thus phonons are well defined in amorphous solids--the main qualitative difference from a crystal is that amorphous phonons do not have a well defined $\underline{k}$ vector.

Figure 2 shows a sketch of the potential energy of a solid against some generalised coordinate. Although the amorphous solid may be metastable with respect to the crystalline phase (the diamond cubic structure in the case of Si) it is nevertheless stable against small displacements and a harmonic Hamiltonian like (3) will provide an adequate description. As the amplitude of vibration gets larger, the solid may tunnel through a potential barrier to a more stable structure. This possibility of tunnelling to neighboring structures is an anharmonic effect and so outside our present theory.

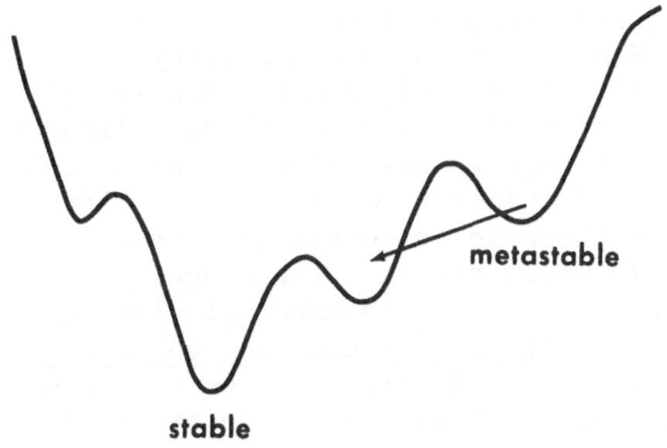

Fig. 2. A sketch of the potential energy of a solid against some generalised coordinate. The metastable states may be crystalline or amorphous and are stable against small displacements. However large displacements or quantum mechanical tunnelling can take the solid to a neighboring minima as indicated by the arrow.

At very low temperatures, a linear specific heat is observed in many amorphous materials. It has been suggested that this is due to low frequency tunnelling models.[7] Of course these modes are not present in the harmonic theory. Whilst it is very likely that these modes do occur in amorphous SiO_2 due to a flexing of Si-O-Si bonds which act as a trigger for the tunnelling, it is much less likely in amorphous Si where the structure is very stable. Experimentally no linear specific heat has been seen in amorphous Si although the very lowest temperatures have not yet been investigated.[8] It is clear that the existence of atoms that make two bonds like O in these covalent networks greatly enhances the tunnelling possibilities.

We have discussed the way in which the potential is set up in a solid. In a crystal we have to solve a matrix whose size is equal to the number of degrees of freedom in the unit cell. As the unit cell gets larger the number of branches increases and the volume of the Brillouin zone gets smaller. An amorphous solid can be thought of as the limit of this process as the unit cell becomes infinite in size and the Brillouin zone becomes just the point $\underline{k} = 0$ so that $\underline{k}$ is no longer useful in classifying the modes. The high pressure phases, Si III and Ge III, provide useful examples of these "intermediate structures." Si III has eight atoms/unit cell and hence 24 phonon branches. There are nine different distinct finite frequencies at $\underline{k} = 0$ as the high symmetry (cubic) makes many modes degenerate. Of these nine frequencies, five are Raman active.[9] This is to be contrasted with diamond cubic Si where there is a single Raman active mode at 520 cm^{-1}. The frequencies in Si III are all below this--roughly speaking we are seeing more of the density of states at $\underline{k} = 0$--eventually when the unit cell becomes infinite in size, we see the complete density of states. It would be wrong, however, to think that we merely relax the $\underline{k}$ selection rules as the unit cell becomes larger; the density of states itself changes slowly. Some features in the density of states remain and some are washed out. This fact, that the density of states looks rather similar in crystalline and amorphous solids, had led a number of authors to attempt to try and describe amorphous solids by applying perturbation theory to crystals. This approach is fundamentally wrong, as there is no continuous structural path between crystalline and amorphous solids. The proper and most satisfying way to approach the problem is to make calculations for amorphous solids from scratch--without invoking any concepts peculiar to crystals like unit cells and Brillouin zones. This is of course difficult in practice, although simple in principle, as it involves the diagonalization of the complete dynamical matrix.

We assume that forces act only between nearest neighbors and the symmetry of a bond in the network is approximately axial so that the potential V_{ij} between a pair of atoms may be written

$$V_{ij} = \frac{3}{2} \beta[(\underline{u}_i-\underline{u}_j)\cdot\underline{r}_{ij}]^2 + \frac{1}{2} (\alpha-\beta)(\underline{u}_i-\underline{u}_j)^2 \qquad (4)$$

where $\underline{r}_{ij}$ is a unit vector joining sites i and j and $\alpha + 2\beta$, $\alpha - \beta$ are the central, non-central force constants. This potential was orginally used by Born[10] in his discussion of the phonons in a diamond crystal.

It is useful to recognise two kinds of disorder that must occur in an amorphous material. The first is quantitative disorder, that is the variation of force constants, like α, β in (4), with site. The environment of every bond in an amorphous solid is slightly different, and so α, β will vary from site to site. This kind of disorder has not been discussed much in amorphous solids, except for localized states in the vicinity of the band gap. Even these states are usually discussed in terms of the crystalline alloy problem. Localized states may occur near band edges where the density of states is very small and arises from unlikely, i.e. improbable, atomic configurations. This subject is still largely unresolved and we shall not pursue it further here. The second kind of disorder is unique to amorphous solids. Topological disorder neglects the variation of force constants from site to site and concentrates on the random network which manifests itself through the summation sign in (3). We shall concentrate on this as it is most important except for these few (<1%) states around the band edges.

It is important to remember that we can only gain a knowledge from the crystal of the forces between certain configurations of atoms. Forces between nearest and next-nearest neighbors in various forms of Si will be about the same if one adopts a local bond picture. However, this will not be so for third nearest-neighbors as this force will surely depend on the dihedral angle. From analyzing experimental data on diamond cubic Si we can at best only determine the forces between third nearest-neighbors in the staggered configuration. We therefore see that in the absence of a knowledge of the general potential (e.g. the screened Coulomb potential), it is necessary to only have very short range forces in our model if progress is to be made.

3. NUMERICAL TECHNIQUES

These techniques have been pioneered by Dean and co-workers and are reviewed in Ref. 5. Most of this work is on SiO_2 and related glasses. Weaire and Alben used the Keating potential (similar to the Born potential (4) but with the non-central force replaced by an angle bending force involving the coordinates of three atoms) with a ratio of angle bending to central force constants of 0.2, similar to the ratio 2/11 in the Born potential previously discussed. It is necessary to have this small non-central piece in the potential to achieve stability. If there are N atoms in the system, there are 3N degrees of freedom. Central forces may be thought of as a constraint on the system involving keeping all the 4 N/2 bonds fixed in length. It can be shown that these constraints are linearly independent and so we are left with N degrees of freedom. This means that there must be N eigenfrequencies at zero frequency for all random networks and crystals with nearest neighbor central forces only. Detailed calculations confirm this (see Fig. 3).

Similar calculations for other random networks with free surfaces confirm the general picture shown in Fig. 3. Present day computers can handle matrices up to about 300 x 300 without using special techniques so that the phonon frequencies in random networks with about 100 atoms may be found numerically. The problem is that the number of surface atoms is high in three-dimensional networks. Even in a network with 500 atoms, about 35% are on the surface. Dean and coworkers[5] have used different boundary conditions to see which parts of the frequency spectrum are surface sensitive. The time involved in diagonalizing a matrix goes roughly as the third power of its size and so doubling the size of the matrix means of factor eight in time. There are special techniques for sparse matrices and also for banded matrices that have been exploited by Dean.[5]

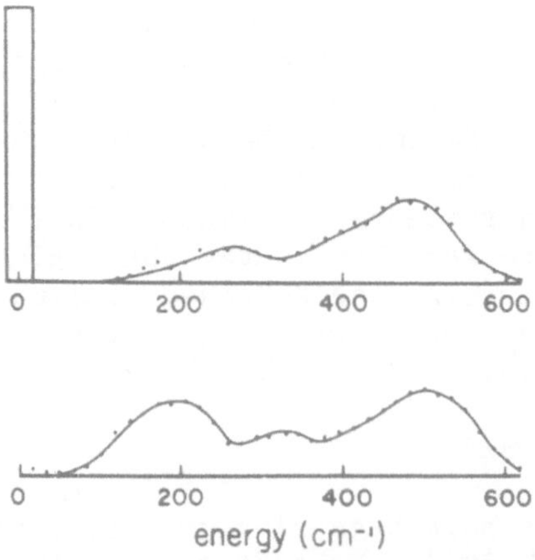

Fig. 3. Calculations of phonon frequencies for Si using Henderson's 61 atom random network with periodic boundary conditions. The upper curve is for nearest neighbor central forces where 1/3 of the weight comes in the delta function at zero frequency and the lower curve is with an additional small angle bending force. From Weaire and Alben.[11]

Comparison with experiment is difficult because the density of states is not measured directly but modulated by matrix element effects that describe the interaction between the probe and the system. The one phonon infrared and Raman scattering from amorphous Ge are shown in Fig. 4.

A comparison with Fig. 4 (with the frequency scale suitably adjusted to go from Si to Ge; this can be done by noting that the optic modes at $\bar{k} = 0$ in diamond cubic Ge and Si are at 301 cm^{-1} and 520 cm^{-1}, respectively) shows that matrix element effects are indeed important particularly in the infrared experiment where the central portion of the spectrum is apparently amplified by matrix element effects.[11] Attempts to include matrix element effects[5,9,11,12] using a local bond interaction have not been terribly successful. Because of this uncertainty we will stay with the density of states as being the more fundamental quantity. The matrix elements for inelastic neutron scattering are known and so this could be used to yield a density of states directly. This is discussed further in section 4.

Numerical calculations have been made by Dean[5,13] for SiO_2 glasses. Unfortunately the disorder is not entirely topological in this case as the Si-O-Si bond angle is thought to vary between about 120° and 180° and this may be thought of as adding quantitative disorder to the network. The Born potential between nearest neighbor Si-O pairs is used (Eqn. 4 with $\beta/\alpha = 14/23 = .61$). Because of the Si,O mass difference the computed spectra fall in several distinct regions as shown in Fig. 5.

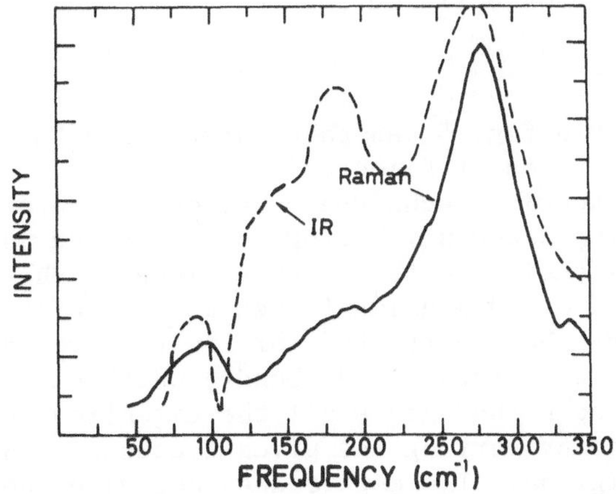

Fig. 4. The infrared absorption and Raman scattering from amorphous Ge.[12] The Raman scattering has the Bose factor $[n(\omega)+1]/\omega$ removed.

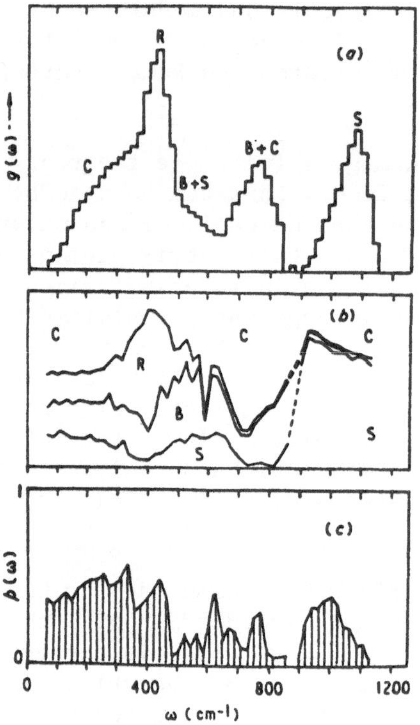

Fig. 5. (a) Frequency spectrum, (b) assignment diagram and (c) participation ratio for an SiO_2 model with fixed end boundary condition. The designations C, R, etc. refer to types of atomic motion described in the text. Note that it is the distance between curves in (b) rather than the heights of the curves themselves, which indicates the proportion of energy arising from different types of motion. From Bell and Dean.[13]

The spectrum in Fig. 5 was calculated for a 330 atom cluster which involves matrices of dimension almost 1000. Bell and Dean have developed a special technique based on the negative eigenvalue theorem for finding frequency histograms of banded matrices. Banded matrices have all their finite elements close to the diagonal and occur naturally in one-dimensional systems. By clever labelling of atoms they also can be constructed for finite three-dimensional networks although with rather wide bands, which naturally increases computer time. The method gives all the eigenfrequencies less than a given one and so by varying the given frequency a histogram is obtained as in Fig. 5a. The method does not give the eigenvectors but "typical" eigenvectors may be found in a separate calculation. From these the <u>character</u> of the <u>vibrations</u> may be found and this is shown in Fig. 5b where C, R, B, S refer to the <u>C</u>ation (Si), and the Rocking, Bending, and Stretching of the anion (0). The participation ratio is defined as

$$p = [3N \sum_{i=1}^{3N} u_i^4]^{-1}$$

at each eigenfrequency where u_i is the eigenvector (normalized to 1) and i goes over all 3N degrees of freedom of the system. For illustration, suppose the eigenvector has 3Nr equal values $(3Nr)^{-1/2}$ and the rest zero; then p = r. Thus p is a reasonable definition of the fraction of atoms participating in a particular mode. It is shown by Fig. 5c for SiO_2 where it can be clearly seen that p is small near the band edges--except the origin, where it approaches 1/3 as one would expect. In a Si random network there is much less quantitative disorder and so localization at the band edges will be much harder to achieve.

A rather more efficient numerical technique involving an equation of motion technique has been used by Beeman and Alben[14] for elemental amorphus semiconductors. They have studied the effects of varying the amorphous structure on the density of states. Matrix elements can be included in a rather straight forward way in this technique.

4. STRUCTURAL POTENTIAL APPROXIMATION (OR BETHE LATTICE)

Although the numerical work described in the previous section provides us with a lot of information, there are two objections that can be raised. The first is that the influence of the surface is rather unclear, and second, little understanding is achieved as to why certain spectral features in the crystal remain and some disappear. Some answers to these questions can be achieved using the structural potential approximation (S.P.A.). This is rather analogous to the coherent potential approximation in alloys which treats the quantitative disorder when two atomic species are randomly arranged on a regular lattice. The S.P.A. tries to understand the role of structure in determining densities of states. In Fig. 6 we show a piece of random network. The forces along bonds are described by a Born potential (4). However, the surface atoms by definition do not have their full complement of bonds and therefore vibrate with too large an amplitude. In order to reduce this amplitude, we place the surface atoms in a potential well

$$V_s = \frac{1}{2} \underline{u}_s \underline{\underline{K}}_s \underline{u}_s$$

where $\underline{u}_s$ is the displacement of a surface atom.

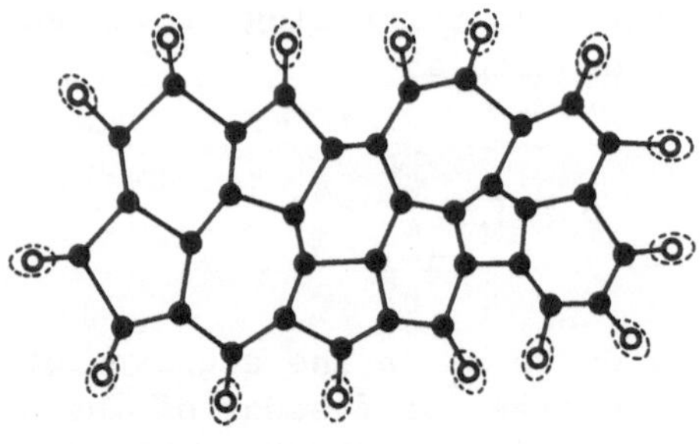

Fig. 6. A piece of random network with interior atoms (solid circles) and surface atoms (open circles). The dashed ellipses on the surface atoms represent the self-consistent potential.

The elements of the symmetric tensor $\underline{\underline{K}}_s$ are chosen so that the mean-square displacement of each surface atom is equal to the average of the mean-square displacements of all the interior atoms at each frequency--appropriate account being taken of the symmetry present in the problem. These conditions determine all the elements of $\underline{\underline{K}}_s$ and the density of states is defined as the average of the local density of states over the whole cluster. This procedure is straightforward in principle but leads to a set of self-consistent equations for the elements of $\underline{\underline{K}}_s$ which must be solved at each frequency ω. However, by adopting this procedure, the abrupt distinction between the surface atoms and the interior atoms is lost as they are all moving with the same mean-square amplitudes. This simulates the medium outside the cluster which can be regarded as having been replaced by an effective medium.

The mean square amplitude of vibration is most easily calculated from the Green function

$$G = (M\omega^2 - V)^{-1} \tag{5}$$

If we write the displacement corresponding to the u_i degree of freedom as $|i\rangle$, then the local density of states $\rho_i(\omega)$ and the contribution to the mean square amplitude $\langle u_i^2\rangle_\omega$ at a frequency ω can be easily derived from G

$$\begin{aligned} \rho_i(\omega) &= -(2M\omega/\pi)\,\mathrm{Im}\,\langle i|G|i\rangle \\ &= -(2M\omega/h)\,\langle u_i^2\rangle \end{aligned} \tag{6}$$

The relations (6) are true at zero temperature. However, it is trivial to include the temperature dependence as the excitations are Bosons. Thus by achieving self-consistency at zero temperature, below, we automatically achieve it at all temperatures.[15] The self-consistency condition is

$$\frac{1}{N_i}\sum_i \langle u_i^2\rangle_\omega = \frac{1}{N_s}\sum_s \langle u_s^2\rangle_\omega \qquad (7)$$

where N_i, N_s are the number of interior surface atoms. The elements of the tensor $\underline{\underline{K}}_s$ are complex and frequency dependent. Thus the cluster potential is non-Hermitian--this can lead to difficulties as the Green function may not have the right analytic properties. This "problem" is also encountered in the C.P.A. In the case of a single tetrahedron, see Fig. 7, the S.P.A. is equivalent to constructing a Bethe lattice, that is a lattice with no closed loops. One can show in this case that the Green function does have the right analytic properties.

The simplest application of the S.P.A. is to the five Si atoms shown in Fig. 7. By symmetry the surface potentials are all equivalent and are ellipsoids with one parameter describing motion along the bond and another motion perpendicular to it. The mean square vibrations of the interior atom are spherically symmetric and by demanding (Eq. 7) that the surface vibrations are also spherically symmetric and equal to those of the interior atom--we obtain two conditions. Thus both parameters in the surface potential may be eliminated and an equation formed for the Green function G

$$\frac{4}{3}\{1+[2(\alpha-\beta)G]^2\}^{1/2} + \frac{2}{3}\{1+[2(\alpha+2\beta)G]^2\}^{1/2} = 1+G(M\omega^2-4\alpha) \qquad (8)$$

This equation may be multiplied out to give a quartic in G. The roots are real except for one complex pair in a certain frequency range. The imaginary parts gives the density of states (Eq. 6) and is shown in Fig. 8.

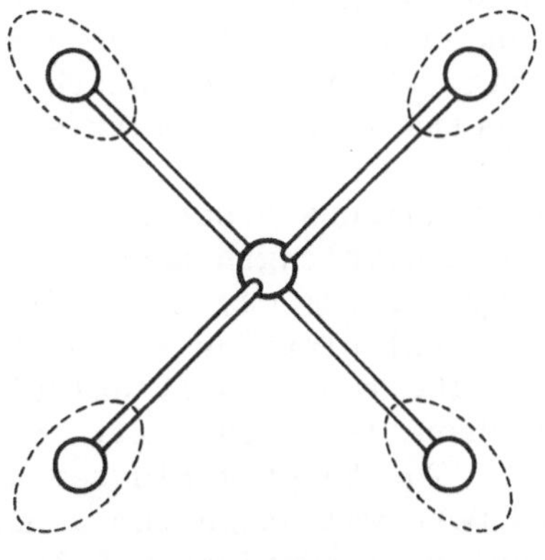

Fig. 7. A single tetrahedron with one interior and four surface atoms. The dashed ellipsoids represent the surface potentials.

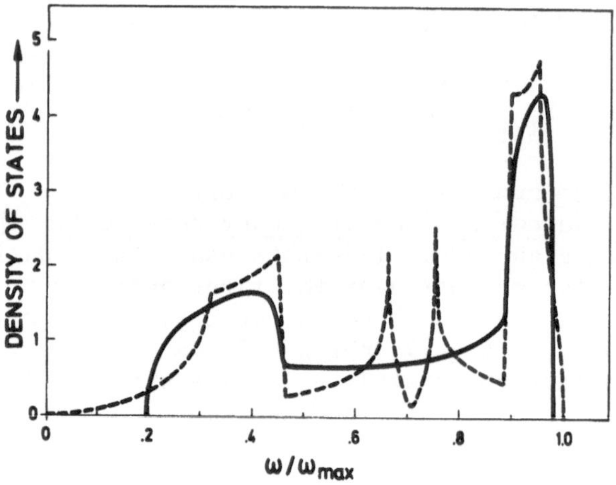

Fig. 8. Comparison of the density of states for Si/Ge in the diamond cubic structure (dashed line) and for a tetrahedral cluster (solid line). The same Born potential is used in both cases.

It can be seen that the broad TA and the sharp TO peaks are retained in the cluster (although the designations TA, TO lose their meanings). The two sharp longitudinal peaks in the center of the crystal spectra are totally absent in the cluster. If we examine the origin of the crystalline peaks in $\underline{k}$ space we can see qualitatively what is happening. The TA, TO peaks have contributions from all over the Brillouin zone. The longitudinal peaks, on the other hand, arise from states extremely close to the hexagonal faces on the Brillouin zone boundary. As the zone boundary is produced by long-range structural order, the states on and very close to the zone boundary should be very dependent on long-range order. The cluster calculation shows very nicely that the TA, TO peaks are produced by the four-fold local coordination and should therefore be present in the amorphous solid. This is consistent with the experimental results (Fig. 4) if we accept that the peak around 175 cm^{-1} in the I.R. spectra is due to matrix element effects.[12]

The calculational techniques described in these two sections fail to give the very long wavelength modes. It is clear that these can not be defined in any natural way for a finite system. A composite picture of various calculations for Si type solids is shown in Fig. 9. In comparing the two cluster calculations it can be seen that the spectral shape is essentially completely determined by the tetrahedral unit. The 85 atom cluster has a broadened peak at high frequencies and more weight in the central region--this is probably due to the angular distortions. Calculations show that the peak at high frequencies is sharper in models with smaller angular distortions.[14]

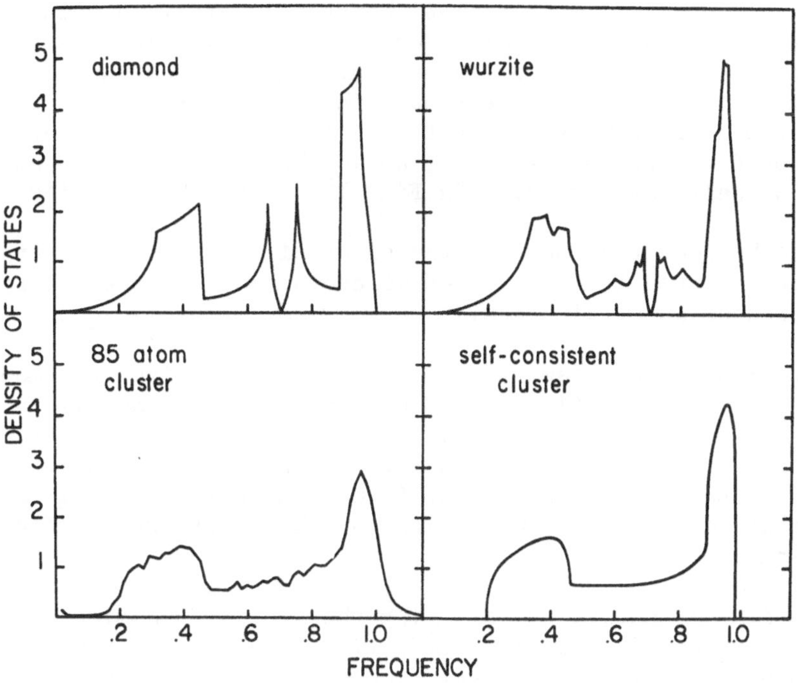

Fig. 9. The density of states for Si with the maximum frequency in diamond cubic normalized to 1. All calculations are for the Born model with β/α = .6 which is equivalent to a ratio of 2/11 between the noncentral and central forces, except for the 85 atom cluster which uses the Keating force constants with a ratio of 2/10 between the angle bending and bond stretching force constants (R. Alben, private communication). This cluster uses "quasi-periodic" boundary conditions in which surface atoms are connected to other surface atoms so that every atom is four-fold coordinated. The diamond and self-consistent cluster are the same as in Fig. 8. Wurtzite is a hexagonal crystal with 4 atoms in each unit cell.

In treating the vibrational modes of multicomponent glasses, Lucovsky[16] has tried to isolate typical atomic configurations containing a few atoms. The vibrational spectra of these "molecules" is then calculated or obtained from experiments on similar molecules in the gaseous phase. This technique has been rather successful particularly in the chalcogenide glasses like As_2Ge_3. The main objection is that the bonds between the "molecule" and the rest of the matrix are neglected. In many cases, these bonds are as strong as the bonds within the "molecule" and so one has no feeling for the width (i.e. localization in a loose sense) of the "molecular band." This difficulty may be largely overcome by using the S.P.A. We have done this for an SiO_2 tetrahedron like Fig. 7. but with four extra oxygen atoms on the Si-Si bonds. In order to illustrate the broadening of the molecular modes, the Si-O-Si units are assumed to be linear. The forces used are the same as in Fig. 5. Ellipsoidal potentials are placed upon the surface Si atoms and the same conditions are applied as in the Si tetrahedron--the O atoms being included via the equations of motion but otherwise not entering into the self-consistency. In Fig. 10 we show the results.

It can be seen that some of the spectral features are sharp, as would be expected from localised molecular motions, but there are also broader spectral bands.

The assignment diagram Fig. 11, which characterizes the modes, is analogous to that of Fig. 5b, except that there is no distinction between R and B for linear Si-O-Si units. It can be seen that most of the character of the modes is determined by entirely local considerations. Because of the linear Si-O-Si unit it is unfair to make a direct comparison between Figs. 5 and 10 although similarities are striking.

The S.P.A. is equivalent to doing the lattice dynamics exactly on a Bethe lattice.[15] This is a tree-like structure that has no rings of bonds. It is a very useful concept when looking at amorphous solids because a) it is soluble and b) the solution gives the density of states directly without first obtaining dispersion relations $\omega(\underline{k})$. Indeed a $\underline{k}$ vector cannot easily be defined. Because there is no k vector, no long-range order and no Brillouin zone; there are no Van Hove singularities in the density of states.

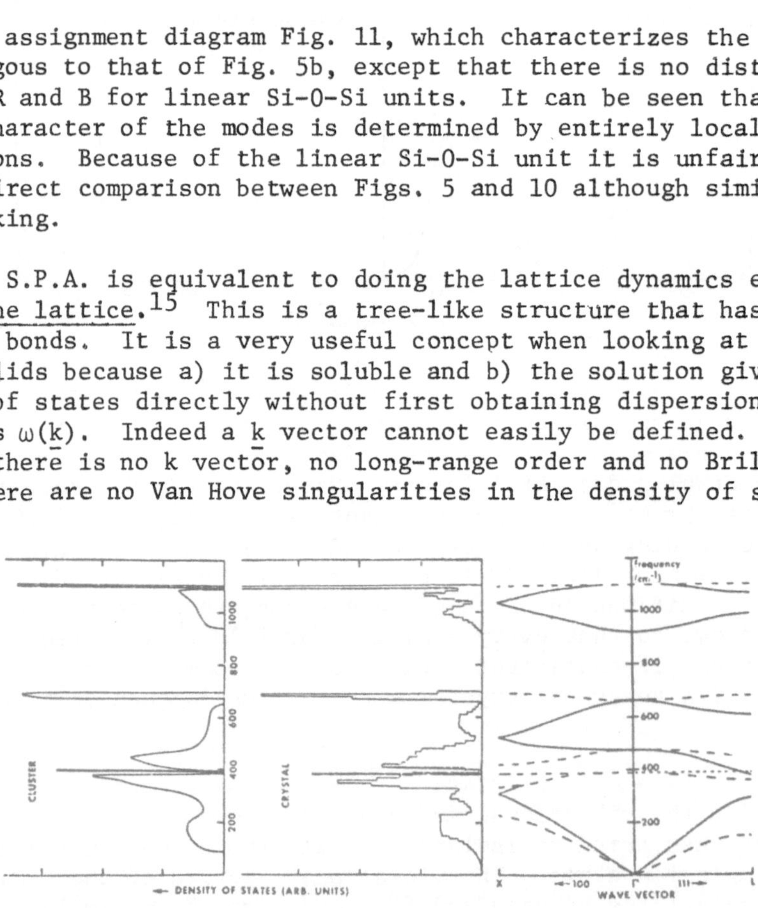

Fig. 10. The density of states for the cluster and for SiO_2 crystobalite (the Si atoms form a diamond cubic lattice and the Si-O-Si bonds are straight) and the phonon dispersion curves for crystobalite (the solid, dashed, dotted lines are singly, doubly and triply degenerate modes respectively). The flat band at 387 cm^{-1} leads to a delta function in the density of states that contains 1/9 of the total weight.

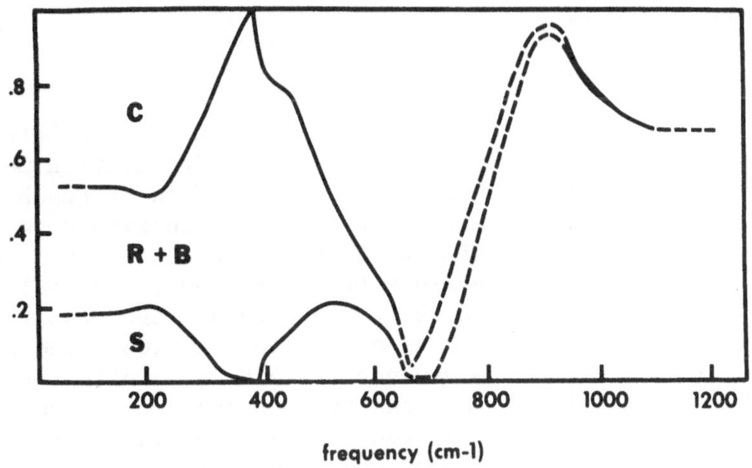

Fig. 11. The assignment diagram where C, R, B, S are explained in the text. There is no distinction between R and B when the Si-O-Si bond is straight. The dotted lines are for the guidance of the eye and are not calculated or even defined.

This approach has been developed much further by Joannopoulous and coworkers.[17] They have made extensive calculations on SiO_2 where the bond angle Si-O-Si is not constrained to be 180° but is given the more reasonable value of 138°. The algebra becomes more complex in this case and the solution proceeds numerically via a transfer matrix technique. The results of these calculations are shown in Fig. 12 together with the inelastic neutron scattering cross section $S(q,\omega)$ in the large q limit. This measures the density of states but with different mass weighting for the Si and O masses as shown in 12(b). This reduces the weight in the peak near the middle of the band to give good agreement with experiment. Laughlin and Joannopoulous have also calculated the infrared absorption and Raman scattering. The infrared absorption agrees will with experiments, but the agreement is less good in the case of Raman scattering because of the uncertainty in the matrix elements.

5. SOME SPECTRAL THEOREMS

The theoretical techniques described so far are quite successful in obtaining actual spectra. However the large amount of computation involved tends to obscure the physics. For this reason Sen and Thorpe[19] have examined a very simple model that contains most of the important physics in a very transparent way.

There is a large class of glasses such as GeS_2 and SiO_2 whose structure is generally believed to be a three-dimensional random

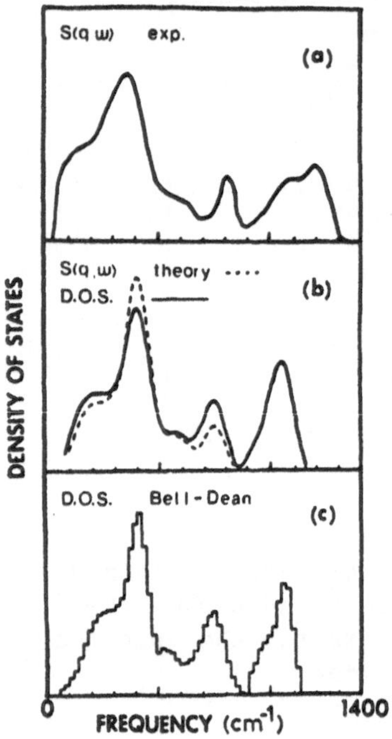

Fig. 12. Comparison of calculated density of states with neutron measurement. (a) Neutron data of Leadbetter and Stringfellow,[18] showing density of states of silica modulated by a fairly uniform matrix element. (b) Solid curve is the calculated average density of states of silicon dioxide Bethe lattice broadened by 75 cm^{-1}. Dotted curve is the density of states of silicon dioxide weighted as discussed in the text for comparison with neutron data. (c) Theoretical density of states for silica from random-network model of Bell and Dean. (From Laughlin and Joannopolous[17]).

network made of the basic AX_4 tetrahedral units. It is also generally believed that these basic tetrahedra largely retain their integrity in the various crystalline phases of these materials as well as in the glassy forms. We expect, therefore, AX_4 molecular modes to play an important role in determining the vibrational spectra of these systems. Indeed, this role is analogous to that of the sp^3 orbitals in tetrahedral networks[20] (i.e., Si and Ge). The AXA angle in general is random. For example, in glassy SiO_2, Mozzi and Warren[21] estimate this angle to lie between 120° and 180° with the maximum in the distribution around 140°. This angle determines to a large extent how the solid-state or collective effects are brought about as it forms the connecting bridge between AX_4 tetrahedra. In this section we introduce a simple model using a nearest-neighbor central force (α) between A and X atoms. This allows us to study the metamorphosis of the molecular-like modes into extended bandlike modes as a function of the bond angle θ in the AXA bridge, which for simplicity we take to be constant for a given network. This model provides a useful framework to study the interplay between the local tetrahedral order and the solid-state effects. The noncentral force constants are generally small, and so the high-frequency optic modes are well represented by this model. However, the network has no resistance to certain kinds of shear motion, and so it not possible to discuss the low-frequency modes without the inclusion of some other forces such as a near-neighbor noncentral force β.

Let α be the central force constant and M and m the masses of the cation (A) and anion (X), respectively. The potential energy is then given by

$$V = \frac{\alpha}{2} \sum_{<ij>} [(\underline{u}_i-\underline{u}_j)\cdot\underline{r}_{ij}]^2 , \qquad (9)$$

where the summation goes over all nearest-neighbor pairs <ij> consisting of A and X atoms. The displacement $\underline{u}_i$ refers to either type of atom, and $\underline{r}_{ij}$ is a unit vector along the AX bond. We examine the equations of motion for a particular bond and eliminate the degrees of freedom of the bridging X atom, shown as x and y in Fig. 13. (Note that motion perpendicular to the bond has no restoring force associated with it.)

$$(M\omega^2-\alpha)u_i = -(\alpha\sin\tfrac{1}{2}\theta)x - (\alpha\cos\tfrac{1}{2}\theta)y,$$

$$(m\omega^2-2\alpha\sin^2\tfrac{1}{2}\theta)x = -(\alpha\sin\tfrac{1}{2}\theta)u_1 + (\alpha\sin\tfrac{1}{2}\theta)u_2 ,$$

$$(m\omega^2-2\alpha\cos^2\tfrac{1}{2}\theta)y = -(\alpha\cos\tfrac{1}{2}\theta)u_1 - (\alpha\cos\tfrac{1}{2}\theta)u_2 . \qquad (10a)$$

We find that

$$[M\omega^2-\alpha-(\alpha\sin\tfrac{1}{2}\theta)/(m\omega^2-2\alpha\sin^2\tfrac{1}{2}\theta)$$
$$-(\alpha\cos\tfrac{1}{2}\theta)^2/(m\omega^2-2\alpha\cos^2\tfrac{1}{2}\theta)]u_1$$
$$= [(\alpha\cos\tfrac{1}{2}\theta)^2/(m\omega^2-2\alpha\cos^2\tfrac{1}{2}\theta)$$
$$-(\alpha\sin\tfrac{1}{2}\theta)^2/(m\omega^2-2\alpha\sin^2\tfrac{1}{2}\theta)]u_2 . \qquad (10b)$$

We can now define an effective force constant α' that connects the u_1 and u_2 displacements shown in Fig. 13

$$\alpha' = -m\omega^2\alpha^2\cos\theta/[m\omega^2-\alpha)^2-\alpha^2\cos^2\theta] . \qquad (11a)$$

This provides a complete description of the dynamics of each AXA bond. A diagonal term α_d is also introduced into the dynamical matrix at each end of the bond

$$\alpha_d = \alpha + \frac{\alpha^2(m\omega^2-\alpha^2\sin^2\theta)}{(m\omega^2-\alpha)^2-\alpha^2\cos^2\theta} \qquad (11b)$$

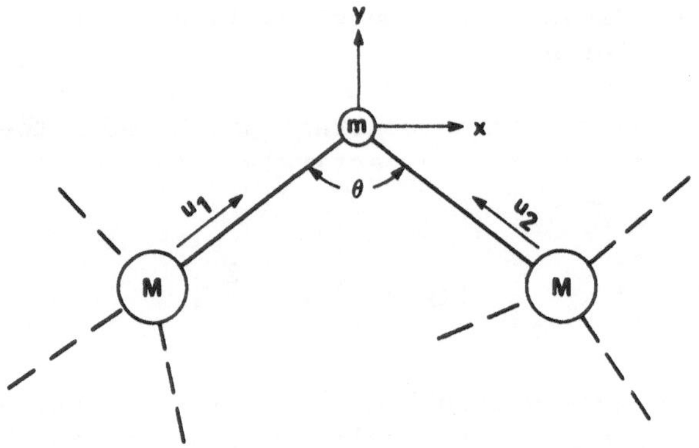

Fig. 13. AXA bond showing the bond angle θ and the various displacements used in the text.

This then maps the complete problem involving the AX_2 network onto one just involving the cation motion. This problem has been studied by Weaire and Alben[11] for Si in a tetrahedral network. They find that the vibrational modes for a Si with only nearest-neighbor central forces are given by

$$M\omega^2 = \frac{4}{3}\,\alpha(1-\varepsilon) \ , \tag{12}$$

where

$$-1 \leq \varepsilon \leq 1 \ . \tag{13}$$

The quantity ε is in fact an eigenvalue of the electronic s-band problem[20] onto which many of these simple random network problems map.[22] The complete spectrum for (4) also contains δ functions at $M\omega^2 = 0$ and $8/3\ \alpha$. We will consider these δ functions in our problem later. With the mapping discussed previously, Eq.(12) becomes

$$M\omega^2 = \frac{4}{3}\,(\alpha_d-\alpha'\varepsilon) \tag{14}$$

which, using (11), gives the frequencies explicitly as

$$\omega^2 = \left(\frac{2\alpha}{3M}+\frac{\alpha}{m}\right) \pm \left[\left(\frac{2\alpha}{3M}\right)^2 + \left(\frac{\alpha\cos\theta}{m}\right)^2 + \frac{4\alpha^2\varepsilon\cos\theta}{3Mm}\right]^{1/2} \ . \tag{15}$$

This leads to two bands whose edges are given by $\varepsilon = \pm 1$, i.e.,

$$\omega_1^2 = (\alpha/m)(1+\cos\theta), \quad \omega_2^2 = (\alpha/m)(1-\cos\theta),$$

$$\omega_3^2 = \omega_1^2 + 4\alpha/3M \ , \qquad \omega_4^2 = \omega_2^2 + 4\alpha/3M \ . \tag{16}$$

These allowed energy bands are shown in Fig. 14 as a function of $\cos\theta$. Note that as $\theta \to 90°$, the two bands degenerate into δ functions at $\omega^2 = \alpha/m$, $\alpha/m + 4\alpha/3M$. These are just the modes of an isolated AX_4 molecule shown in Fig. 15, where the lower frequency α/m is the breathing mode (singlet A_1 mode) and the upper frequency is the triplet (F_2 mode) in which all the motion is in either the x, y, or z direction. It is clear that for bond angles of 90°, there is no coupling between the molecular modes and that as the angle increases from 90°, so does the effective coupling.

There are total of nine modes per AX_2 unit. It is easy to see that five of these are at zero frequency. This is because it is possible to move the network in such a way that all the AX bond lengths are unchanged if we impose four constraints per AX_2 unit-- leading to 9 - 4 = 5 modes at zero frequency. There are also δ functions, each with weight 1, at ω_3 and ω_4, as shown in Fig. 14. These modes are sketched in Fig. 16. By arguments similar to the ones given above involving constraints, it is easy to show that there is weight 1 in each of these two δ functions. The remaining two modes are, of course, the band modes. As the angle is increased from 90°, we go from the molecular A_1 and F_2 modes with weights 1 and 3 to a crossover at $\cos\theta_c = -2m/3M$. For larger angles than this, we have two bands each with weight 2. Eventually the bands

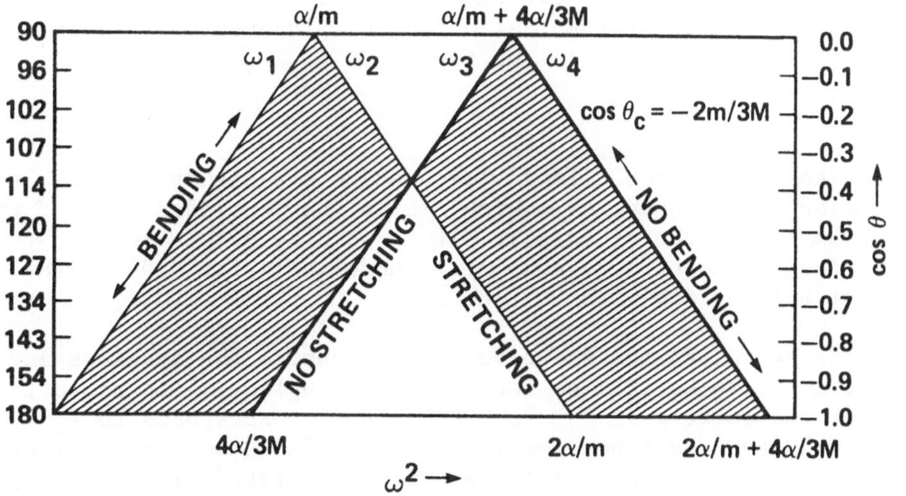

Fig. 14. Allowed frequency regions are shown by the shaded area as a function of $\cos\theta$ (right scale) and θ (left scale). The horizontal axis is linear in ω^2, and a mass ratio $m/M = 16/28$ appropriate to SiO_2 has been used. The character of the band edge modes is shown as are the limiting frequencies for $\theta = 90°$ and 180°. The two heavier lines give the position of δ functions, and the angle at which the two bands just touch (θ_c) is indicated.

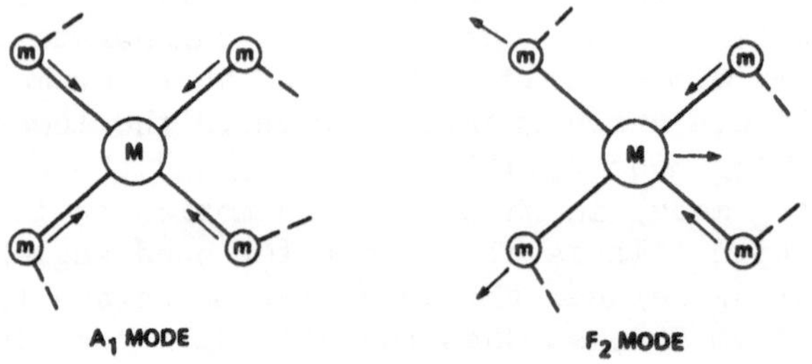

Fig. 15. Eigenmodes of an isolated AX_4 tetrahedron showing the singlet (A_1) and triplet (F_2) mode. The other modes lie at zero frequency because there is no restoring force associated with the motion with only a nearest-neighbor central force.

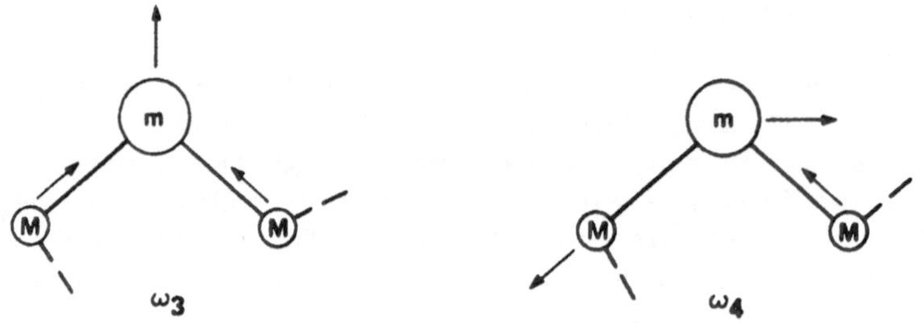

Fig. 16. Eigenmodes associated with the δ functions at ω_3 and ω_4 are made up of linear combinations of displacements on individual bonds of the kind shown in the figure.

spread apart, and the allowed frequencies are higher or lower than the molecular frequencies, which lie in the forbidden part of the spectrum if $\theta > \cos^{-1}(-4m/3M)$. It is pointed out in Ref. 23 that the molecular modes were completely unreliable for SiO_2 in the β-crystobalite form with $\theta \sim 180°$.

The A_1 and F_2 modes are very analogous to the atomic s and p states.[20] From Eq.(11) we see that the coupling α' is proportional to $\cos\theta$ and is zero if $\theta = 90°$. This case corresponds to $V_2 = 0$ of Ref. 20 and describes completely decoupled atoms with a singly degenerate s state and triply degenerate p state. Thus for systems where the AXA angle θ is close to 90° and the noncentral forces are weak, we expect the system will be well described by the molecular modes. Notice that for $\theta < \theta_c$, the positions of the centers of the two spectral bands are given exactly by the molecular frequencies. When $\theta > \theta_c$, the separation of the two bands is greater than predicted by the molecular model and the weights are no longer 1:3 but become 2:2.

Having explored this simple model in some detail, we are in a position to make contact with experiment. Of course, we are looking for a qualitative comparison as the potential energy in a real AX_2 solid is more complex than the one we have used. The best available experiments are intrared absorption and Raman scattering; however, these measure the density of states modified by a matrix element. This is an additional complication as it is very difficult to include matrix element effects. It is, therefore, useful to regard the computer simulations of Bell and Dean[5] as "experiments." We will discuss BeF_2, $GeSe_2$, SiO_2, GeS_2, and GeO_2 which have mass ratios m/M of 2.11, 1.09, 0.57, 0.44, and 0.22, respectively. There is no critical angle for BeF_2, and one is always in the upper part of Fig. 14 with two bands with weights 1:3. For $GeSe_2$, SiO_2, GeS_2, and GeO_2, the critical angles are 136°, 117°, 107°, and 98°, respectively.

We are now in a position to discuss the spectra of these various AX_4 solids.

BeF_2. As noted, there is no critical angle, and we are, therefore, in the upper part of Fig. 14, where the two high-frequency bands are derived from the molecular A_1 and F_2 bands. The work of Bell and Dean[5] (more details are given in Ref. 24) shows clearly that, indeed, the triplet F_2 band (at ~700 cm^{-1}) is split off from the rest of the spectrum and is rather symmetrical. This we would expect because, in the absence of noncentral forces, there would be δ functions at both sides of this band. The A_1 mode has merged with the rest of the band but is still a rather pronounced shoulder at around 350 cm^{-1}. BeF_2 is unusual because of the light Be mass; and although it is in no sense molecular, its spectrum is clearly evolved from the molecular spectrum in the sense that we are in the upper part of Fig. 14 and the upper band has evolved directly from the F_2 molecular mode. We would, therefore, predict that this upper band would be seen very strongly in infrared absorption but have almost no Raman activity. As far as we know, there are no optical data available.

GeS_2 and $GeSe_2$. These materials have been studied extensively experimentally.[25] The experimental spectra have been interpreted very successfully in terms of a molecular model. The measured spectra of GeS_2 have two strong peaks, which are a Raman active A_1 mode at 342 cm^{-1} and an infrared active F_2 mode at 367 cm^{-1}. These modes are surprisingly sharp. This is clear and unmistakable evidence that we are in the upper part of Fig. 14 with $\theta < \theta_c$. For GeS_2, θ_c = 107°, and so we conclude that the bond angle in GeS_2 is less than 107°. This is a very useful conclusion and is what would be expected from chemical bonding considerations.[26] Direct structural information on the bond angle is difficult to obtain from x-ray diffraction. A similar situation probably holds in $GeSe_2$, where the bond angle must be less than the critical value of 136° if the molecular interpretation is to work.

SiO_2 and GeO_2. For SiO_2 and GeO_2, we have the situation where $\theta < \theta_c$. For SiO_2, it is generally believed that there is a fairly broad distribution in bond angles, with a maximum around 140°.[21] The situation is probably similar in GeO_2. We are, therefore, in the lower part of Fig. 14, where solid state effects predominate, and it makes more sense to use the AXA bond as a starting point rather than the AX_4 molecule. The work of Bell and Dean[5,24] shows that there is a high-frequency band (around 1000 cm^{-1}) with weight 2 that has more strength on the high-frequency side as we would expect from the evolution of the δ function. The lower doublet has merged with the lower frequency modes but is still apparent as a wing ~700 cm^{-1}. It is more difficult in this case to make predictions about the optical activity, but we would expect the high-frequency band to be both Raman and infrared active because of the mixing of the molecular A_1 and F_2 modes.

We have deliberately desisted from giving a more quantitative comparison between theory and experiment, as we believe that the usefulness of the central-force model is in giving an overall understanding of the behavior of the upper part of the frequency spectrum. However some success has been achieved in making more quantitative comparisons recently by Galeener.[27]

6. CONCLUSION

In these notes I have tried to discuss the general principles involved in understanding the vibrational spectra of amorphous solids. Of necessity I have only covered a few areas with a concentration on those aspects that have been of particular interest to me and previously published in Refs. 15, 19 and 23. The problem is much more difficult than in molecules and crystalline solids. As a consequence spectra can rarely be calculated with high precision--rather general features have to be identified.

In these lectures I have concentrated on Si and SiO_2 type glasses as these are probably the easiest to understand theoretically. However there has been a great deal of experimental work on other systems.[28,29] In the future more theoretical effort will undoubtedly go into understanding these more complex systems. More inelastic neutron scattering experiments at high q would also be useful as there is not the problem with matrix elements that complicates the interpretation of infrared and Raman data.

These lectures are partly based on work supported by a grant from the National Science Foundation (DMR-77-05983).

REFERENCES

1. G. Herzberg, Molecular Spectra and Molecular Structure, New York, Van Nostrand, 2nd edition, 1964.
2. An elementary introduction may be found in C. Kittel, Introduction to Solid State Physics, J. Wiley and Sons, Inc., New York, 1971, 4th Ed. More exhaustive treatments may be found in M. Born and K. Huang, Dynamical Theory of Crystal Lattices, Oxford University Press, London and New York, 1954; and in Solid State Physics, Supplement 3, 1971, 2nd ed., A. A. Maradudin, E. W. Montroll, G. H. Weiss and I. P. Ipatova, Theory of Lattice Dynamics in the Harmonic Approximation.
3. W. H. Zachariasen, J. Am. Chem. Soc. 54, 3841 (1932), D. E. Polk, J. Non-Cryst. Solids 5, 365 (1971).
4. W. Paul and G.A.N. Connell in Physics of Structurally Disordered Solids, N.A.T.O. Advanced Study Institute Series, Plenum Press, 1976. Edited by S. S. Mitra, p. 45.
5. P. Dean, Rev. Mod. Phys. 44, 127 (1972).
R. J. Bell, Rep. Prog. Phys. 35, 1315 (1972).
6. For example, see the discussion in Ref. 4.
7. P. W. Anderson, B. I. Halperin and C. M. Varma, Phil. Mag. 25, 1 (1972); W. A. Phillips, J. Low Temp. Phys. 7, 351 (1972).
8. C. N. King, W. A. Phillips and T. P. DeNeufville, Phys. Rev. Lett. 32, 538 (1974).
9. R. J. Kobliska, S. A. Solin, M. Selders, R. K. Chang, R. Alben, M. F. Thorpe, and D. Weaire, Phys. Rev. Lett. 29, 725 (1972).
10. M. Born, Ann. Phys. Lpz. 44, 605 (1914).
11. D. Weaire and R. Alben, Phys. Rev. Lett. 29, 1505 (1972); R. Alben in Proc. of Conf. on Tetrahedrally Bonded Amorphous Semiconductors, Yorktown Heights, 1974, AIP Conf. Proc. 20, p. 249.
12. R. Alben, D. Weaire, J. E. Smith and M. Brodsky, Phys. Rev. B11, 2271 (1975).
13. R. J. Bell and P. Dean, Disc. Faraday Soc. 50, 55 (1970).
14. D. Beeman and R. Alben, Advances in Physics 26, 339 (1977).
15. M. F. Thorpe, Phys. Rev. B8, 5352 (1973); see also M. F. Thorpe in Physics of Structurally Disordered Solids, N.A.T.O. Advanced Studies Institute Series, ed. by S. S. Mitra, Plenum Press, p. 623 (1976).
16. G. Lucovsky in the Proc. of the 5th Intl. Conf. on Amorphous and Liquid Semiconductors, Garmish-Partenkirchen, (Taylor and Francis, Lond, 1974), Vol. 2, p. 1099.
17. R. B. Laughlin and J. D. Joannopolous, Phys. Rev. B16, 2942 (1977), Phys. Rev. B17, 2790 (1978).
18. A. J. Leadbetter and M. W. Stringfellow, Neutron Inelastic Scattering, Proceedings of the Grenoble Conference (IAEA), Vienna, p. 501, 1972. (Unpublished).
19. P. Sen and M. F. Thorpe, Phys. Rev. B15, 4030 (1977).

20. D. Weaire and M. F. Thorpe, Phys. Rev. B4, 2508 (1971); M. F. Thorpe and D. Weaire, Phys. Rev. B4, 3518 (1971.
21. R. L. Mozzi and B. E. Warren, J. Appl. Cryst. 2, 164 (1969).
22. D. Weaire and M. F. Thorpe, in Computational Methods for Large Molecules and Localized States in Solids, edited by F. Herman, A. D. McLean and R. K. Nesbet (Plenum, New York, 1972).
23. K. Kulas and M. F. Thorpe, AIP Conf. Proc. No. 31, 251 (1976). Also, M. F. Thorpe, Physics of Structurally Disordered Solids, edited by S. S. Mitra (Plenum, New York, 1976), p. 623.
24. R. J. Bell, N. F. Bird and P. Dean, J. Phys. C1, 799 (1968); R. J. Bell, P. Dean and D. C. Hibbins-Butler, J. Phys. C3, 2111 (1970).
25. See, for example, G. Lucovsky, in Amorphous and Liquid Semiconductors, edited by J. Stuke and W. Brenig (Taylor and Francis, London, 1974), p. 1099; G. Lucovsky, J. P. deNeufville, and F. L. Galeener, Phys. Rev. B9, 1591 (1974).
26. G. Lucovsky (priviate communication).
27. F. L. Galeener, Phys. Rev. B19, 4292 (1979).
28. See for example, C. Perry in these proceedings.
29. See for example, G. Fisher in Physics of Structurally Disordered Solids, N.A.T.O. Advanced Study Institute Series, ed. by S. S. Mitra (Plenum, New York, 1976), p. 703.

ORDER-DISORDER PHASE TRANSITIONS IN SOLIDS

Clive H. Perry

Physics Department, Northeastern University

Boston, Massachusetts 02115, U.S.A.

1 INTRODUCTION

Basic research in phase transitions has the momentum of a decade or more of significant accomplishment in problems ranging from fundamental theory and experiment to the application of many new exciting areas such as order-disorder in alloy design, martensitic transformations and critical phenomena. Infrared (IR), Raman (R) and inelastic neutron (N) spectroscopy have been shown to be powerful tools and much work has been carried out, especially in verification of the "soft mode" idea and to detail studies of "soft mode" behavior at displacive structural phase transitions. Excellent reviews on this topic have been given by Scott (1), Blinc and Zeks (2) and by Dorner and Comès (3).

The temperature dependence of a soft mode does often provide the most direct information on the generalized susceptibility of the order parameter and thus the dynamics of structural phase transitions. One concern has been with extracting critical components from the spectra. However, these relatively simple quantities often represent less information than is contained in the spectra itself and in general can only be obtained unambiguously in rather simple (e.g., quasi-harmonic) situations (4).

For instance, soft mode behavior may not be the only striking change in a spectrum around a phase transformation. In some cases new peaks arise in the spectrum due to the activity of modes in the low temperature (lower symmetry or ordered phase) which are forbidden in the high temperature, higher-symmetry (often disordered) phase. In others, the degeneracy of modes can be totally or partially removed in the lower symmetry phase. Qualitatively it is clear that all changes in mode strength and frequency are somehow related to the change in phase (whether order-disorder, purely structural or a mixture of both) and can be related to the spontaneous value of the order parameter below the transition temperature. The quantitative connection with the phase transition mechanism has been partially investigated experimentally for several cases (5-11) with IR, R and N techniques. Some of these examples have been analyzed with varying success by Petzelt and Dvořák (12) using general formulae and group theoretical means to determine phenomenologically the strength changes and frequency shifts due to a particular phase transition.

All effects of fluctuations are expected to play a greater role for order-disorder phase transitions, than for strictly structural phase transitions as fluctuations of the spontaneous order parameter, η can be comparable with its saturated value η_s. The usual second order processes which lie in the frequency region close to $\omega_i(q)$ contribute strongly and it is consequently particularly difficult to predict temperature dependences close to the transition temperature.

If the order-parameter involves a reorientation of an ion group (e.g., NH_4^+, $^-NO_2$, $^-NO_3$) or there are two equivalent positions on each O...H - O hydrogen bond (e.g., in KDP and its isomorphs), the disorder is, at first sight, of a particularly simple nature. Some understanding of the vibrations in these systems have been made using a perturbation approach. However, many phase transitions of the order-disorder type simultaneously undergo a displacive transition from one ordered arrangement to another. These materials are not amorphous in the usual sense (13) but are characterized by a special type of disorder on one sublattice and from one point of view they are similar to an AB_xC_{1-x} mixed crystal. The mixing is not of different species with different mass but of one species with two kinds of orientations or positions resulting in two kinds of force constants.

Some specific examples of order-disorder phase transitions which have been studied by IR and inelastic (or quasi-elastic) light scattering and neutron scattering spectroscopy are discussed in the following sections.

2 TERBIUM VANADATE

Order-disorder phase transitions are common in magnetic systems. The only second-order phase transition thus far examined which exhibits a central peak in neutron experiments is $PrAlO_3$ (14). The rare-earth vanadates, arsenates and phosphates exhibit similar cooperative Jahn-Teller transitions which involves a continuous tetragonal distortion to an orthorhombic structure (6,15). The transition is driven by a soft exciton which transforms like B_{1g}

for $DyVO_4$ and B_{2g} for $TbVO_4$ at q = 0 but produces no soft optic mode. The exciton couples linearly to an elastic strain (acoustic phonon) of the appropriate symmetry which is the ultimate soft mode of the system. These materials are paramagnetic due to the unpaired electrons of the partly filled 4f shell of the rare earth ion. They belong to the tetragonal space group D_{4h}^{19} at $T > T_D$ with the point symmetry D_{2d} for the rare earth ion. The resultant crystal field is such that it favors large magnetic quadrupole moments with axes in the basal plane. As a consequence of this, there are near degeneracies in the lowest electronic states in the high temperature phase due to the four fold symmetry about the c-axis.

The basic mechanism for the cooperative Jahn-Teller effect is brought about by small displacements of the ligand ions which can change the crystal field acting on the rare earth ions. The ligand displacements themselves interact with each other because of the elastic properties of the lattice and this provides an effective interaction between the quadrupoles of the neighboring ions. The entire crystal can then become unstable with respect to the proper distortion under the influence of the interaction between the groups of ligands. As the disorder-order phase transitions proceeds, the magnetic quadrupoles align themselves and the low-lying states shift and split. The J-T transition occurs at $\sim 33^oK$ in $TbVO_4$ and the results of the temperature dependence of the soft-electronic modes and the associated energy level scheme are shown in Fig. 1.

The application of a magnetic field along one of the quadrupole moments in the basal plane in the disordered tetragonal state can also drive the J-T transition (16,17) by aligning the quadrupoles directly. (A similar result can be obtained by applying an external uniaxial stress to the crystal (15).) The influence of the magnetic field on the electronic modes and the subsequent crystal distortion depends on the direction of application of the field in the basal plane (6). With H//y(100) the field is applied at 45° to one of the easy axes and the electronic modes simply suffer a Zeeman splitting such that

$$E = \pm \{W_2{}^2 + \frac{\Delta^2}{2} + H^2 \pm [(4W_2{}^2 + \Delta^2)H^2 + (\frac{\Delta^2}{2})^2]^{\frac{1}{2}}\}^{\frac{1}{2}}. \quad (2.1)$$

Where $W_2 = (\lambda + \mu)\langle S^z\rangle$, $\lambda + \mu = 25\ cm^{-1}$ (μ is the strain, λ is the mode (acoustic) coupling), $\Delta = 9\ cm^{-1}$ (half the splitting between the two exterior singlets), H is the applied field and $\langle S^z\rangle$ measures the quadrupole moment. The splittings are symmetric as shown in Fig. 2(a) and no lattice distortion takes place (18,19).

If the field is applied along one of the easy axes (i.e., H//($\bar{1}$10)), then the energy levels can be found by solving a cubic

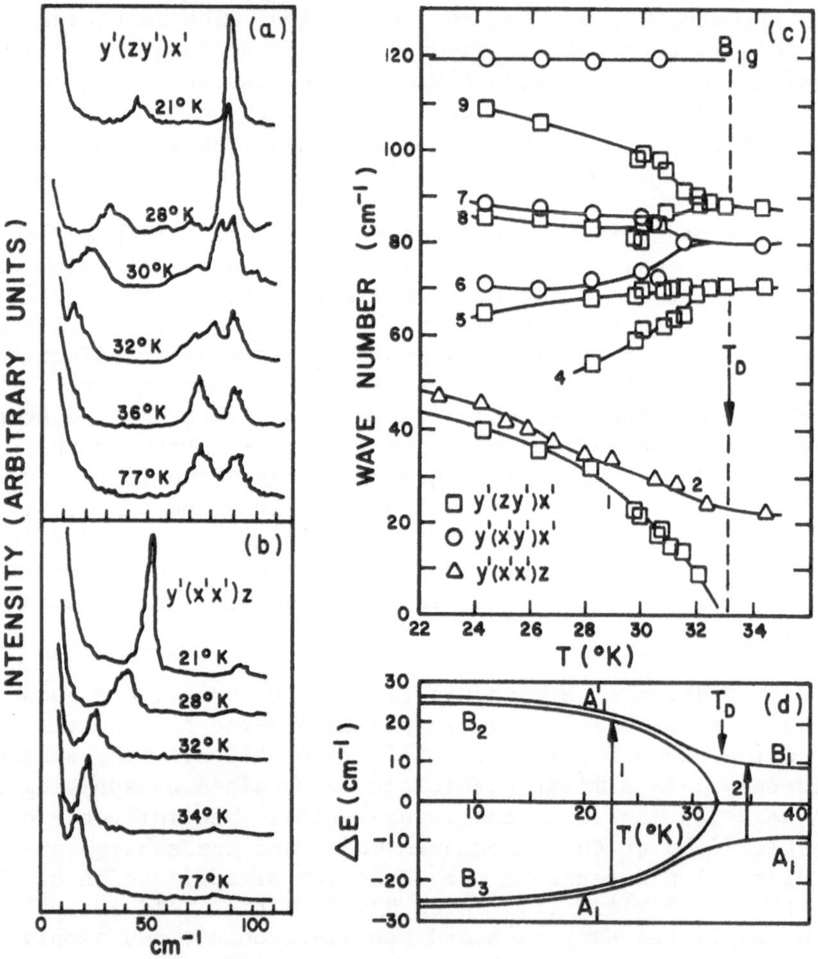

Fig. 1. (a) and (b) Raman spectra of the low-lying electronic transitions showing the soft mode behavior.
(c) Electronic energy levels and observed Raman transitions as a function of temperature.
(d) The $TbVO_4$ energy level scheme.

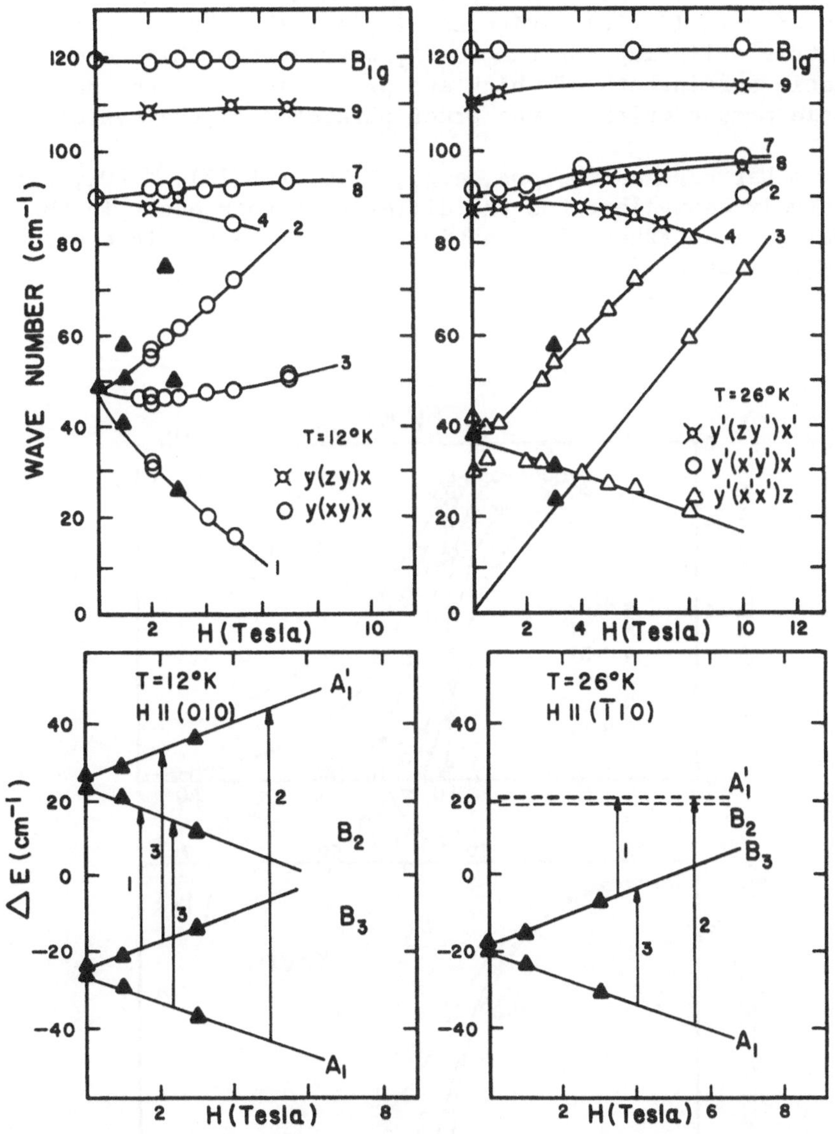

Fig. 2. Observed electronic Raman transition as a function of magnetic field at 12°K. (a) H//(100), (b) H//($\bar{1}$00). The associated energy level scheme is shown below. ▲ Calculated values from Ref. 18.

equation as the $|E_{y'}\rangle$ does not couple to the other three (18). The electronic mode behavior shows a soft exciton and the crystal suffers a tetragonal distortion (Fig. 2(b)). This distortion can be monitored by measuring the E_g phonon mode splitting as a function of H field (16). Figures 3(a) and (b) show the comparison between this splitting for the VO_4 symmetric stretching mode and the elastic strain, both of which are proportional to the nett quadrupole moment which is the order parameter of the system.

The other rare earth ions, e.g., $DyVO_4$ (16,17), $TmPO_4$, etc. are qualitatively equivalent and the differences that occur in the application of an external magnetic field can be attributed to the differing g factors (19).

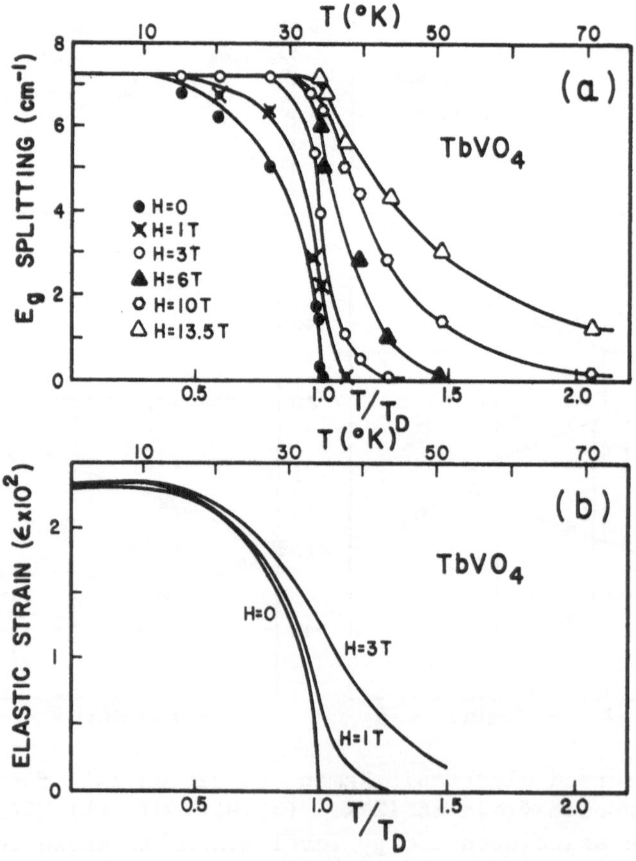

Fig. 3(a). Temperature and magnetic field dependence of the splittings of the "830 cm^{-1}" E_g phonon mode.
(b). The calculated elastic strain using molecular field theory (18).

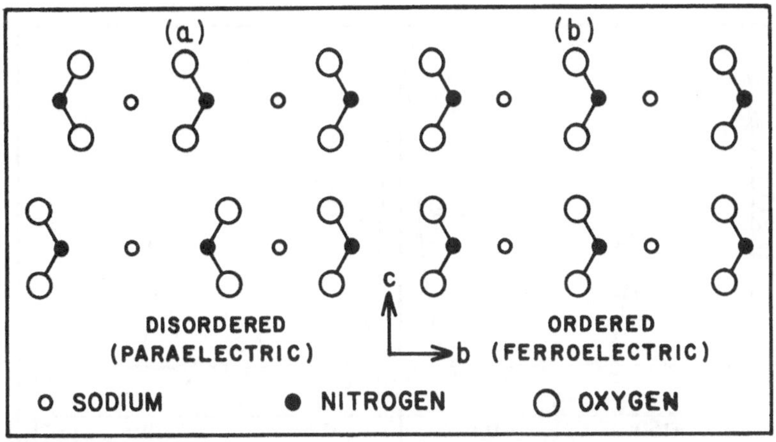

Fig. 4. Orientations of the $^-NO_2$ groups in $NaNO_2$ showing the (a) disordered and (b) ordered phases.

3 SODIUM NITRITE

Sodium nitrite possesses disorder-order, paraelectric (D_{2h}^{25}) to ferroelectric (C_{2v}^{20}) phase transition at ~ 163°C. (Actually a small region exists between ~ 163 and 168°C at atmospheric pressure where it is antiferroelectrically ordered). The normal modes of vibration in the various phases have been studied by infrared (20) and Raman spectroscopy (21) and by inelastic neutron scattering (22). These were found to show very little temperature dependence so that the phase transition does not result from an instability of the crystal against one of the normal modes describing the small displacements of the atoms.

Scattering from the ferroelectric fluctuations resulting from an ordering of the nitrite groups and the consequent large displacements of the Na^+ and $^-NO_2$ ions (~ 0.5Å) produce quasi-elastic scattering which occurs more slowly than those of the long wavelength fluctuations. The disorder-order transition is consequently first order. In the paraelectric phase the energy width of the quasi-elastic peak is much less than the experimental resolution and τ is assumed to be greater than 6×10^{-10} sec.

The ionic positions of the $^-NO_2$ ion in the paraelectric phase can be described as disordered between two equivalent positions (Fig. 4(a)). A spin coordinate was introduced by Yamada and Yamada (23) and by Sakurai et al. (22) to specify the orientation. It was also assumed in their models that either the motion of the Na ion

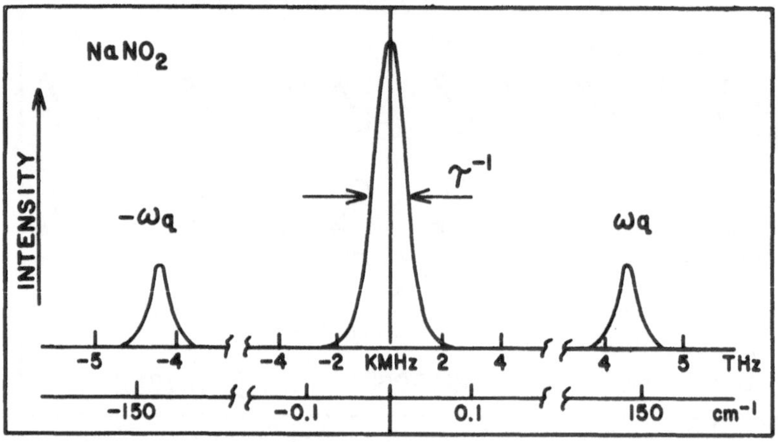

Fig. 5. Spectral response for a quasi-harmonic phonon with an emerging additional central component.

closely followed the NO_2 group (23) or that it was small (22). However, the restoring forces involved in the ferroelectric fluctuations are obviously quite different from those giving rise to the normal fluctuations and any model based upon the normal modes of vibration is inadequate for describing the quasi-elastic scattering.

The potential in which the oxygen atoms move can be simplistically viewed as a double well. At low temperatures the $^-NO_2$ ions lie on the same side of the double well (Fig. 4(b)). The system may be considered to have two characteristic frequencies. One is the normal mode of vibration with a quasi-harmonic frequency, ω_q and the other is associated with a fluctuation from well to well and is given by τ^{-1}. (Fig. 5)

Dielectric studies of Hatta et al. (24) showed that the low frequency response could be characterized by a single Debye relaxation time of $\sim 10^{-8}$ sec. near the transition. Scott (1) quotes a value of $\tau \approx 10^{-10}$ sec for the neutron scattering estimate of the relaxation time at room temperature ($\Gamma \sim 0.05$ cm^{-1}). By comparison, the Debye relaxation time for the NH_4^+ tetrahedral reorientation in the ammonium halides is estimated to be $\sim 9 \times 10^{-11}$ sec from Rayleigh scattering technique (25) and from Raman studies of the O...H - O motion in KDP type materials, it is $\sim 5 \times 10^{-13}$ sec (26). On the other hand, the dwell time of a mobile ion at a given site in a superionic conductor at high temperature is $\sim 10^{-11}$ sec whereas the hopping or jump time is $\sim 10^{-12}$ sec (27). The use of conventional infrared or Raman spectroscopy is generally limited to the study of systems with $\tau < 10^{-12}$ sec so that the peak in the response occurs at frequencies greater than 1 or 2 cm^{-1}. High resolution inelastic neutron scattering (28) and time of flight spectrometers operating from a cold source (29) have improved this resolution by about an order of magnitude. However, one must resort to Brillouin spectroscopy (30) or correlation techniques (31) for the study of systems having even lower frequency ($\tau^{-1} \sim 10^8$ - 10^{10} Hz) acoustic relaxation processes which may occur in molecular solids and liquids.

The response in all these systems can sometimes be characterized by a single relaxation time which is generally considered the soft mode of the system as for example in $NaNO_2$. This mode is not a group theoretically predicted optical phonon and it is often the peculiarity of these order-disorder phase transitions. No normal external lattice vibration (expected from group theory which assumes infinitesimal amplitudes for each normal coordinate) transforms according to the symmetry of the order-parameter. For example, it can involve an indirect coupling, say through the elastic strains (acoustic phonons) and this can often become the ultimate soft-mode of the transition. One must be concerned with the "soft mode" self-energy function, which contains the effects of the soft-mode interactions with all other degrees of freedom in the crystal.

4 AMMONIUM HALIDES

The ammonium and deuteroammonium halides are well-known to undergo a number of structural modifications as a function of pressure and temperature. The various crystallographic structures and transition temperatures are given in Table 1 at atmospheric pressure (32). In the disordered phase CsCl Phase II, the ammonium ions are randomly distributed between two possible energetically equivalent orientations. (In the NaCl Phase I there are six equivalent orientations.) A small distortion along one of the cubic symmetry axes in Phase II causes a transformation to the tetragonal structure (Phase III) which contains two molecules per unit cell. The ammonium ions become parallel ordered along the axis but are anti-parallel ordered in the xy plane perpendicular to the tetragonal axis. In Phase IV, a cubic CsCl ordered structure is obtained in which the NH_4 ions have parallel ordering along each principal symmetry axis and the crystal exhibits piezoelectric properties as the crystal possesses no center of inversion. Comparison of the three "CsCl" structures are shown in Fig. 6. It is now well established that the order-disorder phenomena in the ammonium halide crystals must involve two competing types of interaction. There is a direct coupling between ammonium ions, due mainly to octupole-octupole electrostatic interaction, which favors parallel ordering (33). In addition, there is an indirect interaction which

Table 1. Phase transformations in the ammonium halides

	Phase I (α)	Phase II (β)	Phase III (γ)	Phase IV (δ)
Lattice type (NH_4) ions	NaCl (Disordered)	CsCl (Disordered)	(Tetragonal) (Antiparallel ordering)	CsCl (Parallel ordering)
Structure	O_h^5(Fm3m)	O_h^1(Pm3m)	D_{4h}^7(P4/nmm)	T_d^1(Pm3m)
NH_4Cl	α ⟷ 457.7^{o}K	β ⟷ 242.9^{o}K (256°K; 1.5 Kbar)		δ
ND_4Cl	α ⟷ 348.4^{o}K	β ⟷ 249.4^{o}K (249.4^{o}K; 1 bar)		δ
NH_4Br	α ⟷ 411.2^{o}K	β ⟷ 235°K	γ ⟷ 78°K	δ
			γ ⟷ 105°K	δ
		⟵(215.8^{o}K; 3.2 Kbar)⟶		
ND_4Br	α ⟷ 405°K	β ⟷ 215°K	γ ⟷ 158°K	δ
			γ ⟷ 165°K	δ
		⟵(215°K; 2.2 Kbar)		
NH_4I	α ⟷ 255.8^{o}K	β ⟷ 231.8^{o}K	γ	

The Phase II-Phase IV multicritical point, p^*, T^* is given in parentheses.

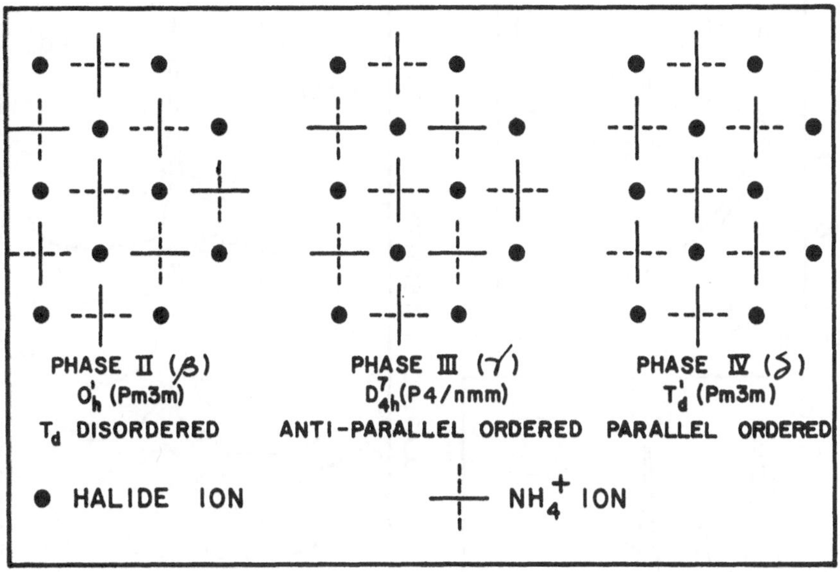

Fig. 6. Orientation of the NH_4^+ groups in the ammonium halides in the (a) disordered, (b) antiparallel ordered and (c) parallel ordered phase, as viewed down the z axis. The --|-- represents the tetrahedral structure of the NH_4^+ ion.

favors antiparallel ordering. Currently, there are two theoretical models for explaining the source of the indirect interactions: (i) an electrostatic model involving induced dipoles caused by distortion or polarization of intermediate halide ions (34); and (ii) a pseudo-spin-phonon coupling model (35).

The transitions in the pure materials and the $(NH_4)Cl_{1-x}Br_x$ mixed crystals have been studied using a variety of techniques but the principal spectroscopic investigations have included Rayleigh (25) and Brillouin scattering (36), infrared (32,37,38), Raman (8-11,38-41) and inelastic neutron scattering (42-46). The phase diagram due to Jahn and Neumann (47) is shown in Fig. 7(a).

The low frequency Raman spectrum of single crystals of NH_4Cl, NH_4Br and NH_4I were measured by Couzi et al. (8) in the disordered phases I and II. Particular attention was paid to the polarization selection rules in the cubic CsCl phase II. The space group of this phase is not exactly O_h^1 but is "T_d^1 disordered". The temperature dependence of the intensity of the disorder induced spectrum on entering the antiparallel ordered phase III or the parallel

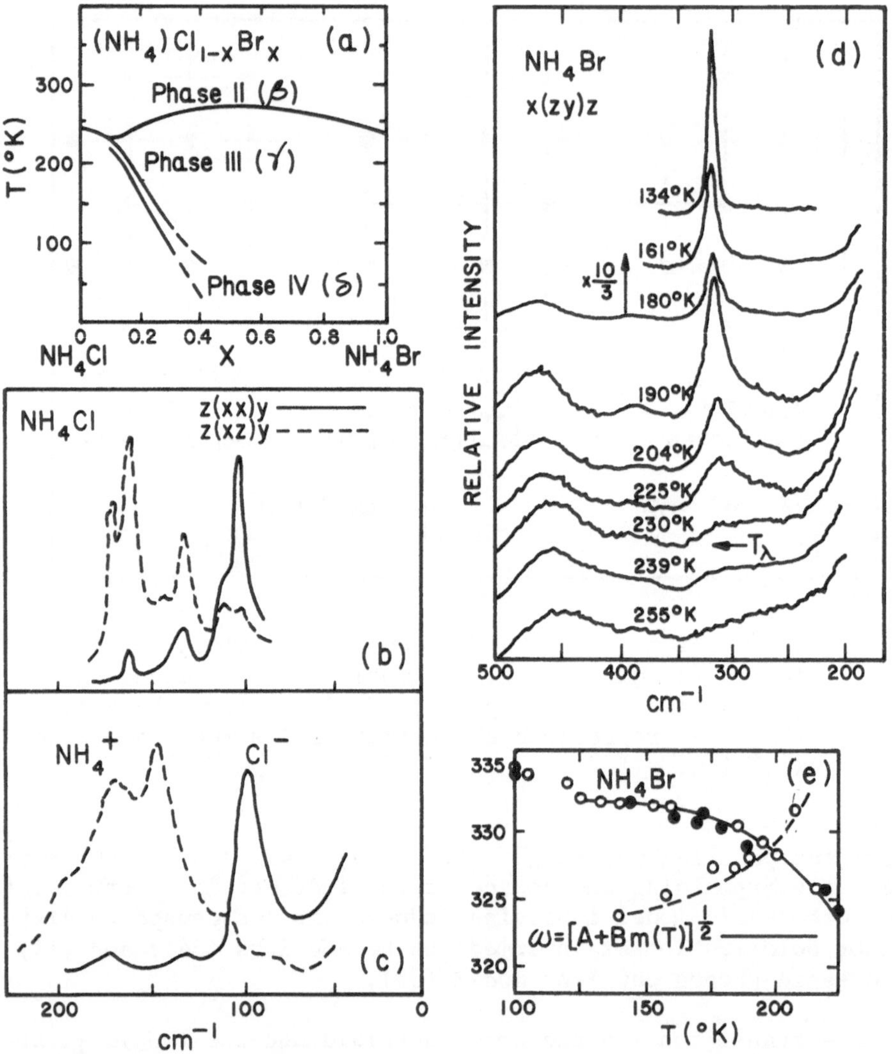

Fig. 7(a) Phase diagram for $NH_4Cl_{1-x}Br_x$ at 1 bar (38,46,47).

(b) Calculated spectra for NH_4Cl in its disordered phase.

(c) Measured spectra in NH_4Cl for comparison with Fig. &(b) showing the crossed (NH_4^+) and uncrossed (Cl^-) vibrations (8).

(d) The temperature dependent E_g librational mode in NH_4Br (11).

(e) Librational frequency ω (cm^{-1}) versus T (^oK). ○ Garland sample (37,50), ● Jahn sample (38,47), □ Wang's data (7). Solid curve is $\omega = [A + Bm(T)]^{\frac{1}{2}}$ (11).

ordered phase IV was studied in some detail. The spectra were interpreted by performing a calculation of the Raman spectrum, based upon the theory of Loveluck and Sokoloff (48) of Raman scattering from phonons in a disordered system using the lattice dynamical results of Teh (49) (See Fig. 7(b)). Excellent agreement was obtained with the major features of the observed polarization selection rules. The two polarization configurations were shown to pick out either NH_4^+ vibrations (crossed polarization) or halide ion vibrations (uncrossed polarization) individually (See Fig. 7(c)). The implications for phase III in NH_4Br were also discussed in the light of these calculations and a brief discussion of NH_4I in phase I was also given. Geisel and Keller (10) considered the influence of the orientation of NH_4^+ ions to analyze the peculiar temperature dependence of the Raman scattering intensities of the "56 cm^{-1}" band (7,40) in the ordered phase III in NH_4Br using a similar theory.

Recently Buhay et al. (11) investigated the polarized Raman scattering from single domain NH_4Br in the frequency region of the librational mode at temperatures near but below the order-disorder phase transition temperature $T_\lambda = 235°K$. The libronic excitation frequency decreases as T_λ is approached from below as shown in Fig. 7(d) in agreement with previous studies by Wang and Wright (7,40). The line-width was found to exhibit an Arrehenius-type temperature dependence with an activation energy of 1.07 kcal/mol. The libronic frequency shift was analyzed by including a contribution to the ammonium ion torsion constant due to the disorder dependent NH_4^+-NH_4^+ interaction. The interaction was assumed to have the Ising form

$$\sum_{ij} J(R_i - R_j)\sigma_i\sigma_j \ , \tag{4.1}$$

where $\sigma_j = \pm 1$ signify the two orientations of the jth ammonium ion. The sign of $J(R_i - R_j)$ was chosen so as to give phase III ordering. The most important contribution was the "self"-torsion constant and the order dependent interaction contribution was then given by

$$\frac{\partial^2}{\partial\theta^2} \sum_j J(-R_j)\sigma_o\sigma_j \ , \tag{4.2}$$

where θ is a rotation of the NH_4^+ ion at the origin. $\sigma_o = +1$ for phase III ordering and σ_j is replaced by m. Then,

$$J(-R_j)\sigma_o\sigma_j \approx -|J(-R_j)|m \ , \tag{4.3}$$

where $m = \langle\sigma_j\rangle$ (i.e., doing mean field theory). The order dependent contribution to the torsion constant is of the form (IBM) where

$$BI = \left|\frac{\partial^2}{\partial\theta^2} \sum_j J(-R_j)\right| \qquad (4.4)$$

and I is the moment of inertia of the ion. Thus the librational frequency (being proportional to the square root of the torsion constant) is given by $\omega^2 = A + Bm$, where $\sqrt{A}$ is the frequency in the absence of this order dependent interaction. The librational frequency versus T is plotted in Fig. 7(e). The solid line is a plot of $\omega = [A + Bm(T)]^{\frac{1}{2}}$ with $\sqrt{A} = 312 \pm 5$ cm^{-1} and $B \approx 100{,}000$ cm^{-2}. The temperature dependent order parameter was taken as $m(T) = (1 - T/T_\lambda)^\beta$. β was determined to be 0.27 ± 0.04 which was in good agreement with the value 0.25 ± 0.02 determined from birefringence and neutron measurements (47) Geisel and Keller (10) obtained $\beta = 0.34$ from Wang's intensity data (40) of the 56^{-1} line.

The polarized measurements also indicated a strong temperature dependent, broad (~ 70 cm^{-1}) band at 279 cm^{-1} at low temperatures possibly due to disordered induced first order Raman scattering.

Bauhofer et al. (38) measured the optical phonons and phase transition in $NH_4Cl_{1-x}Br_x$ mixed crystals using IR and Raman spectroscopy. At 300°K and for all x, the $NH_4Cl_{1-x}Br_x$ system is in the "T_d^1 disordered" phase II. Below ~ 240°K for x = 0.08 to x = 1 the system transforms to the antiparallel ordered tetragonal γ phase (D_{4h}^7, 2 molecules/unit cell). This is accompanied by a reduction of the Brillouin zone and certain zone-edge phonons are folded back to the zone center without significantly changing their frequencies or eigenvectors. At low temperature for $x < 0.4$ and $x \approx 1$ the crystals become parallel ordered (space group T_d^1) (See Fig. 7(a)).

A pure E_g spectrum can be obtained from using $y'(x'y')z$ where x = (100), y = (010), z = (001), x′ = (110) and y′ = ($\bar{1}$10). In the β phase (phase II), a y(xx)z scattering geometry gives an $A_{1g} + 4E_g$ Raman spectra. Predominantly $T_2A(M)$ phonon modes were observed corresponding to the Br^- and Cl^- ions whose strengths varied as x and 1-x respectively. For y(xy)z which provides modes of T_{2g} symmetry, the most prominent features were interpreted as being primarily associated with the NH_4^+ ion corresponding to $T_2O(M_2')$ and $T_1O(M_5')$ zone boundary phonons (8) (see Fig. 8).

In the γ phase (phase III) group theory predicts (32) the following external lattice modes $A_{1g} + B_{2g} + 2E_g + A2_u + E_u$. The four Raman active modes are connected with four M-point modes of the cubic phase denoted by M_4', M_5', M_2' M_5' in analogy with the notation for CsCl (53). The B_{2g} and the A_{1g} modes correspond to vibrations of the NH_4^+ and the anion sublattices respectively in the direction of the tetragonal axis while the other sublattice is at rest. For the E_g modes, both ionic species are vibrating. Sample spectra are shown in Fig. 9.

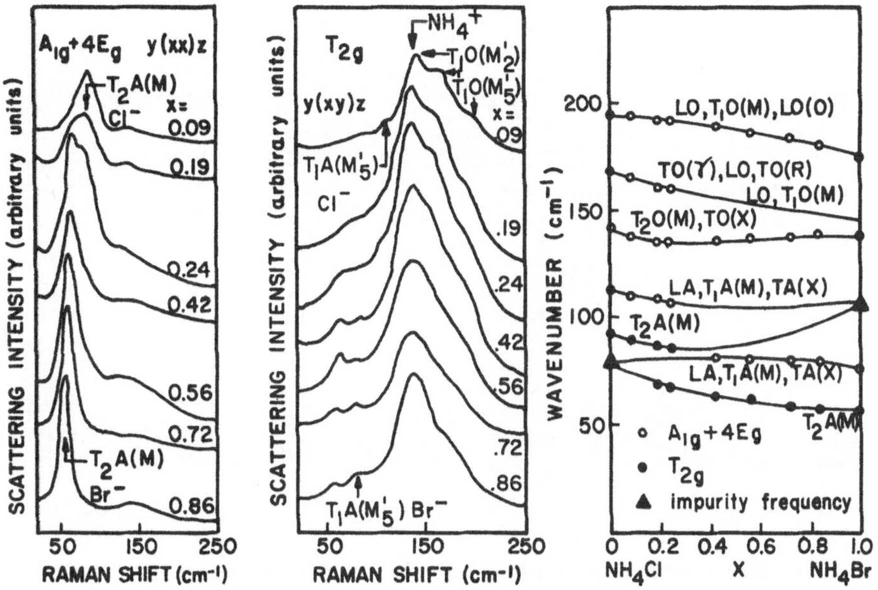

Fig. 8. Raman spectra of $NH_4Cl_{1-x}Br$ in the disordered phase II showing that symmetric scattering gives predominantly the zone boundary T_2A halide vibrations with two mode behavior while the NH_4^+ from T_2O and T_1O occurs in T_{2g} and is essentially concentration independent.

For x = 0.72 (Fig. 10(a)), as well as four main peaks that occur in NH_4Br, an additional A_{1g} mode due to the Cl^- ion is also observed. The mode strengths of the two A_g modes vary as x and 1-x respectively. The NH_4^+ librational E_g mode at about 350 cm^{-1} splits at lower temperatures as can be seen in the spectra taken at 20°K.

For x = 0.19 (see Fig. 10(b)) two successive phase transitions occur (β-γ at 250 and γ-δ at 160°K) (see Fig. 7(a)). This sample shows a certain amount of domain formation on traversing the first transition and some depolarization is observed. Through the γ-δ transition, all the spectral features of the tetragonal phase disappear. In this phase both the TO and LO phonons are observed (32).

The results of a simple extension of the REI model (54) for lattice vibrations for $q \neq 0$ in mixed crystals has been applied to this system (38,55) and are shown in Fig. 10(a). The experimentally determined concentration dependence of the mode frequencies in the tetragonal phase are shown in Fig. 10(b).

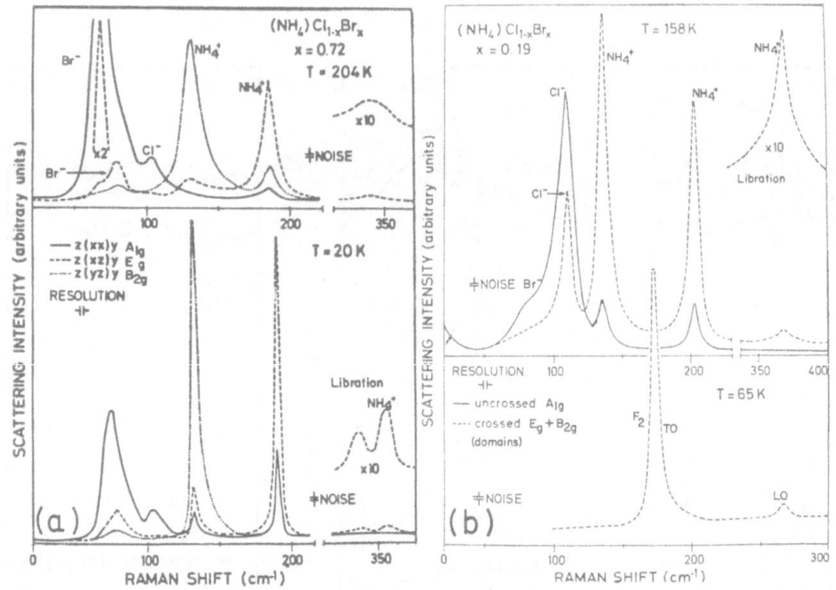

Fig. 9. Temperature dependence of the Raman spectra for (a) x=0.72 and (b) x=0.19 in the antiparallel ordered phase. The strength of the phonon modes corresponding to Br^- and Cl^- ions vary as x and 1-x respectively. Note the splitting of the E_g mode at low temperatures for x=0.72.

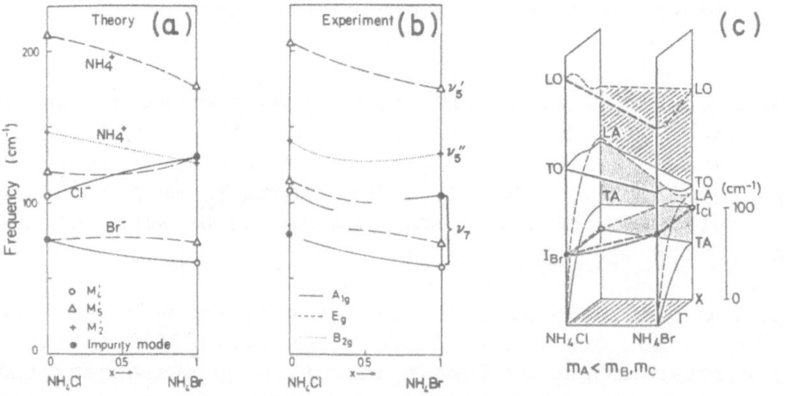

Fig. 10(a) Concentration dependence of the Raman frequencies of $NH_4Cl_{1-x}Br_x$ in the tetragonal phase III at 230°K showing the calculated M point frequencies. The model parameters were obtained from fits to the Raman frequencies for x=0 and x=1.

10(b) Experimental data for comparison with Fig. 10(a).

10(c) Proposed phonon mode scheme for $NH_4Cl_{1-x}Br_x$ from Γ to X (45,55).

Taking into account the simplicity of the model, the agreement with the experimental data is remarkably good. The agreement is acceptable for the Br^- impurity modes but is too high by $\sim 25\%$ for Cl^-. Such deviations may be expected as the impurity modes are only described by simple Einstein oscillators with the frequency $\omega_I = \sqrt{\frac{F}{m_I}}$ where F is the short range force constant for the pure end member and m_I is the mass of the impurity.

The Raman spectra in the β-phase, which reflect a weighted density of states, indicate that the high symmetry points of the Brillouin zone of the pure crystals remain points of a high density of states in the mixed crystals, too (8). The concentration dependence of the peaks of the polarized spectra leads to the conclusion that the system $(NH_4)Cl_{1-x}Br_x$ reveals an acoustical two-mode behavior, which is in good agreement with the results in the tetragonal phase. However, in the γ-phase new interesting features are observed in the acoustical frequency region, which are not present in the spectra of the pure end members (7,32,39). The Raman-active modes in the tetragonal phase of pure NH_4Br can be identified with M-point phonons of the cubic phase. We assume this to be also true for the mixed crystals. Thus, the Raman measurements show, in fact, the concentration dependence of zone-edge phonons. This assumption is supported by qualitative results of a calculation based on the REI model. The classification of mixed crystals into "optical two-mode," "acousto-optical mixed mode" and "acoustical two-mode" systems would account better for the experimental facts than a classification of "optical one-mode" and "optical two-mode" systems. Buhay (56) has also performed measurements of the concentration dependence of the librational mode, which for $x \gtrapprox 0.4$, is observed to split at low temperatures. The two features show that clustering effects may be important as they correspond more closely to NH_4^+ librations in predominantly Cl^- and Br^- environments.

The study of the lattice vibrations in mixed crystals has been restricted mainly to IR and Raman spectroscopic techniques. Such systems are of considerable interest as they are intermediate between a perfect crystal and a fully disordered amorphous material. As many mixed crystals form a miscible series over the whole concentration range between the pure end members, it is possible to observe the behavior of local- or impurity-modes and how they emerge and broaden into bands as the ion composition is changed (13). The long-wavelength optical phonons can generally be well characterized into various types of mode classes and several models have been used to successfully explain the observed behavior (54). However, several questions arise when one is treating a mixed crystal. To what extent can the phonon eigenstates be described by plane waves and how far can these concepts be projected into a Brillouin zone? The effect of mixing strongly influence the amount of disorder and this in turn will determine how well-defined the modes may appear in a particular experiment. The long-wavelength phonon lifetimes generally decrease in mixed crystals but the modes can usually be followed over a large portion of the composition range. The same group-theoretical symmetry rules associated with the end members can be applied to the mixed crystals and any effect of disorder in the interatomic force constants does not play a dominant role for the $q \approx 0$ modes although this observation may be expected to break down for shorter wavelength modes.

Perry et al. (45) obtained the dispersion curves for $NH_4Cl_{1-x}Br_x$ system for x = 0.25, 0.54 and 0.72 using the triple-axis spectrometers IN2 and IN3 at the High Flux Reactor, Grenoble, and can be compared with those obtained previously for NH_4Cl (42) and ND_4Br (44). One mode behavior of the TO and LO modes is observed at the Γ and the M points while two mode behavior is observed for the M point acoustic phonons (38). Genzel and Bauhofer (55) extended the simple REI model concept to the zone boundary phonons and in Fig. 10(c) is shown their proposed phonon schemes for the $NH_4Cl_{1-x}Br_x$ mixed crystal system from Γ to X.

For x = 0.72 the [ξ00]TA and [ξξ0]TA branches were well-defined almost to the zone boundaries and had identical slopes to within the experimental error. The chlorine impurity mode was weakly observed as a broad band ~ (100 cm^{-1}) but as no measurements were taken for $\zeta < 0.3$ it is not known whether this mode can be followed to the Γ point. The M ($\simeq$ X) point Raman frequencies are shown at the Γ point where they were observed as peaks in the one phonon density of states due to the disorder of the NH_4^+ ions. Both frequencies are in reasonable agreement with the zone boundary X point values.

There was a slight increase in the slope for the [ξ00]TA branch for the x = 0.54 sample. Towards the zone boundary the branch merged with a broad peak situated at ~ 66 cm^{-1} which was present across the whole zone; a similar band was observed at ~ 90 cm^{-1}. This latter mode coincided reasonably well with the predicted transverse "impurity" branch (see Fig. 10(b)) but it is not possible to ascertain whether this peak is caused by a one phonon density of states observed due to the disorder or whether it does in fact represent a rather dispersionless branch. Again the zone boundary values are close to the observed Raman frequencies for the two modes (see Fig. 9).

The results for the 0.25 sample indicated that the onset of disorder due to mixing caused a considerable broadening in the region where there should be a lifting of the band mode crossing with the TA and LA acoustic branches. Both the TA and LA branches disappeared into the broad incoherent peak due to the density of phonon states near 80 cm^{-1} (again in good agreement with the Raman data (38). This peak and the broad peak at ~ 93 cm^{-1} appeared for all $\underline{q}$ and in both longitudinal and transverse configurations.

There was no indication of the upper longitudinal branch and only a broad peak at $\sim$ 163 cm^{-1} corresponding to the TO mode could be observed. Apart from a slight increase in the slope of the TA branch and a softening of the TA(X) frequency no measurable difference in the incoherent background was observed in passing from the $\beta \rightarrow \delta$ phase.

The disorder in the ammonium ions did not effect the inelastic neutron scattering studies of the phonons but the large incoherent scattering cross section of hydrogen may well have contributed generally to the difficulty of resolving the impurity band modes for some concentrations. Despite some of these inherent limitations in the study of undeuterated $NH_4Cl_{1-x}Br_x$ mixed crystals, well-defined TA and LA phonons were observed over the composition range investigated. The measured phonon branches from the neutron, IR and Raman data are summarized in Fig. 11 and some typical neutron intensity plots for various $\underset{\sim}{q}$ values are shown in Fig. 12 and the data are representative of all the samples. The model calculations and simple predictions of Genzel and Bauhofer (55) are in qualitative agreement with the experimental data but the effect of mixing and subsequent disorder in the mixed crystals causes the "impurity" branches to appear as rather broad bands. This unfortunately reduced the possibility of observing the lifting of the mode crossings and the ability to separate incoherent scattering due to a phonon density states from a branch with little dispersion.

The NH_4-halide spectra show some interesting features that can be summarized below.

The order-disorder transitions are caused by the reorientation of the NH_4 ions where the reorientation times are long compared to the probe, whether it be IR, Raman or neutron spectroscopy, so that direct information on the transitions dynamics is difficult to obtain. Dielectric studies are not possible as no dipole moment is involved.

In NH_4Cl-rich samples the Phase II-Phase IV transition which occurs at 242°K (1 atmos.) in NH_4Cl is ferrodistortive, non-ferroelectric and non-ferroelastic. It is caused by an A_{2u} order parameter which has the peculiarity that no normal external lattice modes transform with this symmetry. (E.g., the NH_4 libration has F_{1g} symmetry in the disordered phase and is IR and Raman inactive.) The temperature dependence of the F_2 mode strengths where studied in the ordered phase in the infrared (32) and Raman (7,32,39). The results were in qualitative agreement with the variation of the order parameter but are difficult to analyze quantitatively.

The phase II-III transition that occurs in $NH_4Cl_{1-x}Br_x$ samples for $x \geq 0.12$ and occurs at 235°K in NH_4Br at 1 atmos. is antiferrodistortive, non-ferroelectric and improper ferroelastic. The

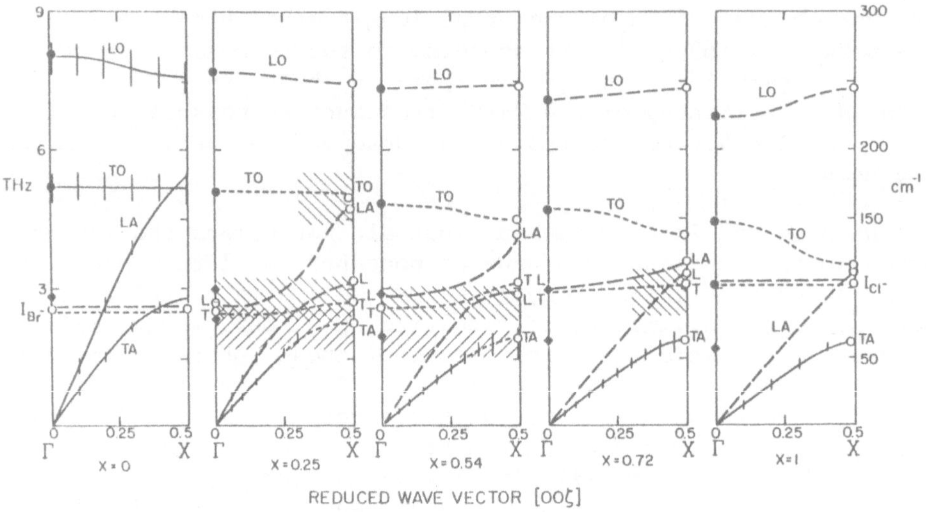

Fig. 11. Phonon dispersion in $NH_4Cl_{1-x}Br_x$ mixed crystals as a function of composition. Symbols: | neutron peaks; \\\ neutron bands; ● IR(Γ) Raman; —— measured curves.

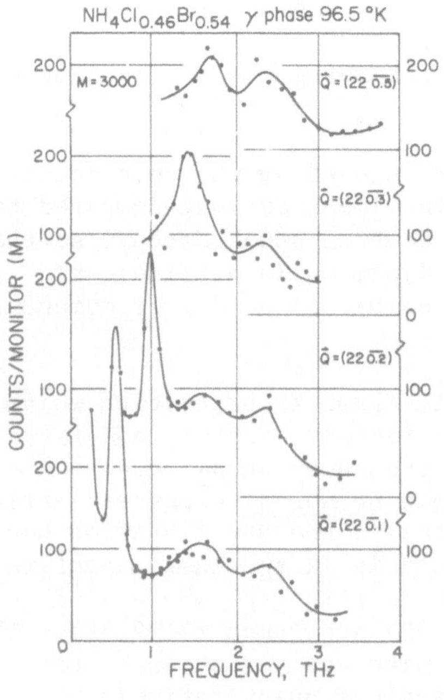

Fig. 12. Neutron intensity data as a function of Q for x = 0.54.

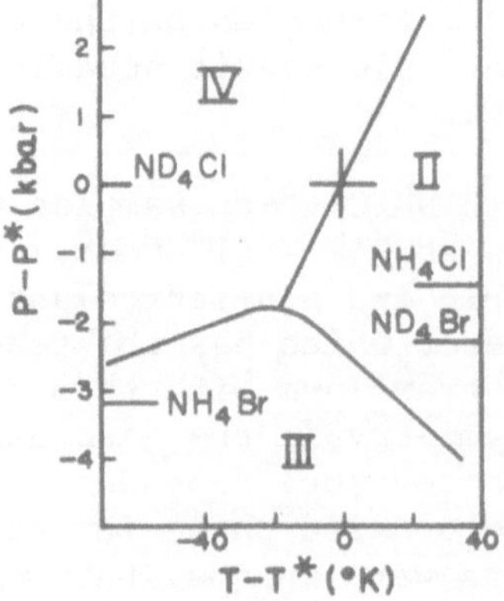

Fig. 13. Master phase diagram for ammonium chloride and bromide (50).

NH_4I II-III transition occurs at 235°K. The phase IV transition can only be induced by the application of pressure (41,51). The order parameter is the same NH_4 reorientation motion as in NH_4Cl but occurs at the M-point and can couple with NH_4-halide translational motion. As mentioned before the M_3' point acoustic mode at $\sim 56\ cm^{-1}$ in NH_4Br has received particular attention (10,40). The results of Perry and Lowndes (32) show that the IR selection rules are not effected for the lattice modes and almost no discontinuity in the F_{1u} mode linewidth was observed in crossing into the tetragonal phase (i.e., no A_{2u}-E_u splitting was observed).

Strong, polarized Raman spectra occurs in the disordered phases for all the ammonium halides and the reader is referred to Refs. 7, 8, 10, 48 for further discussion. There is a softening of the librational modes (11,40) and low frequency translation modes (7,40) as the temperature approaches T_λ from below. The linewidth of the modes change rapidly near T_λ due to the gradual evolution of disorder. Presumably, short range ordering of the NH_4 orientations persist above T_λ but long range ordering (comparably to the probe wavelengths in the IR and visible) decreases rapidly as T approaches T_λ.

Recent results of Hochheimer et al. (51) on the Raman scattering in NH_4Br and NH_4I under hydrostatic pressure have shown that the librational frequency increases with increasing pressure due to the decrease in the lattice parameters. However, close to T_λ, one might expect the NH_4^+-NH_4^+ interactions to also increase with a strong deviation from linear behavior. No marked deviations were found and it is their conclusion that the dominant mechanism for the strong softening with temperature is not the disorder dependent NH_4^+-NH_4^+ interaction as suggested by Buhay et al. (11) but it is the destabilizing effect of the repulsive overlap valence forces that are more important as the lattice parameters decrease. As these authors (51) point out, this is another example where purely electrostatic forces are insufficient to describe the rotational behavior of NH_4^+ ions in the crystal.

Garland et al. (52) in another recent paper on hydrogen bonding and order-disorder phenomena in ammonium halide crystals strongly advocate the spin-phonon coupling mechanism (35) for favoring antiparallel ordering. They point out that the strength of this coupling depends on the mass of the halide ion, the frequency of the M point acoustic mode and the gradient ΔV^1 in the potential for the hydrogen bond between NH_4^+ and X$'$. The obtain empirical values for ΔV^1 and the variations with changes in halide ion, pressure, deuteration and substitution of NH_4^+ by alkali ions are consistent with expectations for systems with weak N-H_4...X hydrogen bonding. Further experimental and theoretical work on the ammonium halides appear necessary to explain the effects of these indirect interactions and their relative contributions as this is obviously a "model" system that can be applied to many other orientationally disordered materials.

A master phase diagram for ammonium chloride and bromide is shown in Fig. 13 due to Leung et al. (50). The disorder-order multicritical point p^*, T^* (indicated by + symbol) is given in Table 1.

5 HYDROGEN BONDED FERROELECTRICS: KDP AND ISOMORPHS

KDP (KH_2PO_4) and its isomorphs have received extensive theoretical and experimental attention. They are of practical interest as their birefringent and piezoelectric properties are utilized in harmonic generation and in electro-optic modulators. The ferroelectric behavior which occurs at a disorder-order phase transition is associated to a large degree with proton tunneling and hydrogen bonding as deuteration has a marked effect on the Curie temperature. At T_c the heavy metal ions (K, Rb, Cs) simultaneously undergo a displacive transition from a D_{2d}^{12} ($I\bar{4}2d$) to a C_{2v}^{19} (Fdd2) structure so the phase transition is of a mixed character.

Descriptions of the phase transition have generally invoked a cooperative ordering of protons in a double-well potential along with each hydrogen bond. deGennes (57) originally showed that the individual tunneling of protons between a double minima potential well would lead to collective tunneling excitations in the presence of proton-proton interactions, and Brout et al. (58) and Tokunaga (59) showed that such collective tunneling excitations would have a soft-mode temperature dependence. Kobayashi (60) extended this concept to include a coupled interaction between a collective proton tunneling mode and a transverse-optic lattice vibration polarized along the c-axis of the crystal and showed that one of the coupled modes, the ferroelectric mode, would have a soft temperature dependence. Dvorák (61) has modified the Kobayashi theory by taking into account the interaction of the a-polarized TA phonon both with the proton tunneling mode and the TO phonon polarized along the c-axis; in this theory, as the transition is approached in the paraelectric phase, the ferroelectric mode drives the acoustic mode towards zero via the piezoelectric interaction (62). Blinc et al. (63) and Reese et al. (64) have pointed out that the proton tunneling model has its limitations because at high temperatures the approximations used in the linear equations of motion (60) break down and lead to diffusional rather than oscillatory motion. Theoretical discussions have examined the consequences of interactions between the ferroelectric mode and fluctuations in the phonon density (65,66) and have yielded a response function for the ferroelectric mode which predicts, under certain circumstances, the existence of a quasi-elastic central component as well as a different form for the condensation of the ferroelectric mode. Lage and Stinchcombe (67) have developed an effective-medium theory for the substitutionally disordered transverse Ising model above its transition and have used it to discuss the pressure and deuteration dependence of the static and dynamic properties of KDP-KD*P mixed system.

Experimentally most spectroscopic studies have concentrated on the condensation of the ferroelectric mode (68-72) and on searches of the central component (73). In the latter case there was some evidence of a central component in KDP but not in CsDA (CsH_2AsO_4).

The possibility of a separate determination of the magnitude of the relaxing self-energy has been pointed out by Young and Elliott (66). The inclusion of a spin-two phonon coupling in their Hamiltonian predicts a splitting of the E modes in the ferroelectric phase in addition to any reststrahlen effects and some evidence of this effect was observed in the KDP spectra of Aggrawal and Perry (69). More extensive studies have been made of KDP, KDA, RbDP and CsDP and their deuterated isomorphs by Leung et al. (74) and their results are used to discuss the temperature dependence of the central component amplitude in these materials.

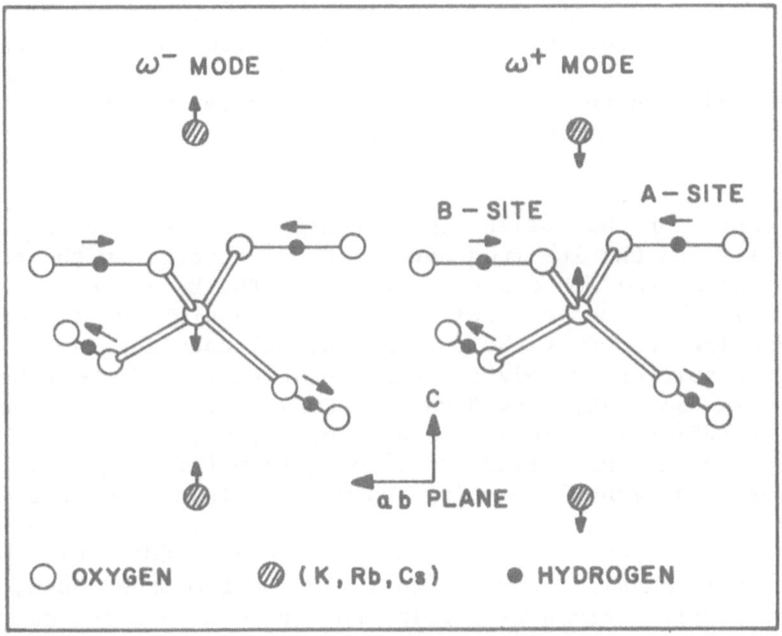

Fig. 14. Eigenvectors for the ω^- and ω^+ modes in KDP and its isomorphs (Ref. 60).

Pressure measurements up to ~ 10 Kbars have been reported for KDP by Peercy (75), RbDP by Peercy and Samara (76) and KDA, RbDA and CsDA by Leung et al. (77). These studies reveal a significant pressure dependence of the ferroelectric mode, which becomes underdamped with increasing pressure. Consequently the low frequency spectrum cannot be described by a Debye relaxation and a damped harmonic oscillator would appear more appropriate in this analysis.

The four tunneling protons in the primitive unit cell belong to a B_2, E and A_2 species. The B_2 and E proton modes are Raman active and would be expected to couple to phonons of the same species. As the proton displacements are nearly perpendicular to the spontaneous polarization, the B_2 (xy) coupled mode spectra has received the most attention. Generally the low frequency xy spectra have been fitted with two (ω_+ and ω_-) coupled damped harmonic oscillators in order to determine the FE mode (ω_-) and its temperature dependence. This has been accomplished with varying degrees of success for extracting the collective proton tunneling frequency, Ω_p and the pure phonon, ω_q from $\omega_\pm^2 = \frac{1}{2}(\omega_q^2 + \Omega_p^2) \pm \{[\frac{1}{2}\omega_q^2 - \Omega_p^2]^2 + G^2\}^{\frac{1}{2}}$ where G^2 is an interaction term. In the Kobayashi system $G = \Delta^2$ (real coupling). In the absence of the lattice coupling, the collective tunneling frequency is given by (60) $\Omega_p^2 = 4\Omega^2 - \Omega J\tanh(\Omega/kT)$ where Ω is the individual tunneling frequency and J is the dipolar proton-proton interaction. A representation of the eigenvectors for the ω_+ and ω_- modes are shown in Fig. 14.

Evidence for the collective protonic or deuteronic motions is observed in both the low frequency B_2 and E spectra in the paraelectric phase (see for example Fig. 15). The B_2 spectra can be "fitted" with two coupled oscillators representing the ferroelectric mode and a low frequency phonon. Additional coupling can exist between the soft optic mode and certain acoustic modes and has been discussed in some detail by Brody and Cummins (62) and by Coombs and Cowley (65). As pointed out by Scott et al. (70) in the limit of large damping, the relaxation time $\tau_p(T) = \Gamma_p(T)/\Omega_p^2(T)$. T/τ_p is the more fundamental quantity and varies linearly with temperature. However, it extrapolates to zero well below the transition temperature with real coupling (Kobayashi model (60)) which provides less realistic physical parameters (70). On the other hand, imaginary coupling gives T/τ_p going to zero within a few degrees of the transition temperature (72) and is the more appropriate choice in discussing the anharmonic theories of Cowley and Coombs (65).

Recently Kweicien et al. (78,79) have concentrated their pressure and temperature studies on the low frequency E-symmetry modes (Fig. 15(a) and (c)). The rather featureless spectra also indicate an interference shape which had previously been interpreted by Scott and Wilson (80) as a scattering from the one phonon

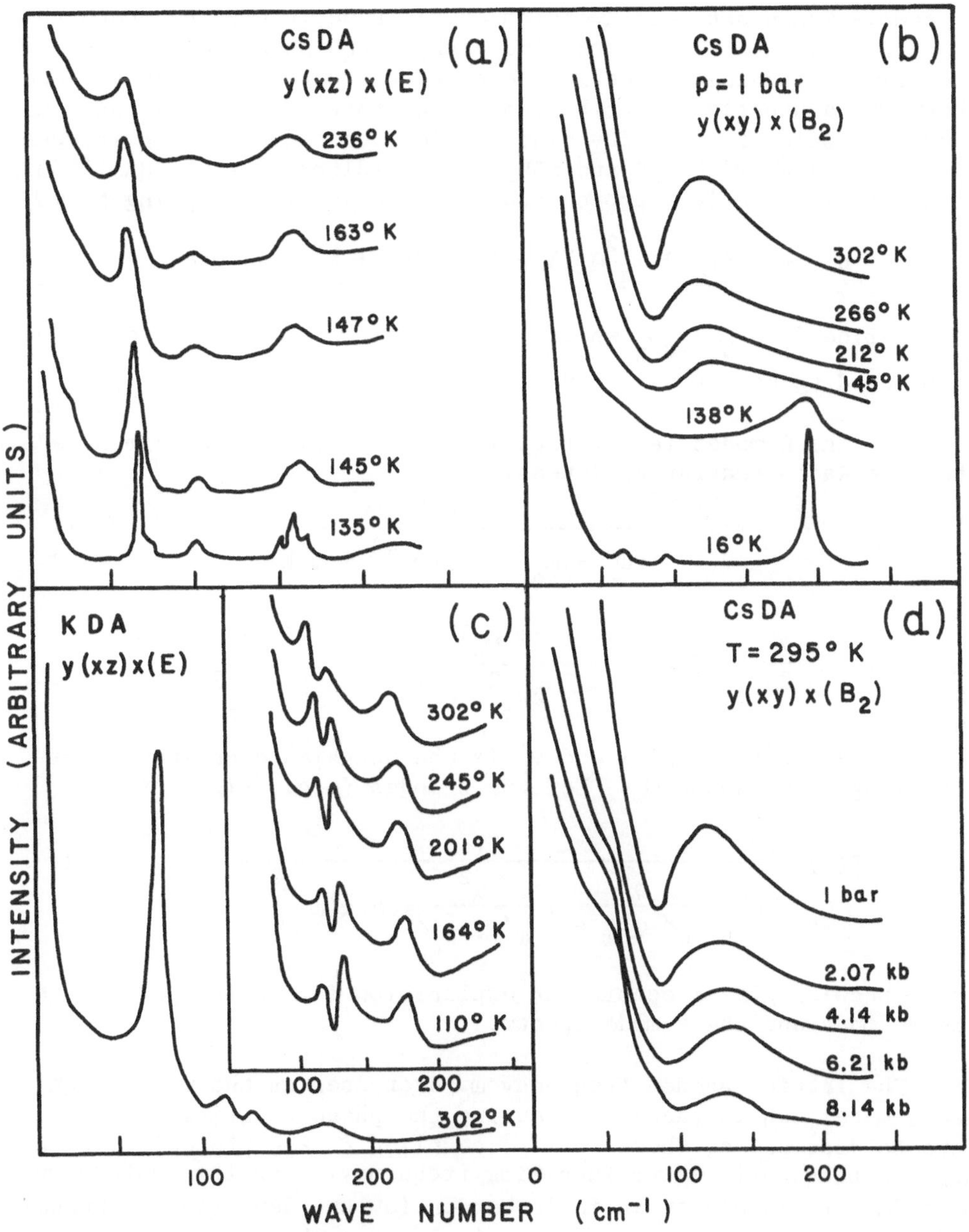

Fig. 15 (a) and (b) comparison of the E and B_2 mode spectra through the paraelectric-ferroelectric phase transition at $\sim$ 137°K in CsDA (77-79). (c) E mode spectra of KDA showing the strong interference (coupled mode behavior) at $\sim$ 125 cm^{-1}. (d) Raman spectra of CsDA as a function of pressure showing the hardening of the low frequency "soft" mode shoulder.

density of states where disorder above T_c would allow $q \neq 0$ modes to become Raman active. Kweicien et al. have interpreted both their E and B_2 mode spectra by assuming each proton sees a different tunneling depending on the influence of the particular lattice mode. They take a Gaussian distribution of the thermal displacements and a corresponding probability distribution for the tunneling frequencies by analogy with NMR and NQR studies (81,82). Their Hamiltonian consists of a lattice, proton and a proton-lattice coupling terms,

$$H = \tfrac{1}{2} \sum_q (p^2 + \omega_q^2 Q_q^2) + \tfrac{1}{2} \sum_i \Omega_p \sigma^x + \tfrac{1}{2}\hbar \sum \lambda_{iq} Q_q \sigma_i \; .$$

They define $\Omega_i = \sum_q \lambda_{iq} Q_q$ where λ_{iq} is the coupling parameter. $\langle \sigma_i \rangle$ is assumed to have a Gaussian distribution around Ω_p.

For the E-modes (e.g., Fig. 15(b) and (c)) they obtain a result that the Raman scattering intensity,

$$I(\omega) \propto \frac{\Gamma(\omega)}{[\omega^2 - \omega_q^2 - \omega\Delta(\omega)]^2 + \omega^2\Gamma^2} \; , \qquad (5.1)$$

$$\text{where} \quad \Gamma = \left(\frac{\sqrt{\pi}}{2\nu_p}\right) \lambda^2 \left[\exp. - \frac{(\omega + \Omega_p)^2}{2\nu_p^2} + \exp. - \frac{(\omega - \Omega_p)^2}{2\nu_p^2}\right] . \qquad (5.2)$$

$\Delta(\omega)$ is Dawson's integral and ν_p is the Gaussian distribution of tunneling times about Ω_p. For the B modes (e.g., Fig. 15(a)),

$$I(\omega) \propto \frac{\Gamma(\omega)}{\left[1 - \frac{\lambda}{\omega^2 - \omega_q^2} + \frac{\lambda^2}{\omega^2 - \omega_q^2}(\omega \mathrm{Re}\chi(\omega)\right]^2 + \omega^2\Gamma^2} \; . \qquad (5.3)$$

Their results give a consistent explanation for the shape of both the B_2 mode and the E mode spectra.

The lattice phonon frequency ω_q decreases as the mass of the heavy metal ion increases to causing the phase transition temperature to decrease as there is more coupling to the protonic motion and a subsequently lower tunneling frequency. For the deuterated materials T_c is higher as it is harder for the deuterons to tunnel and the tunneling frequency is again lower. The effect of pressure is to decrease the O-H-O distance and it is easier to tunnel (the well narrows) and Ω_T becomes harder and underdamped (see Fig. 15(d)). The phase transition temperature from the ferroelectric to the paraelectric phase consequently decreases to lower temperatures with increasing pressure.

6 SUPERIONIC CONDUCTORS

The class of materials often called superionic conductors are solid electrolytes. They exhibit anomalously high ionic diffusion coefficients due to a large number of mobile ions moving among an even larger number of sites (83). The dwell time at a given site is $\sim 10^{-11}$ sec at high temperatures but the jump time is assumed to be considerably shorter (about the same order as the microscopic atomic motions) so that the ionic diffusion constants are comparable to those observed in liquids.

The vibrational modes yield the usual absorption in the IR spectrum while the relaxation of the diffusing ions can produce Debye-type dielectric loss peaks in the long wavelength far infrared region. Similar behavior is exhibited in the Raman spectrum except that the diffusive motions yield quasi-elastic scattering intensity centered at zero frequency. The infrared and Raman activity of the diffusive motion will depend on the local site geometry in a unit cell.

Quasi-elastic neutron scattering spectroscopy has also been applied to the study of the dynamic fluctuations in AgI (84), $RbAg_4I_5$ (85) in order to determine the motion of the Ag ions.

Light scattering in particular has been used to study ionic motion in AgI (86,87), $RbAg_4I_5$ (88), CuCl, CuBr, CuI (89), BaF_2, SrF_2, PbF_2 (90) and ZrO_2-Y_2O_3 (91) systems. The spectra in many cases have been compared in the ordered, disordered and melt phases. Table 2 shows the temperature ranges of the phase transformations in some of the cationic conductors.

Table 2. Temperature Ranges (^{o}C) of the Solid Phases of the Copper Halides and AgI.

	$F\bar{4}3m$-T_d^2 Zinc blende (γ)	$P6_3mc$-C_{6v}^4 Wurtzite (β)	$Fm3m$-O_h^5 α-phase
CuCl	< 407	407-422	-
CuBr	< 385	385-469	469-488 (M.P.)
CuI	< 369	369-407	407-600 (M.P.)
AgI	-	< 148	148-555 (M.P.) ($Im3m$-O_h^9)

The observed spectra can be analyzed by considering a system of non-interacting particles diffusing in a lattice. However, a more complete description of ionic motion includes correlations among interacting mobile atoms.

The high temperature phases of these materials exhibit high ionic conductivities but low electronic conductivities. The absence of free carrier absorption is helpful in obtaining low frequency light scattering spectra and low frequency infrared reflectivity data.

In the copper halides, the low temperature phase has the ordered zinc-blende structure. It is, however, the highly conducting "disordered" β and α phases and the melt phase that are particularly interesting.

The β phase of all three materials is the wurtzite or wurtzite-like structure where the halogen atoms form an HCP lattice. The α phases of CuI, CuBr and AgI exhibit slightly different structures. In α CuI the I atoms form an FCC lattice while in CuBr and AgI the halide ions form a BCC lattice.

In the copper halides the Cu ions predominantly occupy "tetrahedral" sites with a small occupation (~ 10%) of octahedral sites. On the other hand, in α-AgI, the cation sites are highly distorted tetrahedra which also effectively makes the cation lattice disordered in a fairly rigid HCP halide lattice (87). For example, β-AgI even in the wurtzite phase appears to possess static disorder as, in addition to the TO peak, there is a broad intense "disordered" scattering. This has been interpreted by Burns et al. (87) as due to harmonic motion about "fixed" deviations from ideal positions. The side peak at ~ 80 cm^{-1} is possibly due to Ag^+ ions in interstitial octahedral sites whereas the ~ 40 cm^{-1} may be the zone boundary TA as this persists in all phases. Simple harmonic lattice dynamics which incorporates the structural disorder that takes place at high temperatures has been used to explain the phonon data (87).

The IR and Raman data of AgI taken from Refs. (86,87) are shown in Fig. 16(a) indicating the β to α phase transition.

In order to gain a perspective of the low frequency contribution to the light scattering spectra, it is necessary to observe the complete // // and ⊥ // Raman spectra over a broad frequency range (at least a few hundred cm^{-1}). The uncorrected Raman spectra (from Ref. 89) of β (wurtzite), α (superionic) and melt phases of the copper halides and AgI are shown in Fig. 16(b). Similar results have been observed for the ionic conductors BaF_2, $SrCl_2$, PbF_2 (90) and stabilized ZrO_2 (91). All the spectra exhibit features above 100 cm^{-1} and strong temperature dependent scattering at low frequencies.

If I is the experimentally observed Stokes Raman data, then a reduced spectrum corresponding to $I_{red} = \omega I/(n+1) \propto g(\omega)$, where n

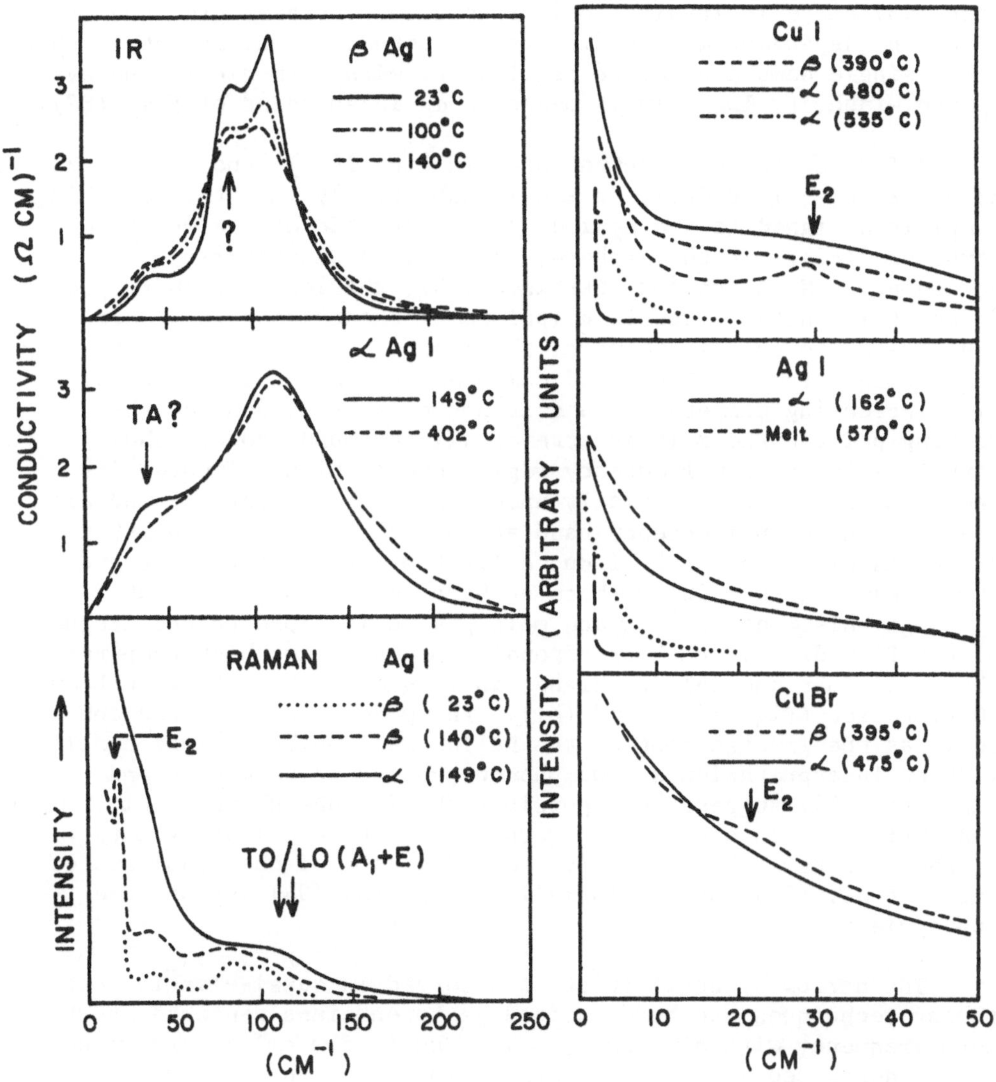

Fig. 16(a) Comparison of the conductivity (IR data) and Raman data of AgI (86,87) showing the β (wurtzite) to α (superionic) phase transition.

Fig. 16(b) Low Frequency spectra of the β, α and melt phases in the Cu halides and AgI (89) —— —— laser line, ... narrow component.

is the Bose-Einstein factor $(\exp.\frac{\hbar\omega}{kT} - 1)^{-1}$, is possibly a preferred way of presenting the data. This format was first used in displaying the Raman spectra of glasses (92) and amorphous semiconductors (93) since it involves only a density of states $g(\omega)$ and frequency dependent matrix elements. Another useful aspect is that for a single damped harmonic oscillator, with harmonic frequency, ω_o and width, Γ, I_{red} always peaks at ω_o independent of Γ/ω_o (87).

Referring to the spectra shown in Fig. 16(b), the mode at 24, 22 and 28 cm^{-1} in β-CuCl, Br and I respectively and at 17 cm^{-1} in β-AgI is assigned to the E_2 mode characteristic of the wurtzite structure (87). In the α-phase, the E_2 mode disappears, but, according to Nemanich and Mikkelsen (89), the low frequency spectra in the $/\!/ \perp$ configuration is comprised of two components. Here $\perp$ and $/\!/$ refer to polarizations of the light $\perp$ and $/\!/$ to the scattering plane respectively. The broad component decreases in intensity with increasing temperature and is not clearly evident at the melting point. The melt spectra of all compounds do not show structure in the low frequency region but at higher frequencies the spectra remain essentially unchanged in all materials and all phases. The high frequency band around 120 cm^{-1} has been interpreted as either disorder induced first order scattering or two phonon Raman scattering. These authors contend however that although this mode in β, α and melt phases is close to the frequency of the TO mode (Γ), the corresponding modes of the high temperature phases of CuBr and CuI are higher in frequency than the zinc blende TO vibration (Fig. 16(a) and (b)). The modes are polarized and scale as the inverse root of the halide mass which is more consistent with an interpretation of two phonon processes. Nemanich and Mikkelsen (89) suggest that possibly the feature shown in Fig. 16(a) and (b) is a first order Raman response due to oscillatory motions of the Cu ions. The suggestion embodies some of the arguments of Burns et al. (87) but the formalism of Geisel (94) may be more appropriate.

The narrow spectral function observed in quasi-elastic scattering techniques can be described as Lorentzians centered about zero frequency with a width of $\sim 10^{10}$ Hz (a few cm^{-1}). The results can be described in terms of correlation functions of fluctuations in the polarizability tensor, $\delta\alpha$ such that

$$S_{ij}(q,\omega) = \int dr\, dr'\, dt\, e^{iq(r-r')-i\omega t} \langle \delta\alpha_{ij}(r,t)\delta\alpha_{ij}(r',0)\rangle \; . \tag{6.1}$$

Here q and ω are the wave vector and frequency differences and ij the polarization indices of the incident and scattered light. For systems which exhibit a single relaxation mode and obey linear rate equations, Eq. (6.1) becomes

$$S_{ij}(q,\omega) = \langle|\alpha_{ij}(q)|^2\rangle\ \Gamma(q)/\pi\ (\omega^2 + \Gamma(q)^2)\ , \qquad (6.2)$$

where Γ(q) is the relaxation rate.

For liquids and gases the simplest example is density fluctuations from a dilute system of isotropic molecules or atoms. A single Lorentzian is observed in the ⊥ ⊥ (or // //) configuration with $\Gamma(q) = q^2D$ where D is the diffusion constant. If the molecules are anisotropic and their rotational motion is characterized by an angular diffusion constant, Φ, then an additional component (with a width of $q^2D + \Phi$) is observed in both ⊥ ⊥ and ⊥ // spectra. Hence, if $\Phi > q^2D$, the additional component will be independent of q or the scattering geometry. On the other hand, in a system of deformable but isotropic atoms or molecules, there is a similar additional component due to the anisotropic distortion when the atoms collide. The characteristic collision time is observed in both ⊥⊥ and ⊥ // spectra. Analogous situations often arise in the disordered phases in ionic conductors.

The samples necessary for light scattering studies in superionics must be of high purity to reduce the parasitic and the wings of the elastically scattered light which often obscures the region of interest (89). Right angle scattering is preferable but in many instances, where it is essential to contain the sample in a sealed ampoule for high temperature studies, a back-scattering geometry has to be employed. In colored samples and in those exhibiting fluorescence, the red or near infrared lines of Kr^+ laser are often preferable over those normally used with an Ar^+ laser.

Small spectral slit widths ($< 0.5\ cm^{-1}$), holographic gratings and the addition of a third monochromator allows measurements of spectra as low as 2 cm^{-1} without resorting to the use of iodine or sodium vapor filters which can only be employed with a single moded Ar^+ or a suitably tuned dye laser. The contribution due to elastically scattered light can usually be determined by comparison with the scattered light from a roughened aluminum surface.

The spectra of α-CuI (Fig. 16(b)) shows most clearly that the central component has two contributions, one with $\Gamma < 15\ cm^{-1}$ and a broader one with $\Gamma > 30\ cm^{-1}$. The narrow component was centered at zero and is described by Eq. (6.2).

The broader component and the E_2 component have been fitted (89) with a damped harmonic oscillator function of the spectral form which in the high temperature limit can be described by

$$I_{ij} = \frac{\Gamma}{(\omega^2 - \omega_o^2)^2 - \Gamma^2\omega_o^2}\ . \qquad (6.3)$$

A possibly more adequate fit for the broad component can be obtained by assuming a Lorentzian centered at zero as described by Eq. (6.2). Consequently there may be some ambiguity in the interpretation of the broad component. This treatment has been adopted in the most recent work of Nemanich (89) and also seems the most appropriate for the ZrO_2-Y_2O_3 system (91).

Application of a simple jump-diffusion model has been considered by Klein (27), Geisel (94) and Dieterich et al. (95).

The spectra are represented by sum of Lorentz oscillators with q independent widths of $\Gamma = f\,\tau^{-1}$ where τ^{-1} is the ionic hopping rate and f is a factor depending on the site symmetry.

In order to account for the depolarized scattering in α-CuI, low symmetry configurations must be involved. Nemanich et al. (89) have suggested that interactions of the mobile ions which are close enough to cause anisotropic polarizability must be included. They separate out the contributions to the polarizability and rewrite Eq. (6.1) so that

$$S_{ij}(q,\omega) = \int dr\, dr'\, e^{iq(r-r')-i\omega t}(\langle\delta\alpha_{ij}^{(o)}(r,t)\delta\alpha_{ij}^{(o)}(r',0)\rangle$$

$$+ \langle\delta\alpha_{ij}^{(1)}(r,t)\delta\alpha_{ij}^{(1)}(r',0)\rangle + 2\langle\delta\alpha_{ij}^{(o)}(r,t)\delta\alpha_{ij}^{(1)}(r',0)\rangle). \quad (6.4)$$

$\delta\alpha^{(o)}$ is the contribution due to a single particle in the lattice formed by the halogen ions and the mobile ions that are isotropically distributed. $\delta\alpha^{(1)}$ is the additional contribution caused by the mobile ions.

The first term in (6.4) is simply single ion correlations. The second term is due to correlation effects of more than one mobile ion. The third term describes the coupling between the two contributions. They point out that scattering from α-CuI and α-AgI is specially interesting to consider because $\delta\alpha_{\perp/\!/}^{(o)} = 0$ and the depolarized scattering is due to terms like

$$S_{\perp/\!/}(q,\omega) = \int dr\, dr'\, dt\, e^{iq(r-r')-i\omega t}\langle\delta\alpha_{\perp/\!/}^{(1)}(r,t)\delta\alpha_{\perp/\!/}^{(1)}(r',0)\rangle . \quad (6.5)$$

For nearest neighbor pairs, the resulting flucations transform as a non-degenerate $\Gamma_1(A_{1g})$ corresponding to total density pairs (// // scattering) and a doubly degenerate Γ_3^+ (E_g) which gives completely depolarized scattering due to differences in densities of the different types of ions (96). The decay rate of the pair fluctuation (narrow component) is approximately the sum of the hopping rates of the two atoms ($\Gamma \sim 2/\tau$) as seen in Table 3 taken from Ref. (89).

Table 3. The narrow (Γ_n) and broad (Γ_b) linewidths observed in the low frequency light scattering spectra (89) and the hopping rate (τ^{-1}) determined from conductivity measurements (89).. ℓÅ is the correlation length.

Material	Phase	T^oC	Γ_n (cm^{-1})	Γ_b (cm^{-1})	τ^{-1} (cm^{-1})	ℓÅ
CuI	Melt	650	4 ± 1	35 ± 8		
	α	535	3 ± 1	45 ± 8	1.3	2.68
	β	395	2 ± 1			2.63
CuBr	α	475	12 ± 2			1.63
	β	395	8 ± 2			2.50
CuCl	β	415	13 ± 2			~2.4
AgI	Melt	570	6 ± 1	25 ± 8	5.2	
	α	162	3 ± 1	30 ± 8	2.1	

The depolarized scattering corresponds to collision induced scattering in gases and liquid and the collision time τ^{-1} is equivalent to the formation of nearest neighbor pairs. In the melt phase Γ_n increases only slightly which shows that local atomic order and the dynamics of the fluctuations are similar in the α and melt phases.

The broad component is also depolarized and is observed in both the α and melt phases. Nemanich et al. (89) suggest that this feature is due to relaxation after or during the diffusive motions. In a jump diffusion model the ionic motion in the solid phase, the width of the broad component (~ 30 cm^{-1}) would be related to τ^{-1} and the data again suggest similar phenomena in the melt.

In the above discussion analogies have been made from charge density fluctuations, rotational diffusion and collision induced scattering. Each of these cases have been studied in a fluid systems and "superionic" conductors are often described in terms of a liquid-like picture although a correlated long-lived site model is more appropriate than a simple liquid model for highly conducting materials.

ACKNOWLEDGMENTS

Portions of this work at Northeastern University were supported in part by NSF grants #DMR75-06789, DMR79-06371, the Research Corporation and NATO Research Grant #1360. Part of the equipment used was provided under NASA Cooperative Equipment Agreement NCAw22-011-079. The facilities provided by the National Magnet Laboratory, M.I.T., the Max Planck Hochfeld Magnetlabor, Grenoble and the Institut Laue-Langevin are gratefully acknowledged. I would like to thank Gertrude Tang for typing the manuscript.

REFERENCES

(1) J. F. Scott, Rev. Mod. Phys. 46, 83 (1974).

(2) R. Blinc and B. Zeks, Soft Modes in Ferroelectric and Antiferroelectrics, North-Holland, Amsterdam (1974).

(3) B. Dorner and R. Comès in Dynamics of Solids and Liquids, S. W. Lovesey and T. Springer, ed., Springer-Verlag, Berlin-Heidelberg-New York, pp. 127-196 (1977).

(4) P. A. Fleury in Light Scattering in Solids, ed. by M. Balkanski, R. C. C. Leite and S. P. S. Porto (Flammarion, Paris, 1976), p. 747.

(5) W. D. Johnson, Jr., and I. P. Kaminow, Phys. Rev. 168, 1045 (1968).

(6) R. J. Elliot, R. T. Harley, W. Hayes and S. R. P. Smith, Proc. Roy. Soc. A238, 217 (1972).

(7) C. H. Wang and R. B. Wright, J. Chem. Phys. 57, 4401 (1972); ibid. J. Chem. Phys. 58, 1411 (1973).

(8) M. Couzi, J. B. Sokoloff and C. H. Perry, J. Chem. Phys. 58, 2965 (1973).

(9) F. Gervais and B. Pirion, Phys. Rev. B11, 3944 (1975).

(10) T. Geisel and J. Keller, J. Chem. Phys. 62, 3777 (1975).

(11) H. Buhay, J. B. Sokoloff and C. H. Perry, J. Chem. Phys. 68, 5139 (1978).

(12) J. Petzelt and V. Dvorák, J. Phys. C: Sol. State Phys. 9, 1571 (1976); ibid., J. Phys. C: Sol. State Phys. 9, 1587 (1976).

(13) A. S. Barker, Jr., and A. J. Sievers, Rev. Mod. Phys. 47, 161 (1975).

(14) R. J. Birgeneau, J. K. Kjems, G. Shirane and L. Ct. Van Uitett, Phys. Rev. B10, 2512 (1974).

(15) G. A. Gehring and K. A. Gehring, Rep. Prog. Phys. 38, 1 (1975).

(16) R. T. Harley, C. H. Perry, D. Cardarelli, and W. Richter, Proc. 6th Raman Conference, Vol. 2, ed. by E. D. Schmid et al. (Heydon) 1978, p. 340.

(17) R. T. Harley, C. H. Perry and W. Richter, J. Phys. C: Sol. State Phys. 10, L187 (1977).

(18) J. W. McPhersons and Yung-Li Wang, J. Phys. Chem. Solids 36, 493 (1975).
(19) D. Cardarelli, Ph.D. Thesis, Northeastern University 1979 (unpublished).
(20) J. D. Axe, Phys. Rev. 167, 573 (1968); M. K. Barnoski and J. M. Ballantyne, Phys. Rev. 174, 948 (1968) and references therein.
(21) E. V. Chisler and M. S. Shur, Sov. Phys.-Solid State 9, 796 (1967).
(22) J. Sakurai, R. A. Cowley and G. Dolling, J. Phys. Soc. Japan 28, 1426 (1970).
(23) Y. Yamada and T. Yamada, J. Phys. Soc. Japan 21, 2167 (1966).
(24) I. Hatta, T. Sakudo and S. Sawada, J. Phys. Soc. Japan 21, 2162 (1966).
(25) P. D. Lazay, J. H. Lunacek, N. A. Clark and G. B. Benedek, Light Scattering Spectra of Solids, ed. G. B. Wright (Springer-Verlag, N.Y.), p. 593.
(26) R. S. Katiyar, J. F. Ryan and J. F. Scott, Phys. Rev. B4, 2635 (1971).
(27) M. V. Klein in Light Scattering in Solids, ed. by M. Balkanski, R. C. C. Leite and S. P. S. Porto (Flammarion, Paris, 1976), p. 351.
(28) M. Birr, A. Heidemann and B. Alefeld, Nucl. Instr. Meth. 95, 435 (1971).
(29) P. Aldebert, A. J. Dianoux and J. P. Traverse, J. de Physique, 1979 (in press).
(30) N. Lagakos and H. Z. Cummins, Phys. Rev. B10, 1063 (1974).
(31) K. B. Lyons, R. C. Mockler and W. J. O'Sullivan, J. Phys. C6, L420 (1973).
(32) C. H. Perry and R. P. Lowndes, J. Chem. Phys. 51, 3648 (1969) and references therein.
(33) T. Nagamiya, Proc. Phys. Math. Japan 24, 137 (1942); 25, 540 (1943).
(34) A. Huller, Z. Phys. 254, 456 (1972); 270, 343 (1974); V. G. Vaks and V. E. Schneider, Phys. Stat. Sol. (a) 35, 61 (1976); A. A. Vlasova, E. E. Tornan and V. E. Schneider, Soc. Phys. Solid State 20, 497 (1978).
(35) Y. Yamada, M. Mori and Y. Noda, J. Phys. Soc. Japan 32, 1565 (1972).
(36) M. Gross and D. Gerlich, Int. Conf. on Lattice Dynamics, ed. M. Balkanski (Flammarion, Paris, 1977), p. 500.
(37) C. Garland and N. Schumaker, J. Phys. Chem. Sol. 28, 799 (1967).
(38) W. Bauhofer, L. Genzel, C. H. Perry and I. R. Jahn, Phys. Stat. Sol. (b) 63, 385 (1974); 63, 465 (1974); Proc. 12th European Congress on Molecular Spectroscopy, Strasbourg, France, July 1-4, 1975 (Elsevier Publishing Co., Amsterdam 1976), p. 263.
(39) C. H. Wang and P. Fleury, Proc. Conf. on Light Scattering Spectra of Solids, G. B. Wright ed. (Springer-Verleg, N.Y.

1969), paper H-4.
(40) C. H. Wang, Phys. Rev. Lett. 26, 122 (1971); C. H. Wang and R. B. Wright, J. Chem. Phys. 56, 2124 (1972); ibid 58, 2934 (1973).
(41) H. D. Hochheimer and W. Dultz, Sol. State Comm. 14, 475 (1974); M. L. Shand, H. D. Hochheimer and C. D. Walker, Sol. State Comm. 20, 1043 (1976).
(42) H. G. Smith, J. G. Taylor and W. Reichardt, Phys. Rev. 181, 1218 (1969).
(43) H. C. Teh and B. N. Brockhouse, Phys. Rev. B8, 3928 (1973).
(44) Y. Yamada, Y. Noda, J. D. Axe and G. Shirane, Phys. Rev. B9, 4429 (1974).
(45) W. Press, J. Eckert, D. E. Cox, C. Rotter and W. Kamitakahara, Phys. Rev. B14, 1983 (1976).
(46) C. H. Perry, I. R. Jahn, V. Wagner, W. Bauhofer, L. Genzel and J. B. Sokoloff, Int. Conf. on Lattice Dynamics (ed. M. Balkanski, Flammarion, Paris, 1977), p. 419.
(47) G. Egert, I. R. Jahn and D. Renz, Sol. State Comm. 9, 775 (1971); I. R. Jahn and E. Neumann, Sol. State Comm. 12, 721 (1973).
(48) J. M. Loveluck and J. B. Sokoloff, J. Phys. Chem. Solids 34, 869 (1973).
(49) H. C. Teh, Canadian J. of Phys. 50, 2807 (1972).
(50) C. W. Garland, R. C. Leung and C. Zahradnik, J. Chem. Phys. (to be published), and R. C. Leung, C. Zahradnik and C. W. Garland, J. Chem. Phys. (to be published).
(51) H. D. Hochheimer, M. L. Shand, C. T. Walker and A. Hüller, J. Chem. Phys. (in press).
(52) C. W. Garland, K. J. Lushington and R. C. Leung, J. Chem. Phys. (in press).
(53) J. L. Warren, Rev. Mod. Phys. 40, 38 (1968).
(54) L. Genzel, T. P. Martin and C. H. Perry, Phys. Stat. Solids (b) 62, 83 (1974).
(55) L. Genzel and W. Bauhofer, Z. Physik. B25, 13 (1976).
(56) H. Buhay (private comm.).
(57) P. A. deGennes, Sol. Stat. Comm. 1, 132 (1963).
(58) R. Brout, K. A. Müller and H. Thomas, Sol. Stat. Comm. 4, 507 (1966).
(59) M. Tokunaga, Prog. Theor. Phys. 36, 857 (1966).
(60) K. Kobayashi, J. Phys. Soc. Japan 24, 497 (1968).
(61) V. Dvorák, Czech. J. Phys. B20, 1 (1970).
(62) E. M. Brody and H. Z. Cummins, Phys. Rev. Lett. 21, 1263 (1968); 23, 1039 (1969).
(63) R. Blinc, V. Dimic, J. Petkovsek and E. Pirkmajer, Phys. Lett. A26, 8 (1967).
(64) R. L. Reese, I. J. Fritz and H. Z. Cummins, Phys. Rev. B7, 4165 (1973).
(65) G. J. Coombs and R. A. Cowley, J. Phys. C: Sol. State Phys. 6, 121 (1973); 6, 143 (1973).

(66) A. P. Young and R. J. Elliott, J. Phys. C: Sol. State Phys. 7, 2721 (1974).

(67) E. J. S. Lage and R. B. Stinchcombe, J. Phys. C: Sol. State Phys. 9, 3295 (1976); 9, 3681 (1976).

(68) I. P. Kaminow and T. C. Damen, Phys. Rev. Lett. 20, 1105 (1968).

(69) D. K. Aggrawal and C. H. Perry, 2nd Int. Conf. on Light Scattering in Solids, Paris 1971 (Flammarian), p. 429; H. Hammer, p. 245; R. S. Katiyar, J. F. Ryan and J. F. Scott, p. 436.

(70) R. S. Katiyar, J. F. Ryan and J. F. Scott, Phys. Rev. B4, 2635 (1971).

(71) C. Y. She, T. W. Broberg, L. S. Wall and D. F. Edwards, Phys. Rev. B6, 1847 (1972).

(72) R. P. Lowndes, N. E. Tornberg and R. C. Leung, Phys. Rev. B10, 911 (1974).

(73) N. Lagakos and H. Z. Cummins, Phys. Rev. B10, 1063 (1974); Phys. Rev. Lett. 34, 883 (1975).

(74) R. C. Leung, N. E. Tornberg and R. P. Lowndes, J. Phys. C: Sol. State Phys. 9, 4477 (1976).

(75) P. S. Peercy, Phys. Rev. Lett. 31, 379 (1973).

(76) P. S. Peercy and G. A. Samara, Phys. Rev. B8, 2033 (1973).

(77) R. C. Leung, N. E. Tornberg and R. P. Lowndes, J. Phys. C: Sol. State Phys. 10, 4855 (1977).

(78) J. Kweicien, Ph.D. Thesis, Northeastern University (1979).

(79) J. Kweicien, A. Widom and R. P. Lowndes (private comm. and to be published).

(80) J. F. Scott and C. M. Wilson, Sol. State Comm. 10, 597 (1972).

(81) R. Blinc in Structural Phase Transitions and Soft Modes (ed. by E. J. Samuelsen, E. Anderson and J. Feder (Universitetsforlaget, Oslo, 1971), p. 97.

(82) M. A. Moore and H. C. W. L. Williams, J. Phys. C:Sol. State Phys. 5, 3169 (1972).

(83) W. Van Gool, Fast Ion Transport in Solids (North-Holland, Amsterdam, 1973).

(84) G. Eckold, K. Funke, J. Kalus and R. Ehechner, J. Phys. Chem. Solids 37, 1097 (1976).

(85) S. M. Shapiro and D. Semmingsen and M. Salamon, Int. Conf. on Lattice Dynamics (ed. M. Balkanski, Flammarian Sciences, Paris, 1977), p. 538.

(86) G. Burns, F. H. Dacol, M. W. Shafer and R. Alben, Sol. Stat. Comm. 24, 753 (1977).

(87) G. Burns, F. H. Dacol and M. W. Shafer, Phys. Rev. B16, 1416 (1977); R. Alben and G. Burns, Phys. Rev. B16, 3746 (1977). See also Chap. VII, Int. Conf. on Lattice Dynamics, ed. M. Balkanski, Flammarian Sciences, Paris, 1977, pp. 519-553.

(88) D. Gallagher and M. V. Klein, J. Phys. C9, L687 (1976).

(89) R. J. Nemanich and J. C. Mikkelsen, Jr., Physics of Semiconductors (ed. B. L. H. Wilson) Inst. Phys. Bristol 1978, p. 661. R. J. Nemanich, R. M. Martin and J. C. Mikkelsen, Jr.,

Sol. State Comm. (to be published). R. J. Nemanich (private communication).
(90) R. T. Harley, W. Hayes, A. J. Rushworth and J. F. Ryan, Int. Conf. on Lattice Dynamics (ed. M. Balkanshi, Flammarian Sciences, Paris, 1977), p. 346.
(91) A. Feinberg and C. H. Perry (unpublished).
(92) R. Shuker and R. W. Gammon, Phys. Rev. Lett. 25, 222 (1970).
(93) J. E. Smith, Jr., M. H. Brodsky, B. L. Crowther, M. I. Nathan and A. Pinczuk, Phys. Rev. Lett. 26, 642 (1971). R. Alben, D. Weaire, J. E. Smith and M. H. Brodsky, Phys. Rev. B11, 2271 (1975).
(94) T. Geisel, Sol. State Comm. 24, 155 (1977).
(95) W. Dieterich, T. Geisel and I. Peschel, Z. Physik. B29, 5 (1978).

THE EFFECT OF MULTIPHONON PROCESSES ON VIBRATIONAL BAND SHAPES IN MOLECULAR SOLIDS

Vincenzo Schettino

Istituto di Chimica Fisica
Università di Firenze
Via G. Capponi 9 - 50121 Firenze

INTRODUCTION

The purpose of this note is to discuss multiple excitations which occur when a photon is absorbed (or scattered) and result in the excitation of two or more vibrational quanta on different molecules. The study of multiple excitations becomes of particular significance in ordered solids because of the existence of the translational symmetry. In these solids the two vibrational excitations are characterized by their frequencies ω_i and ω_j and wavevectors $\underline{k}_i$ and $\underline{k}_j$ .In the following,without much loss of generality,we shall refer only to absorption processes but the conclusions obtained will apply to scattering processes as well. For an absorption process the energy and momentum conservation requires that

$$\omega_o = \omega_i + \omega_j \tag{1}$$

and

$$\underline{q} = \underline{k}_i + \underline{k}_j + \underline{G} \tag{2}$$

where ω_0 and $\underline{q}$ are the photon frequency and wavevector and $\underline{G}$ is a reciprocal lattice vector. From the conservation law (2) it is seen that in a two-phonon band we generally have contributions from phonons of all wavevectors $\underline{k}_i$ and $\underline{k}$ within the Brillouin zone.Therefore two-phonon bands become of interest as possible sources of information on the phonon densities of states and on the intermolecular forces responsible for the phonon dispersion in crystals.

In molecular solids it is convenient to consider different types of two-phonon bands depending on the phonon modes involved in the transitions. The most commonly observed two-phonon bands are those involving two internal vibrations. If the two vibrations have an appreciable dispersion the two-phonon band will appear as a broad band whose structure is related to the combined density of two-phonon states. In general it has been found that in molecular crystals the intensity of these two-phonon bands relative to the intensity of fundamentals is very close to that found in the gas phase[1-5]. This clearly suggests that the excitation mechanisms in the two phases should be similar.

A second type of two-phonon bands of common occurrence in molecular crystals involve an internal and an external mode[7]. These transitions give rise to broad features appearing on both sides of the narrow internal fundamental bands (phonon side bands). Since the internal modes generally have a dispersion small compared to the width of the low frequency phonon band, the structure of phonon side bands will simply depend on the one-phonon density of states

The third type of two-phonon transitions involve two low frequency external vibrations and give rise to very broad bands in the low frequency region. These bands have been observed more rarely in molecular crystals[8].

A two-phonon band can be excited by two different coupling mechanisms[1,9,10]. These are depicted in Fig. 1. In mechanism a) a direct coupling of the photon to the two phonons occurs through the non linear terms of the electric dipole moment (electrical anharmonicity). This coupling mechanism involves the mutual polarization of the mo-

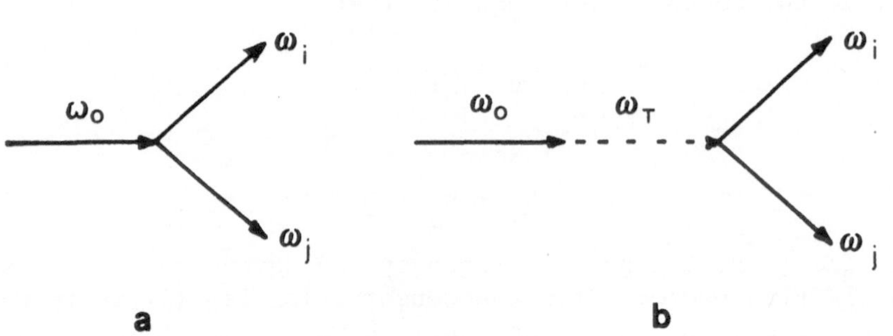

Fig. 1 - Mechanisms of two-phonon excitation.

lecules in the crystal. In mechanism b) there is an indirect coupling to the photon. A photon is absorbed and a transverse optical phonon is first produced. This intermediate virtual phonon rapidly decays into two phonons by the third order terms in the mechanical potential.

The resulting profile of a two-phonon band will depend on the one hand on combined density of states and on the other hand on the relative contribution of the two coupling mechanisms.

TWO-PHONON DENSITY OF STATES

As noted in the previous section the structure of the two-phonon bands is related to the combined density of states. This is therefore the basic quantity needed for a quantitative study of two-phonon bands. The main features can be readily obtained using the harmonic approximation and standard lattice dynamics techniques. Corrections due to anharmonic interactions can be considered in a later stage of refinement.

In the case of phonon side bands it can generally be assumed that the internal vibration has a dispersion small compared to that of the external modes. In these cases the two-phonon density of interest is coincident with the one-phonon density of states [11]. Qualitative comparison of calculated densities of states with infrared and Raman two-phonon bands shows that in the majority of cases (CO_2 [1], OCS [11], N_2O [2], naphtalene [12]) densities of states and band shapes may differ considerably. In some other cases however ((benzene [13]) the phonon side bands seem to reproduce more closely the theoretical density of states.

In the case of transitions involving two internal vibrations the calculation of the two-phonon density of states is greatly simplified. In fact each internal mode can generally be considered independently of all other crystal modes and the exciton model can be used [1]. If f and g denote two internal modes, which may be degenerate, we define one-site excitons

$$|ft;\underline{k}\rangle = L^{-1/2} \sum_n \phi^f_{nt} \exp(i\underline{k}\,(\underline{n} + \underline{t})\,)\ \Pi'\phi^o \qquad (3)$$

where L is the number of unit cells, $\underline{k}$ the wavevector, n and t define the unit cell and site in the crystal and ϕ is a harmonic oscillator wavefunction. The frequencies of the crystal states ω_α are obtained by diagonalization of the perturbation matrix H with elements

$$H^f_{ts}(\underline{k}) = \langle ft;\underline{k}|V|fs;\underline{k}\rangle \qquad (4)$$

where V is a suitable intermolecular potential which is expanded in the appropriate normal coordinates. For molecular crystals V is generally expressed in terms of atom-atom or multipole interactions.

In the study of two-phonon infrared bands involving only internal vibrations V_5 has generally been taken as a resonant dipole-dipole interaction[1-5,14-16].

In a way similar to (3) we may define two-phonon states describing the excitation of two internal vibrations at different lattice sites[1]. After appropriate symmetrization two-phonon states can be simply expressed in terms of the one-phonon states (1) as

$$|ft,gs;\underline{k},-\underline{k}> = |ft;\underline{k}>|gs;-\underline{k}> \tag{5}$$

where a normalization factor has been omitted. In the harmonic limit the elements of the two-phonon perturbation matrix are given by

$$H^{fg}_{tt',ss'}(\underline{k}) = H^{f}_{tt'}(\underline{k}) + H^{g}_{ss'}(-\underline{k}) \tag{6}$$

Therefore the two-phonon energies $\omega^{fg}_{\alpha\beta}(\underline{k})$ can be easily obtained once the one-phonon energies are known from diagonalization of (4)

$$\omega^{fg}_{\alpha\beta}(\underline{k}) = \omega^{f}_{\alpha}(\underline{k}) + \omega^{g}_{\beta}(-\underline{k}) \tag{7}$$

and the harmonic combined density of states is

$$n^{fg}(\omega) = \sum \delta(\omega - \omega^{f}_{\alpha}(\underline{k}) - \omega^{g}_{\beta}(-\underline{k})) \tag{8}$$

In several cases two-phonon bands have been reported where one of the modes has a negligible dispersion ($H_{ss'} = 0$)[1-5]. In these circumstances the two-phonon density of states simply reduces to the density of mode f. When on the contrary the two modes both have an appreciable dispersion the combined density of states may differ completely from that of the individual modes[5]. Special attention is necessary when overtone bands are treated as it has been discussed for crystalline SiF_4[14] and HCl[15].

TWO-PHONON ABSORPTION COEFFICIENT

a) Internal vibrations

The two-phonon absorption coefficient for two-phonon bands involving internal vibrations can be calculated readily using a Green function approach[17]. In fact the Green functions of interest in this case can be obtained explicitily by solution of the equation of motion. Formally it is found that the problem closely resembles that of point defects in crystals, the role of the impurity being played in the present case by the intramolecular anharmonicity.

The absorption coefficient α is given by the Fourier transform of the crystal dipole moment $\underline{M}$

$$\alpha(\omega) \sim \int dt\, e^{-i\omega t} <\underline{M}(t)\ \underline{M}(0)> \tag{9}$$

For the calculation of the contribution to the absorption coefficient due to the indirect mechanism of Fig. 1b the dipole moment is expanded to first order in the normal coordinates and the Green function of interest is then the one-phonon Green function $G(n,m)$ describing the probability that mode f is annihilated at site m and later produced at site n

$$G^{f}(n,m) = -i\theta(t) \langle\langle a_{n}^{\dagger f}(t) ; a_{m}^{f}(0) \rangle\rangle \qquad (10)$$

n and m are composite indices for the cell and site in the crystal and $a^{\dagger}$ and a are the creation and annihilation operators. It is necessary to calculate the renormalization of the Green function (10) due to the intramolecular anharmonic terms

$$H_{an} = \sum_{n} k_{ffg}\, q_{n}^{f}\, q_{n}^{f}\, q_{n}^{g} \qquad (11)$$

The second order moment contains localized and delocalized terms of the type

$$M^{(2)} = \sum_{n} M_{n}^{fg}\, q_{n}^{f}\, q_{n}^{g} + \sum_{nm} M_{nm}^{fg}\, q_{n}^{f}\, q_{m}^{g} \qquad (12)$$

Contributions to the absorption coefficient due to the localized second order moment will depend on the two-phonon Green function

$$G^{fg}\binom{nn'}{mm'} = -i\theta(t) \langle\langle a_{n}^{\dagger f}\, a_{m}^{\dagger g} ; a_{n'}^{f},\, a_{m'}^{g} \rangle\rangle \qquad (13)$$

which is renormalized by the anharmonic Hamiltonian

$$H_{A} = \sum_{n} x_{fg}\, \bar{n}_{f}(n)\, \bar{n}_{g}(n) \qquad (14)$$

where the $\bar{n}$'s are the occupation numbers.

The imaginary part of the Green function (13) gives the renormalized two-phonon density of states $\rho(\omega)$ which will depend on the value of the anharmonicity constant x_{fg}. When x_{fg} is sufficiently small the Green function (13) has poles only within the continuum of two-phonon states and the renormalized density of states is simply slightly deformed with respect to the harmonic case. The deformation increases for larger anharmonicity and if x_{fg} is sufficiently large new poles appear outside the continuum. These new poles give rise to bound states and will correspond to sharp peaks in the absorption spectra. The behaviour of the renormalized density of states as a function of the value of the anharmonicity constant is shown in Fig.2 in a typical case. According to this model it is possible to show that the two-phonon absorption coefficient due to the localized moments is simply proportional to the renormalized two-phonon density of states.

The main contribution of the non localized second order moment

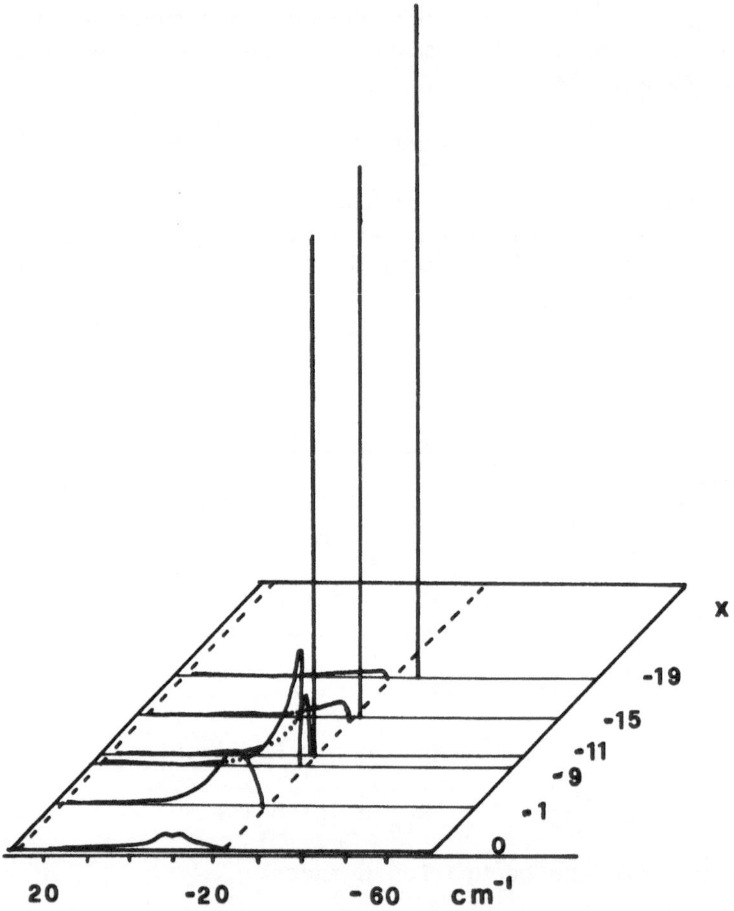

Fig. 2. Renormalized density of states for the $\upsilon_1 + \upsilon_3$ mode of CO_2. The broken lines define the limits of the two-phonon continuum.

which corresponds to the direct coupling mechanism of Fig. 1a, is given by

$$\alpha^{a}(\omega) = \sum_{\underline{k}\alpha\beta} M^{fg}_{o,n}\, e^{i\,\underline{k}\,\underline{n}}\, B^{f}_{\alpha}\, B^{g}_{\beta}\, \delta\left(\omega - \omega^{f}_{\alpha} - \omega^{g}_{\beta}\right) \tag{15}$$

where B^{f}_{α} and B^{g}_{β} are the appropriate eigenvectors obtained from the diagonalization of (4).

In the molecular crystals investigated sofar it has been found that in the majority of cases the main contribution to the two-phonon absorption coefficient is due to the indirect mechanism b .

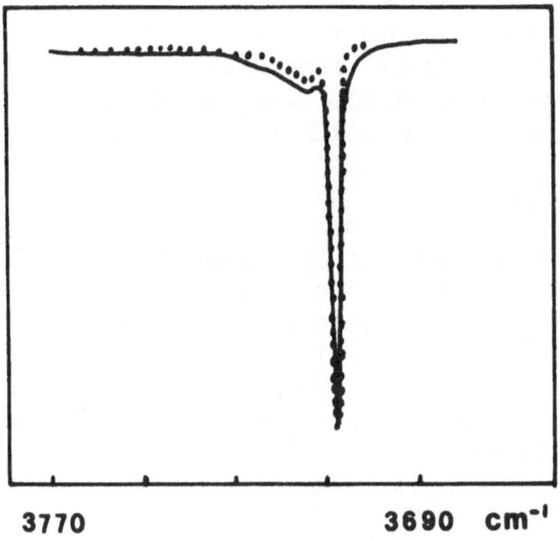

Fig. 3 - Calculated (......) and observed (———) transmission in the $\nu_1 + \nu_3$ two-phonon region of crystalline CO_2.

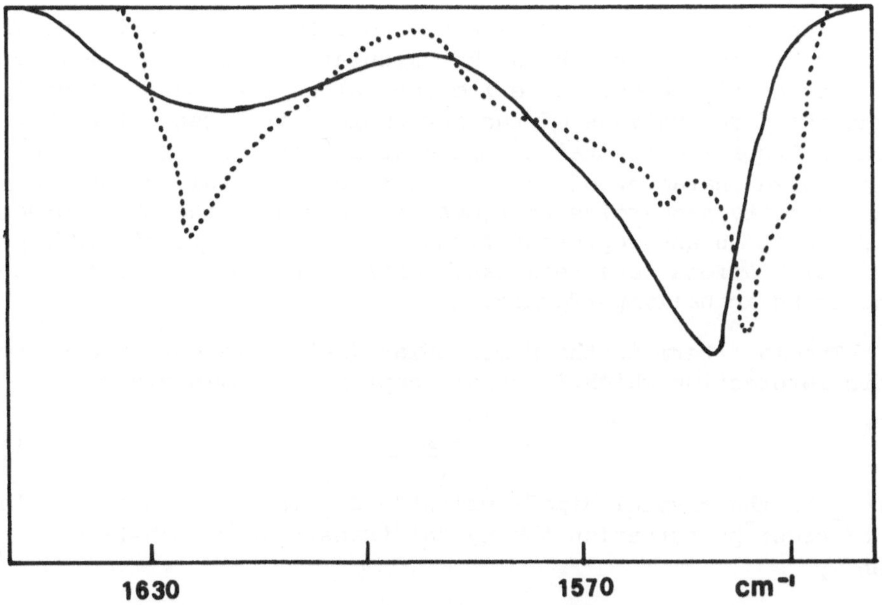

Fig. 4 - Calculated (......) and observed (———) transmission in the $\nu_2 + \nu_3$ two-phonon region of crystalline SF_6.

This compares with the results obtained for combination bands in the gas phase. Using gas phase molecular properties it has been found that a good agreement can be obtained between the calculated and observed spectra. As an example in Figs 3 and 4 results of calculation in the case of occurrence of bound states and resonances are shown. In a few cases, however, it has been found that the contribution of the direct coupling mechanism can be relevant [2,17].

b) Phonon side bands

Complete calculations of the absorption coefficient for phonon side bands have been performed only for the OCS crystal [11,17]. In this particular case it has been found that the contribution of the direct mechanism a) is very small. The contribution of the indirect mechanism b) can be evaluated using second order perturbation theory. The total Hamiltonian of the system (crystal and radiation field) can be written as

$$H = H_o + H_3 + H_i \tag{16}$$

In (16) H_o includes the harmonic crystal Hamiltonian and the free photon Hamiltonian and H_3 is the third order crystal potential

$$H_3 = \sum B(\underline{k},\underline{k}',\underline{k}'')(a_{\underline{k}}+a^{\dagger}_{-\underline{k}})(a_{\underline{k}'}+a^{\dagger}_{-\underline{k}'})(a_{\underline{k}''}+a^{\dagger}_{-\underline{k}''}) \tag{17}$$

In (17) unnecessary indices have been omitted for simplicity. The third order coefficients B can be calculated starting from an atom atom (or multipole-multipole) potential as the Fourier transforms of the third derivatives of the intermolecular potential with respect to the appropriate crystal normal coordinates. Taking advantage of the pairwise character of the interaction potential and of the various symmetry properties that must be obeyed by the third order coefficients in any representation it is possible to simplify practical computations to a reasonable extent and reduce the lattice sums to be actually performed.

The last term in the Hamiltonian (16) represents the photon-phonon interaction which, in dipole approximation, is given by

$$H_i = - \underline{M}\,\underline{E} \tag{18}$$

where $\underline{M}$ is the crystal dipole moment and $\underline{E}$ is the electric field. By second order perturbation theory the transition probability is given by

$$W_{f\leftarrow i} = \frac{2\pi}{h}\left|\frac{\langle f|H_3|m\rangle\langle m|H_i|i\rangle}{E_i - E_m} + \frac{\langle f|H_i|m\rangle\langle m|H_3|i\rangle}{E_i - E_m}\right|^2 \delta(E_i - E_f)$$

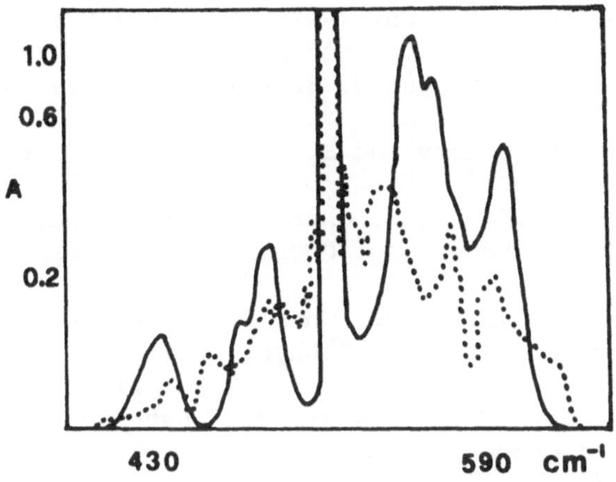

Figure 5 Calculated (••••••) and observed (———) absorbance in the ν_2 region of crystalline OCS

where the summation over the intermediate states has been omitted. Expanding the crystal dipole moment to first order, and substituting (17) and (18) in (19), explicit expressions for the absorption coefficient for sum and difference bands are obtained.

In the case of OCS crystal,application of this theory has shown that the phonon side bands differ greatly from the density of states. In particular, the high frequency part of the density of states does not contribute to the two-phonon absorption coefficient, a feature that has been qualitatively observed in many other cases. It has also been found that the third order coupling coefficients greatly depend on both the phonon energy and wave vector. On the whole, a reasonable agreement is obtained between the calculated and observed band shapes, as can be seen from Figure 5. This demonstrates that the atom-atom potential is able to account for most of the anharmonic interactions in molecular crystals.

REFERENCES

(1) D.A. Dows and V. Schettino, J.Chem.Phys., 58, 5009 (1973)
(2) V. Schettino and P.R. Salvi, Spectrochim.Acta, 31A, 399 (1975)
(3) V. Schettino and P.R. Salvi, Spectrochim.Acta, 31A, 411 (1975)
(4) R. Righini, P.R. Salvi and V. Schettino, Mol.Cryst.Liq.Cryst., 43, 223 (1977)
(5) P.R. Salvi and V. Schettino, Chem.Phys., 40, 413 (1979)
(6) V. Schettino, Chem.Phys., Letters, 18, 535 (1973)

(7) S.S. Mitra, J.Chem.Phys., 39, 3031 (1963)
(8) G. Zumofen and K. Dressler, J.Chem.Phys., 64, 5198 (1976)
(9) M. Lax and E. Burstein, Phys.Rev., 97, 39 (1955)
(10) B. Szigeti, Proc.Roy.Soc. (London), A258, 377 (1960)
(11) P.R. Salvi, R. Righini and V. Schettino, J.Phys., C 10, 11 (1977)
(12) V.L. Broude, L.M. Umarov and E. Sheka, Phys. Stat. Sol.(b) 78, 325 (1976)
(13) S.D. Colson and P.B. Klein, Chem.Phys., Letters, 34, 17 (1975)
(14) D.P. Craig and V. Schettino, Chem.Phys., Letters, 23, 315 (1973)
(15) V. Schettino and P.R. Salvi, Chem.Phys., 41, 439 (1979)
(16) F. Bogani and V. Schettino, J.Phys., C 11, 1275 (1978)
(17) F. Bogani, J.Phys., C 11, 1283 (1978); 11, 1297 (1978)
(18) R.G. Della Valle, P.F. Fracassi, V. Schettino and S. Califano, Chem.Phys., in press

NEUTRON SPECTROSCOPY STUDIES ON ROTATIONAL STATES IN MOLECULAR SOLIDS

H. Stiller

Institut für Festkörperforschung der Kernforschungsanlage Jülich, Postfach 1913, 517o Jülich, Germany

This lecture presents first a brief introduction to the information which neutron spectroscopy can provide on condensed particle systems and, secondly, some examples of information obtained by neutron spectroscopy: a) on single particle rotational motions in molecular solids, b) on collective motions in disordered orientational coordinates. By collective motions we mean motions with phase relations in the displacements of neighbouring particles; such phase relations usually are expected for ordered systems only. By single particle motions we mean motions without such phase relations.

1. NEUTRON SPECTROSCOPY

In general, neutron spectroscopic measurements are done with monochromatic radiation like Raman or Brillouin scattering experiments. The incident wave is characterized by a frequency ω_o and a wave vector $\vec{k}_o$, with $|\vec{k}_o| = 2\pi/\lambda_o$, if λ_o is the wavelength. We then measure scattered radiation, the scattered wave characterized by frequencies ω' and wave vectors $\vec{k}'$, such that the scattering itself can be characterized by a distribution of frequency changes

$$\omega = \omega_o - \omega', \tag{1a}$$

and wave vector changes (called scattering vectors)

$$\vec{Q} = \vec{k}_o - \vec{k}'. \tag{1b}$$

Figure 1 shows such a measurement schematically. The probability for scattering into a solid angle element $d\Omega$ and into a frequency interval $d\omega'$ is

$$P(\theta,\omega')d\Omega d\omega' = 1-\exp\left[-Nl \frac{d^2\sigma}{d\Omega d\omega'}\right] d\Omega d\omega' \approx N\cdot l\cdot \frac{d^2\sigma}{d\Omega d\omega'} d\Omega d\omega'.$$

N is the density of scattering centers, l the effective neutron path in the sample. $d^2\sigma/d\Omega d\omega'$ is the so-called double differential scattering cross section. It is defined as the flux $\vec{j}'$ of scattered neutrons passing with a velocity $\vec{v}'$ (corresponding to their energy $h\omega'$) through an area $r^2 d\Omega$ in distance r from the sample, relative to the incident flux $\vec{j}_o$:

$$\frac{d^2\sigma}{d\Omega d\omega'} = r^2 \frac{|\vec{j}'|}{|\vec{j}_o|} \tag{2}$$

There are two important differences with respect to the scattering of light. In the first place, the relation between wave number and frequency is different. It is $\omega_o = c\, k_o$ and $\omega' = c\, k'$ for light (and also X-rays), with c the light velocity, but

$$\omega_o = \frac{\hbar}{2m} k_o^2 \quad , \quad \omega' = \frac{\hbar}{2m} k'^2 \tag{3}$$

for neutrons, with m the neutron mass. This is the reason that with neutrons we can simultaneously measure frequency shifts $\omega = \omega_o - \omega'$ of a magnitude comparable to what one measures with Raman and Brillouin spectroscopy, and scattering vectors of a magnitude $|\vec{Q}| = (k_o^2 + k'^2 - 2 k_o k' \cos\theta)^{1/2}$ comparable to what one measures with X-rays.

The second important difference is that the neutrons are scattered not by electrons but by the atomic nuclei (by electrons only in as far as the electron spins are unpaired; this scattering shall not be considered here). The scattering by the individual nucleus is isotropic and hence can be characterized by a mere number, the scattering length a |1|. Moreover, it can be described in first Born approximation, so that the cross-section (2) becomes |1|

$$\frac{d^2\sigma}{d\Omega d\omega} = \frac{k'}{k_o} S(\vec{Q},\omega) \tag{4}$$

$$S(\vec{Q},\omega) = \sum_{no} p_{no} \sum_{n} \Big| \sum_{j} \langle n | a_j e^{i\vec{Q}\vec{r}_j} | n_o \rangle \Big|^2 \delta\left(\omega - \frac{\varepsilon_{no}-\varepsilon_n}{\hbar}\right) \tag{5}$$

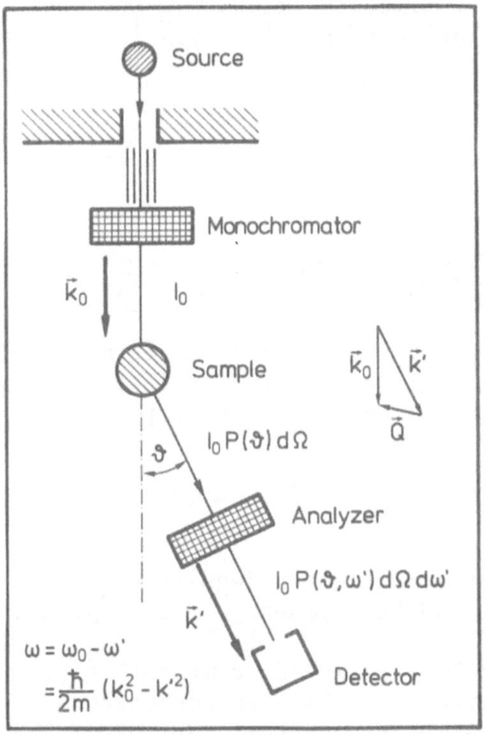

Fig. 1. Schematic representation of a neutron scattering experiment. The monochromator selects neutrons of wave vector $\vec{k}_o$, the analyser selects from the scattered radiation neutrons of wave vector $\vec{k}'$. If I_o is the neutron intensity behind the monochromator, then the intensity between sample and analyser is proportional to $I_o P(\theta) d\Omega_\theta$, and behind the analyser $I_o P(\theta, \omega') d\Omega_\theta\, d\omega'$ ($P(\theta) d\Omega_\theta$ = probability for scattering into a solid angle element $d\Omega_\theta$ at scattering angle θ; $P(\omega')d\omega'$ = probability for scattering into the frequency interval between ω' and $\omega' + d\omega'$).

$\sum_j$ goes over all scattering nuclei, $\vec{r}_j$ is the position of nucleus j; $|n\rangle$ is the final, $|n_o\rangle$ the initial state of the scattering system, p_{no} the statistical weight of $|n_o\rangle$.

To illustrate the usefulness of the symmetry-adapted harmonics as employed in the lectures by Robert Pick and Karl Michel, consider a system of molecules with $\vec{R}_m$ the position of the center of gravity of molecule m and $\vec{r}_{jm}$ the position of nucleus j in molecule m relative to $\vec{R}_m$. $\sum$ over j in equation (5) then can be written

$$\sum_m f_m(\vec{Q})e^{i\vec{Q}\vec{R}_m}$$

with

$$f_m(\vec{Q}) = \sum_{j=1}^{M} a_{jm} \, e^{i\vec{Q}\vec{r}_{jm}}$$

if M is the number of nuclei in molecule m. It has been shown |2| how much for instance a structural analysis for systems with large rotational amplitudes can be simplified if the molecular form factor $f_m(\vec{Q})$ is written

$$f_m(\vec{Q}) = \int \rho_a(\vec{r}) \, e^{i\vec{Q}\vec{r}} \, d\vec{r}$$

and then the scattering length density $\rho_a(\vec{r})$ expanded into functions adapted to the symmetry of the system.

The scattering law (5) is the simplest possible scattering law. The coupling of neutron radiation to condensed matter is simple, in particular much simpler than the coupling of light in infrared absorption as well as in Rayleigh, Brillouin and Raman scattering. As a consequence, the interpretation of neutron scattering data, especially the evaluation of intensities, often can be carried much farther than the one of optical data. Moreover, as mentioned above, in contrast to light scattering for which $\vec{Q} \simeq 0$, neutrons are scattered with finite scattering vectors $\vec{Q}$ ($10^{-3} \leq |\vec{Q}| \leq 10 Å^{-1}$). The information hence is also much more detailed. It relates to the behaviour of the scattering particles not only in time but also in space.

This is seen in the most general way by the transformation |3|:

$$\int e^{i(\omega t-\vec{Q}\vec{r})} \, S(\vec{Q},\omega) \, d\vec{Q}d\omega \equiv G_a(\vec{r},t).$$

The function $G_a(\vec{r},t)$ thus defined is with equation (5)

$$G_a(\vec{r},t) = \frac{1}{(2\pi)^{3/2}} \left\langle \sum_{j,j'} a_j a_{j'} \int d\vec{r}' \, \delta(\vec{r}+\vec{r}_j(o)-\vec{r}')\, \delta(\vec{r}'-\vec{r}_{j'}(t)) \right\rangle ;$$

$\vec{r}_j(t)$ being the position of nucleus j at time t. $G_a(\vec{r},t)$ is a density correlation function in space and time, weighted with the scattering amplitudes a_j. If there are different kinds of nuclei in the system, in a disordered arrangement, then only average amplitudes enter the scattering:

$$\langle a^2 \rangle = \frac{1}{N} \sum_j a_j^2 \quad ; \quad \langle a \rangle^2 = \frac{1}{N^2} \left(\sum_j a_j\right)^2 \qquad (6)$$

A disordered mixture of differnt kinds of nuclei can originate not only from the presence of different kinds of atoms but also from the presence of different isotopes and from random orientations of nuclear spins, since the scattering lengthes a_j depend on the relative orientation of the nuclear spin to the spin of the neutron. The scattering law then becomes |3|

$$S(\vec{Q},\omega) = a_{inc}^2 \, S_{inc}(\vec{Q},\omega) + a_{coh}^2 \, S_{coh}(\vec{Q},\omega) \qquad (7)$$

with

$$a_{inc}^2 = \langle a^2 \rangle - \langle a \rangle^2 , \quad a_{coh}^2 = \langle a \rangle^2 , \text{ and} \qquad (8)$$

$$S_{inc}(\vec{Q},\omega) = \int e^{i(\vec{Q}\vec{r}-\omega t)} \, G_s(\vec{r},t) \, d\vec{r}dt,$$
$$S_{coh}(\vec{Q},\omega) = \int e^{i(\vec{Q}\vec{r}-\omega t)} \, G(\vec{r},t) \, d\vec{r}dt. \qquad (9)$$

Now

$$G_s(\vec{r},t) = \frac{1}{(2\pi)^{3/2}} \left\langle \sum_j \int d\vec{r}' \, \delta(\vec{r}+\vec{r}_j(o)-\vec{r}') \, \delta(\vec{r}'-\vec{r}_j(t)) \right\rangle \qquad (1o)$$

is the self-correlation function, the probability that the same particle which at time 0 was at $\vec{r} = 0$ will be at $\vec{r}$ at time t, and

$$G(\vec{r},t) = G_s(\vec{r},t) + G_d(\vec{r},t), \quad \text{where} \qquad (11)$$

$$G_d(\vec{r},t) = \frac{1}{(2\pi)^{3/2}} \left\langle \sum_j \sum_{j' \neq j} \int d\vec{r}' \, \delta(\vec{r}+\vec{r}_j(o)-\vec{r}') \, \delta(\vec{r}'-\vec{r}_{j'}(t)) \right\rangle \qquad (12)$$

is the pair-correlation function, the probability that if a nucleus was at time 0 at 0 another nucleus will be at $\vec{r}$ at time t. $S_{inc}(\vec{Q},\omega)$ is called the incoherent, $S_{coh}(\vec{Q},\omega)$ is called the coherent scattering law. The reason for these names is immediately obvious if for a moment we consider $S(\vec{Q}) = \int S(\vec{Q},\omega)d\omega$. From equation (5) this is simply

$$S(Q) = |\sum_j a_j e^{i\vec{Q}\vec{r}_j}|^2$$

which becomes with equations (6) and (8)

$$S(\vec{Q}) = N a^2_{inc} + a^2_{coh} \sum_{j,j'} e^{i\vec{Q}(\vec{r}_j - \vec{r}_{j'})} .$$

Consider a very simple example |4| : a nucleus bound to a dumb-bell molecule which has a fixed orientation $\vec{s}$ and does nothing but jump from time to time instantaneously by 180°, so that the nucleus has two equilibrium sites $\vec{r}_1$ and $\vec{r}_2$ with $\vec{s} = \vec{r}_2 - \vec{r}_1$ (fig. 2).

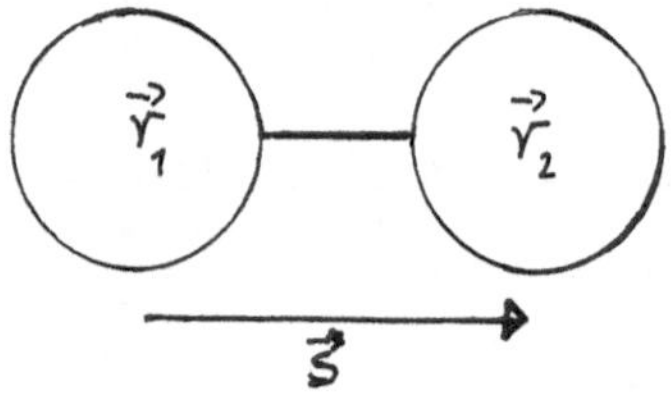

Fig. 2

Then

$$G_s(\vec{r},t) = \delta(\vec{r}-\vec{r}_1)p_1(t) + \delta(\vec{r}-\vec{r}_2)p_2(t),$$

with $p_\nu(t)$ the probability that at time t the nucleus is at site ν; $p_1 + p_2 = 1$. We have two possible initial conditions: a) $p_1(0) = 1$, b) $p_2(0) = 1$.

$$G_s(\vec{r},t) = \frac{1}{2}(G_s{}^a(\vec{r},t) + G_s{}^b(\vec{r},t)).$$

If

$$\frac{dp_1}{dt} = \frac{1}{\tau}(-p_1+p_2),$$

then

$$p_1(t) = p(t) = \frac{1}{2}(1+e^{-2t/\tau}),$$

$$G_s(\vec{r},t) = \delta(\vec{r})\frac{1}{2}\left\{1+e^{\frac{-2t}{\tau}}\right\} + \frac{1}{4}\left\{\delta(\vec{r}-\vec{s}) + \delta(\vec{r}+\vec{s})\right\}\left\{1-e^{\frac{-2t}{\tau}}\right\};$$

$$S_{inc}(\vec{Q},\omega) = \delta(\omega)\frac{1}{2}\left\{1+\cos(\vec{Q}\vec{s})\right\} + \frac{1}{2\pi}\frac{2\tau^{-1}}{\omega^2+(2\tau^{-1})^2}\left\{1-\cos(\vec{Q}\vec{s})\right\}.$$

The elastic scattering, $\delta(\omega)$, in S_{inc} originates from the fact that $G_s(\vec{r},t)$ is finite for $t \to \infty$. The term $1 + \cos(\vec{Q}\vec{s})$ represents a diffraction originating from the nucleus being distributed in the time average onto the two sites of distance $\vec{s}$.

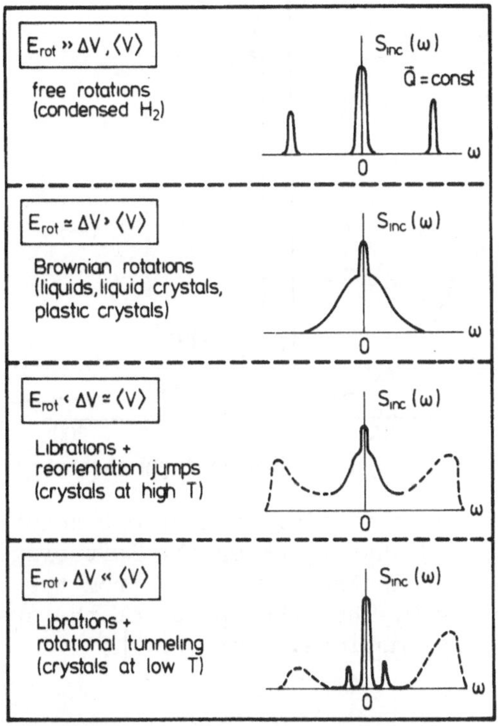

Fig. 3. Schematic representation of spectra S_{inc} $(\omega, \vec{Q} = \text{const})$ for different relative strengths of the mean orientation-dependent interaction $\langle V \rangle$ between molecules, the fluctuating part ΔV of the interaction and the kinetic rotational energy E_{rot} which the molecules may take up. The broken parts of the spectra shall indicate densities of state for librations. Contributions from translational motions are not shown. The indicated sharp elastic lines, at $h\omega = 0$, appear in solids only.

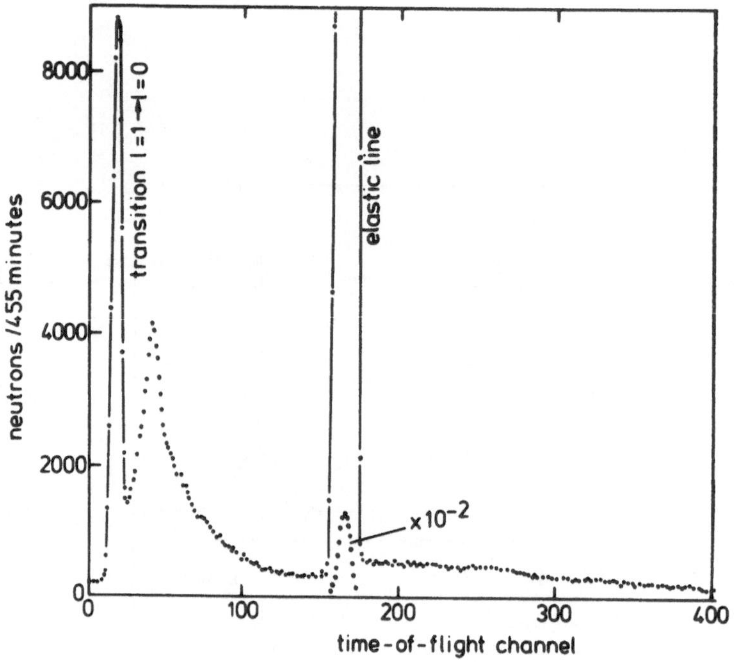

Fig. 4. Wavelength distribution of neutrons scattered from solid hydrogen. The sharp peak on the left originates from free rotations. It is the transition from the rotational state l = 1 to l = 0. The other parts of the spectra are due to translational motions. From ref. |5|.

2. OBSERVATIONS ON SINGLE PARTICLE ROTATIONS IN MOLECULAR SOLIDS

In condensed systems, the molecules move under the influence of their mutual interactions. The forces experienced by a molecules fluctuate:

$$V(t) = \langle V \rangle + \Delta V(t).$$

If we disregard coupling of internal and external coordinates, then - if V is an orientation dependent interaction - the character of the rotational motion will be governed by the strengths of $\langle V \rangle$ and ΔV relative to the kinetic rotational energies E_{rot} which the molecule may take up. Fig. 3 represents four simple cases, together with the expected corresponding spectra $S_{inc}(\omega, \vec{Q} = \text{const})$.

Free rotations (with the angular momentum a good quantum number) are a rare case in condensed systems. Only two examples, liquid and solid hydrogen and the α phase of solid methane, are known. For

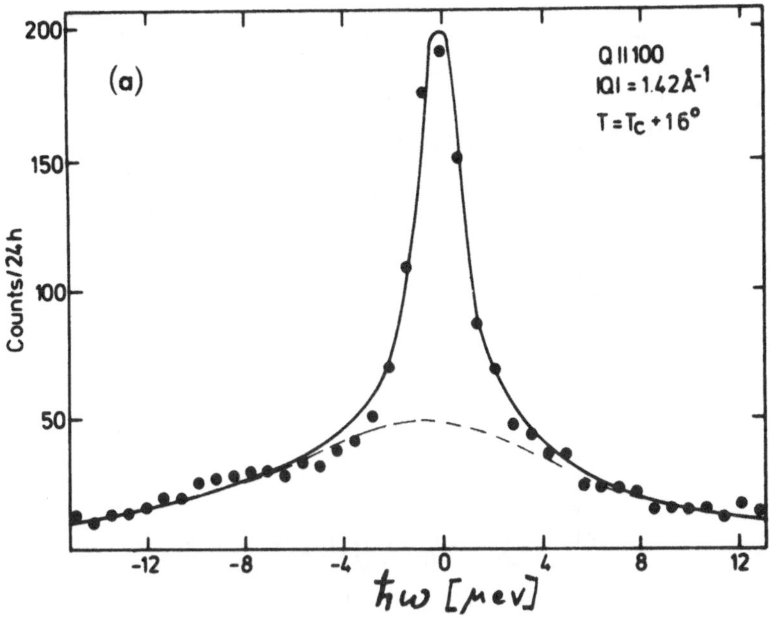

Fig. 5a. Typical quasi-elastic part of the energy distribution of neutrons scattered from β $-NH_4Cl$. The broadened component originates from reorientation jumps of the NH_4 tetrahedra ($\langle V \rangle \simeq \Delta V > E_{rot}$). There are 12o^{o} reorientations around the 3-fold as well as 9o^{o} reorientations around the 2-fold axes of the tetrahedra. The corresponding jump rates could be determined individually with measurements for different orientations of a single crystal, exploiting the $\vec{Q}$-dependences of the elastic line and of the three Lorentzians to which the jump mechanism gives rise |6|. These $\vec{Q}$-dependences are diffraction patterns as illustrated with a simple example in the preceding section. From ref. |7|.

each of the other three types of motion presented in fig. 3 very many examples have been studied neutron spectroscopically. Figures 4 to 6 show for each case one example.

The classical motions, Brownian rotation and rotation jumps, can be visualized classically also in space, of course. Thus the Q-dependence of the elastic line for instance in the case of Brownian rotation of a tetrahedron, as in the case of $(NH_4)_2$ Sn Cl_6 at high temperatures (fig. 6), is given simply by the square of a zero-order spherical Bessel function, j_o^2 (QR), with R the intramolecular N-H distance. This is the diffraction from a hollow sphere: with Brownian rotation, for $t \rightarrow \infty$, each proton is evenly distributed on the surface of such a sphere. With quantum-mechanical motions, like free rotations and rotational tunneling, the behaviour of the individual

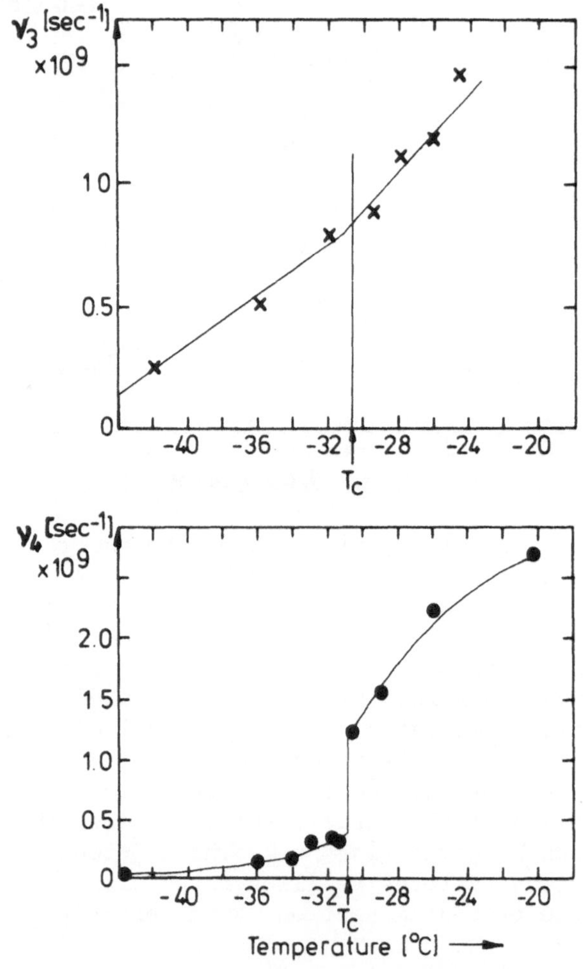

Fig. 5b. The jump rates $V_3 = \tau_{120°}^{-1}$ and $V_4 = \tau_{90°}^{-1}$, for reorientations in NH_4 Cl as a function of temperature. From ref. |7|.

nucleus in space is less easily visualized. It is determined by the rotational wavefunctions. The Q-dependence of the scattering can be particularly interesting in these cases. Especially in the case of rotational tunneling its determination would provide very detailed information on the potential |9|. However, to my knowledge, such measurements have not yet been carried out successfully.

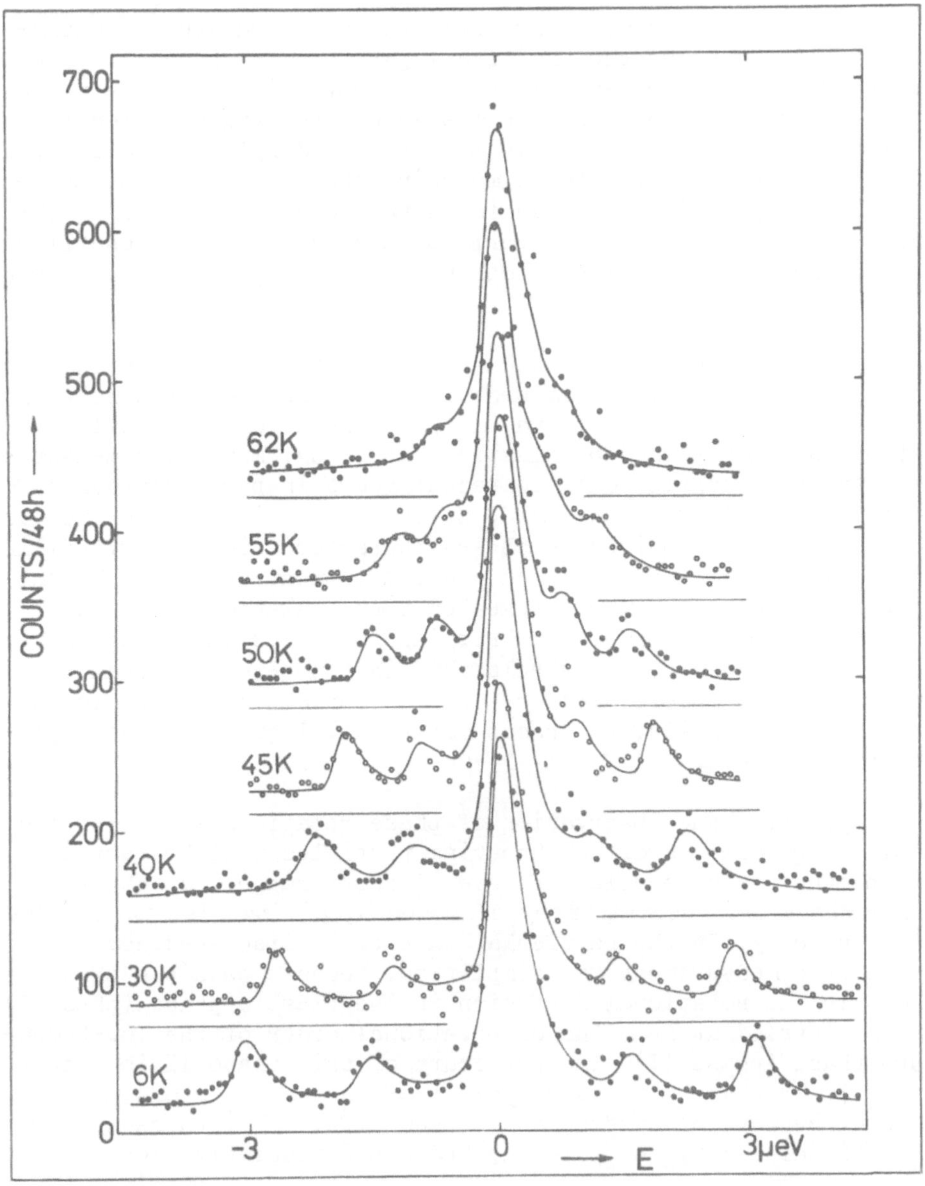

Fig. 6. Energy distribution of neutrons scattered from $(NH_4)_2 \cdot SnCl_6$. The side maxima originate from transitions within the tunnel splitting of the librational ground state of the NH_4 tetrahedra ($\langle V \rangle \gg \Delta V$, E_{rot}). With increasing temperature, that is increasing ΔV, the tunneling rotations go over into a classical Brownian rotational motion ($\langle V \rangle \lesssim \Delta V \simeq E_{rot}$). From ref. |8|.

3. COLLECTIVE MOTIONS IN DISORDERED COORDINATES

With incoherent scattering, only the behaviour (the assembly-averaged behaviour) of the individual particle is observed, independent of whether the particle actually moves individually or collectively. Periodic motions, for instance, are always collective. The displacement of one particle affects the displacements of others in a regular way; the particles move in phase. In fact, it can be very misleading to interpret an incoherent neutron scattering observation of such motions in terms of a single particle oscillation, for instance as a libration in a fixed single particle potential.

Phase relations in periodic displacements of neighbouring particles actually can be determined with coherent inelastic neutron scattering, in much the same way as phase relations in equilibrium positions are determined by elastic scattering. Many such measurements have been carried out, in recent years in particular also for librations in molecular crystals. However, in this lecture, we will not consider these motions which are characteristic for crystalline solids only. I will rather discuss a measurement which, with coherent neutron scattering, has revealed phase relations also in a non-periodic motion, namely in Brownian rotations in a plastic solid. The question of phase relations in non-periodic dynamic displacements is interesting, of course, for all kinds of disordered systems; liquids, amorphous materials, liquid crystals, plastic crystals, etc.

In general, the observation of phase relations in non-periodic dynamic displacements of neighbouring particles will be very difficult, because such correlations are of short range, so that their Fourier transforms contain many components, giving rise to diffuse scattering only. In the particular case to be discussed here, solid methane, the measurement was simplified, because possible phase relations in the rotational diffusion in the plastic phase (phase I) could be guessed at from the orientational order of the low-temperature phase (phase II). The structure of this phase II is shown in fig. 7. In phase I the arrangement of the carbon atoms, the molecular centers of masses, is the same (fcc); but there is no orientational order, the molecules rotate diffusively |10|.

The orientational order of phase II, as shown in figure 7, gives rise to superlattice reflections halfway between the reciprocal lattice points of the fcc lattice of phase I. Figure 8 shows the elastic neutron scattering observed in phase I around a point of reciprocal space where in phase II such a superlattice reflection appears. The observations made with this scattering are summarized by the formulae and numbers given in figure 8.

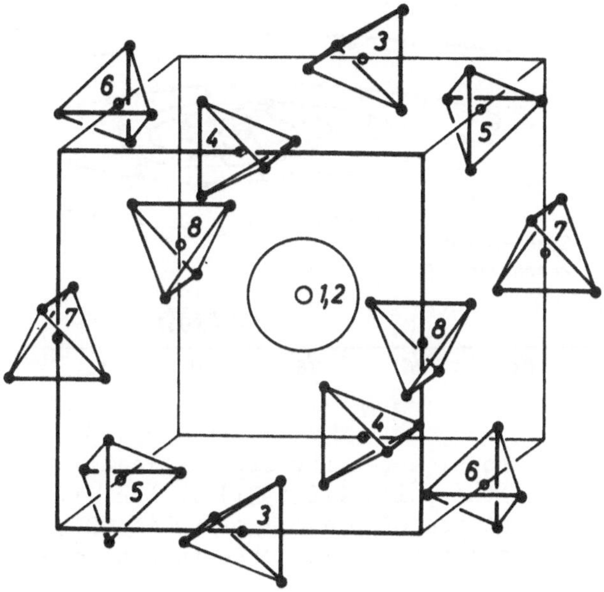

Fig. 7. Structure of phase II (27.0 > T > 22.1K) of solid CD_4. The lattice of the carbon atoms is fcc. As to the orientation of the molecules, 2 out of 8 have preferred orientations. Each of these unoriented molecules sits in a cage of orientationally ordered ones. The orientational order is of anti-ferro type. From ref. |10|.

Figure 9 shows the part around $\omega = 0$ of an energy distribution of neutrons scattered in phase I with the scattering vector $\vec{Q}$ pointing to the center of the intensity distribution as shown in figure 8. All observations made in this way are summarized in the equations and numbers given with the figure. The data show: each molecule performs rotational diffusion, nearly independent of temperature as long as the phase transition is not approached very closely; yet clusters of orientational order are formed of sizes and lifetimes which increase critically when the temperature is lowered towards the phase transition. The molecules partly rotate in phase; the phase relations approximate the orientational order of phase II.

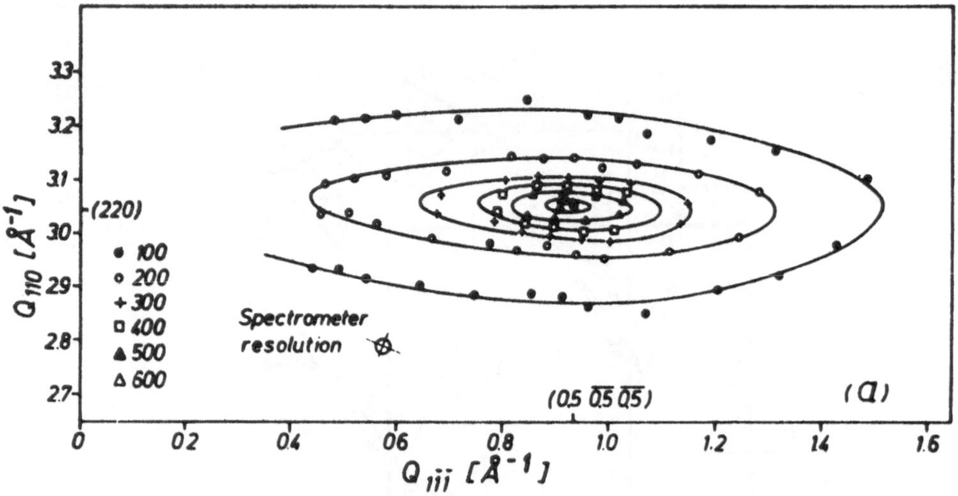

Fig. 8. Contours of equal intensity of elastically scattered neutrons around the point 1/2 $(5\bar{3}\bar{1})$ in the $[1\bar{1}2]$-zone of the reciprocal lattice of phase I of solid CD_4, 0.6° above the transition to phase II. The observations can be summarized as follows:

$$S_{coh}(\vec{Q},\omega = 0) = \frac{\chi_s}{1+\xi_{\parallel}^2 q_{\parallel}^2 + \xi_{\perp}^2 q_{\perp}^2} ;$$

$\vec{Q} = \vec{q} - \vec{g}_s$ with $q_{\parallel}$, $q_{\perp}$ the components $\parallel$ and $\perp$ to $\langle 111 \rangle$;

$\vec{g}_s$ = position of superlattice reflection in phase II.

The ξ are the corresponding correlation lengths:

$$\xi_{\parallel}, \xi_{\perp} \propto \left(\frac{T - T_c^*}{T_c^*}\right)^{-\nu} \quad \text{with } \nu = 0.5 \pm 0.03;$$

$$\chi_s \propto \left(\frac{T - T_c^*}{T_c^*}\right)^{-\gamma} \quad \text{with } \gamma = 1.13 \pm 0.10. \text{ From ref. } |11|.$$

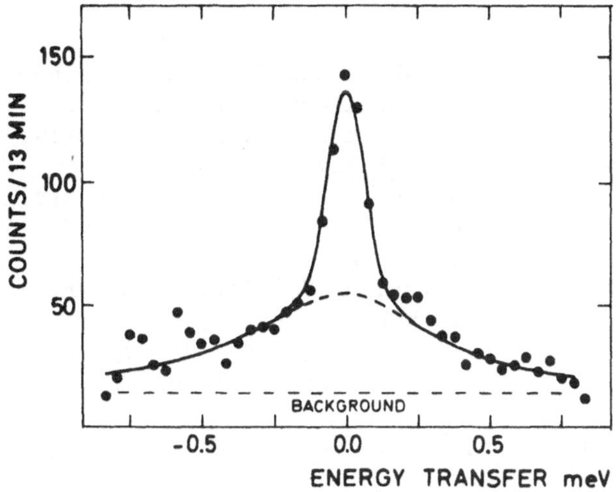

Fig. 9. Energy distribution of neutrons scatterd in phase I of solid CH_4, 11.2° above the transition to phase II, with $\vec{Q}$ pointing to the center of the contours shown in fig. 8. The energy distribution is composed of three parts: a central one with no natural width (broadened by resolution only) and two others of finite and very different widths. Of these two only one is shown in the figure, the third one being so broad that it appears as a "background" only on the scale of the figure. The unbroadened and the very broad component could be observed everywhere in reciprocal space, with Q-independent width, whereas the component of intermediate width appeared in regions of intense coherent scattering (like in fig. 8) only, with a width

$$\Gamma = \Lambda(1 + \xi_{\parallel}^2 q_{\parallel}^2 + \xi_{\perp}^2 q_{\perp}^2)$$

$\vec{q}$ as defined in the text to fig. 8. This component could be described with

$$S_{coh}(\vec{Q},\omega) = S_{coh}(\vec{Q},\omega = 0) \cdot \frac{\Gamma(Q)}{\omega^2 + \Gamma^2(\vec{Q})}$$

Obviously, the unbroadened and the very broad component in the energy distribution originate from incoherent scattering. The width of the very broad component in fact agrees with the single particle rotational diffusion constant. The width of the intermediate (the coherent) component is the reciprocal lifetime of clusters of a local orientational order corresponding to the orientational structure of phase II. From ref. |12| .

REFERENCES

/1/ see for instance W. Marshall and S. Lovesey: "Theory of Thermal Neutron Scattering", Clarendon Press, Oxford 1971
/2/ W. Press and A. Hüller: Acta Cryst. A 29, 252 (1973)
/3/ L. Van Hove: Phys. Rev. 95, 249 (1954)
/4/ see for instance T. Springer: "Quasielastic Neutron Scattering for the Investigation of Diffusive Motions in Solids and Liquids", Springer Verlag 1972
/5/ A. Bickermann, H. Spitzer, H. Stiller, H. Meyer, R. Lechner and F. Volino: Z. Physik B 31, 345 (1978)
/6/ K.H. Michel: J. Chem. Phys. 58, 1143 (1973)
/7/ J. Töpler, D. Richter and T. Springer: J. Chem Phys. 69, 317o (1978)
/8/ W. Prager, W. Press, B. Alefeld and A. Hüller: J. Chem Phys. 67, 5126 (1977)
/9/ A. Hüller: Phys. Rev. B 16, 1844 (1977)
/1o/ W. Press: J. Chem. Phys. 56, 2597 (1972)
/11/ A. Hüller and W. Press: Phys. Rev. L. 29, 266 (1972)
/12/ W. Press, A. Hüller, H. Stiller, W. Stirling and R. Currat: Phys. Rev. L. 32, 1354 (1974)

ROTATION-TRANSLATION COUPLING IN COLLECTIVE MODES OF MOLECULAR SOLIDS AND LIQUIDS

G. Jacucci and M.L. Klein*

Dipartimento di Fisica, Libera Università degli
Studi di Trento, 38050 Povo (Trento), Italy
*Chemistry Division, National Research Council of Canada,
Ottawa K1A OR6, Canada

1 INTRODUCTION

This article is concerned with rotation-translation coupling in molecular solids and liquids. We first review some of the computer simulation work carried out on solid and liquid N_2 using atom-atom potentials (1-4). We show that the collective modes, particularly the shear modes, can be strongly influenced by molecular reorientation. Next we recall briefly some experimental data on depolarized light scattering from supercooled fluids which show evidence of transverse hydrodynamic modes (5,6). The coupling of shear waves to reorientational motion in a molecular dynamics (MD) computer simulation study of liquid H_2O is then considered (7). Finally, by analogy with the dynamics of a two-component ionic fluid (8) a method is suggested for studying the coupling of shear waves to molecular reorientation in a more systematic fashion.

2 CONDENSED NITROGEN

2.1 Liquid-N_2

This is the simplest molecular fluid for which extensive MD calculations have been reported (1,2). The calculations utilized atom-atom potentials and were carried out for a system of 500 molecules under conditions close to the triple point. Information on molecular reorientational motion is contained in the correlation function appropriate to depolarized light scattering:

$$P(t) = \sum_{i,j} \langle P_2(\vec{u}_i(o) \cdot \vec{u}_j(t)) \rangle /N$$

where $\vec{u}_i$ is a unit vector along the molecular symmetry axis of molecule i and $P_2(x) = \frac{1}{2}(3x^2 - 1)$. P(t) can be partitioned into a self term (i = j) and a distinct part $P_d(i \neq j)$. The results of the MD calculation (2) are shown in Figure 1.

We see from this figure that the depolarized light scattering from liquid N_2 will be dominated by the self part and collective effects as embodied in P_d will be very difficult to detect.

The collective modes were studied via the dynamical structure factor $S(Q,\omega)$ which is the Fourier transform of the correlation function of the density operator:

$$F(\vec{Q},t) = \langle \rho(\vec{Q},o)\rho(-\vec{Q},t) \rangle /N$$

with:

$$\rho(\vec{Q},t) = \sum_j \exp(i\vec{Q} \cdot \vec{r}_j(t))$$

It should be stressed that here the summation runs over atoms (1). Figure 2 shows $S(\vec{Q},\omega)$ for $Q \sim 0.2$ Å^{-1}. The spectrum shows the typical Rayleigh-Brillouin structure, the longitudinal sound wave being reasonably well resolved. For this value of Q the spectrum is almost indistinguishable from that of the CM motion. The overall conclusion from the liquid studies (1,2) was that specific effects due to the anisotropic intermolecular forces were difficult to detect. The system behaved largely as if the CM motion and molecular reorientation were independent.

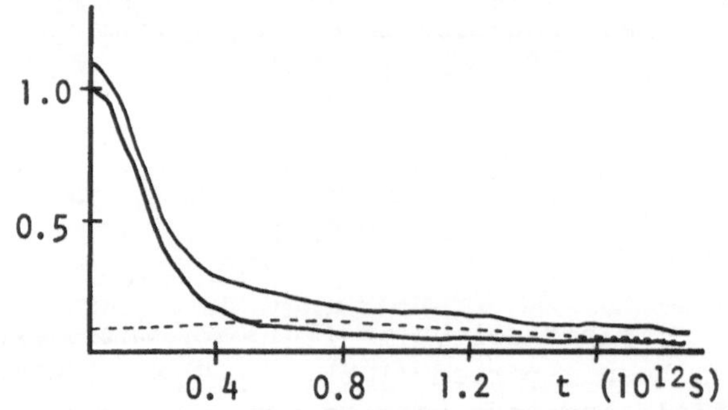

Figure 1 Orientational correlation function P(t) (upper curve) and its self part (lower curve) $P_2(t)$. The dashed curve is the distinct part

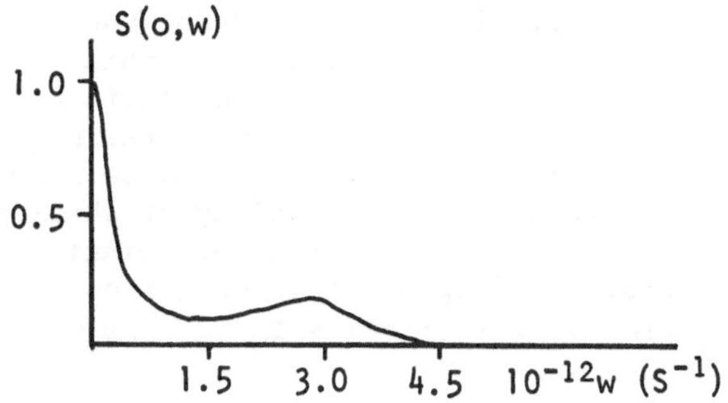

Figure 2 Dynamical structure factor (normalized) for $Q \sim 0.2$ Å^{-1}

2.2 Solid α-N_2

In solid α-N_2 at sufficiently low temperatures the molecules execute small amplitude oscillations about well defined equilibrium positions, the molecular CM forming an fcc lattice. As the crystal is heated up reorientational motion sets in. The effect of this motion on collective modes in the solid was studied by MD (3) and a dramatic softening of the shear modes was found. Figure 3 shows $S(\vec{Q},\omega)$ for a transverse acoustic mode propagating in the <001> direction of this solid both above and below the orientational order→disorder transition (T_d) calculated by MD (3). The arrows in the figure indicate the results of a lattice dynamical calculation using the same state conditions and intermolecular potential. At low temperature the agreement between lattice

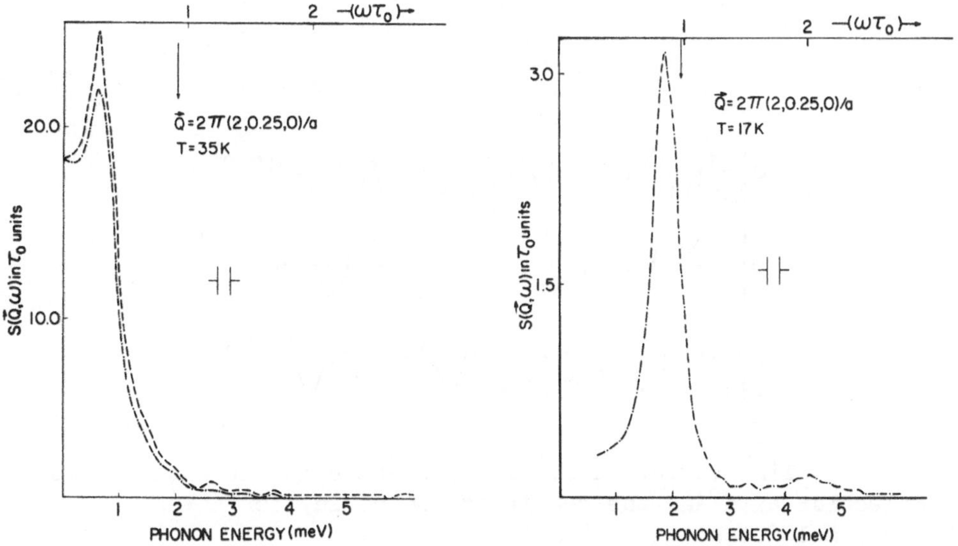

Figure 3 $S(\vec{Q},\omega)$ data for $\vec{Q} = 2\pi(2,0.25,0)/a$ for solid α-N_2

dynamics and the MD results is not too bad but once molecular reorientation sets in the enhanced softening of this shear mode is readily seen. For longitudinal modes the lattice dynamical calculations are in reasonable agreement with the MD data both above and below T_d. Accordingly, we interpret these results as demonstrating the importance of the coupling of shear modes to reorientational motion. The ratio of longitudinal sound speed to that of the transverse sound speed for propagation in the <001> direction changes from $v_\ell/v_t \sim 2$ for $T < T_d$ to $v_\ell/v_t \sim 5$ when $T > T_d$.

2.3 Solid β-N_2

In solid nitrogen when the orientational order is lost ($T > T_d$) the solid transforms into a structure in which the CM occupy an hcp lattice. We will therefore now discuss the MD results for this phase (4). The MD study carried out for 288 molecules, under conditions that correspond roughly to $T \sim 48$ K, $V \sim 26$ cm^{3}mol^{-1}, indicated that the single molecule motion in β-N_2 is rather complicated. Figure 4 shows the reorientational motion of three randomly chosen molecules. It can be seen that the molecules undergo frequent rapid reorientation, often through 180°. There seems to be no evidence of any coherent precession about the c-axis. The time covered in this figure is about 5000 MD steps or about 25 ps. $P_2(t)$, the single molecule reorientational auto correlation function of $P_2(\cos\theta)$ and its associated power spectrum $S_2(\omega)$ are shown in Figure 5. The value of $\tau_2 = \int_0^\infty P_2(t)dt$ was found to be about 10% larger than the experimental value derived from nuclear spin lattice relaxation data. For future reference we note that $S_2(\omega)$ has decayed almost to zero when $\omega\tau_0 \sim 3$ ($\tau_0 \sim 16$ cm^{-1}, 0.50 THz, or 2 MeV).

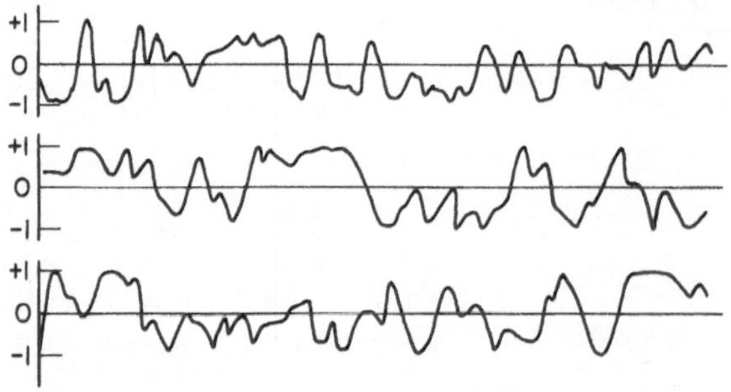

Figure 4 Time variation of cos θ, where θ is the angle between the molecular axis and the c-axis of the solid, for three arbitrary molecules

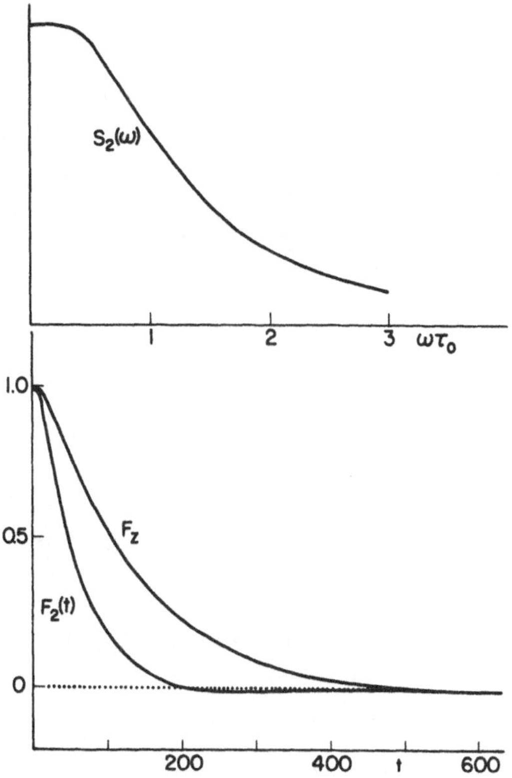

Figure 5 Orientational autocorrelation function of $P_2(\cos\theta)$ and its associated power spectrum

Some examples of collective modes in β-N_2 are shown in Figures 6 and 7. These figures display the dynamical structure factor $S(\vec{Q},\omega)$ for both longitudinal and transverse modes propagating along the crystal c-axis. Phonon energies calculated from a sphericalized intermolecular potential (and shown as vertical lines in the figures) agree reasonably well with longitudinal phonons but disagree substantially with the transverse modes. This behavior is similar to that reported for α-N_2 (when $T > T_d$) in Section 2.2. In the β-phase the ratio of the longitudinal to transverse sound speed derived from the smallest $\vec{Q}$-response functions is rather large ($v_\ell/v_t \sim 3$).

Comparison of $S_2(\omega)$ shown in Figure 5 with the (005) large Q crystal response shown in Figure 6 reveals the dominance of single molecule reorientational motion. However, since the spectrum is far from zero at $\omega\tau_0 \sim 3$, collective effects must make some contribution also.

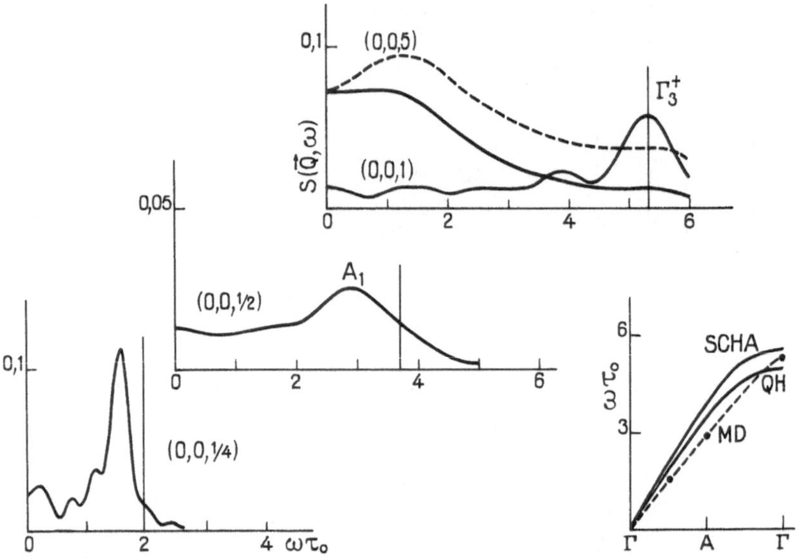

Figure 6 The dynamic structure factor $S(\vec{Q},1)$ for longitudinal phonons propagating along the c-axis of β-N_2.

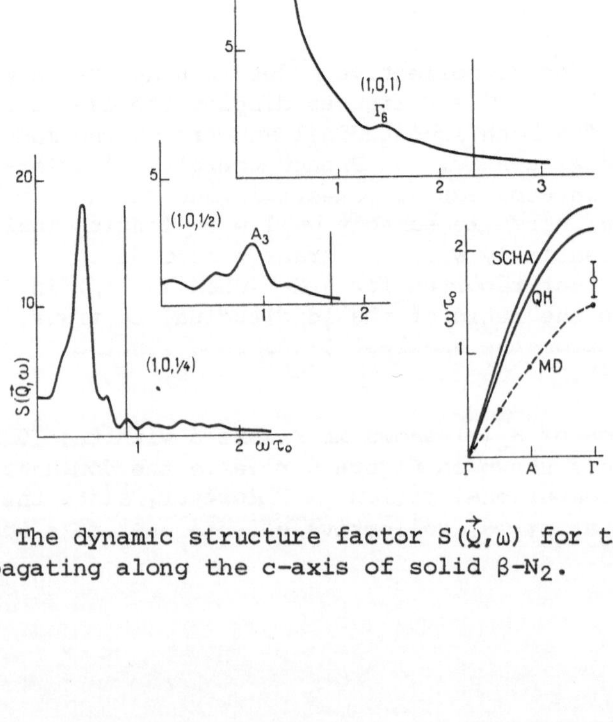

Figure 7 The dynamic structure factor $S(\vec{Q},\omega)$ for transverse phonons propagating along the c-axis of solid β-N_2.

3 ROTATION-TRANSLATION COUPLING

3.1 Historical Review

The coupling of shear waves to molecular reorientations was first observed in the quasi-elastic depolarized light scattering from high viscosity fluids. Fabelinskii et al (5) noted that upon supercooling certain molecular liquids showed well defined peaks corresponding to transverse hydrodynamic modes. Similar results have also been observed in simpler molecular fluids such as CS_2, by Stoicheff and his collaborators (6).

This coupling was also revealed directly using computer simulation of a model system (7). The system chosen was liquid H_2O. This seemed to be appropriate because this is a highly structured fluid for which a reasonable model exists, namely the ST2 model derived and extensively used by Rahman and Stillinger (9).

3.2 Reorientation Response to an External Shear

Here we describe in some detail the simulation experiment (7) because of its pedagogical value. Consider first a perturbation having the character of an acoustic mode which is applied to the MD system. In practice this can be achieved by applying an external force to the centre of gravity of each molecule, of the type:

$$F_y(x_i) = F_o \cos(2\pi x_i/L)$$

where x_i is the x-component of the position vector of particle i

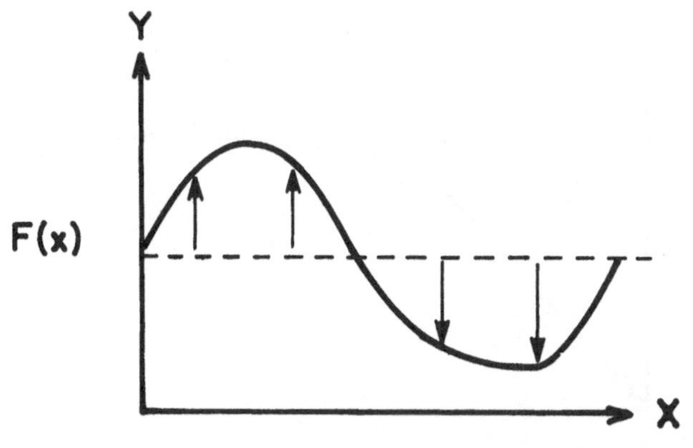

Figure 8 Spatial pattern of the applied force

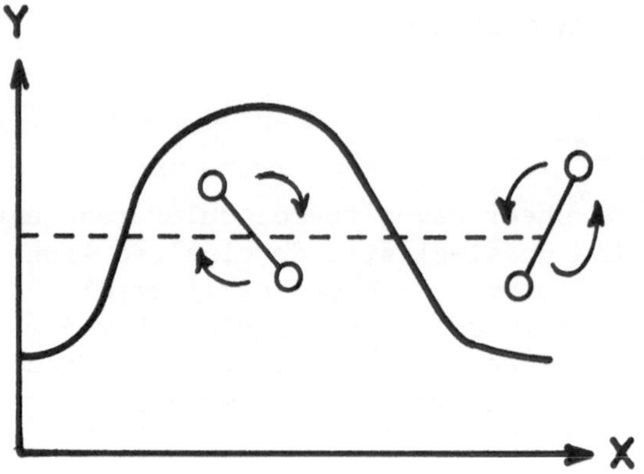

Figure 9 Spatial pattern of the transverse current response

and L is the box length. This force which is applied, after some suitable elapse time t, to a previously equilibrated system, is held constant while the shear current response is evaluated.

This response has the form of a transverse current wave whose time dependent amplitude is given by the stationary wave:

$$S_y(t) = \int dx \Delta v_y(x) \rho(x) \cos(2\pi x/L)$$

Here the particle density:

$$\rho(x,t) = \sum_i \delta(x - x_i(t)\cdot)$$

and Δv is the incremental velocity of the centre of gravity arising

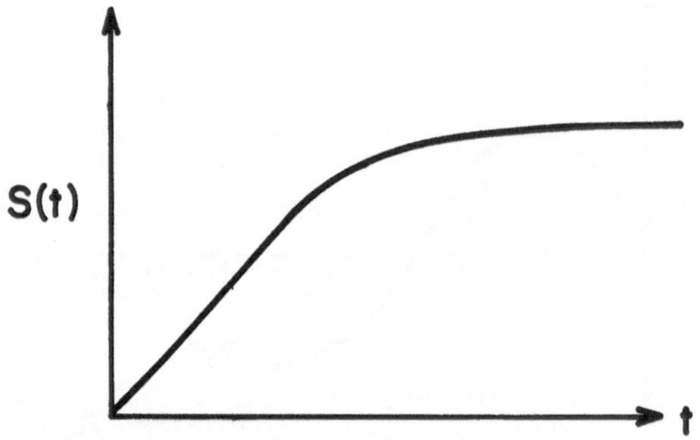

Figure 10 Wave amplitude of the transverse current response

from the application of the external shear. The system responds to the external wave by generating a transverse acoustic current where spatial pattern is a standing wave polarized in the y direction with wave vector in the x direction, as for the applied force (see Figure 8).

The time dependence of S(t) is sketched in Figure 10. Note the linear rise as a function of t at short times and the asymptotic plateau. This short time linear behaviour is nothing more than the free particle response dictated by the ratio of the magnitude of the applied force divided by the appropriate molecular mass. At long times the asymptotic value of this current can be used to evaluate the viscosity of the fluid (10,11). In addition to this response of the CM of the molecules, there is also a reorientational response of the molecules (see Figure 9). A molecule sitting at a node of the shear wave is rotated, while those sitting at the maximum amplitude of the shear wave do not rotate at all. This rotation occurs about an axis perpendicular to the plane of the applied shear wave (in this case the z axis). The reorientational response is nothing more than the z component of the angular momentum density of the molecules and appears as a standing wave polarized in the y direction with wave vector along x but phase shifted by $\pi/2$ with respect to the applied shear force. The response in this case is a wave of amplitude:

$$L_z(t) = \int dx \Delta L_z(x) \rho(x) \sin(2\pi x/L)$$

It should be noted that (Figure 11) $L_z(t)$ at short times is quadratic in time. This arises because the torque exerted on the molecules grows with the velocity gradient, whose own rise is

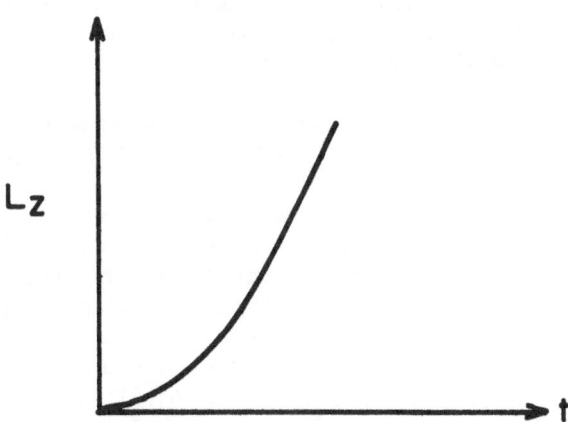

Figure 11 Angular momentum density response when F is applied to CM

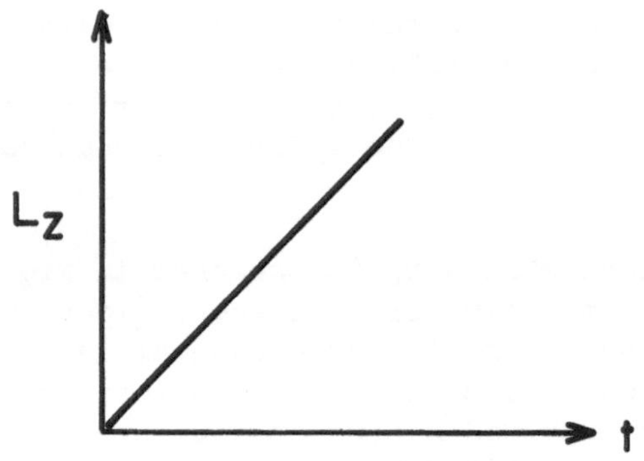

Figure 12 Angular momentum density response when F is applied to atoms

linear at short time. We stress that this phenomenon only exists because of the anisotropy of the intermolecular potential.

3.3 Coupled Currents

If we had applied the shear force to the individual atoms, rather than the CM, we could find a linear short time rise in the L_z response (see Figure 12), because in effect we are now applying a torque directly to each molecule in addition to the CM displacement. Of course, the moment of inertia will now come into play and the molecules will respond to the applied torque accordingly. If the moment of inertia is small, as in H_2O, then rotation of the molecule will be larger than is needed to minimize "internal friction". It is then clear that in order to create a shear current in the liquid (with the equilibrium value of this "internal strain") causing the minimum reaction for the orientational degrees of freedom, one will have to apply simultaneously a shearing force to the CM and a suitably chosen torque.

We suggest that in general the dynamics of molecular systems should be analyzed by looking at the appropriate linear combinations of these two types of current (S and L) of the same $\vec{Q}$ and polarization with coefficients that are also $\vec{Q}$-dependent. There will, of course, be two types of current, one predominantly translational in character, the other rotational (at least at long wavelength). Such ideas are analogous to those adopted in the analysis of collective modes in molten salts (8,11). In the latter case, starting from current densities $J+(\vec{Q},t)$ and $J-(\vec{Q},t)$, one constructs four correlation functions $C_{\alpha\beta}(\vec{Q},t)$, where α and β label + or − ions. One then obtains a matrix of the second moments of the power spectra of these currents $C_{\alpha\beta}(\vec{Q},\omega)$, which in general is not diagonal. This means that for a given Q value, the correlation function will contain optic and acoustic features.

For each Q value a suitable choice of two linear combinations of $J+(\vec{Q},t)$ and $J-(\vec{Q},t)$ will yield power spectra where second moment matrix is now diagonal. This particular choice separates out a pair of uncoupled modes for all wavelengths. For long wavelengths these go over to the usual optic and acoustic modes. The diagonal elements of this new matrix are now better separated with the "optic" mode lying higher in frequency and the "acoustic" lower than before this "diagonalization" is performed.

In the experiment on H_2O mentioned earlier, such a procedure was found to lower the second moment of the transverse current spectrum by about a factor of 2 with respect to that derived from the CM motion alone. Clearly a more systematic investigation of this type could be useful. Moreover, in our opinion, this type of analysis should be explored further and applied to the analysis of molecular motion in condensed phases. Suitable candidates for study abound. For example, those crystals showing orientationally disordered (plastic crystal) phases such as CH_4, CCl_4, DCl, CO etc would appear to provide good systems on which to explore these ideas further as well as the corresponding liquids.

REFERENCES

(1) J.J. Weis and D. Levesque, Phys. Rev., A13, 450 (1976)

(2) D. Levesque and J.J. Weis, Phys. Rev., A12, 2584 (1975)

(3) J.J. Weis and M.L. Klein, J. Chem. Phys., 63, 2869 (1975)

(4) M.L. Klein and J.J. Weis, J. Chem. Phys., 67, 217 (1979)

(5) V.S. Starunov, E.V. Tiganov and I.L. Fabelinskii, Zh. Eksp. Teor. Fiz. Pis'ma Red., 5, 317 (1967) (JETP Lett., 5, 260 (1967))

(6) G.I.A. Stegeman and B.P. Stoicheff, Phys. Rev. Letters, 21, 202 (1968); Phys. Rev., A7, 1160 (1973); E.D. Enright and B.P. Stoicheff, J. Chem. Phys., 60, 2536 (1974); ibid 64, 3658 (1976)

(7) G. Jacucci and A. Rahman, Workshop on Simulation of Long Time Phenomena, CECAM (1974)

(8) C. Abramo, M. Parinello and M. Tosi, J. Phys., C7, 4201 (1974) M. Parinello and M. Tosi, Rev. Nuovo Cimento, 2 (1979)

(9) A. Rahman and F.H. Stillinger, J. Chem. Phys., 55, 3336 (1971)

(10) E.M. Gosling, I.R. McDonald and K. Singer, Mol. Phys., 26, 1475 (1973)

(11) G. Cicotti, G. Jacucci and I.R. McDonald, Phys. Rev., A13, 426 (1976)

PARTICIPANTS

Directors

S. Bratos	Laboratoire de Physique Théorique des Liquides, Université Pierre et Marie Curie, Paris, France
R.M. Pick	Département de Recherches Physiques, Université P. et M. Curie, Paris, France

Organizing Committee

G. Birnbaum	National Bureau of Standards, Washington, USA
J.P. Boon	Service de Chimie Physique 2, Université de Bruxelles, Belgium
S. Bratos	Laboratoire de Physique Théorique des Liquides, Université P. et M. Curie, Paris, France
C. Brot	Laboratoire de Physique de la Matière condensée, Université de Nice, Nice, France
S. Califano	Laboratorio di Spettroscopia Molecolare, Università di Firenze, Florence, Italy
J. Lascombe	Laboratoire de Spectroscopie IR, Université de Bordeaux, Talence, France
R.M. Pick	Département de Recherches Physiques, Université P. et M. Curie, Paris, France

Lecturers and Round Table leaders

C.A. Angell	Purdue University, Lafayette, USA
G. Birnbaum	National Bureau of Standards, Washington, USA
S. Bratos	Laboratoire de Physique Théorique des Liquides, Université P. et M. Curie, Paris, France
D. Buckingham	University Chemical Laboratory, Cambridge, UK
S. Califano	Laboratorio di Spettroscopia Molecolare, Università di Firenze, Florence, Italy
G. Jacucci	Dipartemento di Fisica, Università degli Studi di Trento, Povo, Italy
J. Janik	Instytut Fizyki Jadroweg W. Krakovie, Krakow, Poland
J. Lascombe	Laboratoire de Spectroscopie IR, Université de Bordeaux, Talence, France
A. Laubereau	University of Bayreuth, Bayreuth, FRG
K. Michel	Theoretische Physik, Universität des Saarlandes, Saarbrücken, FRG
D. Oxtoby	James Franck Institute, University of Chicago, Chicago, USA
C.H. Perry	Northeastern University, Boston, USA
R.M. Pick	Département de Recherches Physiques, Université P. et M. Curie, Paris, France
G. Robertson	Department of Physics, University of Cape Town, Rondebosch, South Africa
V. Schettino	Università di Firenze, Istituto di Chimica Fisica, Florence, Italy
D. Sette	Istituto di Fisica, Facoltà di Ingegneria, Università di Roma, Roma, Italy
W. Steele	Department of Chemistry, The Pennsylvania State University, University Park, USA

H. Stiller	Institut für Festkörperforschung des Kernforschungsanlage Jülich, Jülich, FRG
M.F. Thorpe	Department of Physics, Michigan State University, East Lansing, USA
J. Vincent-Geisse	Laboratoire de Spectroscopie IR, Université Pierre et Marie Curie, Paris, France
R. Zwanzig	University of Maryland, Institute for Physical Science and Technology, College Park, USA

Participants

M. Alves Marques	Centro de Fisica da Materia Condensada, Instituto de Fisica e Matematica, Lisbon, Portugal
M. Allavena	Centre de Mécanique Ondulatoire Appliquée, Paris, France
F. Baglin	Department of Chemistry, University of Nevada, Reno, USA
R. Bailey	Department of Chemistry, University of Strathclyde, Glasgow, UK
J.L. Beaudouin	Laboratoire de Recherches Optiques, Université de Reims, France
A. Behrens	Fachbereich Physikalische Chemie, Universität Marburg, Marburg, FRG
S. Besnainou	Centre de Mécanique Ondulatoire Appliquée, Paris, France
D. Bougeard	Physikalische und Theoretische Chemie, Universität Essen, Essen, FRG
S. Brueck	Lincoln Laboratory, M.I.T., Lexington, USA
B. Bussian	Anorganisch-Chemisches Institut der Universität Heidelberg, Heidelberg, FRG
V. Chandrasekharan	Laboratoire des Interactions Moléculaires et des Hautes Pressions, Université Paris-Nord, Villetaneuse, France

S.H. Chen	Nuclear Engineering Department, M.I.T., Cambridge, USA
J. Chesnoy	Laboratoire d'Optique Quantique, Ecole Polytechnique, Palaiseau, France
J. Clouter	Department of Physics, University of Toronto, Toronto, Canada
L. Colombo	Institute "Rudjer Boskovic", Zagreb, Yugoslavia
L. Dalmolen	Solid State Physics Laboratory, Groningen, Netherlands
S. Dattagupta	Hahn Meitner Institut, Berlin, FRG
G. Döge	Institut für Physikalische Chemie der Technischen Universität Braunschweig, Braunschweig, FRG
T. Dorfmüller	Fakultät für Chemie, Universität Bielefeld, Bielefeld, FRG
M. Dubs	Institute for Physical Chemistry, ETH-Zurich, Zurich, Switzerland
W. Dultz	Fachbereich Physik, Universität Regensburg, Regensburg, FRG
C. Dreyfus	Département de Recherches Physiques, Université P. et M. Curie, Paris, France
J. Fivez	Theoretische Physik, Universität des Saarlandes, Saarbrücken, FRG
J. Gea	Facultad de Ciencas, Universidad Autonoma de Madrid, Cantoblanco Madrid, Spain
A. Gharbi	Faculté des Sciences de Tunis, Département de Physique, Belvedere Tunis, Tunisia
M. Gross	Department of Physics and Astronomy, Tel-Aviv University, Ramat-Aviv, Israël
K. Hoshino	School of Mathematics and Physics, University of East Anglia, Norwich, UK
J. Kieffer	Laboratoire des Interactions Moléculaires et des Hautes Pressions, Université Paris-Nord, Villetaneuse, France

D. Kirin	Institut "Rudjer Boskovic", Zagreb, Yugoslovia
K. Kobashi	Institut für Festkörperforschung Kernforschungsanlage, Jülich, FRG
H. Krasser	Institut für Festkörperforschung Kernfor-Schungsanlage, Jülich, FRG
P. Krueger	University of Calgary, Calgary, Canada
R. Kruse	Institut für Physikalische Chemie I, University of Karlsruhe, Karlsruhe, FRG
G. Lucazeau	CSP Université Paris-Nord, Villetaneuse, France
W. Luck	Fachbereich Physicalische Chemie Philipps, Universität Marburg, Marburg, FRG
T. Luty	Institute of Theoretical Chemistry, University of Nijmegen, Nijmegen, Netherlands
R. Lynden-Bell	University Chemical Laboratory, Cambridge, UK
G. Mariotto	Dipartimento di Fisica, Università egli Studi di Trento, Povo, Italy
F. Marsault	Laboratoire de Physique Expérimentale Moléculaire, Université P. et M. Curie, Paris, France
N. Meinander	University of Helsinki, Accelerator Laboratory, Helsinki, Finland
P. Menger	Laboratorium Voor Fysische Chemie, Universiteit Van Amsterdam, Amsterdam, Netherlands
K. Menn	Institut für Theoretische Physik III, Universität Giessen, Giessen, FRG
M. Merrian	Laboratoire d'Optique Appliquée, Ecole Polytechnique ENSTA, Palaiseau, France
M. Nardone	Istituto di Fisica G. Marconi, Rome, Italy
V.T. N'Guyen	Laboratoire d'Infrarouge, Université Paris-Sud, Orsay, France
N. Nielsen	Department of Chemistry, University of Oslo, Oslo, Norway

F. Owens	Energetic Materials Laboratory, Armement Research Development, Command Dover, USA
R. Paul	Chemistry Department, The University of Calgary, Calgary, Canada
M. Perrot	Laboratoire de Spectroscopie Infrarouge, Université de Bordeaux, Talence, France
E. Praestgaard	Chemistry Laboratory, Ørsted Institute, Copenhagen, Denmark
P. Prasad	Chemistry Department, State University of New York, Buffalo, USA
J. Raich	College of Natural Sciences, Colorado State University, Fort Collins, USA
H. Romanowski	Institute of Chemistry, Wroclaw University, Wroclaw, Poland
A. Saint-Pierre	Saint-Vincent Archabbey, Latrobe, USA
M. Sampoli	Istituto di Fisica, Università di Venezia, Venezia, Italy
M. Sanquer	Département de Physique Cristalline & Chimie Structurale, Université de Rennes-Beaulieu, Rennes, France
J.L. Sauvajol	Université de Lille, Villeneuve d'Ascq, France
C. Schutte	Department of Chemistry, University of South Africa, Pretoria, South Africa
R. Seelinger	Physikalisches Institut, Universität Stuttgart, Stuttgart, FRG
D. Siapkas	Second Laboratory of Physics, University of Thessaloniki, Thessaloniki, Greece
A. Stepanescu	Istituto Elettrotecnico Nazionale,Galileo Terraris, Torino, Italy
S. Yip	Nuclear Engineering Department, M.I.T., Cambridge, USA
K. Szczepaniak	Institute of Experimental Physics, Warsaw University, Warsaw, Poland

G. Tarjus	Laboratoire de Physique Théorique des Liquides, Université P. et M. Curie, Paris, France
J. Teixeira Dias	The Chemical Laboratory, University of Coimbra, Coimbra, Portugal
D. Tildesley	Physical Chemistry Laboratory, Oxford University, Oxford, UK
C.W. Van der Lieth	Anorganisch Chemisches Institut des Universität Heidelberg, Heidelberg, FRG
H. Van Elburg	Laboratorium voor Fysische Chemie, Amsterdam, Netherlands
D. Varshneya	Vitreous State Laboratory, Department of Physics, Catholic University, Washington DC, USA
M.J. Velutini	Laboratoire Dymor, Nice, France
S. Venugopalan	Raman Research Institute, Bangalore, India
H. Versmold	Institüt für Physikalische Chemie der Universität Würzburg, Würzburg, FRG
C.H. Wang	Department of Chemistry, University of Utah, Salt Lake City, USA
R. Wertheimer	Physik Department der Technischen Universität München, Munich, FRG
H. Wieldraayer	Van der Waals Laboratorium, University of Amsterdam, Amsterdam, Netherlands
R. Wilde	Department of Chemistry, Texas Technical University, Lubbock, USA
J. Yarwood	Department of Chemistry, University of Durham, Durham City, UK
M. Yvinec	Département de Recherches Physiques, Université P. et M. Curie, Paris, France
H. Zelsmann	CEA-CEN Grenoble, Résonance Magnétique, Grenoble, France

AUTHOR INDEX

ABRAMO C. et al., 431, 440
ADELMAN S.A. et al., 63
ADELMAN S.A. et al., 94
AGGRAWAL D.K. et al., 368-369
AKHAMOV S.A. et al., 175
ALBEN R. et al., 348-349, 354, 360
ALBEN R. et al., 393-396
ALDEBERT P. et al., 375
ALFANO R.R. et al., 180
AMORIM DA COSTA A.H. et al., 93
AMOS R.D., 25
AMOUREUX J.P. et al., 316
ANDERSON P.W., 336
ANDERSON P.W. et al., 346
ANDREAE J.H. et al., 206
ANGELL C.A. et al., 191-194, 197-198
ARNDT R. et al., 47
ARNDT R. et al., 131, 133, 139
ASSELIN M. et al., 101
AXE J.D., 373
AXILROD B.M. et al., 31
AZIZ R.A. et al., 1

BAILEY R.T., 43, 56, 167
BAILEY R.T., 62-63
BALKANSKI M. et al., 375, 398
BANSAL M. L. et al., 134
BANSAL M. L. et al., 328
BARKATT A. et al., 197
BARKER A.S. et al., 368, 383
BARKER J.A. et al., 1, 31, 32
BARNOSKI M.K. et al., 373
BAROJAS J. et al., 93
BARTOLI F.J. et al., 44-46, 131, 132
BARTOLI F.J. et al., 116, 121, 128, 130
BASS R. et al., 206
BATTAGLIA M.R. et al., 48
BAUDOUR J.L. et al., 306
BAUHOFER W. et al., 377-381, 384
BECK H., 279
BELL R.J. et al., 349-350, 364
BENEDEK G.B. et al., 208
BERNE B.J. et al., 208-209, 222, 224
BERTONCINI P.J. et al., 30
BESNARD M. et al., 133
BICKERMANN A. et al., 422
BIJVOET J.M. et al., 305
BIRGENEAU R.J. et al., 368
BIRNBAUM G., 62, 64
BIRNBAUM G., 148, 151
BIRR M. et al., 375
BJARNASON J. et al., 73
BLINC R., 53
BLINC R. et al., 367, 388, 392
BLIOT F. et al., 84
BOGANI F. et al., 238, 408
BOGANI F. et al., 412-413
BOLDESKUL A.E. et al., 133
BONADEO H. et al., 238, 244, 249
BORN M. et al., 3
BORN M. et al., 221, 263
BORN M. et al., 273, 276
BORN M. et al., 347
BOSE R. et al., 193
BOUACHIR M. et al., 133
BOURNAY J. et al., 108, 109, 112
BOYS S.F. et al., 9
BRADLEY C.J. et al., 280, 282, 312
BRADLEY D.J. et al., 167

BRATOS S. et al., 43, 53, 56
BRATOS S. et al., 54, 102, 107, 112, 184
BRATOS S. et al., 45-46, 52, 91, 95, 117, 186
BRATOS S. et al., 44-45, 91, 95, 121, 128
BREALY G.J. et al., 34
BREUILLARD-ALLIOT C. et al., 124, 132
BRINDEAU E. et al., 45
BRIQUET J.C. et al., 135, 138
BRODBECK C. et al., 133
BRODY E.M. et al., 388, 390
BROT C. et al., 63
BROT C. et al., 297
BROUDE V.L. et al., 417
BROUT R. et al., 388
BRUECK S.R. et al., 175
BRUINING J. et al., 123
BUCKINGHAM A.D., 12, 20, 22, 24, 26, 28, 37
BUCKINGHAM A.D., 158, 186
BUCKINGHAM A.D., 161
BUHAY H. et al., 368, 377-379, 383, 387
BULANIN M.O., 123, 127, 148, 156
BUNKER P.R., 4
BURNS G. et al., 393-396
BUTLER J.A.V., 3

CABANA A. et al., 127
CAHILL J.E., 248
CALAWAY W.F. et al., 175
CALIFANO S., 222, 228, 242
CALLENDER R. et al., 302
CALOIN M. et al., 217
CALOINE B., 123, 129
CAMPBELL J.H. et al., 131-133
CARDARELLI D., 369, 372
CERTAIN P.R., 30
CHANDRAIAH et al., 160
CHESNOY J. et al., 158
CHEUNG P.S.Y. et al., 94
CHISLER E.V. et al., 373
CICOTTI G., 441
CLARKE J.H.R. et al., 148, 163
CLARKE J.H.R. et al., 182, 184, 188, 189
CLAVERIE P., 9
CLAYDON M.F. et al., 102
CLEMENTS W.R.L. et al., 97, 175
COHEN S.S. et al., 47
COLLES M.J. et al., 178
COLSON S.D. et al., 407
CONSTANT M. et al., 131, 134, 137
CONWAY A., 31
COOMBS G.J. et al., 388, 390
COOPER P.A., 305
COSTINES M.E. et al., 125
COULSON C.A. et al., 55
COUZI M. et al., 368, 377-378, 380, 383, 387
COX T.I. et al., 161-163
CRAIG D.P. et al., 408

DAHLER J.S. et al., 12
DALGARNO A., 28, 32
DANIELMEYER H.G., 205
DAVID J.G., 134-135
DAVIES G.J. et al., 163
DAVIDSON D.W. et al., 192
DEAN P., 343, 348-349, 363-364
DECIUS J.C. et al., 237
DEL BENE J. et al., 37
DELLA VALLE R.G. et al., 241, 248, 413
DE MARIA A.J. et al., 167
DEMARTINI F. et al., 180
DE RAEDT B. et al., 280, 282, 296 297, 300
DE RAEDT H. et al., 299-300
DEZWAAN J. et al., 133
DIANOUX A.J. et al., 296, 331, 338
DIESTLER D.J., 43, 45
DIESTLER D.J. et al., 47
DIESTLER W. et al., 398
DIJKMAN F.G. et al., 47
DIJKMAN F.G. et al., 138
DOGE G. et al., 137
DOGE G. et al., 136-138, 140, 184
DOGE G. et al., 49, 140
DOLLING G., 279
DONOVAN B. et al., 222
DORNER B. et al., 367
DORVAL P. et al., 136
DOWS D.A. et al., 236, 238, 406, 408-409
DREYFUS C. et al., 125, 132
DURAND D. et al. 302

DVORAK V., 388
DYKE T.R. et al., 34, 36

ECKOLD G., 393
ECONOMOU E.N. et al. 333-334, 336
EDGELL W.F., 195
EGERT G., 377-378, 380
EIGEN M. et al., 197
EINSTEIN A., 297
ELLIOTT R.J., 368-369
ENRIGHT E.D. et al., 431, 437
EVANS J.C., 102, 112
EVANS M.W., 80, 85
EWING G.E., 123

FABELINSKII L., 208
FEINBERG A. et al., 397-398, 402
FEYNMAN R.P., 5
FEYNMAN R.P., 170
FISHER G., 364
FISCHER S.F. et al., 47, 139, 184
FISCHER S.F. et al., 185
FLEURY P.A., 367
FLUBACHER P., 32
FONTAINE D. et al., 287, 290, 313, 320
FONTAINE D. et al., 320, 325, 333
FONTAINE H. et al., 306
FROST B.S., 153
FUJIYAMA T. et al., 129
FURRY W.H., 297

GALATRY L. et al., 122, 158
GALEENER F.L., 364
GALLAGHER D. et al., 393
GANGEMI F.A., 4
GARLAND C.W. et al., 377-378, 386-388
GEHRING G.A. et al., 368-369
GEISEL T. et al., 368, 377, 379, 380, 387, 396, 398
GENZEL L. et al., 381-385
GERRATT J., 30
GERVAIS F. et al., 368, 377
GIBBS J.H., 191
GILBERT M. et al., 127, 140-141, 184, 338
GILBERT M. et al., 75, 134, 140
GILBERT M. et al., 310, 328
GILL E.B. et al., 123
GILLEN K.I. et al., 76-77, 134
GIORDMAINE J.A., 169, 180
GIUA R. et al., 236
GOLSTEIN M., 192
GORDON E.I. et al., 217
GORDON R.G., 44, 169
GORDON R.G., 79
GORDON R.G., 84, 326
GORDON R.G., 44, 62
GOSLING E.M. et al., 431, 439, 441
GOTZE W., 279, 292
GOULAY-BIZE A.M. et al., 125-126
GRIFFITHS J.E., 8, 136
GRIFFITHS J.E., 130-134
GROSS M. et al., 377
GUILLOT B. et al., 87
GUISSANI Y., 123, 129

HAIDA O. et al., 191
HAMMER H. et al., 389
HANKINS D. et al., 37
HARADA J. et al., 249
HARLEY R.T. et al., 369, 372, 397-398
HARRIS C.B. et al., 52
HATTA I. et al., 374
HELLMANN H., 5
HERZBERG G., 3
HERZBERG G., 341
HERZFELD K.F. et al., 204
HESP H.M.M. et al., 175
HESTER R.E. et al., 196
HILLS B.P. et al., 49, 52
HIRSCHFELDER J.O. et al., 6, 30
HOCHHEIMER H.D. et al., 387
HOLZER W. et al., 151
HOOGE F.N. et al., 155-156
HOOTON D.J., 273, 277
HORNER H., 278
HOWELL F.S. et al., 195
HULLER A. et al., 296
HULLER A. et al., 377
HULLER A. et al., 420, 428
HUNTER J.L. et al., 206, 217
HUONG P.V. et al., 121-122
HYNES J.T., 63

IVANOV E.N., 297

JACOB J. et al., 134

JACUCCI G., 325
JACUCCI G., 431, 437
JAHN J.R. et al., 377, 378, 380
JANSEN L., 18
JONES D.R. et al., 131-132, 135

KADANOFF L.P. et al., 269
KAMINOV J.P. et al., 389
KASTLER A. et al., 244
KATIYAR R.S. et al., 374, 390-391
KAUZMANN W., 191
KEESOM W.H., 8
KETELAAR J.A.A. et al., 155-156
KEYES T. et al., 63, 87
KING C.N. et al., 346, 348
KITTEL C. et al., 341
KIVELSON D. et al., 87
KIVELSON D. et al., 163
KLEIN M.L., 431, 434
KLEIN M.V., 375, 398
KNAUSS D.C., 49, 51-52
KOBAYASHI K., 389-390
KOBLISKA R.J. et al., 346, 349
KOLLMAN P.A. et al., 35, 36
KOBS W. et al., 3
KREEK H. et al., 19
KRISHNA MURTI G.S.R., 305
KROLL D.H. et al., 296
KUBO R., 49
KUBO R., 62
KUBO R., 104
KUBO R., 183
KUBO R., 270
KULAS K. et al., 364
KUNG A.H., 167
KURKI-SUONIO, 282-283
KWEICIEN J. et al., 391-392
KWOK J. et al., 147, 158-159

LADANYI B.M. et al., 63
LAFAIX M. et al., 140
LAGAKOS N. et al., 375, 389
LAGE E.J. et al., 388
LAMB J., 204
LANDAU L.D. et al., 44
LANDSBERG G.S. et al., 53
LANGHOFF P.W. et al., 28
LARSSON K.E.J. et al., 326, 329
LAUBEREAU A., 8
LAUBEREAU A., 97
LAUBEREAU A., 167-173, 175-176, 179-180, 182-187
LAUGHLIN R.B. et al., 357-358
LAUGHLIN W.T. et al., 193
LAX H. et al., 406
LAZAY P.D. et al., 374, 377
LEADBETTER A.J. et al., 306
LEADBETTER A.J. et al., 358
LE DUFF Y. et al., 123
LEGAY D. et al., 125-126, 137
LEIBFRIED G. et al., 263
LEICKNAM J.C. et al., 45
LEUNG R.C. et al., 389-391
LEVANT R., 47
LEVANT R., 137
LEVESQUE D. et al., 98
LEVESQUE D. et al., 431-432
LEVI G. et al., 83
LEVY H.A. et al., 279
LINDER B., 33
LITOWITZ T.A., 207
LIV B. et al., 30
LIVINGSTON R.C. et al., 297
LIVINGSTON R.C. et al., 328
LOISEL J. et al., 136
LONDON F., 8, 28
LONG C.A. et al., 154
LONG D.A., 63
LONGUET-HIGGINS H.C., 1, 3, 10, 15
LOVELUCK J.H. et al., 379, 387
LOWNDES R.P. et al., 389-390
LUCOVSKY G., 355, 363
LUTY F., 296
LYNDEN-BELL R.M., 49-52
LYNDEN-BELL R.M., 75
LYONS K.B. et al., 375

MACEDO P.B. et al., 195
McCLUNG R.E.D., 127-128, 130
McINTYRE D. et al., 208
McKELLER A.R.W., 2
McLACHLAN A.D., 33
McLAUGHLIN D.R. et al., 30
McPHERSONS J.W. et al., 369, 371, 372
MADDEN P.A. et al., 49, 174
MAIER M. et al., 169
MAKER P.D., 73
MARADUDIN A.A. et al., 237, 260
MARADUDIN A.A. et al., 263

MARADUDIN A.A. et al., 341
MARECHAL Y., 53-54
MARGENAU H. et al., 5, 232
MARKAM J.L., 212
MARSAULT J.P., 123, 130-131
MARSAULT-HERAIL F. et al., 127
MARSHALL W. et al., 416
MARTIN R.M., 337
MARYOTT A.A. et al., 83
MAZZA CURATI V. et al., 196, 199
MEAL J.H. et al., 74
METIU H. et al., 47
MICHEL K.H. et al., 291, 296
MICHEL K.H. et al., 325
MICHEL K.H. et al., 426
MIKLAVC A. et al., 186, 187
MIRONE P. et al., 65
MITRA S.S., 406
MIZUSHIMA S. et al., 2
MONTROSE C.J. et al., 208, 214-215
MOORE M.A. et al., 392
MORADI-ARAGNI A. et al., 139
MORAWITZ H. et al., 45
MORE M. et al., 316
MORI H., 291
MOROKUMA K., 36
MOUNTAIN R.D., 208
MOZZI R.L. et al., 358, 364
MULLER K. et al., 52, 85-86
MURREL J.N., 31

NAFIE L.A. et al., 45, 67, 116, 121, 128, 169
NAGAMIYA T., 375
NARAYANAMURTI V. et al., 297
NEMANICH R.J. et al., 393-399
NEMETHY G. et al., 3
NETO N. et al., 221-222, 224, 228, 231, 233-234
NICHOLS W.H., 214
NIELSEN H.H., 74

O'DELL J. et al., 88
ONSAGER L., 33
OSTLUND N.S. et al., 9
OUILLON R. et al., 124, 135, 137, 139
OXTOBY D.W. et al., 43, 45, 49, 56
OXTOBY D.W. et al., 62, 65
OXTOBY D.W. et al., 45, 47-48, 186
OXTOBY D.W. et al., 49-50, 52, 94, 97
OXTOBY D.W. et al., 94-96, 139
OZAWA K., 156

PAUL R. et al., 45
PAUL W. et al., 343
PAULING L. et al., 29
PEERCY P.S., 390
PERCHARD J.P. et al., 121-122
PERRY C.H., 364
PERRY C.H., 375, 377-378, 380, 381, 383, 385, 387
PETZELT J. et al., 368
PHILLIPS W.A. et al., 346
PICK R.M., 286
PICK R.M., 311, 316, 322, 338
PIMENTEL G.C. et al., 56
PINAN-LUCARRE J.P. et al., 135, 136-137
PLASS K.G., 204
POLK D.E., 341-342
POPKIE H. et al., 36
PRAGER W. et al., 429
PRESS W. et al., 282, 284, 310, 313, 315, 316, 418
PRESS W. et al., 293
PRESS W. et al., 296, 429
PRESS W. et al., 377, 382
PRESS W. et al., 426-427
PULLMAN B. et al., 33

QUENTREC B. et al., 87, 93

RAHMAN A. et al., 91
RAHMAN A. et al., 437
RAICH J.C. et al., 98
RAICH J.C. et al., 278
RAKOV A.V., 119
REDDY S.P. et al., 147-148
REESE R.L. et al., 388
REID C. et al., 199
REYNOLDS P.A., 233
RIEHL J.P. et al., 52, 94
RIGHINI R. et al., 232-233, 237, 239, 241-242, 261-262
RIGNY P., 230
ROBERTSON G.N. et al., 54, 103,105

ROMANOWSKI H., 54
ROOTHAAN C.C.J., 30
ROSE M.E., 67-68
RÖSCH N., 54
ROTHSCHILD W.G., 47, 186
ROTHSCHILD W.G., 47, 133
ROTHSCHILD W.G., 136, 138
ROWE J.M. et al., 302
ROWE J.M. et al., 313

SAIGHI L. et al., 154
SAKURAI J. et al., 373-374
SALVI R. et al., 406-408, 413
SAMSON R. et al., 161
SANDORFY C., 53, 101
SARE J.M. et al., 189-190
SATIJO S.K. et al., 136
SCHETTINO V. et al., 256, 404
SCHETTINO V. et al., 254
SCHETTINO V. et al., 406-408, 413
SCHIFF L.I., 26
SCHMIDT W. et al., 167
SCHNEPP O., 237
SCHNEPP O., 278
SCHROEDER J. et al., 133,138-140
SCHULZ J., 134
SCHUSTER P. et al., 34
SCHWARTZ M. et al., 133
SCOTT J.F. et al., 367, 374, 390
SCOTTO M., 123, 177
SEILMEIER A. et al., 182
SELOUDOUX R. et al., 133
SEN P. et al., 357, 364
SETTE D., 204, 206
SEYMOUR R.S. et al., 282, 284
SHAND M.L. et al., 377, 387
SHAPIRO M.M. et al., 153
SHAPIRO S.M., 393
SHE C.Y. et al., 389
SHEKA E., 241
SHENG T., 153
SHEPPARD N., 53
SHUGARD M. et al., 94
SHUKER et al., 396
SINANOGLU O., 33
SINGER K., 87
SKOLD K., 397
SMITH A.L. et al., 150, 153-154
SMITH H.G. et al. 377, 384
SMITH J.E. et al., 396
SMYTH C.P., 194
SOUSSEN-JACOB J. et al., 126, 133
SPANNER K., 182-183
SPRINGER T., 420
STARKSCHALL G., 19
STARUNOV V.S. et al. 431, 437
STEELE W.A. et al., 62, 71, 79, 82-83, 87,
STEELE W.A. et al., 127, 131, 134
STEGEMAN G.I.A. et al., 215, 216
STEGEMAN G.I.A. et al., 431, 437
STEPANOV B.I., 53
STEVENS W.J., 30
STEWART E.S. et al., 204
STOCKMEYER R., 297
STREIB W.E. et al., 305
SUNDER S. et al., 128, 133-134
SZIGETI B., 406

TABISZ G.C., 148, 160
TADDEI G. et al., 228
TAKAGI K., 217
TANABE K., 134, 139
TANFORD C., 3
TANG K.T. et al., 28
TAPIA O., 33
TEACHOUT R.R., 30
TEH H.C. et al., 377, 379
THIBAUDIER C., 330
THORPE M.F. et al., 346, 349, 352 356, 358, 360, 362, 364
TOKUHIRO T. et al., 49
TOKUNAGA M., 388
TOPLER J., 427-428
TRANCHANT F. et al., 305
TREFLER M. et al., 4
TURNBULL D. et al., 189
TURRELL G., 122

VAKS V.G. et al., 377
VALIEV K.A., 45, 49, 140
VAN GOOL W., 393
VAN GUNSTEREN W.F. et al., 94
VAN HOVE L., 418-419
VAN KAMPEN N.G., 49
VAN KONYNENBURG P. et al., 125
VAN KONYNENBURG P. et al., 45, 129, 162
VAN KRANENDONK J., 148-149, 156, 158

VAN WOERKOM P.C.M. et al.,49-50, 52, 93, 140
VAN WOERKOM P.C.M. et al., 140-141
VENKATARAMAN G. et al., 279
VERLET L., 91
VINCENT-GEISSE J., 56
VINCENT-GEISSE J. et al., 83, 127
VINCENT-GEISSE J. et al., 123, 126-127, 129, 140
VIOSSAT G. et al., 123
VLASOVA A.A. et al., 377
VODAR B. et al., 148, 155, 157, 158
VON DER LAGE F.C. et al., 281-282
VON DER LINDE D. et al., 170, 182

WAGNER E.L. et al., 279
WALL T.H., 134
WALLACE D.C., 354, 356
WANG C.H. et al., 49, 51-52
WANG C.H. et al., 368, 377-379, 383, 385, 387
WANG S.C., 8
WARREN J.L., 380
WATANABE A. et al., 2
WEAIRE D. et al., 348-349, 358, 360, 362
WEGDAM G.H. et al., 87
WEIDEMANN E.G. et al., 54
WERTHEIMER R.K., 49, 51-52, 186
WEIS J.J. et al., 431-433
WELSH H.L., 148
WIGNER E., 286
WILDE R.E. et al., 47
WILLIAMS D.E., 232
WITKOWSKI A. et al., 52
WOFSY S.C. et al., 4
WONG J. et al., 189, 192
WOODCOCK L.V. et al., 194

YAMADA Y. et al., 373-374, 377, 384
YAMADA Y. et al., 311, 377, 387
YAMAMOTO T. et al., 293
YARWOOD J. et al., 54, 106
YARWOOD J. et al., 133, 137
YOUNG A.P. et al., 388-389
YVINEC M. et al., 313-315, 320, 337-338

ZACHARIASEN W.H., 341-342, 358
ZINTH W. et al., 174
ZUBAREV D.N., 269
ZUMOFEN G. et al., 406-407
ZWANZIG R.W., 291
ZYGAN-MAUS R. et al., 187

SUBJECT INDEX

Absorption coefficient, 204 ff, 211 ff, 217
Absorption process, 405 ff
Amorphous solids, 341-364
Anderson localization, 336
Anharmonic coefficient, 409
Anharmonic effects, 37 ff, 54 ff, 64, 74, 98, 253, 273, 278, 295, 405
  electrical, 406
  frequency shift, 259 ff

Band intensity, 133, 146 ff, 149, 153, 158, 160, 243
Band shapes
  amorphous solids, 354
  Brillouin in liquids, 207
  glasses, 194-196, 200
  hydrogen bond, 57, 61, 91, 96, 106, 112, 117, 121, 137
  viscous liquids, 189, 194-197
  totally symmetric modes, 337
Band widths
  Brillouin in liquids, 210, 216
  hydrogen bond, 57
  glasses, 194-196
  molecular crystals, 101, 107, 112, 117, 120, 125 ff, 176 ff, 253, 259 ff
  viscous liquids, 194-196
Bethe lattice (*see also* structural potential approximation), 352, 353, 356-358
  surface atoms, 348, 351-353, 355
Born Oppenheimer approximation, 3 ff, 343, 344
Bragg reflection spectroscopy, 217, 218
Brillouin spectra, 206, 210, 217
Brillouin zone, 238-260, 267, 346, 354, 356
Bulk viscosity, 209, 210

Central peak, 368
Coherent potential approximation, 351, 353
Coherent scattering law, 173 ff, 180, 209-210, 308, 335 ff, 416-420, 432
Collision induced processes, 44, 50 ff, 58, 62, 76, 118, 121, 125, 148 ff, 170, 185
Combination bands, 253
Commutations relations, 255 ff
Computer simulation (*see also* molecular dynamics), 50, 52, 61, 67, 83, 87, 91 ff, 431 ff
Coordinates
  internal normal, 45, 54, 64 ff, 93, 96, 151, 160, 169, 224, 226, 227
  molecular, 223, 233, 246, 247
Coriolis coupling, 52, 62, 74 ff, 128, 132, 182, 310, 323, 328
Correlation function, 46, 52, 67, 68, 91, 93, 129, 162, 203 ff, 270, 285, 416, 420, 431
  rotational, 46, 52, 62 ff, 68,

Correlation function (continued)
rotational (continued), 72 ff, 82, 93, 117 ff, 291 ff, 296 ff, 307, 310, 320 ff, 435 ff
vibrational, 46, 52, 93 ff, 117 ff, 197
Correlation time, 107 ff, 117-129
rotational, 77 ff, 120, 123, 129 ff, 308-337
vibrational, 95, 97 ff, 120, 128
Cubic harmonics, 282
Crystals
electric moment, 244 ff
polarizability tensor, 244

Decoupling approximation, 276, 279, 316, 317, 335
Density fluctuations, 207-210, 212-222
Density of states
one phonon, 199, 238-239, 262, 309, 336, 338, 346-349, 351-358, 363, 406
two phonon, 260, 407
Dephasing processes, 46 ff, 55, 147, 169, 175
Depolarization ratio, 119, 161
Dielectric constant, 207, 209
Double well potential, 374
Dynamical coupling (*see also* correlation function, vibrational), 309, 318, 323, 336
Dynamical matrix, 222, 227, 268 ff, 346, 359

Electrostatic interactions, 233, 238-239
multiple expansion of, 6, 15, 20 ff, 152, 234-235, 237, 242, 245, 375
Euler angles, 69 ff, 280, 306, 313 ff ff
Excitation process, 168 ff
Exciton (molecular), 407-408

Fermi resonance, 55 ff, 110 ff, 182
Ferroelectrics, 373, 388 ff
Fluctuation dissipation, 270
Force constant disorder, 347
Free rotation, 421 ff

Glasses, 194, 196, 200, 355, 358
Glass transition, 189-195
Green function, 254, 269 ff
anharmonic, 408

Harmonic approximation, 268, 271 ff
Hydrodynamic singularity, 279
Hydrogen bond, 34 ff, 53 ff, 101 ff, 122, 133, 368, 371, 388 ff

Incoherent scattering law, 323, 416-424
Induction energy, 7, 14, 26 ff
Infrared tensor (*see also* transition moments), 43, 244 ff, 308
Irreducible Raman spectrum, 319, 324, 335
Irreducible spherical tensor, 69 ff, 318

Jahn Teller effect, 368-369

Landau Placzek peak, 279
Lattice modes, 55, 223, 240-241, 249, 253, 260, 264 ff
Libration, 326, 328, 332
Librational modes, 280, 296, 379 ff
Linear molecules, 290 ff, 305-311 ff, 320-321, 325, 332
Linear response, 270
Liouville operator, 292
Local field effects, 50 ff, 62, 80 ff, 129, 158, 160, 309, 323, 333, 334, 336
Localized states, 347, 356
Local modes, 384, 385
Long range order, 354

Magnetic interactions, 9
Memory functions, 47, 62, 80 ff, 125, 131, 292

Mixed crystal, 377
Modulation speed, 46, 48 ff, 119, 137 ff, 183
Molecular dynamics, 431 ff
Molecular orientation, 279, 306, 325 ff
Molecular reorientation, 280, 325 ff, 368, 373 ff, 431, 433
Mountain model, 208, 215, 218
  mode, 210-212, 214-217, 431, 437
Multiple moments, 17 ff

Normal coordinates, 307

Order parameter, 294, 296, 367ff
Oriented gas model, 244-250
Oscillators (model of coupled), 10 ff
Overtones, 154, 253, 408

Paraelectrics, 369, 373
Perturbation method, 254-256 ff
Phase transition
  displacive, 367 ff
  order disorder, 367 ff, 433
Phonon
  amorphous solids, 345
  dispersion curves, 223, 237-238, 241, 255
  dynamical matrix, 222, 227, 268 ff
  lifetime, 259 ff
Picosecond spectroscopy, 136, 139, 167 ff
Piezoelectrics, 375
Plastic (orientational disorder) phase, 278-279, 293 ff, 305 ff, 441
Polarizability, 22
  tensor (*see also* Raman tensor), 63, 65 ff, 70, 91, 130, 160, 207, 244 ff
  hypertensor, 161
Polarization energy, 236, 238, 240
Population lifetime, 94, 169, 175, 180 ff
Potential energy, 32 ff, 227-232
Potential energy (continued)
  atom-atom, 347, 351-363, 412, 431
  between internal modes, 307, 334, 337
  dispersion, 98
  intermolecular pair-wise, 98, 227, 231 ff, 260
  van der Waals, 231-240
Probability distribution function, 282 ff, 306, 310 ff, 325
Pseudo spin, 170, 310
Pseudo spin phonon coupling, 377, 387

Rakov's method, 119 ff, 125, 129, 135
Raman intensity, 248, 249
Raman tensor (*see also* polarizability tensor), 43-44, 169, 244 ff, 307, 318 ff
Random network, 342-343, 348, 352, 358-361
Relaxation function, 292
Relaxation processes, 183, 189-192, 197
  double, 206, 217-218
  rotational, 121 ff, 296, 374
  vibrational, 43 ff, 64, 103 ff, 136 ff, 204-207, 209, 214
Relaxation time (*see also* correlation time), 105, 169
  energy (*see also* population lifetime), 47 ff
  rotational, 374
  vibrational, 189-192, 197, 204-206, 209-212, 214, 216-217
Renormalized harmonic approximation, 277 ff
Residence time, 309, 311, 332, 334 ff, 374, 393 ff
Response function, 270, 292, 437
Restoring forces matrix, 292
Rigid body approximation, 223
Rotation matrix, 225-247
Rotational current, 440

Rotational diffusion coefficient, 81, 120, 130 ff
Rotational models, 85, 124, 136, 325 ff
  Debye (or diffusion), 81, 83, 119, 132, 296-297, 331, 421-422 ff, 426, 429
  Gordon (or extended diffusion), 84, 124 ff, 326-332
  jump, 297, 329, 421-424
Rotation translation coupling, 203, 431 ff
Rotation vibration coupling, 53

Second quantization, 255
Secular (non secular) variables, 291
Sen and Thorpe model, 357-363
Short range order, 343
Side band, 406-407, 410 ff
Simultaneous transitions (*see also* two phonon bands), 152, 155 ff
Soft mode, 367 ff
Solvent effects, 33 ff
  frequency shift, 137-158
Sound velocity, 207, 209-211, 217
  frequency dependence, 209, 211, 213, 214, 217
  longitudinal, 344-345
  transverse, 434-435
Spectral moments, 86, 88, 117, 119, 127, 129, 133, 138
Spherical harmonics, 278, 311
Static susceptibility, 291, 294
Structural potential approximation (*see also* Bethe lattice), 351-357
Superionic conductor, 374, 393 ff
Symmetry adapted functions, 280ff, 293, 312, 321 ff, 418

Tetrahedral units
  amorphous solids, 353-358
  viscous liquids, 197-199
Three body interactions, 149, 150 ff
Topological disorder, 347
Torques (mean square), 123, 125, 127, 131
Transition moments (*see also* IR tensor), 44, 73, 151
  second order dipole moment, 411-412
Transition dipole coupling, 128, 136, 140 ff, 185, 237, 238, 338
Transport coefficient, 295
Tunnelling, 296, 388 ff
Two phonon bands (*see also* simultaneous transition), 405 ff

Ultrasonic measurements, 204-207, 215, 217

Van der Waals forces, 314, 323
Vibrational models, 47, 138
Viscous liquids, 189-200, 314 ff, 431, 437

Wigner functions, 67, 293, 314 ff

SUBSTANCE INDEX

AgI, 393 ff
Ar, 190
As_2Ge_3, 355

BaF_2, 393 ff
BeF_2, 363
Br_2, 82 ff, 147, 155 ff, 205

CBr_4, 306, 315 ff
$C(CH_3)Cl_3$, 170, 182, 184 ff
$C(CH_3)_2(CN)_2$, 306
$C(CH_3)_3Br$ (t-butylbromide), 134
$C(CH_3)_3Cl$ (t-butylchloride), 134, 306, 315, 320, 325
$C(CH_3)_3CN$ (t-butylcyanide), 306
$C(CH_3)_4$ (neopentane), 306
CCl_4, 190, 215, 441
$CDCl_3$, 133
CD_3CN, 140
CD_3H, 130 ff
CD_3I, 138
CD_3OH, 108 ff
CD_4, 126 ff, 292, 310, 316, 426ff
C_2D_2, 126
C_6D_6, 178 ff
CF_3Cl, 306
CF_4, 127, 140 ff, 306, 328, 338
$CHCl_3$, 71, 73, 75 ff, 133
CHF_3, 133
CH_2Br_2, 135
CH_2Cl_2, 141, 199, 206, 217
CH_3Cl, 206
CH_3CN, 132, 140, 190
CH_3D, 130
CH_3I, 131, 137 ff, 140, 181 ff
CH_3OH, 112 ff, 178
CH_3X (methylhalides), 131
$(CH_3)_3N$ (trimethylamine), 133
CH_4, 126 ff, 290, 422, 429, 441
C_2H_2, 126
C_2H_4, 71 ff
$C_2H_4N_2S$ (1,2,5-thiodiazole), 135
$(CH_2OH)_2$(ethylene glycol), 177 ff
C_2H_5OH, 191 ff
C_3H_4 (allene), 133
$C_4H_2N_2$ (succinonitrile), 306
C_4H_4O (furane), 136
C_4H_4S (thiophen), 135 ff, 305
C_6H_6, 76 ff, 79, 83, 134, 190, 217, 249, 407
$C_6H_5CH_3$, 190, 194 ff
C_6H_5OD, 106
C_6H_5OH, 106 ff
C_6H_{12}, 134, 305
C_9H_7N (quinoline), 136
$C_{10}H_8$ (naphtalene), 240, 407
$C_{10}H_{16}$ (adamantane), 311, 316
$C_{10}H_{18}$ (decaline), 199
$C_{18}H_{14}$ (paraterphenyl), 306
Cl_2, 123
ClO_3F, 134
CN^-, 297 ff
CO, 123, 441
CO_2, 125 ff, 147, 154, 161, 238, 407, 410, 413
$Co(CO)_4^-$, 196
CS_2, 147, 154, 161, 162, 206
CsH_2AsO_4(CsDA), 389 ff
CsH_2PO_4(CsDP), 389 ff
CuBr, 393 ff
CuCl, 393 ff
CuI, 393 ff

D_2, 123

DCl, 441
DF, 121-122
D_2O, 134
$DyVO_4$, 369, 372

GaAs, 344
Ge, 346, 349, 354
GeO_2, 363 ff
GeS_2, 358, 363
$GeSe_2$, 363

H, 345
H_2, 123, 148, 155, 422 ff
HCl, 121 ff, 148, 157, 190, 408
HF, 34, 121-122
HOD, 134
H_2O, 36, 134, 431, 437
He, 345

KCN, 305 ff, 311, 313, 320, 325
KH_2AsO_4(KDA), 389
KH_2PO_4(KDP), 345, 388 ff

N_2, 93, 97 ff, 123, 139, 147, 153, 155, 170, 175 ff, 189 ff, 305, 431 ff
N_2(solid α, β), 433, 434
NaCN, 305, 313, 320, 332
$NaNO_2$, 373 ff
$NaNO_3$, 305
NaOH, 199
ND_4Br, 375 ff
ND_4Cl, 375 ff
NF_3, 75, 134, 140
NH_3, 133, 239, 249, 260
NH_4Br, 375 ff
NH_4Cl, 279 ff, 375 ff, 424 ff
$NH_4Cl_{(1-x)}Br_x$, 377 ff
NH_4I, 375 ff
$(NH_4)_2SnCl_6$, 423, 425
NO, 160, 206
NO_3^-, 192 ff
N_2O, 123 ff, 139, 407

OH^-, 196
O_2, 123, 153 ff
OCS, 125 ff, 407, 412

PH_3, 133
PbF_2, 393 ff
$PrAlO_3$, 368 ff

Rare earth ions, 368 ff
$RbAg_4I_5$, 393 ff
RbH_2PO_4(RbDP), 389 ff

SF_6, 127, 147, 161, 206, 411
S_2Cl_2, 190
SO_2, 134 ff, 206
Si, 342 ff, 346 ff, 354 ff
$Si(CH_3)_4$, 128
SiF_4, 238, 408
SiO_2 (silica crystobalite),342 ff, 355 ff, 357 ff, 362
$SnBr_4$, 175 ff
SrF_2, 393 ff

$TbVO_4$, 369 ff
$TmPO_4$, 372 ff

$ZnCl_2$, 197 ff
ZrO_2-Y_2O_3, 393 ff